AAPG Treatise of Petroleum Geology Reprint Series

The American Association of Petroleum Geologists
gratefully acknowledges and appreciates the leadership and support
of the AAPG Foundation in the development of the
Treatise of Petroleum Geology.

Geophysics II

Tools for Seismic Interpretation

Compiled by
Edward A. Beaumont
and
Norman H. Foster

Treatise of Petroleum Geology
Reprint Series, No. 13

Published by
The American Association of Petroleum Geologists
Tulsa, Oklahoma 74101-0979, U.S.A.

Copyright © 1989
The American Association of Petroleum Geologists
All Rights Reserved

Library of Congress Cataloging-in-Publication Data

Geophysics II.

(Treatise of petroleum geology reprint series; no. 13
Includes bibliographical references.
1. Petroleum—Prospecting.   2. Seismic interpretation.
I. Beaumont, E.A. (Edward A.)   II. Foster, Norman H.
III. Series.
TN271.P4G47   1989   622'.1828   89-18158

ISBN 0-89181-412-4
ISSN 1046-0144

AMERICAN ASSOCIATION OF PETROLEUM GEOLOGISTS FOUNDATION
TREATISE OF PETROLEUM GEOLOGY FUND*

Major Corporate Contributors
($25,000 or more)

Chevron Corporation
Mobil Oil Corporation
Oryx Energy Company
Pennzoil Exploration and Production Company
Shell Oil Company
Union Pacific Foundation

Other Corporate Contributors
($5,000 to $25,000)

Cabot Energy Corporation
Canadian Hunter Exploration Ltd.
Conoco Inc.
Marathon Oil Company
The McGee Foundation, Inc.
Phillips Petroleum Company
Texaco Philanthropic Foundation
Transco Energy Company

Major Individual Contributors
($1,000 or more)

C. Hayden Atchison
Richard R. Bloomer
A.S. Bonner, Jr.
David G. Campbell
Herbert G. Davis
Paul H. Dudley, Jr.
Lewis G. Fearing
James A. Gibbs
George R. Gibson
William E. Gipson
Robert D. Gunn
Cecil V. Hagen
Frank W. Harrison
William A. Heck
Roy M. Huffington
Harrison C. Jamison
Thomas N. Jordan, Jr.
Hugh M. Looney
John W. Mason
George B. McBride
Dean A. McGee
John R. McMillan
Rudolf B. Siegert
Robert M. Sneider
Jack C. Threet
Charles Weiner
Harry Westmoreland
James E. Wilson, Jr.

The Foundation also gratefully acknowledges the many who have supported this endeavor with additional contributions, which now total more than $12,000.

*Contributions received as of November 1, 1989.

Treatise of Petroleum Geology
Advisory Board

W.O. Abbott
Robert S. Agatston
John J. Amoruso
J.D. Armstrong
George B. Asquith
Colin Barker
Ted L. Bear
Edward A. Beaumont
Robert R. Berg
Richard R. Bloomer
Louis C. Bortz
Donald R. Boyd
Robert L. Brenner
Raymond Buchanan
Daniel A. Busch
David G. Campbell
J. Ben Carsey
Duncan M. Chisholm
H. Victor Church
Don Clutterbuck
Robert J. Cordell
Robert D. Cowdery
William H. Curry, III
Doris M. Curtis
Graham R. Curtis
Clint A. Darnall
Patrick Daugherty
Herbert G. Davis
James R. Davis
Gerard J. Demaison
Parke A. Dickey
F.A. Dix, Jr.
Charles F. Dodge
Edward D. Dolly
Ben Donegan
Robert H. Dott, Sr.*
John H. Doveton
Marlan W. Downey
John G. Drake
Richard J. Ebens
William L. Fisher
Norman H. Foster
Lawrence W. Funkhouser
William E. Galloway
Lee C. Gerhard
James A. Gibbs
Arthur R. Green
Robert D. Gunn
Merrill W. Haas
Robert N. Hacker
J. Bill Hailey
Michel T. Halbouty
Bernold M. Hanson
Tod P. Harding
Donald G. Harris
Frank W. Harrison, Jr.

Ronald L. Hart
Dan J. Hartmann
John D. Haun
Hollis D. Hedberg*
James A. Helwig
Thomas B. Henderson, Jr.
Francis E. Heritier
Paul D. Hess
Mason L. Hill
David K. Hobday
David S. Holland
Myron K. Horn
Michael E. Hriskevich
J.J.C. Ingels
Michael S. Johnson
Bradley B. Jones
R.W. Jones
John E. Kilkenny
H. Douglas Klemme
Allan J. Koch
Raden P. Koesoemadinate
Hans H. Krause
Naresh Kumar
Rolf Magne Larsen
Jay E. Leonard
Ray Leonard
Howard H. Lester
Detlev Leythaeuser
John P. Lockridge
Tony Lomando
John M. Long
Susan A. Longacre
James D. Lowell
Peter T. Lucas
Harold C. MacDonald
Andrew S. Mackenzie
Jack P. Martin
Michael E. Mathy
Vincent Matthews, III
James A. McCaleb
Dean A. McGee
Philip J. McKenna
Robert E. Megill
Fred F. Meissner
Robert K. Merrill
David L. Mikesh
Marcus E. Milling
George Mirkin
Richard J. Moiola
D. Keith Murray
Norman S. Neidell
Ronald A. Nelson
Charles R. Noll
Clifton J. Nolte
Susan E. Palmer

Arthur J. Pansze
John M. Parker
Alain Perrodon
James A. Peterson
R. Michael Peterson
David E. Powley
A. Pulunggono
Donald L. Rasmussen
R. Randolf Ray
Dudley D. Rice
Edward C. Roy, Jr.
Eric A. Rudd
Floyd F. Sabins, Jr.
Nahum Schneidermann
Peter A. Scholle
George L. Scott, Jr.
Robert T. Sellars, Jr.
John W. Shelton
Robert M. Sneider
Stephen A. Sonnenberg
William E. Speer
Ernest J. Spradlin
Bill St. John
Philip H. Stark
Richard Steinmetz
Per R. Stokke
Donald S. Stone
Doug K. Strickland
James V. Taranik
Harry TerBest, Jr.
Bruce K. Thatcher, Jr.
M. Raymond Thomasson
Bernard Tissot
Donald Todd
M.O. Turner
Peter R. Vail
Arthur M. Van Tyne
Harry K. Veal
Richard R. Vincelette
Fred J. Wagner, Jr.
Anthony Walton
Douglas W. Waples
Harry W. Wassall, III
W. Lynn Watney
N.L. Watts
Koenradd J. Weber
Robert J. Weimer
Dietrich H. Welte
Alun H. Whittaker
James E. Wilson, Jr.
Martha O. Withjack
P.W.J. Wood
Homer O. Woodbury
Mehmet A. Yukler
Zhai Guangming

*Deceased

IN APPRECIATION...

The American Association of Petroleum Geologists and the AAPG Foundation gratefully acknowledge the contributions of the Society of Exploration Geophysicists to the Treatise of Petroleum Geology Reprint Series volumes on geophysics. The crucial role played by SEG in advancing exploration geophysics is universally recognized in the petroleum industry and is plainly documented by the many papers from *Geophysics* and *Geophysics: The Leading Edge of Exploration* reproduced in these volumes. The spirit of advancement and expansion in the science of geophysical exploration for hydrocarbons as well as the continuing synergism that melds the professions of geology and geophysics is exemplified by the permission granted by SEG to reproduce its papers in this series.

Although SEG and AAPG both have a long history of independent activities and autonomous operation, there have been, and there continue to be, cooperative efforts to improve the professionalism of both geologists and geophysicists. Previous activities, such as joint meetings, research conferences, and publications, exemplify the ongoing cooperative efforts which create results far more valuable than independent efforts by either group would have achieved. SEG's contributions to the earth sciences and to the Treatise of Petroleum Geology Reprint Series volumes on geophysics are gratefully acknowledged by AAPG, the AAPG Foundation, and the Advisory Board of the Treatise of Petroleum Geology.

INTRODUCTION

This reprint volume belongs to a series of that is part of the *Treatise of Petroleum Geology*. The *Treatise of Petroleum Geology* was conceived during a discussion we had at the 1984 AAPG Annual Meeting in San Antonio. When our discussion ended, we had decided to write a state-of-the-art textbook in petroleum geology, directed not at the student, but at the practicing petroleum geologist. The project to put together one textbook gradually evolved into a series of three different publications: the Reprint Series, the Atlas of Oil and Gas Fields, and the Handbook of Petroleum Geology; collectively these publications are known as the *Treatise of Petroleum Geology*. With the help of the Treatise of Petroleum Geology Advisory Board, we designed this set of publications to represent the cutting edge in petroleum exploration knowledge and application. The Reprint Series provides previously published landmark literature; the Atlas collects detailed field studies to illustrate the various ways oil and gas are trapped; and the Handbook is a professional explorationist's guide to the latest knowledge in the various areas of petroleum geology and related fields.

The papers in the various volumes of the Reprint Series complement the different chapters of the Handbook. Papers were selected on the basis of their usefulness today in petroleum exploration and development. Many "classic papers" that led to our present state of knowledge have not been included because of space limitations. In some cases, it was difficult to decide in which Reprint volume a particular paper should be published because that paper covers several topics. We suggest, therefore, that interested readers become familiar with all the Reprint volumes if they are looking for a particular paper.

Geophysics is an indispensable tool for geologists looking for and developing oil and gas fields. Because it lets us "see" into the subsurface, geophysics allows petroleum geologists to build better images of the subsurface than is possible using only surface geology and information from well bores. In the past, geophysics was the domain of the geophysicist, and the geophysicist alone acquired, processed, and interpreted geophysical data. During the past two decades, however, the technology of geophysics has exploded; at the same time, the petroleum industry has been forced to look for more and more subtle traps in more and more difficult terrain. This placed a tremendous burden on geophysicists, and they naturally looked to their colleagues, the geologists, for relief. At first, geologists only helped with interpretation. Today, however, geologists are also involved in helping geophysicists make decisions regarding acquisition and processing of data.

The choice of papers in these geophysics reprint volumes reflects this evolution. The papers were chosen to help geologists, not geophysicists, enhance their knowledge of geophysics. Math-intensive papers were excluded because those papers are relatively esoteric and have limited applicability for most geologists. Many of the papers included do contain mathematical equations, but they were selected because they are germane, and the math is presented at a level that, we trust, the majority of geologists are now comfortable with.

The number and distribution of the papers reprinted in these volumes reflect the current importance and uses of the different geophysical methods described in the papers. We have divided the topic of geophysics into four volumes. The first volume contains papers on Seismic Methods. Papers in this volume are concerned with seismic theory and are grouped into six sections: Seismic Methods, Seismic Rock Properties, Seismic Acquisition, Seismic Processing and Display, Seismic Velocities, and Migration. Volume II is subtitled Tools for Seismic Interpretation. Section titles in this volume are Synthetic Seismograms and Velocity Inversion; Seismic Modeling; Seismic Attributes: Amplitude, Frequency, Phase, Velocity; Shear Waves; Amplitude Variation with Offset; and Vertical Seismic Profiling. Volume III, Geologic Interpretation of Seismic Data, contains sections on Structural Interpretation and Stratigraphic Interpretation. The last volume is on Gravity, Magnetic, and Magnetotelluric Methods. It contains two sections: Gravity and Magnetic Methods, and Magnetotelluric Methods.

We would like to thank the various societies and publishers who gave us permission to reprint these papers, especially the Society of Exploration Geophysicists. We also wish to thank the members of Advisory Board of the Treatise of Petroleum Geology who suggested papers for these volumes, especially R. Randy Ray. Randy Ray is a geophysicist who was trained initially as a geologist. From a large list of proposed papers, he helped us select papers that would be both understandable and useful to a geologist exploring for and developing oil and gas fields.

Edward A. Beaumont
Tulsa, Oklahoma

Norman H. Foster
Denver, Colorado

TABLE OF CONTENTS

GEOPHYSICS II
TOOLS FOR SEISMIC INTERPRETATION

SYNTHETIC SEISMOGRAMS AND VELOCITY INVERSION

The synthesis of seismograms from well log data.   R. A. Peterson, W. R. Fillippone, and F. B. Coker 2

Inversion of seismograms and pseudo velocity logs.   M. Lavergne and C. Willm 25

Well log editing in support of detailed seismic studies.   Brian E. Ausburn 45

SEISMIC MODELING

Stratigraphic modeling: a step beyond bright spot.   E. V. Dedman, J. P. Lindsey, and
M. W. Schramm, Jr. 83

Three-dimensional seismic modeling.   Fred J. Hilterman 88

Interpretive lessons from three-dimensional modeling.   Fred J. Hilterman 106

Stratigraphic modeling and interpretation—geophysical principals and techniques.
Norman S. Neidell and Elio Poggiagliolmi 131

Synthetic seismic sections of typical petroleum traps.   Bruce T. May and Franta Hron 161

SEISMIC ATTRIBUTES: AMPLITUDE, FREQUENCY, PHASE, VELOCITY

Application of amplitude, frequency, and other attributes to stratigraphic and hydrocarbon
determination.   M. T. Taner and R. E. Sheriff 191

Reflections on amplitudes.   R. F. O'Doherty and N. A. Anstey 218

Velocity spectra and their use in stratigraphic and lithologic differentiation.   Ernest E. Cook and
M. Turhan Taner 247

Outlining of shale masses by geophysical methods.   A. W. Musgrave and W. G. Hicks 263

Three-dimensional seismic monitoring of an enhanced oil recovery process.   Robert J. Greaves
and Terrance J. Fulp. 279

SHEAR WAVES

Basis for interpretation of Vp/Vs ratios in complex lithologies.   Raymond L. Eastwood and
John P. Castagna 295

Evaluation of direct hydrocarbon indicators through comparison of compressional- and shear-wave
seismic data: a case study of the Myrnam gas field, Alberta.   Ross Alan Ensley 313

Relationships between compressional-wave and shear-wave velocities in clastic silicate rocks.
J. P. Castagna, M. L. Batzle, and R. L. Eastwood 325

Direct hydrocarbon detection using comparative P-wave and S-wave seismic sections.
James D. Robertson and William C. Pritchett 337

Vp/Vs and lithology.   R. S. Tatham 348

Vp/Vs—a potential hydrocarbon indicator.   Robert S. Tatham and Paul L. Stoffa 357

Amplitude Variation with Offset

Plane-wave reflection coefficients for gas sands at nonnormal angles of incidence.   W. J. Ostrander 373

Vertical Seismic Profiling

Vertical seismic profiling—a measurement that transfers geology to geophysics.   B. A. Hardage 389

The use of vertical seismic profiles in seismic investigations of the earth.   A. H. Balch, M. W. Lee,
J. J. Miller, and Robert T. Ryder 412

Prediction of overpressure in Nigeria using vertical seismic profile techniques.   S. Brun,
P. Grivelet, and A. Paul 425

Offset source VSP surveys and their image reconstruction.   P. B. Dillon and R. C. Thomson 436

Vertical seismic profiling technique emerges as a valuable drilling tool.   R. J. Roberts and J. D. Platt 459

Table of Contents

Geophysics I
Seismic Methods

Seismic Waveforms and Resolution

Aspects of seismic resolution.
R. E. Sheriff.

How thin is a thin bed?
M. B. Widess.

Complex seismic trace analysis of thin beds.
James D. Robertson and Henry H. Nogami.

The limits of resolution of zero-phase wavelets.
R. S. Kallweit and L. C. Wood.

Resolution comparison of minimum-phase and zero-phase signals.
M. Schoenberger.

Seismic Rock Properties

Formation velocity and density—the diagnostic basics for stratigraphic traps.
G. H. F. Gardner, L. W. Gardner, and A. R. Gregory.

Effect of water saturation on seismic reflectivity of sand reservoirs encased in shale.
S. N. Domenico.

Seismic Acquisition

Whatever happened to ground roll?
Nigel A. Anstey.

Field techniques for high resolution.
Nigel A. Anstey.

Vibroseis' gentle massage obtains structural data safely, economically.
N. A. Anstey.

The Vibroseis system of seismic mapping.
Robert L. Geyer.

Vibroseis parameter optimization.
Robert L. Geyer.

Seismic data enhancement—a case history.
R. J. Graebner.

Seismic Processing and Display

Common reflection point horizontal data stacking techniques.
W. Harry Mayne.

Correlation techniques—a review.
N. A. Anstey.

The digital processing of seismic data.
Daniel Silverman.

Seismic data display and reflection perceptibility.
Frank J. Feagin.

Semblance and other coherency measures for multichannel data.
N. S. Neidell and M. Turhan Taner.

Predictive deconvolution: theory and practice.
K. L. Peacock and Sven Treitel.

Estimation and correction of near-surface time anomalies.
M. Turhan Taner, F. Koehler, and K. A. Alhilali.

Seismic signal processing.
Lawrence C. Wood and Sven Treitel.

Seismic Velocities

Seismic velocities from surface measurements.
C. Hewitt Dix.

An analysis of stacking, rms, average, and interval velocities over a horizontally layered ground.
M. Al-Chalabi.

Time-depth and velocity-depth relations in western Canada.
C. H. Acheson.

Apparent velocity from dipping interface reflections.
F. K. Levin.

A velocity function including lithologic variation.
L. Y. Faust.

Seismic data indicate depth, magnitude of abnormal pressures.
E. S. Pennebaker, Jr.

Synthetic sonic logs—a process for stratigraphic interpretation.
R. O. Lindseth.

Velocity spectra—digital computer derivation and applications of velocity functions.
M. Turhan Taner and Fulton Koehler.

The effects of cracks on the compressibility of rock.
J. B. Walsh.

Migration

Migration.
P. Hood.

Two-dimensional and three-dimensional migration of model-experiment reflection profiles.
William S. French.

A simple theory for seismic diffractions.
A. W. Trorey.

Migration of seismic data from inhomogeneous media.
Les Hatton, Ken Larner, and Bruce S. Gibson.

Time migration—some ray theoretical aspects.
P. Hubral.

The wave equation applied to migration.
D. Loewenthal, L. Lu, R. Roberson, and J. Sherwood.

Wave-front charts and three dimensional migrations.
Albert W. Musgrave.

Migration by Fourier transform.
R. H. Stolt.

TABLE OF CONTENTS

GEOPHYSICS III
GEOLOGIC INTERPRETATION OF SEISMIC DATA

STRUCTURAL INTERPRETATION

A process of seismic reflection interpretation.
J. G. Hagedoorn.

Interactive seismic mapping of net producible gas sand in the Gulf of Mexico.
Alistair R. Brown, Roger M. Wright, Keith D. Burkhart, and William L. Abriel.

Interactive interpretation of seismic data.
Anthony C. Gerhardstein and Alistair R. Brown.

Geologic interpretation of seismic profiles, Big Horn basin, part II: west flank.
Donald B. Stone.

STRATIGRAPHIC INTERPRETATION

Inferring stratigraphy from seismic data.
R. E. Sheriff.

Seismic stratigraphy, a fundamental exploration tool.
G. R. Ramsayer.

Seismic facies analysis concepts.
M. M. Roksandic.

Seismic signatures of sedimentation models.
J. C. Harms and P. Tackenberg.

Integration of biostratigraphy and seismic stratigraphy: Pliocene-Pleistocene, Gulf of Mexico.
John M. Armentrout.

The role of horizontal seismic sections in stratigraphic interpretation.
Alastair R. Brown.

Seismic stratigraphy and global changes of sea level, part 10: seismic recognition of carbonate buildups.
J. N. Bubb and W. G. Hatlelid.

Seismic interpretation of carbonate depositional environments
J. M. Fontaine, R. Cussey, J. Lacaze, R. Lanaud, and L. Yapaudjian.

Seismic expression of carbonate build-ups, northwest Java basin.
J. E. Burbury.

Field development with three-dimensional seismic methods in the Gulf of Thailand—a case history.
C. G. Dahm and R. J. Graebner.

Aspects of seismic reflection prospecting for oil and gas.
P. N. S. O'Brien.

Predictive isopach mapping of gas sands from seismic impedance: modeled and empirical cases from Ship Shoal Block 134 field.
Robert D. Woock and Alan R. Kin.

New seismic technology can guide field development.
J. P. Lindsey, M. W. Schramm, Jr., and L. K. Nemeth.

How hydrocarbon reserves are estimated from seismic data.
J. P. Lindsey and C. I. Craft.

Progress in stratigraphic seismic exploration and the definition of reservoirs.
Norman S. Neidell and John H. Beard.

Interpretation of depositional facies from seismic data.
J. B. Sangree and J. M. Widmier.

Seismic stratigraphic model of depositional platform margin, eastern Anadarko basin, Oklahoma.
William E. Galloway, Marshall S. Yancey, and Arthur P. Whipple.

Exploration for oil accumulations in Entrada Sandstone, San Juan basin, New Mexico.
Richard R. Vincelette and William E. Chittum.

Table of Contents

Geophysics IV
Gravity, Magnetic, and Magnetotelluric Methods

Gravity and Magnetic Methods

Gravity and magnetics for geologists and seismologists.
L. L. Nettleton.

Exploring for stratigraphic traps with gravity gradients.
Sigmund Hammer and Rodolfo Anzoleaga.

Measurement of gravity at sea and in the air.
Lucien J. B. LaCoste.

An approximate solution of the problem of maximum depth in gravity interpretation.
D. C. Skeels.

Use of gravity, magnetic, and electrical methods in stratigraphic-trap exploration.
L. L. Nettleton.

The direct approach to magnetic interpretation and its practical application.
Leo J. Peters.

Magnetotelluric Methods

Basic theory of the magneto-telluric method of geophysical prospecting.
Louis Cagniard.

Processing and interpretation of magnetotelluric soundings.
G. Kunetz.

The magnetotelluric method in the exploration of sedimentary basins.
Keeva Vozoff.

Magnetotelluric responses of three-dimensional bodies in layered earths.
Philip E. Wannamaker, Gerald W. Hohmann, and Stanley H. Ward.

Synthetic Seismograms and Velocity Inversion

TECHNICAL PAPERS

THE SYNTHESIS OF SEISMOGRAMS FROM WELL LOG DATA*

R. A. PETERSON†, W. R. FILLIPPONE†, AND F. B. COKER†

ABSTRACT

Under certain simplified but realistic physical assumptions, the basic data from continuous velocity surveys in wells can be used to simulate the variations in acoustic impedance in the ground which give rise to seismic reflections. An analogue computer is described which makes use of the basic well data to produce synthetic seismic records which resemble actual seismograms from shothole explosions. This process provides an interesting insight into the requisite physical conditions, as well as the physical processes, whereby seismic reflections are set up in the earth. The close relationship between seismograms and well logs is brought out. Illustrations are given of field results.

INTRODUCTION

The basic concepts of the reflection seismograph method are relatively simple and straightforward. A charge of dynamite is exploded at A (Figure 1) and the resulting seismic pulse travels outwards as a wave in all directions. The wave is reflected at Horizon B and travels upwards towards the surface. The reflected wave is received by seismometer spread C–C, and recorded by instruments in truck E.

However, as we deal day by day with seismograms recorded from actual explosions in the real earth, it becomes apparent that there are a number of complicating factors which enter into the reflection seismograph process. In order more fully to understand and interpret the data on seismograms it is necessary that we look closely into all phases of the process.

For example, what is the form of the seismic pulse radiated from the shotpoint? How does it change during transmission through the earth? What is the nature of the reflection process in multiple-layered rocks? What are the characteristics of extraneous seismic waves and "noise"? What are the effects of the recording instrument system on the reflection data?

A complete analysis of these factors would encompass the entire field of reflection seismology. The following discussion will be limited more specifically to the reflection process itself in multiple-layered rocks in the light of accurate and detailed velocity information now available from continuous interval-velocity logging equipment. Certain aspects of the shotpoint and recording processes that are directly connected with the over-all problem will also be discussed.

---

* Presented at the Pacific Coast Regional Meeting of the Society in Los Angeles on November 12, 1954 and at the Midwestern Regional Meeting in Dallas on November 18, 1954. Manuscript received by the Editor March 14, 1955.

† United Geophysical Corporation, Pasadena, California.

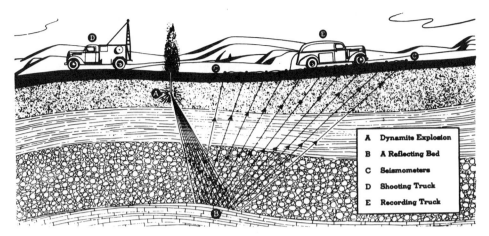

Fig. 1. A schematic diagram of the seismograph method.

FORMATION OF THE SEISMIC PULSE

Consider first the generation of the initial seismic pulse at the shotpoint.

(a) The dynamite explosion itself produces an intense pressure pulse of very short duration, measured in fractions of a millisecond.

(b) The wall rock or material surrounding the explosive charge is subjected to intense stresses exceeding the elastic limit for some distance out from the charge. In most cases the material flows "plastically," i.e., it goes through a non-linear stress-strain cycle (Lampson, 1945). Figure 2 illustrates schematically the behavior of a unit volume of wall rock material originally in physical state V_0, S_0, subjected to stress changes ΔS and volume changes ΔV by the passage of the intense explosion pressure pulse. It should be noted that the physical condition of the rock material suffers a permanent change. The final point of intercept on the $-\Delta V$ axis represents the permanent decrease in volume due to collapse of void spaces in the wall rock material.

(c) The velocities of propagation of various parts of the stress wave are proportional to the corresponding slopes of the stress-strain curve in Figure 2. The initial portion of the wave travels with a velocity characteristic of the medium and proportional to the curve slope near V_0, S_0. However, the high stress levels of the wave travel at a *lower* velocity proportional to the lesser slope near the crest of the stress-strain curve, and hence fall behind the initial part of the pulse. Because of this delay, and other mechanical factors related to the spherical divergence of the wave (Dix, 1949), the initial short duration pressure pulse of the dynamite explosion is ultimately *transformed* into a *seismic pulse* of considerably longer duration. This transformation is illustrated in Figure 3.

(d) It is significant to note that the "time break" as usually recorded on the seismogram corresponds to the instant of explosion and the "front" or "first break" of the traveling pulse. However, in the process of transformation of energy into the primary seismic pulse there is introduced an actual time delay (Figure 3). The "bulk" of the pulse energy is delayed and appears in the large excursions several milliseconds after the instant corresponding to the "time break." The magnitude of this seismic pulse delay is of the order of one half the width of the seismic pulse. This delay,

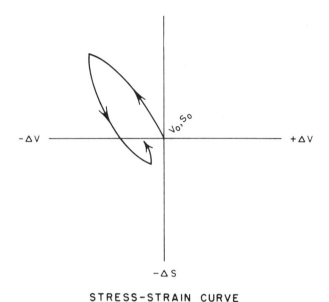

STRESS-STRAIN CURVE

FIG. 2. Stress-strain curve of rock material during passage of intense explosion pressure pulse.

of "mechanical" origin, is in addition to the electrical and electromechanical time delays occurring in the seismograph recording instruments.

(e) The form of this primary seismic pulse or "wavelet" has been carefully studied by Norman Ricker (1953a) and others. In general it will vary in breadth and duration depending on charge size and the type of material in which the charge is detonated. Also, as the pulse travels through the ground, it slowly loses its high-frequency content and increases in breadth. In some cases it may also be followed by an inverted smaller pulse resulting from reflection of the initial pulse from the surface of the ground or base of the weathered layer (Van Melle and Weatherburn, 1953). Nevertheless, for the sake of simplicity we may consider that the primary seismic pulse incident on various rock layers in the ground is ideally of the general shape indicated in Figure 3, with a typical period in the range of 0.010 to 0.030 seconds.

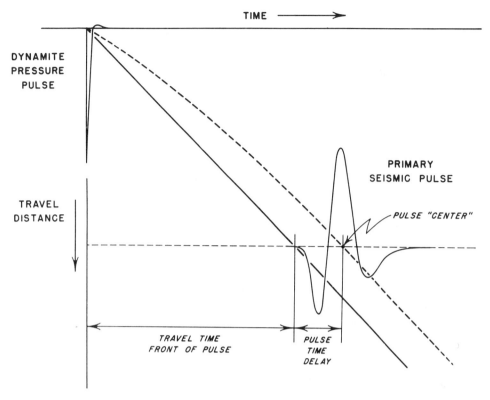

Fig. 3. Diagram illustrating transformation of dynamite pressure pulse into seismic pulse, and the attendant time delay.

THE REFLECTION PROCESS

We consider next the process of reflection of the downward-traveling primary seismic pulse. For the purposes of the present discussion, we will consider that the seismic pulse (once formed) propagates through the earth and reflects from rock strata in accordance with essentially *linear* laws. It will be appreciated later that this assumption allows the earth to be represented by linear analogue apparatus. Also, for the sake of simplicity we will disregard the generally spherical form of the wave front and treat the problem as that of a plane wave traveling vertically downwards in a direction perpendicular to flat-lying rock strata. If the wave is initially traveling in rock of density ρ_1 and seismic wave velocity v_1, and then passes into rock of density ρ_2 and velocity v_2, a portion of the energy in the wave will be reflected at the interface and the remainder will be transmitted. This reflection process is diagrammed schematically in Figure 4.

Seismic wave travel time related to depth in the ground is scaled vertically downwards and elapsed time in the sense of measured time on the seismogram is scaled horizontally. The downward-traveling incident wave is partially reflected by the abrupt change in rock density and velocity. For the case of normal incidence, the reflected and transmitted pulses have identically the same shape and breadth as the incident pulse, but differ in amplitude. The ratio of the amplitude of the reflected wave to that of the incident wave is termed the "reflection coefficient." The reflection coefficient R has the value given by the expression

$$R = \frac{A_r}{A_i}$$

$$R = \frac{\rho_2 v_2 - \rho_1 v_1}{\rho_2 v_2 + \rho_1 v_1}.$$

This expression as written refers directly to "pressure amplitude." In the case when A refers to particle velocity, the polarity of this quantity must be defined in reference to the direction of wave travel. The relationship illustrates that the amplitude of the reflected pulse is determined by the *change* in the density-velocity product between the two rock layers. This product is commonly referred to as the "acoustic impedance" of the rock. Thus we say that the reflection of seismic waves results from variations in acoustic impedance of the rock medium in which the wave travels. Figure 4 illustrates the case of a single-step variation

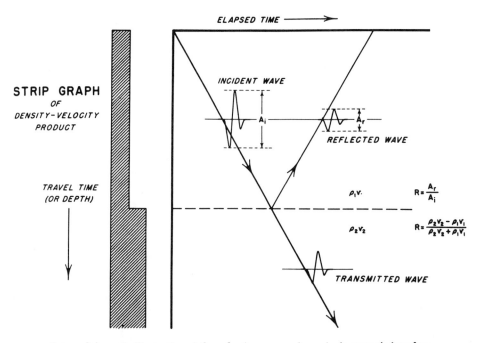

FIG. 4. Schematic illustration of the reflection process for a single acoustic interface.

in acoustic impedance, as indicated in the strip graph at the left hand side.

In addition to specifying the amplitude of the reflected wave, the above expression also indicates that the algebraic sign associated with the reflection coefficient depends upon the relative values of the acoustic impedance of the media adjacent to the interface. In particular, when the incident wave is propagating from a medium of low acoustic impedance into one of higher acoustic impedance, the corresponding reflection coefficient is positive. In this case, the incident, reflected, and transmitted waves will all have the same phase or "polarity." On the other hand, when the wave travels progressively from a medium of higher acoustic impedance into one exhibiting a lower value, the corresponding reflection coefficient is negative. When this situation exists, both the incident and transmitted waves have the same polarity, but the polarity of the reflected wave is reversed.

The Logarithmic Approximation

At this point we can accomplish a significant simplification by introducing an approximate expression for the reflection coefficient, namely

$$\frac{A_r}{A_i} = \frac{\rho_2 v_2 - \rho_1 v_1}{\rho_2 v_2 + \rho_1 v_1}$$

$$A_r \cong \left[\frac{\Delta(\rho v)}{2(\rho v)}\right] A_i$$

$$A_r \cong \left[\frac{1}{2} \Delta \log (\rho v)\right] A_i$$

where
- ρ = rock density
- v = rock velocity
- Δ is "incremental change in."

In using this expresion we may picture the earth as built up of a large number of very thin rock layers, each with a different value of acoustic impedance (ρv). Then the above expression may be interpreted as stating that *the amplitude of the wave reflected by each incremental change or "step" in acoustic impedance is proportional to the corresponding incremental change in the value of the logarithm of acoustic impedance*. For the sake of simplicity, the progressive change in amplitude in the seismic pulse as it travels downwards is disregarded. Although only an approximation, this expression holds very closely, even for quite large-step changes in the value of the acoustic impedance, as illustrated in Figure 5. The utility of the expression will become apparent as discussion proceeds.

Consider next the situation arising from two changes in acoustic impedance, as diagrammed in Figure 6. In this case, the upper "step" change represents an increase in the value of the acoustic impedance, while the lower step corresponds

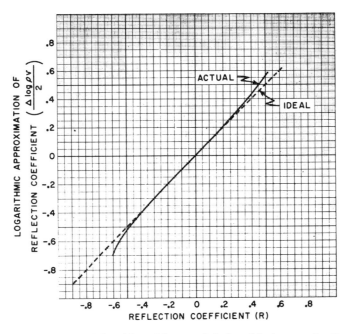

Fig. 5. Illustration of the wide useful range of the logarithmic approximation.

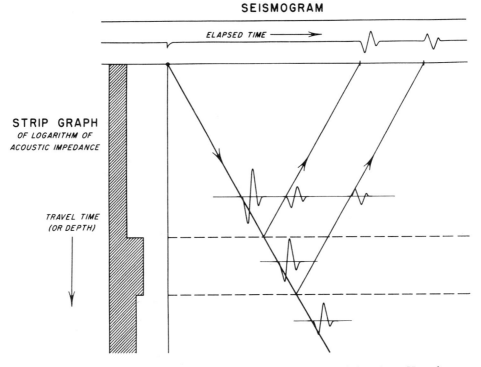

Fig. 6. Schematic illustration of the reflection process for two acoustic interfaces. Note the inversion of the wave reflected from the lower interface.

to a decrease. Consequently, the two resulting reflected pulses have opposite polarities. In addition, other small multiply reflected pulses could be shown, but these are disregarded here for the sake of simplicity.

REFLECTION PROCESS IN THE LAYERED EARTH

In the actual earth we rarely, if ever, have situations as simple as those depicted in Figures 4 and 6. Detailed velocity data available from continuous interval velocity surveys indicate that acoustic impedance varies almost continuously with depth, in direct relation to the type of rock. Figure 7 illustrates a typical relationship between lithologic, velocity, and electrical resistivity well logs. It is apparent that in contrast with the case of a few simple "steps" in

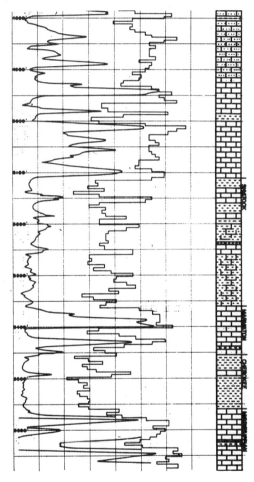

FIG. 7. Comparison of lithologic, interval velocity, and resistivity well logs.

rock velocity, we are confronted with a more complicated problem in analyzing the actual reflection process. The principles involved, however, are the same as discussed before. Even if the graph of the logarithm of acoustic impedance versus travel time or depth is essentially a continuous curve, we may consider it built up of a large number of small steps. In principle these steps may be thought of as occurring at infinitesimally small depth intervals, but in practice they need only be closely spaced compared with the shortest wavelength of interest (Wolf, 1937). Then, as shown in Figure 8, each small step may be considered to give rise to a small reflected pulse of appropriate amplitude and polarity, and the reflected wave energy observed at the surface is the summation of all these small individual reflected pulses. Here again in the interest of simplification we are disregarding the many multiple reflections arising in a layered medium, as well as the progressive change in amplitude in the downward-traveling primary seismic pulse.

In principle, the summation of the many individual layer reflection contributions can be carried through by numerical methods, using adding machines or digital computing devices. Usually, however, the process can be carried out much more conveniently and efficiently by the use of an analogue type computer.

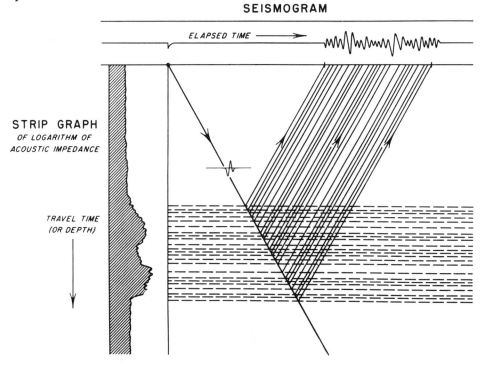

Fig. 8. Schematic illustration of the reflection of a seismic pulse from an acoustic impedance distribution exhibiting continuous variation.

PRACTICAL SYNTHESIS APPARATUS

Figure 9 shows a portion of United Geophysical Corporation's SEISYN COMPUTER, an analogue computer for the synthesis of reflection seismograms from basic well log data.

The several SEISYN components are illustrated schematically in the block diagram of Figure 10. The functions of these components are briefly discussed in the following paragraphs.

The first step in the computing process is the introduction into the SEISYN apparatus of the basic information concerning the variation of acoustic impedance of rock strata with depth, or rather with vertical travel time of seismic waves. This operation can be carried out in several ways, but only one method will be described here. A strip graph is made with the logarithm of acoustic impedance plotted against twice the vertical travel time. Since the acoustic impedance is equal to the numerical product of rock density and rock velocity, detailed well logs of both density and velocity would be desirable. However, it is usually not necessary to have both logs available in order to synthesize seismograms satisfactorily. In fact, if the rock density and velocity are related approximately by any general expression of the form

$$\rho = kv^n$$

FIG. 9. Photograph of a portion of a model of the SEISYN COMPUTER.

then it is apparent that the logarithm of the density-velocity product is equal to a constant plus $(n+1)$ times the logarithm of the velocity alone. Moreover, it is not necessary that the values of k and n remain strictly invariable. It is sufficient that they be reasonably constant over intervals in the well corresponding approximately to the longest wavelengths of interest. Therefore, a detailed continuous velocity log will usually suffice for the purpose of plotting the strip graph.

The graph is photographically reproduced on film with the included area on one side of the curve "blacked in," and the area on the other side transparent.

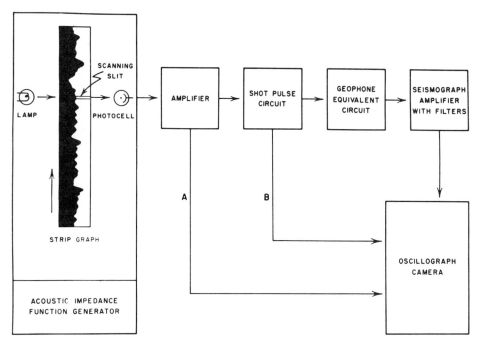

Fig. 10. Block diagram of Seisyn Computer.

The film strip is then photoelectrically scanned at an appropriate speed in the Seisyn apparatus to produce an electrical signal proportional to the logarithm of acoustic impedance. This signal is amplified and passed through a pulse-forming circuit so designed that an applied single voltage step corresponding to a single step in acoustic impedance produces a single seismic-type pulse corresponding to the pulse reflected from a single rock discontinuity. Since the strip graph is considered to be built up of a large number of small steps, each individual step will produce a reflected pulse of correct amplitude and the resulting output of the pulse-forming circuit will be the summation of pulses from all incremental steps in acoustic impedance. Furthermore, since the strip graph is plotted to twice travel time scale, each incremental reflected pulse will have an appropriate time delay corresponding to the reflection time from the shot origin to the point of reflection and return. Thus the output signal of the pulse circuit automatically represents the synthetic or predicted ground motion, usually thought of in terms of ground particle velocity at the surface geophone location. This information can be photographically recorded on channel B of Figure 10.

To incorporate the action of the seismograph recording system, this composite ground motion signal may be used either to drive a shaking table on which geophones are placed, or it may be introduced into an electrically equivalent geophone circuit to accomplish the same purpose. In either case, the resulting geophone output signal is introduced into a seismograph amplifier having appropriate filter characteristics. Then the amplifier output signal is recorded in conventional fashion on an oscillograph record, usually in bridled form on several traces so that it can be compared directly with an actual seismogram taken in the field at the well location.

Examples of Synthetic Seismograms

At the time of writing of this paper, only a limited number of synthetic seismograms have been prepared for comparison with actual field seismograms recorded at the corresponding well locations. In view of the several simplifications and approximations utilized in the process, it would not be surprising if agreement between actual and synthetic seismograms were less than perfect. However, experience to date provides every indication that the principal features of actual field seismograms exclusive of seismic "noise" can be predicted with fair precision by synthesis from continuous interval velocity well data. In cases where reflection events observed on the actual record are not indicated on the synthetic record, the lack of agreement may in itself be significant. The possibility of multiple reflections or other such phenomena may be indicated.

Comparisons of corresponding actual and synthetic seismograms for two field examples are shown in Figures 11 and 12. In each case the upper panel shows portions of the actual field seismogram; the second panel shows the corresponding synthetic seismogram recorded with identical geophones and filters; the third panel of Figure 11 illustrates the lithologic and continuous interval-velocity well logs plotted versus two-way vertical travel time; and the fourth or lower panel shows the synthetic seismogram advanced in time to compensate for the time delay in the over-all reflection seismograph process. The same time delay is present in both field and synthetic seismograms. In the case of the field seismogram, part of the delay is "mechanical" in origin, and originates in the process of transformation of the initial dynamite pulse into the primary seismic pulse. In the synthetic seismogram, the same time delay originates electrically in the "shot pulse circuit." In both cases, the geophones and amplifiers introduce time delays of electrical and electromechanical origin.

In Figure 11 two of the principal reflection events are timed at .358 and .513 seconds. The corresponding times on the synthetic record are .358 and .510 seconds. Particularly in the case of the .358 second reflection event, the similarity in

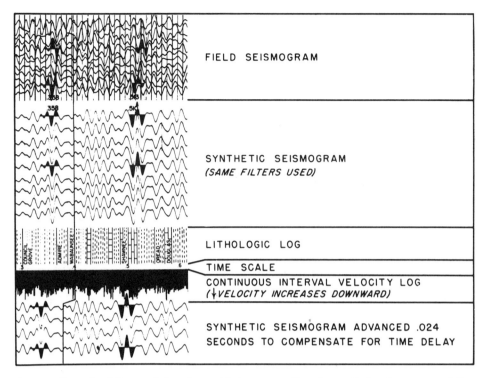

FIG. 11. Example of the correlation observed between a portion of a synthetic seismogram and the corresponding portions of a field seismogram and the lithologic and interval velocity well logs.

waveform details between the field and synthetic seismograms is very close. On the lower panels of Figures 11 and 12, a good degree of correlation exists between the continuous interval velocity log and the synthetic seismogram traces. This result is to be expected within the frequency limits of the seismic pulse spectrum and the recording system transmission characteristics.

The field seismogram of Figure 12 was recorded in an area of poor record quality, due to highly irregular near-surface conditions and a consequent high level of "noise" and scattered energy. Also, the shotpoint and geophone spread were located approximately 2100 feet from the well location in which the continuous interval velocity data were obtained. Nevertheless, there appears to be general agreement for the two principal events, with corresponding times of .655 versus .643 seconds and .807 versus .792 seconds respectively. In the lower panels of Figure 12, the continuous interval velocity log is drawn with polarity opposite to that of Figure 11, and the correlation with the synthetic seismogram traces is best for this choice. This situation will be discussed more fully later in reference to problems of phase characteristics.

Up to this point the principal steps in the process of seismogram synthesis have been briefly described. In the following, the discussion of some of the steps will be expanded, and some general conclusions will be drawn.

Choice of Seismic Pulse Network Characteristics

Consider again the seismic pulse. In field procedure it is formed by an explosion in the shot hole. The shape of the resulting pulse and its duration are largely controlled by the size and depth of charge and by the physical properties of the wall rock. In the SEISYN analogue apparatus the seismic pulse is simulated by passing a step voltage through an electrical network. Normally, this network can be a band-pass filter having adjustable low and high frequency cut-off points and variable attenuation slopes. In principle, then, various shot hole parameters are simulated by adjustment of the several filter parameters. The proper selection of these parameters is, of course, the practical problem.

There are two general approaches to this problem. The ideal situation is one in which it is possible to obtain a "true" pressure recording of the down-traveling seismic pulse simultaneously with the recording of the field seismogram over the well. This approach calls for a pressure geophone at appropriate depths in the well and associated equipment for recording the seismic pulse from surface shot holes. Then in the SEISYN process, the electrical-analogue "seismic-pulse"

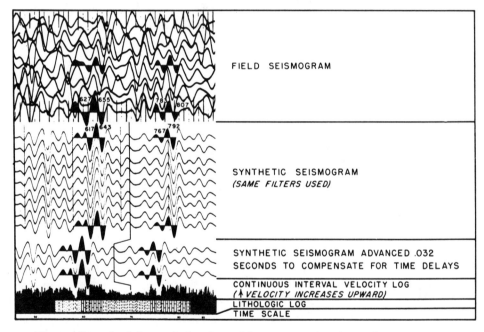

FIG. 12. Example of the correlation observed between a portion of a synthetic seismogram and the corresponding portions of a field seismogram and the lithologic and interval velocity well logs. This area is one of poor record quality but major events appear to correlate.

filter can be adjusted until the output pulse produced by a step voltage input matches the pressure pulse observed in the well.

However, in many cases when the seismic pressure pulse data are not available, another approach must be followed. This alternative method consists of "cut-and-try" selection of the seismic pulse filter parameters until the best "fit" is obtained between the synthetic and actual field seismograms. This procedure appears to work out quite well in practice.

Figure 13 shows a comparison for the same well, between a seismic pulse recorded by a deep-well pressure geophone, and the electrical-analogue seismic pulse incorporated in the synthesis of the seismogram shown in Figure 11. In this case the deep-well pressure geophone recording was made during a conventional well velocity survey, while the field reflection seismogram of Figure

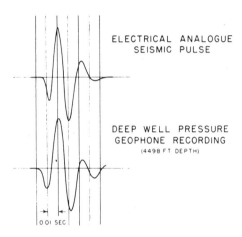

FIG. 13. Comparison of SEISYN pulse and seismic pulse recorded in a deep well.

11 was recorded at another time. Nevertheless, the electrical analogue pulse, arrived at by cut-and-try procedures to give the best fit to the field seismogram, closely resembles the observed pressure pulse.

A More Complete Analogue

In many cases the electrical analogue seismic pulse arrived at by the cut-and-try process is more complicated than actual pressure pulses observed in the well. The reason is probably to be found in the presence of other complicating factors. For example, the weathered layer through which the reflected waves must pass has transmission velocities highly contrasting with the rocks below and the air above. Under certain conditions, strong multiple reflection phenomena can take place, with the result that the weathered layer takes on the characteristics of a frequency selective wave filter (Wolf, 1940). Accordingly, a more complete analogue computer should include a filter unit to simulate the filtering action of the weathered layer. This filter is shown in block diagram form in Figure 14. However, since there are rarely sufficient field data to determine the characteristics of such a filter directly, it is necessary to consider its effects to be combined with those of the shot-pulse filter. The combined units are outlined by a dashed line block labeled SEISYN PULSE NETWORK in Figure 14. In addition, there may be other frequency-selective elements in the complete seismic reflection recording process, such as the mechanical coupling between the geophone

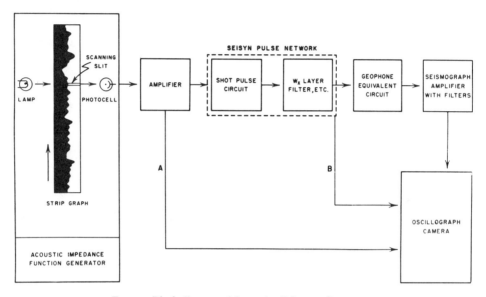

FIG. 14. Block diagram of "complete" SEISYN COMPUTER.

and surface soil. Such effects again may be considered as combined or included with others in the SEISYN PULSE NETWORK, and are thus incorporated in the final result through the cut-and-try procedure previously described.

RELATIONSHIP BETWEEN SEISMOGRAMS AND WELL LOGS

Consider another aspect of the seismogram synthesis process, that is, the close relationship observed between the traces of a reflection seismogram and a well log of lithology versus depth. In the actual reflection seismograph process, we normally think of an over-all "system" in which the seismic pulse generated by a shot-hole explosion is the "input signal." The earth corresponds to a "transmission line" or medium, and the geophones and amplifiers constitute the receiving end of the transmission system.

On the other hand, by utilizing the electrical analogue process presented here, we arrive at the same end result, but with a different arrangement of "components." In particular, the layered earth which is the "transmission line" in the actual seismograph process, becomes the "input signal" in the analogue process. Moreover, since acoustic impedance is one of the physical properties of rocks, we can say, in effect, that the "lithologic" well log, plotted against travel time rather than depth, is the input signal to the reflection seismograph system, and that the reflection seismogram is the output signal. Figure 15 illustrates this viewpoint. In the light of well known electrical circuit behavior, this manner of looking at the reflection seismograph process is particularly helpful in recognizing the detailed and controlling influence of lithologic sequence on the reflection seismogram. It is first of all apparent that the frequency-selective filters of the seismograph amplifier and the frequency spectrum of the primary seismic pulse have a controlling influence on the seismogram. However, it is even more signifi-

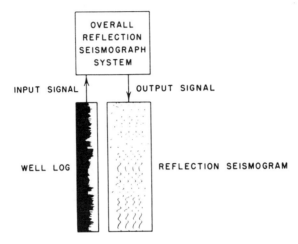

FIG. 15. Illustration of relationship of well log to reflection seismogram.

cant to note that the sequence, timing, periods, and "form" of the reflected waves on the seismogram, i.e., the output signals, are directly controlled by and are correlative with the "lithologic" well log plotted to appropriate time scale, i.e., the input signal. With due allowance for the time delays and frequency discrimination inherent in the recording system and seismic pulse frequency spectrum, it follows that individual "peaks" and "troughs" on the seismogram correspond to individual high or low velocity zones in the lithologic section. Wide peaks or troughs correspond to "thick" beds or velocity zones, and narrow peaks or troughs correspond to "thin" beds or zones.

It is of interest to note that if the over-all reflection seismograph system were completely "high fidelity," that is, if it transmitted all frequencies equally well and introduced no "noise," the output signal would exactly duplicate the input signal. In other words, the reflection seismogram trace would be an accurate facsimile of the well log (logarithm of acoustic impedance plotted to twice vertical travel time). Actually this idealized situation is never realized because it would require that the primary seismic pulse be a simple "step wave," and also, that the recording instrumentation be completely "broad band." In the field operations, the presence of extraneous seismic waves and incoherent "scattered" seismic energy usually requires the use of frequency selective filters in the recording system. However, as we know from experience with other "imperfect" transmission systems, the output signal can still bear a reasonable similarity to the input signal. In this sense the reflection seismogram is a "reasonable facsimile" of the well log, within the limits of the frequency band passed by the over-all system. This fact is borne out in Figures 11, 12, and 15.

In Figure 16 a hypothetical well log has been constructed to illustrate the principles of the relationship between lithology and the corresponding seismogram. Particularly idealized situations have been chosen to show the nature of the relationship in some detail. The early or shallow part of the well log depicts a

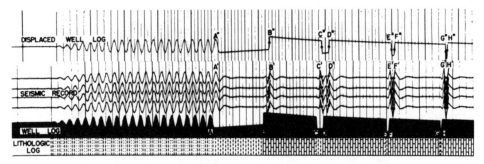

Fig. 16. Hypothetical well log and corresponding seismogram, illustrating principles of relationship.

sustained alternating series of relatively high and low velocity beds, for example, sandstone and shale. This series ends at A, and is followed by an extended low velocity zone to depth B. Below this depth is a continuous high velocity zone, for example, limestone, with low velocity "shale" beds at CD, EF, and GH. Above the hypothetical well log of Figure 16 is plotted the corresponding synthetic seismogram. Relatively wide frequency spectrum seismic pulse and amplifier filters have been employed. It will be noted that the seismic traces are delayed approximately .017 seconds relative to the well log. In typical field practice, total delays of .025 to .055 seconds or more are commonly observed. This total delay represents the sum of the mechanical delay inherent in the primary seismic pulse and the electrical and electromechanical delays in the recording system.

For comparison purposes the well log trace has been displaced .017 seconds to the right and replotted above the seismic record traces. Corresponding features can then be correlated directly. Peaks on this seismic record correspond directly to velocity maxima on the well log, and troughs correspond to velocity minima. It is interesting to note that the seismic reflection wave frequency in the early part of Figure 16 is *independent* of both the frequency characteristics of the seismograph recording system, and the spectrum of the primary seismic pulse. A sinusoidal input signal invariably produces a sinusoidal output signal of identical frequency (assuming linear operation). Since the frequency spectrum of this particular input signal, in the sense of Figure 15, contains essentially only a single frequency component rather than a continuous band of frequencies, no change in "character" can be produced by selective filter action. For example, wavelet contraction and wavelet expansion type filters cannot produce their effects on this type of input signal.

Consider next the low velocity interval AB. The preceding rock velocity variations terminate at A, or A'', and the velocity remains essentially constant to B. Likewise the corresponding reflected waves "terminate" at A', but not as abruptly as in the case of the well log. This effect is due to the "transient" behavior of the seismic amplifier filter and recording system, and to the "oscillatory" form of the primary seismic pulse. Ideally the seismogram would exhibit a quiet zone from A' to B', but in actual practice the interval will be occupied by extraneous seismic waves and "noise."

The sudden increase in rock velocity at B gives rise to the reflected wave at B'. It is interesting to note with reference to the displaced well log trace that the "upsweep" in velocity at B'' is correlative with the "upsweep" from trough to peak marked with the slant line at B', rather than the small "peak" some .010 seconds earlier. This earlier energy may be regarded as part of a "precursor" resulting from filter limitations, in the same way that the energy following the slant line may be regarded as a "tail" similarly resulting from filter limitations. The case of a single "step" in velocity corresponds to an input signal with a very broad frequency spectrum. The frequency pass band of the filter system is by comparison very limited. Hence the "output signal" on the seismogram can only be an imperfect facsimile of the input step wave signal.

The low velocity "shale" beds at CD, EF, and GH are correspondingly represented on the seismogram by "troughs" bordered by slant lines at $C'D'$, $E'F'$, and $G'H'$. Norman Ricker (1953b) has clearly described the principles controlling the degree of resolution obtainable.

Figure 17 provides a further illustration of the relationships between the seismogram and lithology. As in Figure 16 the early part of the log represents alternating high and low velocity beds. The corresponding seismogram and displaced well log are plotted above as in the case of Figure 16. However, a somewhat more narrow filter has been used, and the time delay correction in this case is .027 second. It is to be noted that although the boundaries between successive beds are "sharp" rather than sinusoidal as in Figure 16, the seismogram is nevertheless essentially sinusoidal. The sharp formation boundaries represent a high frequency content in the input signal, which is not passed by the relatively narrow filter system. Hence, essentially only the "fundamental" frequency corresponding to the basic lithologic sequence is recorded. The abrupt changes in bed thicknesses at I and J are approximately depicted on the seismogram at I' and J'.

Tuning and Reinforcement Effects

In addition to general frequency effects, a "tuning" or "reinforcement" phenomenon may occur for certain relationships between bed thickness and the maximum response frequency of the combined seismic pulse filter-geophone-amplifier system. In particular, for the case of beds of alternating high and low velocity, a bed thickness of one half wavelength computed in terms of two-way travel time and peak frequency of the above system will result in a considerable

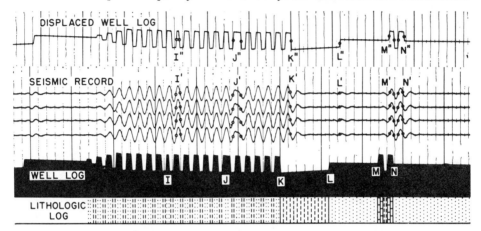

FIG. 17. Hypothetical well log and corresponding seismogram further illustrating principles of relationship.

increase in reflection amplitude. If velocity contrasts are strong, the effect may be further enhanced by multiple reflection phenomena within the rock layers.

This reinforcement or tuning effect is illustrated at $M'N'$ in Figure 17 and is also apparent in Figures 11 and 12, where it will be noticed that the strongest reflection events correspond to alternating sequences of high and low velocity strata of appropriate thickness. In fact, it can probably be assumed that when "noise" is not the controlling factor, the seismograph observer often obtains his "best appearing" records by empirical cut-and-try adjustment of filters and shooting procedures to "tune in" with the thickness of certain sequences of strata of alternating high and low velocities.

Examination of synthetic seismograms leads to the observation that a "continuous train" of reflected waves is usually to be expected. In practice, there are usually no true "beginnings" of individual reflection events, but rather successive "swelling" and "pinching" of envelope amplitude, corresponding to "tuning" effects between bed thickness and the over-all system frequency characteristics. In effect, every peak and every trough represents reflected energy corresponding to variations in rock velocities. In actual field seismograms, extraneous "noise" is also present, and only the stronger reflection events are visible, the lesser events being "lost" in the background of noise.

TIME DELAY AND PHASE DISTORTION CORRECTION

The close relationship between the reflection seismogram and the corresponding well log raises the question: Can this relationship be improved or even made exact through the use of suitable corrective filters and networks? In general, rather severe limitations are imposed on this approach by the presence of extraneous noise. It is apparent that if the signal-to-noise ratio is less than unity in certain attenuated frequency bands, the restoration of response in these bands may only succeed in bringing up the ground noise level relative to the signal. However, interesting possibilities exist in the correction of over-all phase distortion in the system.

The general result of phase distortion is to cause different frequency components to suffer different time delays. In particular, this factor may be considered

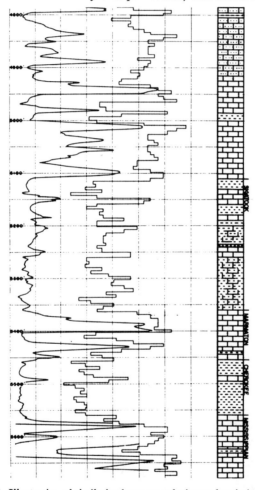

FIG. 18. Illustration of similarity between velocity and resistivity logs.

in the light that reflected energy from thin beds or strata experiences a different time delay than reflected energy from thick beds. Such distortion, of course, weakens and confuses the relationship between the reflection seismogram and the corresponding well log. In making corrections for phase distortion, it is desirable to correct *over-all* distortion including the geophone and amplifier, as well as a phase distortion which may be inherent in the shape of the seismic pulse. Synthetic seismogram studies are particularly helpful in arriving at data necessary for devising methods of *over-all* correction of the phase distortion in the entire system.

In addition to time delay effects, phase distortion can cause confusion as to "polarity" relationships between input and output signals, in this case between the well log and the reflection seismogram. As noted in the discussion of Figures 11 and 12, the closest correlations between seismogram traces and well log required an inversion of polarity (and accompanying change in time delay correction as in the case of Figure 12.) Presumably, this situation results when the tangent intercept of the phase shift curve for the entire system falls near an odd multiple of 180°. It is of interest to note that polarity "inversion" can result from the form of the primary seismic pulse, as well as from the phase characteristics of the recording instruments.

SEISMOGRAM SYNTHESIS FROM OTHER TYPES OF WELL LOG

As a final topic, it is of interest to note the close relationship observed in many instances between rock velocity and electrical resistivity. Figure 18 illus-

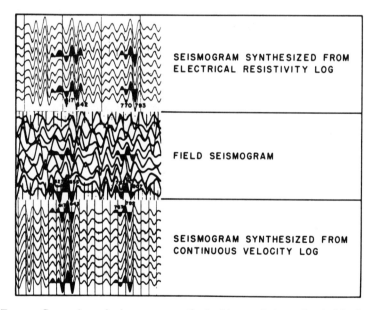

FIG. 19. Comparison of seismograms synthesized from velocity and resistivity logs.

trates the close similarity between the electrical resistivity log and continuous interval velocity log for a particular well. This similarity suggests the possibility of constructing synthetic seismograms from resistivity and other logs as well as continuous velocity logs. Figure 19 compares a synthetic seismogram made in this fashion with the field seismogram of Figure 12. Also shown is a seismogram synthesized from the continuous velocity log of the same well. In this particular case the agreement between the "resistivity seismogram" and the actual field seismogram is substantially as good as the agreement between the "velocity seismogram" and field seismogram. It is believed that refinements in the technique of employing resistivity and other logs may even further extend the usefulness of the seismogram synthesis procedure.

ACKNOWLEDGMENTS

The authors wish to thank the United Geophysical Corporation for encouragement in publishing this paper. They also wish to thank Dr. C. A. Swartz, Mr. E. T. Howes, and Mr. R. L. Duke for valuable help and suggestions.

REFERENCES

Dix., C. Hewitt, 1949, On the minimum oscillatory character of spherical seismic pulses: Geophysics v. 14, p. 17–20.

Lampson, C. W., 1945, Final report on effects of underground explosions: NDRC Report No. A-479, OSRD Report No. 6645.

Ricker, Norman, 1953a, The form and laws of propagation of seismic wavelets: Geophysics, v. 18, p. 10–40.

Ricker, Norman, 1953b, Wavelet contraction, wavelet expansion, and the control of seismic resolution: Geophysics, v. 18, p. 769–792.

Van Melle, F. A. and Weatherburn, K. R., 1953, Ghost reflections caused by energy initially reflected above the level of the shot: Geophysics, v. 18, p. 793–804.

Wolf, Alfred, 1937, The reflection of elastic waves from transition layers of variable velocity: Geophysics, v. 2, p. 357–363.

Wolf, Alfred, 1940, The time delay of a wave group in the weathered layer: Geophysics, v. 5, p. 367–372.

INVERSION OF SEISMOGRAMS AND PSEUDO VELOCITY LOGS *

BY

M. LAVERGNE ** and C. WILLM **

Abstract

LAVERGNE, M., and WILLM, C., 1977, Inversion of Seismograms and Pseudo Velocity Logs, Geophysical Prospecting 25, 231-250.

Pseudo velocity logs can be obtained by seismogram inversion, using true amplitude processing and detailed investigation of move-out velocities. The precision of the results depends on the quality of the seismic data and on the possibility of deconvolving without increasing the noise. An investigation is made of the deformation of pseudo logs due to seismic signal variations and to imperfections of deconvolution.

Both marine and land examples are shown, in some cases with adjustment on well logs. When the dips are large, time sections must be migrated and pseudo velocity logs must be computed from migrated sections. Comparison of sonic logs with pseudo velocity logs obtained in the same area is usually good enough to obtain information on lithological parameter variations by adjustment of pseudo velocity logs on sonic logs. Even when no well is available, pseudo velocity logs can give some indications on the nature of sediments between seismic horizons.

1. Introduction

Geophysical methods for lithology identification and direct detection of hydrocarbons have been developed to a large extent in the last three years.

Information pertaining to lithology and petrophysics can thus be obtained by a detailed investigation of seismic amplitudes. For example, relative amplitudes are used to estimate lithologic variations in sedimentary sequences (O'Doherty and Anstey 1971) and to identify the existence and the extent of some hydrocarbon zones (Larner, Mateker, and Wu 1973; Lindsey and Craft 1973; Sheriff 1973).

Relative amplitude can also be converted into velocities and densities to give a better estimation of lithology and petrophysics. In favourable conditions—such as small dips, low noise, and the absence of multiple reflections—it is possible to compute pseudo acoustic-impedance logs, or pseudo velocity and pseudo density logs from seismic sections with preserved amplitude (Lavergne 1975).

---

* Paper presented at the 37th E.A.E.G. Meeting in Bergen, June 1975.
** Institut Français du Pétrole, 92502 Rueil-Malmaison, France.

2. Principle of the Method

1. *Initial processing with amplitude preservation* (cf. fig. 1)

Amplitude preservation means that a given reflecting horizon should show the same amplitude characteristics on the final section, no matter what type of sequence is encountered in the overlying sediments. It is only partially obtained by the initial processing.

This consists in:

True amplitude recovery

T.A.R. is the process for removing the effects of variable gain in the field recording and for adjusting the amplitude by an appropriate gain program to compensate for spherical divergence and absorption.

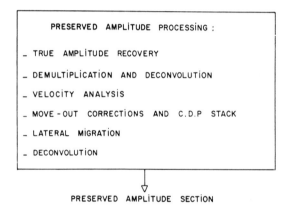

Fig. 1. Initial processing with amplitude preservation.

The processing should be carried out with amplitude preservation, and neither dynamic equalization nor normalization should be applied.

By dynamic equalization, we mean adjusting the gain of different time windows along the trace so that their amplitudes are comparable. By normalization, we mean scaling the amplitudes so that every trace has the same mean energy level.

For divergence correction, a method taking into account the ray curvature is preferably used (Newman 1973). For absorption correction, an average of 0.1 to 0.3 dB per wavelength depending on the nature of the subsurface is generally introduced.

The gain program should be applied only to parts of the record where the signal/noise ratio is greater than 1, in fact from the shallowest identifiable reflection down to a certain depth.

Demultiplication (i.e. removal of multiple reflections) and deconvolution before stacking

When necessary, deconvolution before stacking is performed by a predictive filter in order to eliminate reverberations and peg-leg multiples. The prediction interval should be adapted to the period of the multiple.

Velocity analysis

Very accurate moveout velocity analysis should be made. This information is used both for moveout correction and for pseudo velocity calibration. The interval velocities are used, after appropriate corrections, to introduce low frequency information into the pseudo velocity log.

Moveout corrections and CDP stacking

CDP gathering should be performed without dynamic equalization and normalization. If an equalization should still be absolutely necessary, large enough time-windows should be used to preserve relative amplitudes.

The effect of spread length on the amplitude and phase is difficult to estimate, and therefore all reflections with too large angles of incidence should be eliminated by muting.

Lateral migration

Whenever dips become large, a lateral migration should be performed in order to place events in a proper position and to remove diffraction hyperbolas.

Lateral migration is preferably done by the wave equation method using the downward continuation of seismograms because it preserves the character and the relative amplitudes of reflections (Claerbout and Doherty 1972).

Deconvolution after stacking

Deconvolution after stacking is eventually performed in order to contract the seismic pulse. It must be emphasized, however, that deconvolution is basically a destructive form of filtering (Paige 1973). Minor changes in frequency content can produce great changes in output amplitude. The deconvolution process therefore should not attempt to enlarge too much the frequency spectrum but rather to shorten the pulse and reduce its number of arches.

2. *Pseudo acoustic impedance and pseudo velocity computation*

Pseudo acoustic-impedance and pseudo velocity sections are obtained from the preserved amplitude section by the technique described in fig. 2.

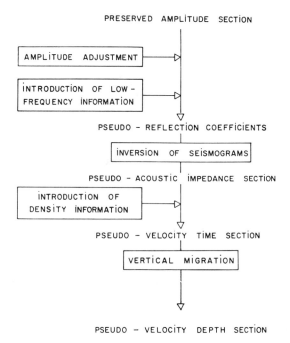

Fig. 2. Processing for pseudo velocity computation.

Amplitude adjustment

The amplitudes of the seismic trace are multiplied by a scaling factor, which transforms the trace into a pseudo reflection coefficient time-series. This is done by calibrating some particular horizons of the seismic section with the corresponding reflection coefficients measured on a sonic log or determined by other methods (cf. 4-1).

Introduction of low-frequency information

Low-frequency information, mainly below 10 Hz, is absent on seismic traces. It must be introduced into the seismic record using the move-out velocity analysis information or other sources of information.

Inversion of seismograms—Computation of acoustic impedances

The computation of the acoustic-impedance log is performed using an algorithm that is inverse of that used to compute the reflection coefficients from the acoustic impedance.

The acoustic impedance Z_{i+1} in the $(i+1)$th layer is computed as a function of the acoustic impedance Z_i in the ith layer and the reflection coefficient k_i at the ith interface by the formula:

$$Z_{i+1} = Z_i \frac{1 + k_i}{1 - k_i} \qquad (1)$$

An initial value must be given to the acoustic impedance Z_1 in the first layer.

Algorithm (1) is derived from the pressure amplitude reflection coefficient for normal incidence:

$$k_i = \frac{Z_{i+1} - Z_i}{Z_{i+1} + Z_i} \qquad (2)$$

The transmission losses are not taken into consideration. When they become important, for example for a layered system of sedimentation, appropriate corrections should be made.

Computation of velocity and density

When pseudo velocity logs are requested, rather than pseudo acoustic-impedance logs, velocity and density must be computed separately. This is possible if a velocity-density dependence can be derived, for example from statistical well analyses, in definite areas and for certain types of sediments. In the Paris Basin and the Aquitaine Basin, an average velocity-density relation can be derived in most cases in shaly sands, shaly limestones, and marls:

$$D = 1.62 + 0.00021 \, V \qquad (3)$$

where D is the density in g/cm³ and V the longitudinal velocity in m/s.

In many cases, however, there is no simple velocity-density relation. For example, in problems involving reflections from oil and gas sand reservoirs, velocities and densities are related by several parameters, including water saturation, porosity, and compressibility (Domenico 1974).

Time to depth conversion

If necessary vertical migration is performed, using the pseudo velocity time function. Pseudo velocity logs are then obtained as a function of depth.

The pseudo velocity sections obtained are tentative sketches of velocity logs computed from the seismic data. They provide a synthesis of the amplitude and of the moveout velocity analysis information. They should be calibrated on wells whenever possible.

Many sources of errors remain:

— imperfections of demultiplication and deconvolution,
— inaccuracies of amplitude preservation and calibration,
— inaccuracy of migration with uncertain velocities,
— absence of high and low frequency information on seismic traces,
— difficulty of separating velocities from densities,
— presence of various kinds of noises on the records,
— errors due to reflections not occurring in the vertical plane.

Only very good seismic data can result in an acceptable level of errors.

3. Tests of the Method

1. Synthetic examples

As a verification of the technique, synthetic pseudo velocity logs have been obtained from synthetic seismograms computed from a sonic log. (fig. 3.1).

The synthetic pseudo velocity log (fig. 3.2) obtained from a spike-pulse synthetic seismogram without multiples is absolutely identical to the initial sonic log (a spike-pulse on a digital record is a pulse $\delta(t)$ such that $\delta = 0$ if $t \neq 0$ and $\delta = 1$ if $t = 0$). The synthetic pseudo velocity log obtained from a spike-pulse synthetic seismogram with multiples shows a good analogy to the initial sonic log, but small differences in details due to the multiple reflections. This is the kind of error one might expect when computing pseudo velocity logs from imperfectly "demultiplied" seismic traces (fig. 3.3).

Fig. 3. Synthetic pseudo velocity logs obtained from spike-pulse and non-spike pulse synthetic seismograms.

(1) Sonic log 1. The vertical scale is converted to vertical propagation time, at a 4 ms sampling rate.
   A) Spike-pulse synthetic seismogram computed without multiples.
   B) Spike-pulse synthetic seismogram computed with multiples.
   C) Synthetic seismogram computed without multiples with a zero mean seismic pulse. The pulse chosen is roughly a 40 ms period, zero mean sinusoidal arch.
(2) Pseudo velocity log obtained from synthetic seismogram A.
(3) Pseudo velocity log obtained from synthetic seismogram B.
(4) Pseudo velocity log obtained from synthetic seismogram C.
(5) Pseudo velocity log obtained from synthetic seismogram C after re-inserting low frequency information. The pseudo velocity log is fitted to the slope of sonic log 1 between 0.57 and 1.12 seconds. Comparison of pseudo velocity logs (4) and (5) shows the velocity modification occurring when low frequency information is introduced. It is interesting to note the increasing velocity contrasts, associated with the increasing velocity values, in the lower part of pseudo velocity log (5). Densities are assumed to be constant.

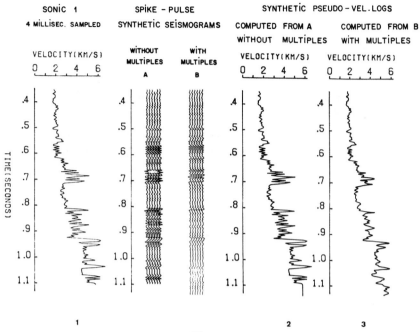

Fig. 3a

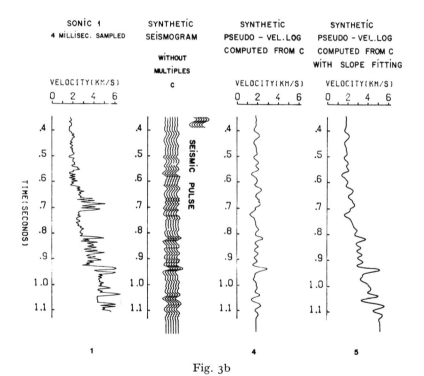

Fig. 3b

Synthetic pseudo velocity logs obtained from synthetic seismograms computed with a pass-band seismic pulse show neither the high nor the low frequency components of the initial sonic log (fig. 3.4). These have been cut off by the filtering effect of the seismic pulse. Some of the low frequency information can be re-inserted by fitting the pseudo velocity values to the average slope of the sonic log (fig. 3.5). The general features of the sonic log are then present on the pseudo velocity log, except for the missing high frequencies and deformations due to the non-spike nature of the seismic pulse.

Synthetic pseudo log 5 gives an idea of the pseudo velocity log obtained when amplitude preservation, calibration, and demultiplication are perfectly performed, when density can be exactly separated from velocity, when no noise is present on the record, when pulse compression is performed up to a two-arch, zero-mean signal and when the low frequency information is introduced into the seismic record.

2. *Field cases*

How does the pseudo velocity log technique perform on *real field cases*?

Fig. 4 shows pseudo velocity logs computed from land seismic traces recorded near wells in Canada (Grau, Hémon, and Lavergne 1975).

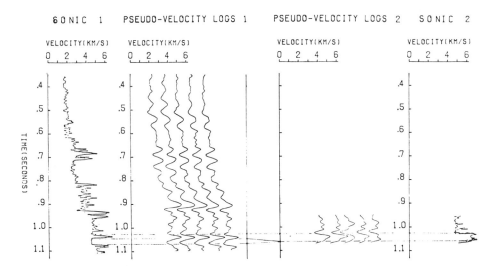

Fig. 4. Comparison of sonic logs 1 and 2 measured in wells 1 and 2, with pseudo velocity logs computed from seismic traces recorded in the vicinity of the wells. Distance between wells is 4 km. Vertical scales have been converted to vertical propagation times, with a 1 ms sampling.

The field technique is twelve fold CDP coverage, 24 traces per shot, 66 m between traces. A certain number of consecutive shots have been selected. The processing steps are:

— amplitude recovery,
— velocity analysis,
— twelve fold CDP stack,
— computation of pseudo velocity logs.

Densities are assumed to be constant. No deconvolution is performed. True amplitude recovery is not performed because the gain information was not preserved during field recording; the amplitude processing is limited to a dynamic equalization of traces. In this case no quantitative significance should be attributed to the absolute values of the velocities obtained. Attention should rather be paid to lateral variations from one pseudo velocity log to the next.

On the left part of fig. 4, sonic log 1 is represented with five adjacent pseudo velocity logs, computed from seismic traces recorded in the vicinity of well 1. On the right, sonic log 2 is represented with five adjacent pseudo velocity logs computed in the vicinity of well 2.

The similarity between sonic and pseudo velocity logs is relatively fair in the lower part of the records. The 5500 m/s high velocity layer at 1050 ms is about 25 to 30 m thick (10 ms) at well 1, and its thickness increases to 100 or 120 m at well 2. This method makes it possible to investigate lateral facies variations using pseudo velocity logs calibrated on wells.

4. Pseudo-Velocity Sections on Deep Sea Drilling Project Sites

Pseudo velocity sections have been obtained from standard seismic sections near drilling site 372 of Deep Sea Drilling Project Leg 42A, in the Western Mediterranean Basin (Scientific Party 1975).

1. *Recording and processing. Wiggle display of the pseudo velocity sections*

Seismic line J 211 is recorded 15 miles east of Minorca, with Flexichoc, twenty-four fold CDP coverage, 24 traces per shot, 100 m between traces.

Processing is performed with amplitude preservation (figs. 5 and 6). Detailed moveout velocity analysis is made, and interval velocities are computed every fourth shot (Fig. 7). A lateral migration is performed after stacking using the wave equation method. No deconvolution is made.

Amplitude adjustment is made on the water bottom reflection coefficient whose intensity is computed from comparisons between first and higher order reverberations. Densities are assumed to be constant. Low frequency information is introduced using interval velocities. Interpolation is performed to obtain one interval velocity function for each trace. Vertical migration is finally performed to convert vertical time to true depths.

From 238 adjacent traces in fig. 6, 238 pseudo velocity logs are computed over a 12 km distance.

The pseudo velocity section obtained is represented by two adjacent three dimensional diagrams in fig. 8a-b in wiggle display.

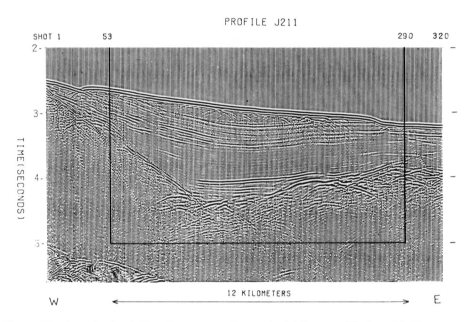

Fig. 5. Portion of seismic line J 211, 15 miles east of Minorca, Western Mediterranean Basin.

Data acquisition:

— M.V. Florence. Flexichoc twenty-four fold CDP coverage.
— Source: 1 Flexichoc, 123 cm diameter, depth 16 m.
— Distance between shots 50 m.
— Streamer twenty-four traces, ninety-nine hydrophones/trace, depth 20 m.
— Distance between traces 100 m.
— Binary gain recording—4 ms sampling.
— Recording filter: 12.5/80 Hz.

Initial processing:

— Demultiplexing.
— True amplitude recovery.
— Velocity analysis.
— Moveout corrections and CDP stack.
— No deconvolution.

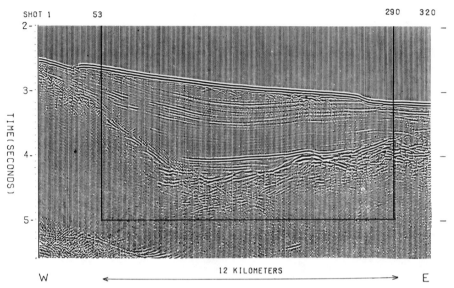

Fig. 6. Portion of seismic line J 211 with lateral migration after stack.

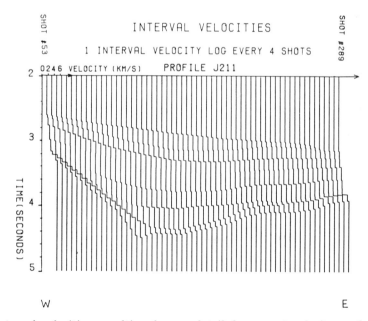

Fig. 7. Interval velocities, resulting from a detailed moveout velocity analysis of line J 211. One interval velocity function was obtained every fourth trace.

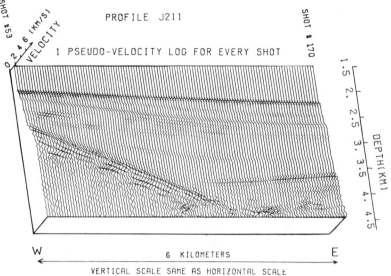

Fig. 8a

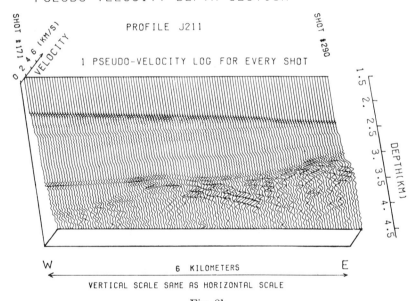

Fig. 8b

Fig. 8. Pseudo velocity section as a function of depth and horizontal distance, obtained from line J 211 between shots 53 and 290. Horizontal and vertical scales are the same The low frequency information is introduced using interval velocities of fig. 7 convolved with a 0-10 Hz low-pass filter.

a - shots 53 to 170,
b - shots 171 to 290.

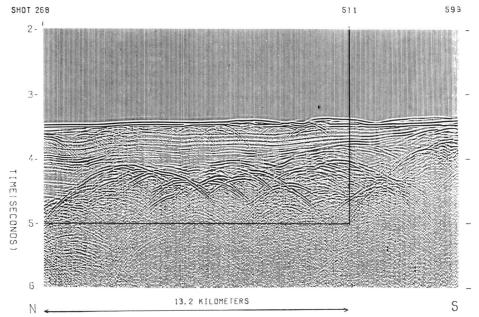

Fig. 9. Portion of seismic line J 212, 25 miles east of Minorca. Data acquisition and initial processing as in fig. 5.

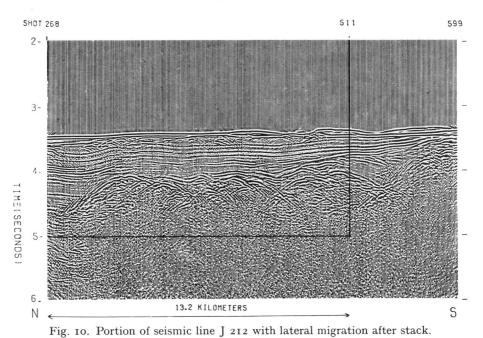

Fig. 10. Portion of seismic line J 212 with lateral migration after stack.

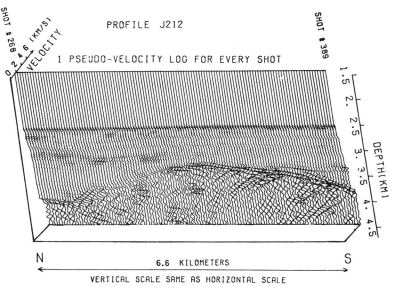

Fig. 11a

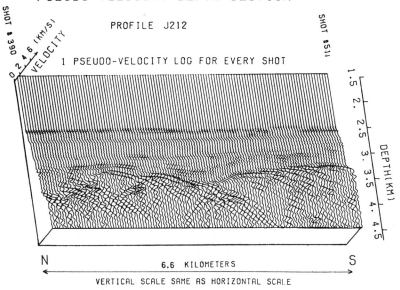

Fig. 11b

Fig. 11. Pseudo velocity section as a function of depth and horizontal distance, obtained from line J 212 between shots 268 and 511. Horizontal and vertical scales are the same. The low frequency information is introduced using interval velocities convolved by a 0-10 Hz low-pass filter.

a - shots 268 to 389,
b - shots 390 to 511.

Line J 212 is recorded south of line J 211 with the same technique. Processing is performed with amplitude preservation (fig. 9). A lateral migration is made (fig. 10) with no deconvolution.

Amplitude adjustment is made on the water bottom reflection coefficient; densities are assumed to be constant; low frequency information is introduced using interval velocities; interpolation is made to obtain one interval velocity function for each trace, vertical migration is finally performed.

From 244 adjacent traces in fig. 10, 244 pseudo velocity logs are computed over a 13.2 km distance.

The pseudo velocity section obtained is represented by two adjacent three dimensional diagrams in fig. 11a-b in wiggle display.

Lithology identification

Site DSDP 372 was drilled a few miles south of Line J 211. The drilling penetrated 884 m of sediments down to lower Burdigalian mudstones and sandstones. Continuous coring was performed. Using this information, an attempt is made to identify lithology on the pseudo velocity sections.

On line J 211, the sediments encountered from the surface down to the basement can be identified at shot 171 as:

— from 0 to 2200 m: water. Velocity 1500 m/s.

— 2200 to 2700 m:
Quaternary-Pliocene and Upper Miocene (?) marl formation. Velocity increasing with depths from 1800 m/s to 2200 m/s.

— 2700 to 3600 m:
Miocene clayey mudstone.
Velocity around 3200 m/s.

— 3600 to 4200 m:
Probable Aquitanian-Oligocene sandstones-mudstone-evaporites (?)
Velocity averaging 4000 m/s.

— Below 4200 m:
Basement.
Velocity averaging 5000 m/s. (?)

On line J 212, the sediment encountered at shot 390 can be identified as:

— from 0 to 2500 m: water. Velocity 1500 m/s.

— 2500 to 3300 m:
Quaternary-Pliocene and Upper Miocene (?) marl formations.
Velocity increasing with depth from 1800 m/s to 2200 m/s.

— 3300 to 3800 m:
  Miocene clayey mudstone. Velocity around 3200 m/s.
— Below 3800 m:
  Possible Aquitanian-Oligocene sandstones-mudstones-evaporites (?) (4000 m/s) and Basement (5000 m/s).

2. Colour display of the pseudo velocity sections

The wiggle displays shown on fig. 8 and 11 are tentative representations of three parameters—depth, horizontal distance, velocity—on a two dimensional sheet of paper. It is clear that they are not the best representations for lithology identification and facies variation detection.

Other types of representations including colour displays were therefore investigated.

Figs. 12 and 13 are pseudo velocity sections of lines J 211 and J 212, obtained as a function of time and horizontal distance, in variable density colour display. These should be compared to fig. 6 between shots 53 and 290 for J 211, and to fig. 10 between shots 268 and 511 for J 212.

The display technique consists in representing the amplitude levels as a certain number of colour levels. Special equipment is available that converts any type of function, whose values have been initially stored in the computer, to different levels of colour. The function displayed can be a geophysical parameter related to lithology, such as amplitude, velocity, density, acoustic impedance, absorption coefficient, dominant frequency; or a lithological parameter such as sand-shale ratio or porosity. The colour display of polarity can also be useful for bright-spot investigations.

Figs. 12 and 13 show that the colour display is a good representation for detecting lateral and vertical facies variations.

An attempt to relate colour to lithology can be made:

— the blue zone below the sea-bottom, with velocities between 1800 m/s and 2200 m/s, could represent Quaternary-Pliocene and Upper-Miocene (?) marl formations;

— the alternate blue and yellow zone, with velocities varying alternately around 2800 m/s, the upper part of the Miocene clayey mudstone;

— the yellow zone with velocities around 3200 m/s, the lower part of the Miocene mudstone;

—- the yellow and brown zone above the basement, with velocities around 4000 m/s, could represent the Aquitanian-Oligocene sandstone.

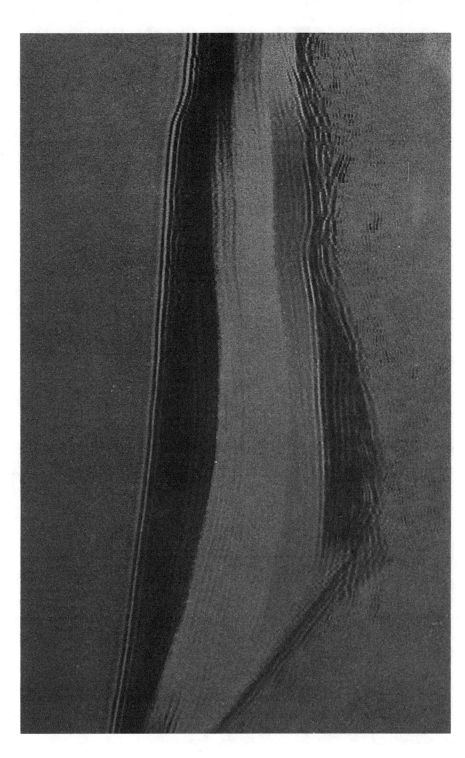

Fig. 12. Pseudo velocity section as a function of vertical time and horizontal distance, obtained from line J 211 between shots 53 and 290. Colour display of the pseudo velocities. The colour velocity scale is represented on the right. On this scale the velocity increases linearly from 1000 to 6000 m/s, from top to bottom. Vertical time and horizontal distance correspond to those of fig. 6.

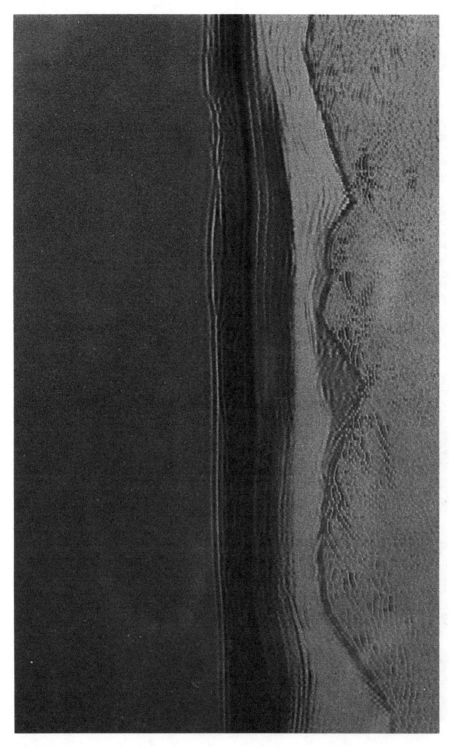

Fig. 13. Pseudo velocity section as a function of vertical time and horizontal distance, obtained from line J 212 between shots 268 and 511. Colour display of the pseudo velocities. The colour velocity scale is represented on the right. On this scale the velocity increases linearly from 1000 to 6000 m/s. from top to bottom. Vertical time and horizontal distance correspond to those of fig. 10.

It is interesting to observe the lateral variations of velocities in the Miocene mudstones and in the Aquitanian-Oligocene sandstones. They could correspond to facies variations in these layers or to variations of compaction with depth.

It is clear that the colour display, neatly showing lateral and vertical facies variations, is well adapted to the representation of geophysical parameters for lithological investigation.

5. Conclusion

Pseudo velocity logs can give useful information on lateral and vertical facies variations, especially when displayed in colour.

The precision of the velocity determination depends very much on the quality of seismic data, the accuracy of amplitude preservation, the possibility of calibration on wells.

At the present stage, the technique works in the two-dimensional domain. This means that the validity of the method depends on the hypothesis that all visible reflections occur from mirrors in the same vertical plane. In the future, it is hoped that three-dimensional techniques will help to develop pseudo velocity logs in three-dimensionally structured regions. The method should make it possible to draw maps of lithological and petrophysical variations.

We are indebted to Compagnie Française des Pétroles Total for making some of their data available to us.

References

CLAERBOUT, J. F., and DOHERTY, S. M., 1972, Downward continuation of moveout-corrected seismograms, Geophysics 37, 741-768.

DOMENICO, S. N., 1974, Effect of water saturation on seismic reflectivity of sand reservoirs encased in shale, Geophysics 39, 759-769.

GRAU, G., HÉMON, CH., and LAVERGNE, M., 1975, Possibilités nouvelles pour la sismique stratigraphique, Panel 9, Development in seismic data handling and interpretation, 9th World Petroleum Congress, Tokyo, 225-234.

LARNER, K. L., MATEKER, E. J., and WU., C., 1973, Amplitude: its information content, in Lithology and direct detection of hydrocarbons using geophysical methods, Symposium of the Geophysical Society of Houston, 17-47.

LAVERGNE, M., 1975, Pseudo-diagraphies de vitesses en offshore profond, Geophys. Prosp. 23, 695-711.

LINDSEY, J. P., and CRAFT, C. I., 1973, How hydrocarbon reserves are estimated from seismic data, in Lithology and direct detection of hydrocarbons using geophysical methods, Symposium of the Geophysical Society of Houston, 119-121, reprinted from August 1, 1973 World Oil.

NEWMAN, P., 1973, Divergence effects in a layered earth, Geophysics 38, 481-488.

O'DOHERTY, R. F., and ANSTEY, N. A., 1971, Reflections on amplitudes, Geophysical Prospecting 19, 430-458.

PAIGE, D. S., 1973, The dark side of the bright spot, in Lithology and direct detection of hydrocarbons using geophysical methods, Geophysical Society of Houston, 186-219.

RICE, R. B., 1962, Inverse convolution filters, Geophysics 18, 4-18.

SHERIFF, R. E., 1973, Factors affecting amplitudes—A review of physical principles, *in* Lithology and direct detection of hydrocarbons using geophysical methods, Symposium of the Geophysical Society of Houston, 3-16. See also Geophysical Prospecting 23 (1975), 125-138.

SCIENTIFIC PARTY, 1975, Glomar Challenger returns to the Mediterranean Sea, Geotimes, August, 16-19.

WELL LOG EDITING

IN SUPPORT OF DETAILED

SEISMIC STUDIES

by

Brian E. Ausburn

J. R. Butler and Company
Houston, Texas

ABSTRACT

Modern seismic processing techniques often permit relatively detailed interpretation. Inherent in many of these detailed seismic applications is the necessity of calibrating seismic information with wellbore data. Consequently, every effort must be made to insure that wellbore data provide the best possible picture of in situ physical properties of the subsurface section.

This paper discusses the need for well log editing and the fact that the editor must not only recognize bad data but must also determine the most appropriate substitution values. Three levels of editing (mechanical, interpretive and modeling) are identified and examples of each are presented.

INTRODUCTION

Sophisticated seismic processing techniques developed in recent years have provided the industry with more meaningful data than could have been predicted by even optimistic forecasters of five to ten years ago. These, in turn, often permit relatively detailed interpretation.[1-5] An example of such a detailed interpretation is shown in Figure 1. This carbonate porosity example, taken from Wittick and Frink's 1976 SEG paper, presents the display that can be obtained with synthetic velocity logs derived from seismic data. Note that with the velocity base line of each trace (velocity bias = 16,000 ft/sec) and the velocity scale (13,333 ft/sec/inch) given, quantitative reservoir information can be obtained.

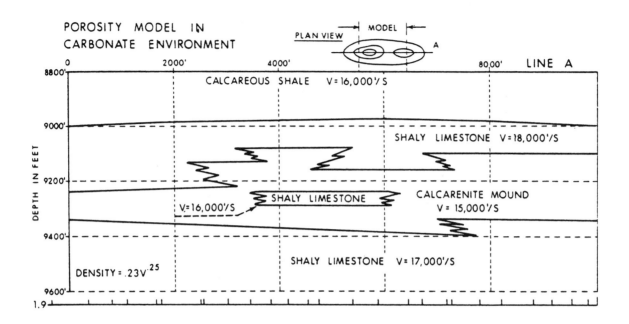

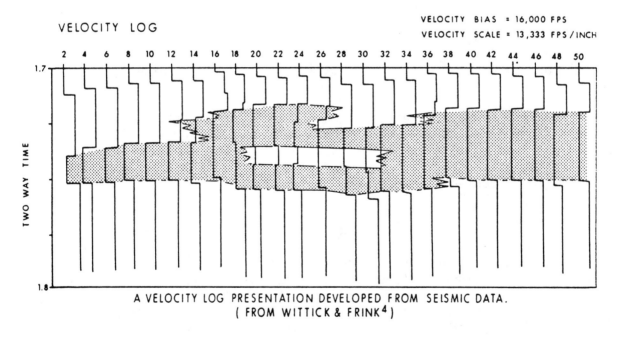

A VELOCITY LOG PRESENTATION DEVELOPED FROM SEISMIC DATA.
(FROM WITTICK & FRINK[4])

FIGURE 1

Inherent in most of these detailed seismic applications is the necessity of calibrating seismic information with wellbore data. Consequently, every effort must be made to insure that wellbore data provide the best possible presentation of in situ physical properties of the subsurface section.

Several things must be done to reduce well log and seismic data to a valid comparison level; e.g., time-depth reconciliation, filtering to account for different frequency contents, integrating so that intervals can be compared to intervals and not interfaces, etc. However, the very important first step before using well log data for geophysical control is log editing.

Everyone who has worked quantitatively with well logs recognizes that it is not uncommon for portions of a well log to contain data that may be significantly different from the true, in situ formation properties. This may be due to either log calibration problems or environmental conditions.

However, people who do not routinely work with quantitative log data may not be familiar with the errors sometimes included in well log measurements. As geologists and geophysicists, hungry for wellbore information to "tie down the seismic", began to use well logs to construct synthetic seismograms[25], well logs were often digitized from "Top to Bottom" in a completely raw, unedited mode. No wonder that the tie from wellbore to seismic was often difficult. Still, it was generally accepted that some gross log editing was required to correct, for example, obvious hole washouts; and, of course, log analysts have always paid careful attention to the quality of log values through pay zones. However, it was not until the advent of the "bright spot (seismic amplitude analysis) technology", or more recently, stratigraphic geophysics (reservoir continuity) that "fine line" editing has been practiced.

As is generally known, almost all formations are altered when a borehole is drilled through them. The more competent formations show an imperceptible change, while the softer formations often suffer significant, obvious alterations.

Shales are altered by exposure to mud filtrate; they cave, erode, absorb water and swell. The degree of weathering is a function of shale type and properties of the mud, such as water loss, filtrate salinity, and weight. Also important are exposure time and other mechanical factors involved in drilling a hole. Sandstones are altered by relaxation (a function of mud weight), erosion (a function of mud weight and bit hydraulics) and invasion with mud filtrate (a function of mud weight and water loss characteristics and bit hydraulics).

Log editing is often an interpretive judgement. Consequently, one runs the risk of editing out valid data, but the risk is usually worthwhile to avoid including invalid information that may cloud the whole picture. Figure 2 presents a severe, but plausible, simulated editing situation. A brief discussion of some of the edits follows:

- 3 -

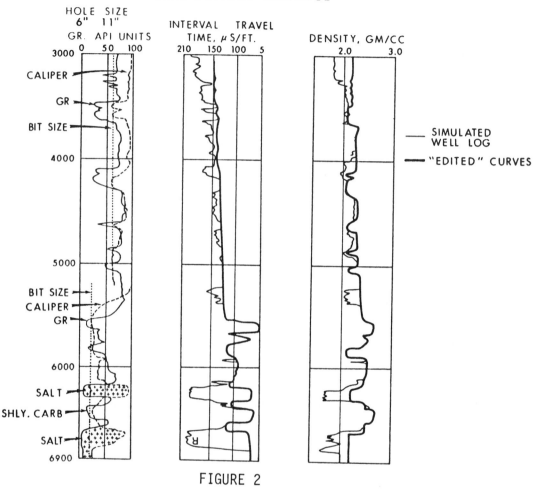

FIGURE 2

Interval 3000-5520

This interval is a sand/shale sequence where most of the shales are washed-out, but many of the sands are not. The combination of hole enlargement and shale alteration can result in apparent shale velocities and densities being markedly less than that <u>judged</u> to be appropriate for the section.

The judgement is often based on trend curves developed from other wells in the area in which the hole conditions for the equivalent stratigraphic section were much better. In this simulated example, the "correction trend" is substantiated by observed shale velocity values in sections where the hole has not washed-out; i.e., 4960-5220 and 5370-5520. It would be postulated that 4960-5220 has not washed out because it was logged relatively soon after it was drilled and, therefore, had less exposure time. Perhaps 5370-5520 has not washed out because the

- 4 -

mud quality was improved and was more easily maintained after the casing string was set at 5220.

Note that the shale interval below the casing shoe, 5220-5370, has washed out. Possibly, the mud quality was not good immmediately upon drilling out of casing but, by the time the drill bit reached 5370, mud properties had stabilized resulting in less hole enlargement. In addition, mud circulation hydraulics (annular turbulence) can sometimes cause washouts beneath casing shoes. If one can determine a mechanical reason for a hole washout, values from nearby "equivalent zones" can be substituted. However, if no mechanical reason can be identified as the cause of hole washout, one should carefully consider the substitution value. Perhaps the reason the zone washed out is because it is different from nearby "equivalent zones".

F

The density log has to be edited for many of the same reasons. Shale alteration and washouts affect the density log, too. Hole washouts, as they pertain to density log readings, can be subdivided into rugosity and enlargement. Rugosity, or roughness, sufficient to cause density log errors (lack of pad contact), can occur sometimes with relatively little hole enlargement. Significant hole enlargement results in a different mechanical conformation of logging tool pad to borehole wall than that used when the standard bulk density-count rate relationship was developed.

Interval 5520-6900

This interval is a mixed section of limestone, sandstone, shale and salt. With exception of a few obvious hole washouts noted on the caliper in shale sections, most of the editing of this lower section is related to the salt sections. It is postulated that the salt layer, 6190-6300, is washed out completely. As a result, both the sonic and density logs are recording essentially the properties of borehole fluids; these are markedly different from salt parameters! Note that the density log has been edited from 6790-6900 even though the hole is in-gauge throughout that section. Due to a different electron/bulk density relationship for salt from that of most other sedimentary rocks, the apparent density log bulk density of salt is not the true bulk density. By contrast, the sonic response in the gauge salt section substantiates the validity of the sonic edits in the washed out salts.

Figure 3 compares synthetic seismograms that would be obtained from raw log data and from edited log data. As would have been predicted from the severity of edits shown on Figure 2, the differences in synthetic seismograms are significant.

Note in the shallow part of the section the synthetic from unedited data has much more character than that from the edited data. These apparent events represent only the observed contrasts in the erroneous information and not any real acoustic impedance contrast in the subsurface. Note also that the last major event from the unedited set occurs at approximately 2.03 seconds, while on the edited version this event occurs at approximately 1.875 seconds. This difference reflects

- 5 -

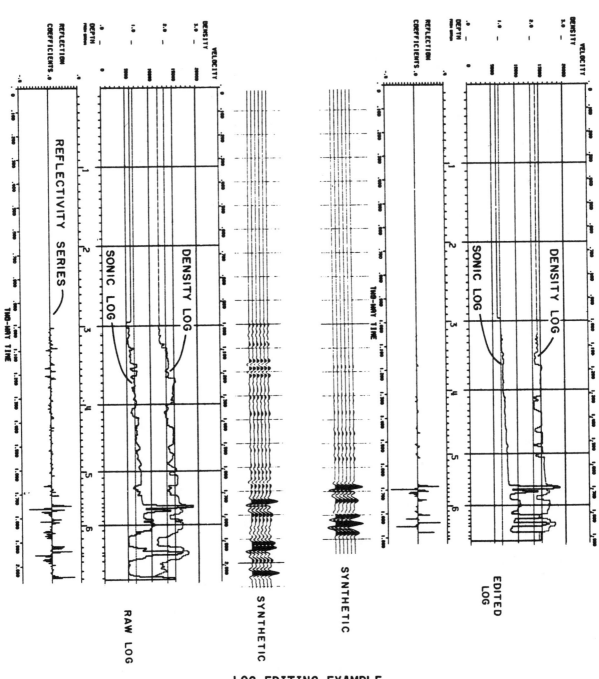

LOG EDITING EXAMPLE

FIGURE 3

the significance of the uphole velocity in achieving depth/time ties between seismic and wellbore data. In this case, the "too slow" velocity observed on the up-hole portion of the raw log could make a difference of 500 to 1,000 feet in the location of the last events.

To conclude this introductory section, it is observed that there are two major facets of log editing:

1. <u>One must be able to recognize that bad data have been recorded.</u> This may be easily determined in the cases of obvious noise (for example, cycle skips) or hole washouts. This may not be easy in the cases of subtle changes in hole conditions, different lithologies, borehole weathering, or undetected or unrecorded log calibration problems.

2. <u>One must try to determine better values to substitute for the bad ones.</u>

THREE LEVELS OF WELL LOG EDITING

Well log editing can be subdivided into three levels; mechanical, interpretive and modeling. There is, of course, some overlap between these levels (particularly between interpretive and modeling), but generally, they would grade from mechanical to modeling as requirements change from simple to complex and information and experience in an area increase from sparse to abundant.

MECHANICAL

Mechanical log editing by definition should be possible on a set of well logs from any locality - even if one were not familiar with the stratigraphic section or the particular drilling and logging problems of that province. The first of these mechanical checks would be the verification of log calibration.

Log Calibration

Because of their importance, a brief review of the operating principles[7,8] and calibration procedures[9,10] of the two common porosity logs that have direct geophysical applications is presented in Appendix A. It is sufficient to say here that log calibration should always be examined. Even though synthetic seismogram development utilizes the contrast in acoustic impedance between zones observed on well logs, calibrated logs are still necessary because 1) calibration errors may not affect the entire data range in a linear fashion, and 2) the determination of absolute physical properties, like overburden gradients and formation velocities, are required.

- 7 -

It is recognized that it is not always possible to obtain perfectly calibrated well logs. The well site geologist or engineer is sometimes under considerable pressure to "accept a mediocre log" due to the cost in rig time to re-log or to the risk of sticking the tool while trying "just one more pass". However, there is seldom any excuse for not having a calibration shown. If a calibration is at least shown, an attempt can be made to rescale the log in order to present the recorded values in a more appropriate data range.

Obvious Instrument/Electronic Noise

Not all mechanical log editing is related to log calibration. There are other log problems that are only indirectly related, if at all, to log calibration.

Figure 4 is a picture of an actual acoustic log exhibiting both cycle skips and noise spikes - they often go together.

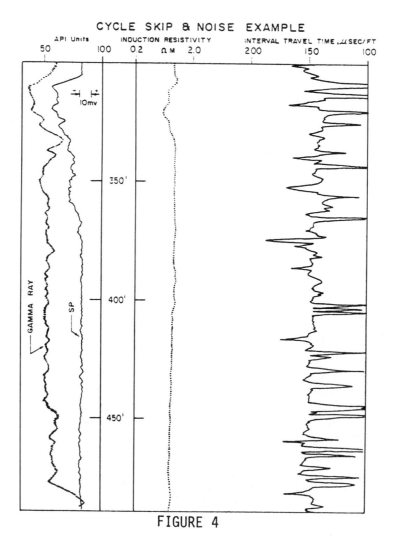

FIGURE 4

Speculation: Perhaps there was gas in the mud which would attenuate the amplitude of the acoustic signal causing cycle skips, or too high an interval travel time. The logging engineer, therefore, reduced the downhole gain selection in an attempt to "pick" the first compressional arrival. By so doing, however, extraneous electronic noise and road noise were picked as legitimate arrivals which are indicated by the noise spikes (too short an interval travel time).

Obvious Hole Related

Figure 5 is a typical example of erroneous density data being recorded due to hole conditions. Note that near the casing shoe, the formation has eroded and slumped so badly that the hole size from the caliper is indicated to be 17" or larger! The density log is "off scale" on the low end indicating an apparent density of less than 1.66 gm/cc (the mud density was 1.32 gm/cc) which is obviously not the formation density.

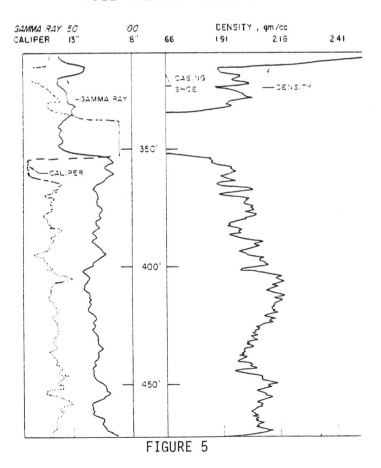

FIGURE 5

Obvious Hydrocarbon Effects

Figure 6 is a typical density log response observed through a high porosity, gas-bearing zone. The true water-bearing density is probably similar to that observed below 200 ft. (fictitious depth), while the correct density in the undisturbed zone is probably even lower than the observed density log value.

Gas effect on an acoustic log is shown on Figure 7. Note through the gas interval, the Δt curve is saturated at approximately 205 μsec/ft (4,900 ft/sec). In other words, the fastest path for the acoustic signal is through the borehole fluid. What is the correct velocity of sound through this gas-bearing interval? It is interesting to note that the acoustic signal is affected even through the gas/water transition zone (168 - 202 feet) where the apparent gas saturation is small.

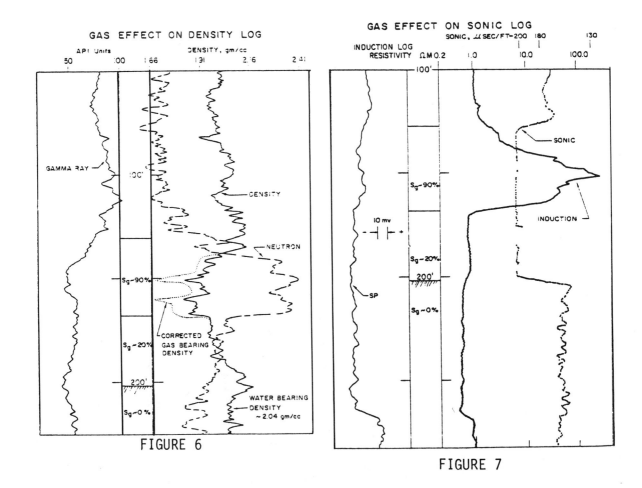

FIGURE 6

FIGURE 7

INTERPRETIVE

A step beyond mechanical log editing is interpretive editing. This type of editing includes making judgements in both "recognizing" bad data and in "substituting" better ones.

Check Shot Surveys

Check-shot or seismic reference surveys are obvious ways to either validate or correct acoustic log information. This is probably the oldest method for tieing seismic times to well depths. Check shot surveys using closely spaced stations in particular can often be a great deal of help in determining interval velocities.

Seismic Information

Sometimes, the seismic information itself can be used to edit the well logs. This would be the case particularly through long, uniform stratigraphic sections where seismically determined velocities are available and unique. At other times, obvious seismic markers or events could be used to judge the contrast in acoustic impedance at lithologic interfaces. From this, some of the physical properties of the formations can be inferred. Although the seismic may only "see" low frequencies and thick beds relative to a well log, it does "see" the formation undisturbed by a borehole; this is a fact worthy of noting and remembering.

Depth Trend Relationships

As wellbore data become available on a province, basin or geologic trend, depth trend relationships are developed for a number of reasons: e.g., to determine the top of abnormal pressures, to project porosity development at depth, or to provide a basis for geophysical modeling. One of the uses for depth trend relationships is to provide a basis for interpretive well log editing.

Figure 8 is a typical composite velocity - depth relationship, while Figure 9 is a typical composite density - depth relationship for a specific province. Note on Figure 9 the variation between the composite relationship from several wells and the relationship for just one well. Recognizing the magnitude of this variation would be very important to consider when making judgements relative to the editing of another well in the province.

Of course, each province and age of formation has a particular relationship. For example, at 10,000 feet, the onshore South Louisiana normal pressured shale

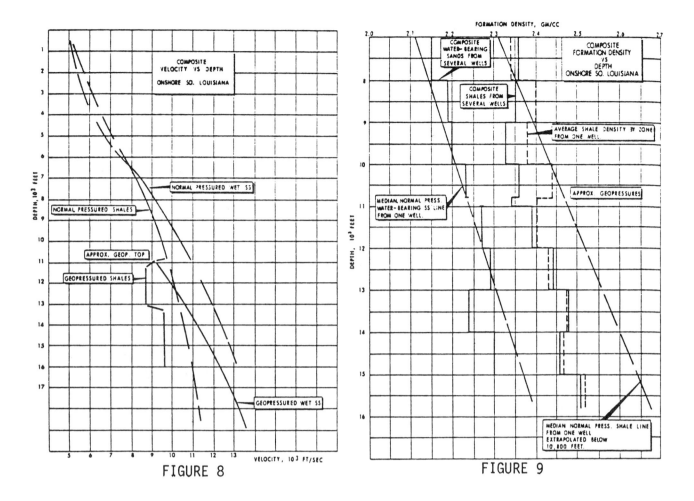

FIGURE 8

FIGURE 9

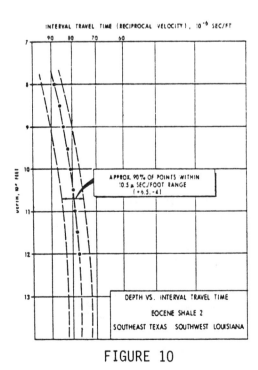

FIGURE 10

(Figure 8) would have a velocity of approximately 9300 ft/sec, while the Eocene Shale from Southwest Louisiana (Figure 10) would have a velocity of approximately 12,300 ft/sec. Likewise, the velocity of normal pressured wet sandstone in South Louisiana (Figure 8) would have a velocity of approximately 10,400 ft/sec, while the Eocene sandstones of Southwest Louisiana (Figure 11) would have a velocity of approximately 13,000 ft/sec. In other words, before making interpretive editing judgements, it should be established that the appropriate relationship is being used!

- 12 -

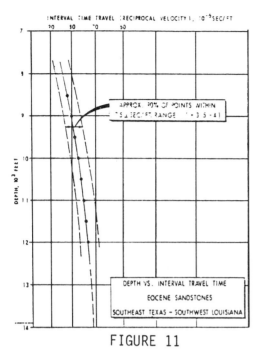

FIGURE 11

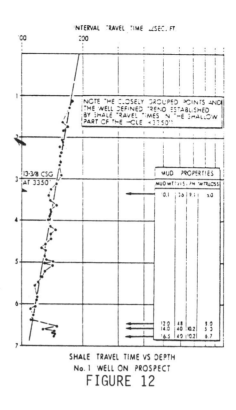

SHALE TRAVEL TIME VS DEPTH
No. 1 WELL ON PROSPECT
FIGURE 12

Figures 12 and 13 are examples of depth trend curves being used specifically for editing purposes. Figure 12 is a conventional plot of interval travel time for clean (pure) shale versus depth. The well represented by these data was the first one on the prospect, and consequently, the mud properties were closely controlled (this is only a speculation, but a plausible one), with low water loss and moderate mud weight. Note the closely grouped points and the well defined trend that can be drawn through the data in the shallow portion of the hole.

Figure 13 is a similar type plot for a subsequent well on the prospect. This follow-up well

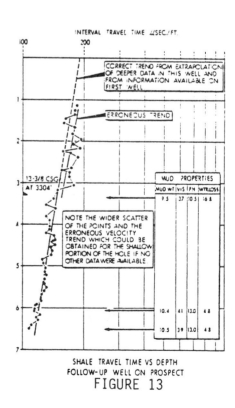

SHALE TRAVEL TIME VS DEPTH
FOLLOW-UP WELL ON PROSPECT
FIGURE 13

- 13 -

is probably only two to three miles removed from the first. In this case, it is speculated that less care was taken with the mud, possibly because the first well had located the pressures and there seemed to be less reason to maintain good mud properties from a drilling standpoint. As a result, the borehole showed more caving (weathering), which in turn, makes all the velocities appear slower. Note the wider scatter in the data points and the erroneous velocity trend that might have been developed in the absence of other information-like data from the first hole and from the deeper data in this hole below the casing shoe. Once casing was set, the mud properties were apparently more closely controlled and the trend developed through the points 3300 - 6500 is quite close to that obtained from the first hole.

Interlog Relationships

Interlog relationships can be extremely helpful in either editing logs or developing estimates of geophysical parameters when the appropriate logs have not been obtained at a critical location. They are particularly valuable in areas where sonic and density variations are not primarily a function of depth. Moreover, it is quite often desirable to estimate the velocity character of a zone from the character, say, of the resistivity log rather then merely substituting some average depth-derived velocity. Table 1 lists seven common interlog relationships.

INTERLOG RELATIONSHIPS

FOR

LOG EDITING PURPOSES AND/OR FOR ESTIMATING VELOCITY OR DENSITY WHEN APPROPRIATE LOGS HAVE NOT BEEN OBTAINED

1. RESISTIVITY - VELOCITY
2. RESISTIVITY - DENSITY
3. DENSITY - VELOCITY
4. GAMMA RAY - VELOCITY
5. GAMMA RAY - DENSITY
6. NEUTRON - VELOCITY
7. NEUTRON - DENSITY

TABLE 1

Figure 14 presents relationships of velocity and formation resistivity developed for three Eocene formations in Southwest Louisiana. Note that each formation has a slightly different relationship. This approach is essentially an "Archie type" plot of sonic-resistivity data; for example, the logarithm of a "porosity parameter" (in this case, Δt-50) was plotted versus the logarithm of formation resistivity. Most formations follow this predictable relationship of decreasing velocity with decreasing resistivity at a constant water resistivity. (An exception to this would be organically-rich source rocks and, of course, hydrocarbon-bearing porous zones.) In addition to being careful that a relationship is not projected across lithology and water resistivity changes, one must ascertan that the resistivity being used to estimate a velocity value is comparable to that used in developing the algorithm. If induction logs are used, a fair amount of variation in hole size and mud resistivity can be tolerated. If electrical devices are used, additional steps to correct the resistivity may be required.

F

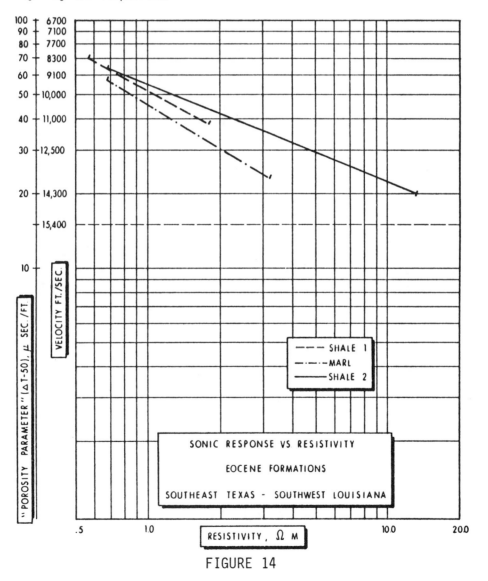

FIGURE 14

Figure 15 is a slightly different presentation of the velocity-resistivity relationship shown on the previous figure. Note that there is very little difference between the normal and geopressured formations.

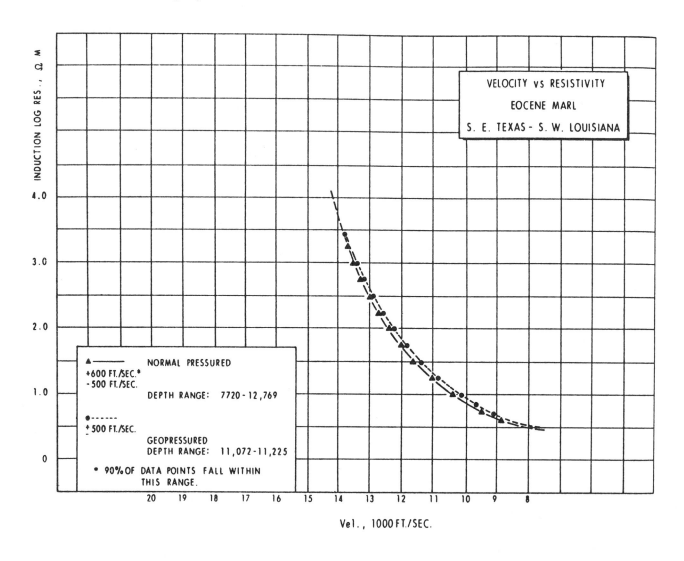

FIGURE 15

Figure 16 is similar to Figure 15 in that it is a plot of velocity versus resistivity on coordinate paper, but for shales rather than marls. Note the velocity-resistivity relationship appears to hold for an interval of almost 6900 feet.

-16-

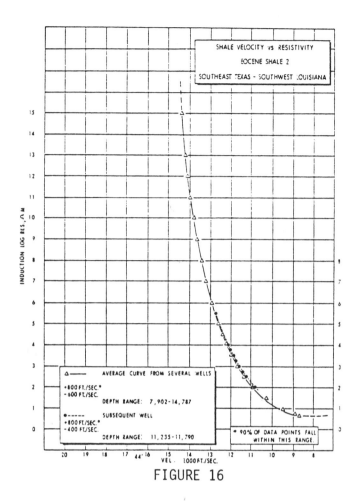

FIGURE 16

Figure 17 presents the observed relationship between density and velocity for an Eocene shale. Gardner's empirical relationship[11] is shown on this figure for comparison. Note that at a velocity of 12,000 ft/sec, there is approximately 0.13 gm/cc difference in density estimate between the two relationships. In view of this discrepancy, the density-velocity relationship for a specific province or formation should be determined whenever possible.

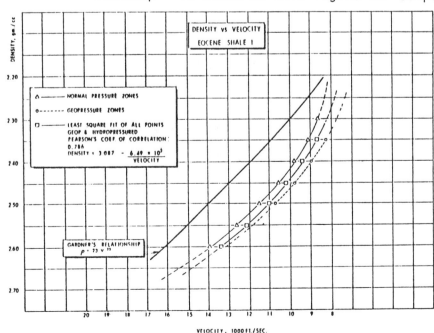

FIGURE 17

Correcting for Intrinsic Differences Between True and Apparent Bulk Density

In density log interpretation, it is commonly assumed that the quantity,

$$2 \left(\frac{\text{Summation of the Atomic Numbers}}{\text{Molecular Weight}} \right)$$

is equal to unity. However, it is recognized that for some substances this relationship does not hold.[6] Consequently, there are notable differences between true bulk densities and apparent bulk densities recorded by a gamma-gamma density log for various substances. These relationships are shown on Figure 18.

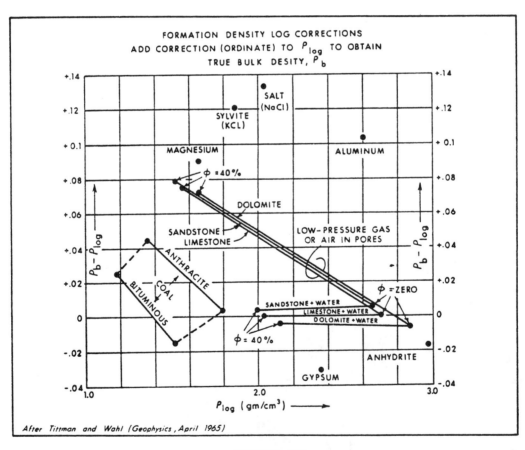

FIGURE 18

Correcting Log Values in Hydrocarbon Zones

The values of density and velocity recorded in hydrocarbon-bearing zones by well logs will almost always differ to some extent from the true, in situ values due to the invasion of mud filtrate. Note, for example, Figure 6. The bulk density recorded appears to have been affected by filtrate invasion. The theoretical density of the formation at the observed gas saturation should be on the order of 1.80 gm/cc, rather than the 1.88 gm/cc average shown on the log. Likewise on the sonic log shown in Figure 7, a theoretical solution would suggest that the true gas-bearing velocity should be on the order of 4500'/seconds (220 microseconds/foot) rather than the borehole fluid velocity actually measured. What would be the properties of this zone if it were completely water filled? This is sometimes difficult to determine. One can <u>assume</u> for this example (Figure 7) that the rock quality is comparable to that observed below the gas-water contact, at say, 210 - 220 feet, and use the observed values from that interval. Sometimes, other log data, e.g., natural gamma radiation, can reinforce such assumptions.

Use of the Latest Logging and Coring Technology

Whenever possible, the latest logging and coring technology should be utilized when preparing to edit well logs for seismic comparison and/or calibration. Discussion of the differences observed between sonic log velocities and check shot surveys has been in the literature since 1959 (Wood)[12], while applications of a long-spaced sonic device also were discussed in 1959 by W. G. Hicks[13]. Figure 19, taken from Hicks' paper[13], presents data that were collected during an experiment in the Anadarko Basin of Oklahoma. It is interesting to note that in

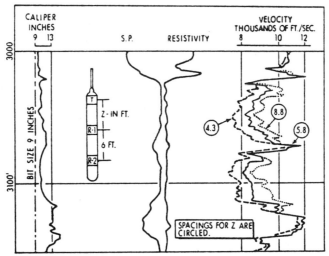

EFFECTS OF BOREHOLE WEATHERING AS DISPLAYED BY DIFFERENT ACOUSTIC LOG SPACINGS

Note that in shaly intervals the shorter distances read velocities that are too low, but in sandy zones all agree.
After Hicks.

FIGURE 19

shaly intervals, the shorter spaced distances (transmitter to the first receiver) read velocities that are too low while in the sandy zones (as reflected by the SP curve), all velocity measurements agree.

Interest in long-spaced acoustic devices has been rekindled in the last four to five years with the advent of more detailed seismic modeling and the desire for better calibration of the seismic with borehole information. Long-spaced acoustic sondes are currently available from Schlumberger and Dresser-Atlas in spacings up to ten feet. Table 2 summarizes the current availability of long-spaced acoustic devices.

TABLE 2

Summary of Available Long-Spaced Acoustic Devices
(November 1976)

COMPANY	SPACING	REMARKS
Birdwell	See remarks.	Birdwell's 3-D Velocity Log provides travel time (not Δt) information on transmitter-receiver spacings from 3 to 100'.
Dresser-Atlas	7'	Currently, four tools are available; one is borehole compensated.
Schlumberger	8-10' 10-12'	Several tools are available; all have one transmitter and two receivers.

Better borehole density data can be obtained by use of the Borehole Gravimeter (Figure 20)[14,15]. Unfortunately, these tools are much more limited in availability than even the long-spaced acoustic logs. In addition, since the current version of Borehole Gravimeters requires stationary measurements, the provinces where this information would be most

- 20 -

BASIC DENSITY EQUATION FOR GRAVIMETER DATA

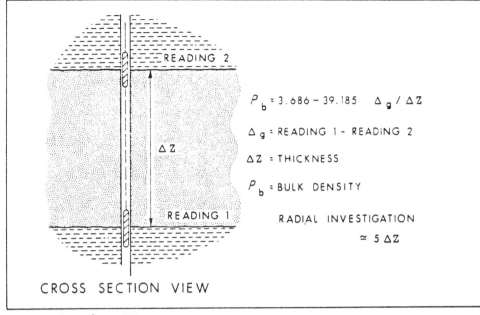

AFTER RASMUSSEN

FIGURE 20

helpful (sequences of soft sands and shales that are susceptible to borehole weathering) are the areas where it would be most hazardous to make the surveys. Nonetheless, in extremely critical calibration situations, consideration should be given to a Borehole Gravimeter Survey. It might be possible, for example, to run a Borehole Gravimeter in a nearby cased hole to provide density information over critical intervals. There may be some areas where helpful shale density information could be obtained by use of a pressure core barrel and/or stressed core analysis.

MODELING

A natural outgrowth of interpretive well log editing is the modeling phase. One of the most common modeling questions is the effect of substitution of different fluids into the subsurface prospect. This can be accomplished by either theoretical relationships or by empirical observations; both methods have problems.

The determination of density from the theoretical relationship is straight-forward and, consequently, is no particular problem as shown by the following:

- 21 -

$$\rho_B = \phi(\rho_f) + (1 - \phi)\rho_g \quad (1)$$

where

ρ_B = formation density

ρ_f = formation fluid density

ϕ = fractional porosity

ρ_g = grain density

where

$$\rho_f = S_w(\rho_w) + (1 - S_w)\rho_{hc} \quad (2)$$

where

S_w = water saturation

ρ_w = density of formation water

ρ_{hc} = density of hydrocarbon.

Formation velocity, however, is not so straight-forward nor easily computed. One of the most common theoretical equations is the Gasmann Equation where the compressional wave velocity[16] is given as:

$$V_\rho = (M/\rho)^{0.5} \quad (3)$$

where

$$M = \overline{M} + \frac{\left(1 - \dfrac{\overline{k}}{k_s}\right)^2}{\dfrac{\phi}{k_f} + \dfrac{(1-\phi)}{k_s} - \dfrac{\overline{k}}{k_s^2}} \quad (4)$$

with

V_ρ = compressional wave velocity, cm/sec.

M = elastic modulus, dynes/cm^2

ρ = formation density, gm/cm^3

k = bulk modulus, dynes/cm^2

ϕ = fractional porosity.

Definition of subscripts:

s = solid of which rock frame is made

f = fluid in the pore space.

Definition of superscripts:

bar = properties of the empty rock frame.

Properties of the saturated rock carry neither subscript nor superscripts. The major drawback to this approach is the difficulty in determining the bulk moduli, particularly those of the empty rock frame. Numerical examples of Gasmann calculations are presented in Appendix B.

The alternative is to use empirical observations of the fluid effects on formation density and acoustic logs. The major drawback of this approach is that invasion of mud filtrate will usually reduce the effect of hydrocarbons on well log measurements. Consequently, empirical observations of hydrocarbon effects are often considered to be minimums with the true effect being something more significant than that observed.

Figure 21 is the frequency distribution of sonic log velocities observed for a sandstone formation in South Louisiana. Distribution of density log readings for the same gas-bearing sandstone in South Louisiana is shown in Figure 22. Both sets of observations are obtained from wireline logs and, yet, it is interesting to note that the difference between gas- and water-bearing velocities is greater (approximately 8%) than the difference observed in gas- and water-bearing densities (approximately 4%). Sonic log measurements in high porosity hydrocarbon zones are often influenced less by invasion than the density log measurements. It has been noted from both empirical observations and theoretical considerations that the majority of the gas effect on formation velocity occurs at low gas saturation[17,18] (similar to the condition of the near-wellbore invaded zone), while the effect on density is linearly proportional to gas saturation.

Lithologic substitution in the modeling phase of log editing (in essence, Stratigraphic Modeling) is a natural outgrowth of depth trend development and other associated studies of the specific stratigraphic section. These types of investigations should provide the appropriate parameters to be substituted when determining the seismic manifestation of sand pinchouts or porosity loss or enhancement.

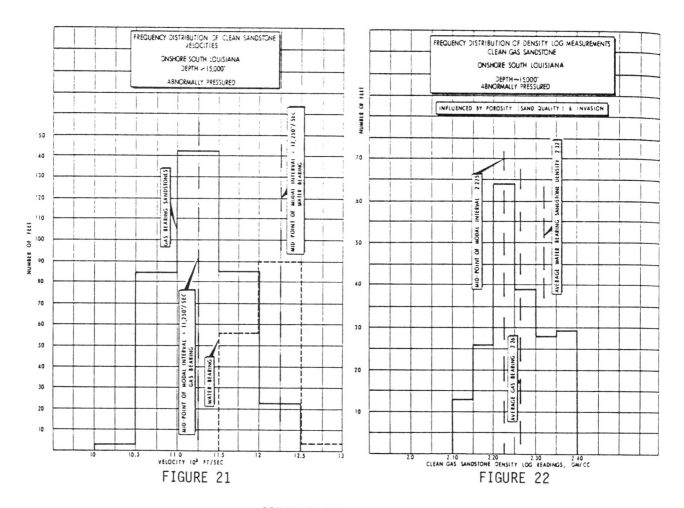

FIGURE 21

FIGURE 22

CONCLUSIONS

Seismic responses from zones of interest are influenced by the magnitude and distribution of earth properties in the total subsurface through which the seimic wave must pass. In order to extract detailed quantitative, or even qualitative, information for specific zones from seismic data, it is necessary to model as best we can the total subsurface section. Well logs are the prime source of subsurface information with which seismic responses can be modeled and/or calibrated. Before they can be used for this purpose, however, they must be subjected to several operations. These include time-depth reconciliation and filtering and/or intergrating techniques. However, the first step is log editing.

Well log editing has two major facets: first, to recognize that incorrect data have been recorded and second, to determine better values to substitute for the incorrect. There are at least three levels of well log editing: mechanical, interpretive, and modeling. These would generally grade from mechanical to modeling as requirements change from simple to complex and information in an area increases from sparse to abundant.

- 24 -

Modern seismic techniques have the potential of allowing us to transfer zones of interest as seen by borehole measurements onto seismic sections and thereby to capitalize on the ability of seismic to provide information on lateral variations in the earth parameters. In order to approach this goal, however, we must do our best to put the subsurface as seen by borehole measurements back into its proper, in situ (undisturbed) condition.

REFERENCES

1. Dedman, E. V., Lindsey, J. P., and Schramm, M. W.: "Stratigraphic Modeling: A Step Beyond Bright Spot," World Oil (May 1975).

2. Lindseth, Roy O., and McLeay, Robert D.: "The Seislog, A New Formation Logging Method," presented at the SPWLA Seventeenth Annual Logging Symposium, Denver, June 9-12, 1976.

3. Lindsey, J. P., Schramm, M. W., and Nemeth, L. K.: "New Seismic Technology Can Guide Field Development," World Oil (June 1976) 59-63.

4. Wittick, T. R. and Frink, A. P.: "Interpretive Methods of Stratigraphic Modeling," presented at the 46th Annual Meeting of the Society of Exploration Geophysicists, Houston, October 1976.

5. Nath, Ashoke K., and Meckel, Lawrence D., Jr.: "Seismic Modeling for Structural and Stratigraphic Interpretation," presented at the 46th Annual Meeting of the Society of Exploration Geophysicists, Houston, October 1976.

6. Tittman, J., and Wahl, J. S.: "The Physical Foundations of Formation Density Logging (Gamma-Gamma)," Geophysics (April 1965).

7. Dresser-Atlas, Log Review I, Dresser Industries, Inc., 1974.

8. Schlumberger, Log Interpretation, Volume I - Principles, Schlumberger, Ltd., New York, 1972.

9. Waller, W. C., Cram, M. E., and Hall, J. E.: "Mechanics of Log Calibration," presented at the SPWLA Sixteenth Annual Logging Symposium, June 4-7, 1975.

10. Cochrane, J. E.: "Principles of Log Calibration and Their Application to Log Accuracy," Journal of Petroleum Technology (June 1966).

11. Gardner, G. H. F., et al,: "Formation Velocity and Density - The Diagnostic Basis for Stratigraphic Traps," Geophysics (1974) Volume 39, Number 6, 770-780.

12. Wood, A. B.: "A Comparison of Well Velocity Methods in South Texas," Geophysics (1959) Volume 24, Number 3, 443-450.

13. Hicks, Warren G.: "Lateral Velocity Variations Near Boreholes," *Geophysics* (July 1959) Volume 24, Number 3, 451-464.

14. Rasmussen, N. F.: "Borehole Gravity Survey Planning and Operations," presented at the SPWLA Fourteenth Annual Logging Symposium, May 6-9, 1973.

15. Jageler, A. H.: "Improved Hydrocarbon Reservoir Evaluation Through Use of Borehole Gravimeter Data," *Journal of Petroleum Technology* (June 1976).

16. White, J. E.: *Seismic Waves: Radiation, Transmission and Attenuation*, McGraw-Hill Bood Company, Inc., New York (1965) 131-133.

17. Domenico, S. N.: "Effect of Water Saturation on Seismic Reflectivity of Sand Reservoirs Encased in Shale," *Geophysics* (1974) Volume 39, Number 6, 759-769.

18. Ritch, H. J., and Smith, J. T.: "Evidence for Low Free Gas Saturations in Water-Bearing Bright Spot Sands," presented at the SPWLA Seventeenth Annual Logging Symposium, Denver, June 9-12, 1976.

19. Crawford, G. E., Hoyer, W. A., and Spann, M. M.: "Frequency Response and Acoustic Resonance in Acoustic Logging," *The Log Analyst* (January-February 1973).

20. Flowers, B. S.: "Overview of Exploration Geophysics - Recent Breakthroughs and Challenging New Problems," *The American Association of Petroleum Geologists Bulletin* (January 1976) Volume 60, Number 1.

21. Neinast, G. S., and Knox, C. C.: "Normalization of Well Log Data," presented at the SPWLA Fourteenth Annual Logging Symposium, Lafayette, May 6-9, 1973.

22. Rudman, A. J., Whaley, J. F., Blakely, R. F., and Biggs, M. E.: "Transformation of Resistivity to Pseudo-Velocity Logs," *The American Association of Petroleum Geologists Bulletin* (May 1976) Volume 60, Number 5, 879-882.

23. Wood, A. B.: *A Textbook of Sound*, (Second Edition), G. Bell & Sons, Ltd., London (1944) 360-362.

24. Clark, Sydney P.: "Handbook of Physical Constants," GSA Memoir 97, (1966).

25. Peterson, R. A., Fillipone, W. R., and Coker, F. B.: "The Synthesis of Seismograms from Well Log Data," *Geophysics* (July 1955) Volume 20, Number 3, 516-538.

26. Amyx, Bass, and Whiting: *Petroleum Reservoir Engineering*, McGraw-Hill Book Company, Inc., New York (1960) 57-64.

APPENDIX A:

Operating Principles[7,8] and Calibration Procedures[9,10]
for
Acoustic and Density Logs

Acoustic Logs

The transmitters shown on Figure A-1 of a compensated acoustic log are pulsed alternately (say 15 times per second). The speed of sound in the sonde and in the mud-filled borehole is less than in the formation. Accordingly, the first arrivals of sound energy at the receivers correspond to sound travel paths in the formation near the borehole wall. Compensation for minor hole washouts and sonde tilt is made by averaging the values obtained from both sets of transmitter-receiver combinations.

The elapsed time of the detection of the first arrival at the two receivers is measured, thereby producing the parameter known as interval travel time: seconds/foot.

Usually, the distance between corresponding receivers is two feet and the time magnitude ranges from 80 microseconds to 400 microseconds. This is expressed in terms of interval travel time as 40×10^{-6} seconds/foot to 200×10^{-6} seconds/foot, or 40 microseconds/foot to 200 microseconds/foot. Interval travel time is, of course, the reciprocal of the common geophysical parameter, velocity, expressed in feet per second.

SCHEMATIC OF COMPENSATED ACOUSTIC LOG

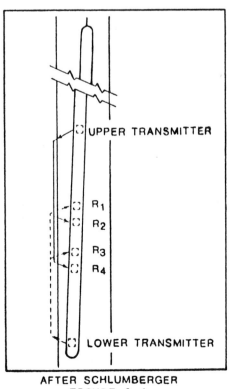

AFTER SCHLUMBERGER
FIGURE A-1
- 27 -

Porosity is determined from relationships such as those shown on Figure A-2. For example:

67 μsec./ft. = 14925 ft./sec. ≈ 8% porosity in a sandstone

83 μsec./ft. = 12048 ft./sec. ≈ 21% porosity in a sandstone

VELOCITY - POROSITY RELATIONSHIP

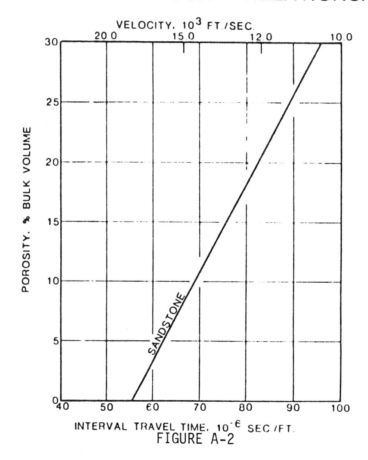

FIGURE A-2

Figure A-3 is a schematic of the BHC Sonic Logging System. The real-time signals for all four receivers are received at the surface and the surface gain control is used to determine the level of amplification.

"Up to this point, all four signals are processed sequentially and exactly alike. No calibration is necessary.

The real-time signals are converted to interval transit-time measurements by digital counting circuits using a crystal-controlled oscillator as their time base. Thus, the only calibration required is the adjusting of the analog circuitry to produce a recorder trace deflection proportional to the digital number representing Δt. Fixed digital numbers representing

- 28 -

interval transit times of 40, 60, 80, 100 and 140 microseconds/foot are fed from the oscillator to the digital-to-analog converter to allow the recorder trace to be adjusted accordingly. The errors of calibration of the BHC Sonic System are primarily related to the accuracy with which the recording galvanometers are adjusted to the calibration standard. The crystal-controlled oscillator to which the calibration standard is based is very stable."[9]

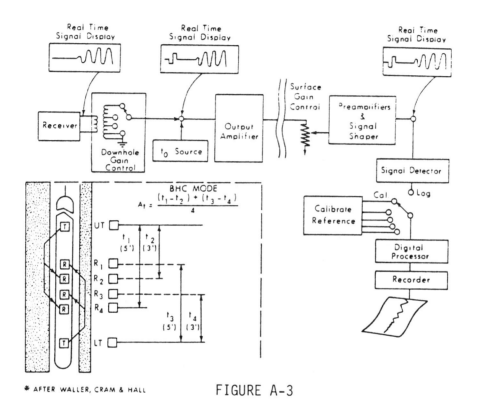

FIGURE A-3

Nonetheless, recorded data are not always a perfect representation of the true interval travel time of the media being investigated. Noise caused by sonde motion is usually the determining factor for downhole gain selection. Too low a gain setting can result in the inclusion of extraneous electronic and "road noises" (sonde motion) which often manifest themselves as high velocity spikes. Too high a gain setting results in the appropriate compressional arrival being missed on, most likely, the far receiver. This manifests itself as a cycle skip (low velocity excursions). Downhole/empirical or in situ calibration points should always be examined if they are present. These are valuable for three reasons:

- 29 -

(a) to check that the conventional techniques were actually and properly done,

(b) to identify changes imposed on the system due to temperature and pressure which can't be identified or compensated for by conventional techniques, and

(c) to identify logging line or bridle connection leaks that only show up when the logging tool is hanging in a mud-filled hole.

Good in situ calibration points for acoustic logs would be casing (57.5 μsec/ft), anhydrite (50.0 μsec/ft), and salt (67.0 μsec/ft).

Density Logs

The density log, as its name implies, measures the density of the near-wellbore formations. A radioactive source, applied to the borehole wall in a shielded skid, emits medium-energy gamma rays into the formation (See Figure A-4). By Compton Scattering, the gamma rays lose some of their energy to the electrons upon each collision. The number of those scattered gamma rays reaching the detector at a fixed distance is a function of electron density. Electron density is proportional to bulk density for most earth materials.

DRESSER ATLAS DENSILOG TOOL

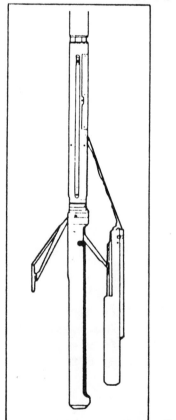

FIGURE A-4

- 30 -

Most modern density log surveys are made with a compensated system (Figure A-5). This method of using measurements made at near and far detectors allows for compensation for minor borehole rugosity and mudcake. Figure A-6 displays a typical density-porosity relationship.

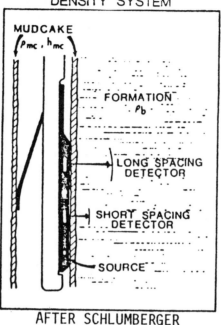

AFTER SCHLUMBERGER
FIGURE A-5

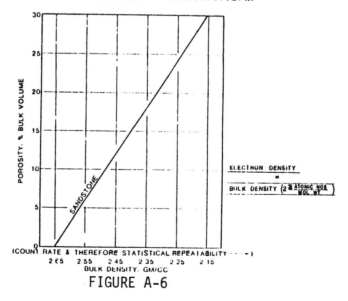

FIGURE A-6

- 31 -

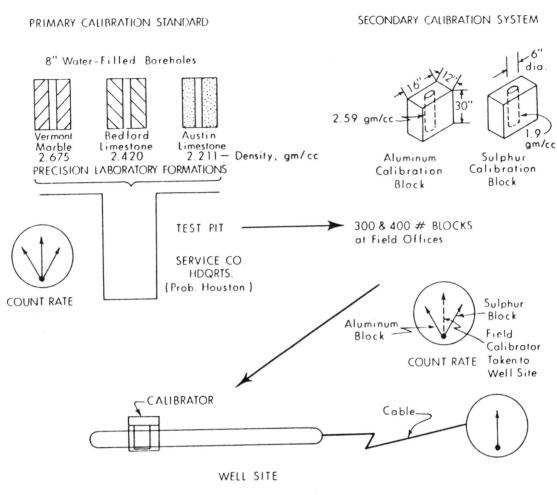

DENSITY LOG CALIBRATION SYSTEM

FIGURE A-7

Each density log source and detector combination are calibrated by documenting count rates observed when the system is in environments of known densities. Figure A-7, a schematic of the calibration sequence, is fairly self-explanatory. The count rate of the density detector in the presence of a portable gamma ray source is documented immediately after the count rate is noted in the two calibration blocks. This procedure does <u>not</u> allow for detection of problems with the gamma ray source used in the logging tool. For extremely critical operations, service companies usually will take calibration blocks to the well site in order to improve calibration accuracy.

Note that since apparent density recorded by the density log is inversely proportional to count rate, the density log has better statistical repeatability at low densities than at high. The same comments relative to the value of in situ calibrations made for acoustic logs also applies to the density. Downhole calibrators for the density log would be anhydrite (2.97 gm/cc) or salt (2.03 gm/cc). Sometimes zones of known porosity and lithology are available for calibration checks.

APPENDIX B

Numerical Examples of Gasmann Calculations

For illustrative purposes, two elementary, numerical examples of Gasmann calculations are presented below.

CASE 1

Hard Rock Example

Given: Dolomitic limestone with 20% porosity

Grain Density (ρg)	=	2.78 gm/cc
Acoustic Velocity*	=	14,000 ft/sec
	=	71.4 μsec/ft.
Bulk Density*	=	2.44 gm/cc
Formation Water	=	200,000 ppm TDS
Reservoir Temperature	=	240° F
Reservoir Pressure	=	4700 psi.

*At 100% Water Saturation.

What percent change in velocity could be expected if 75% of the pore space were filled with methane gas?

STEP 1

Solve equation (3) for M (sometimes referred to as the space

- 33 -

modulus) for 100% water-bearing condition.

$$M = V_p^2 \rho = 44.5 \times 10^{10} \text{ dynes/cm}^2.$$

STEP 2

Determine K_s.
From Reference 24, dolomite = 82×10^{10} dynes/cm^2 and limestone = 67×10^{10} dynes/cm^2.
For this example, K_s was <u>estimated</u> to be 74.5×10^{10} dynes/cm^2.

STEP 3

Estimate $\overline{K}$.

$$C_b = 3.7 \times 10^{-6} \text{ psi}^{-1} \text{ from Reference 26. Assume } \overline{C}_b \simeq C_b.$$

$$\overline{K} = 1/\overline{C}_b = 1.86 \times 10^{10} \text{ dynes/cm}^2.$$

STEP 4

Determine fluid properties at reservoir conditions.
Water compressibility (C_w) was estimated to be 2.26×10^{-6} psi^{-1}.
Water Density = 1.085 gm/cc.

$$K_w = 1/C_w = 3.05 \times 10^{10} \text{ dynes/cm}^2.$$

Gas density was estimated to be 0.157 gm/cc.
Gas compressibility (C_g) was computed from:

$$C_g = 1/\text{pressure} - (1/Z \times dZ/d \text{ pressure}) = 161 \times 10^{-6} \text{ psi}^{-1}.$$

$$K_g = 1/C_g = 0.0428 \times 10^{10} \text{ dynes/cm}^2.$$

Gas-water combination properties.

$$\rho_f = 0.25 (1.085) + 0.75 (0.157) = 0.389 \text{ gm/cc}.$$

K_f from the Woods[23] equation,

$$1/K_f = S_w/K_w + S_g/K_g.$$

$$K_f = 0.0568 \times 10^{10} \text{ dynes/cm}^2.$$

STEP 5

Solve equation (4) for $\overline{M}$ at $S_w = 100\%$.

$$\overline{M} = 44.5 \times 10^{10} - \frac{\left(1 - \frac{1.86 \times 10^{10}}{74.5 \times 10^{10}}\right)^2}{\frac{0.2}{3.05 \times 10^{10}} + \frac{0.8}{74.5 \times 10^{10}} - \frac{1.86}{(74.5 \times 10^{10})^2}}$$

$$= 32.9 \times 10^{10} \text{ dynes/cm}^2.$$

STEP 6

Solve equation (4) for M with combination gas/water conditions.

$$M = 32.9 \times 10^{10} + \frac{\left(1 - \frac{1.86}{74.5}\right)^2}{\frac{0.2}{0.0568 \times 10^{10}} + \frac{0.8}{74.5 \times 10^{10}} - \frac{1.86}{(74.5 \times 10^{10})^2}}$$

$$= 33.2 \times 10^{10} \text{ dynes/cm}^2.$$

STEP 7

Solve equation (1) for gas/water combination.

$$\rho_B = 0.2(0.389) + 0.8(2.78) = 2.30 \text{ gm/cc}.$$

STEP 8

Solve equation (3) for V_p.

$$V_p = \frac{33.2 \times 10^{10}}{2.30}^{0.5} = 3.80 \times 10^5 \text{ cm/sec}$$

$$= 12,500 \text{ ft/sec}.$$

SUMMARY

Utilizing this methodology and the stated assumptions, there would be an approximate 11% change in velocity* from a 100% water saturated zone to one that has identical rock properties but 75% gas saturation in

* $\frac{14,000 - 12,500}{14,000}$

- 35 -

the pore space. As a matter of interest, if the rock frame compressibility were assumed to be 5.0×10^{-7} psi^{-1}, the computed gas/water-bearing velocity would be 13,100 ft/sec. This is only 7% slower than the observed water-bearing velocity and helps demonstrate the sensitivity of the calculation to the assumption of rock frame modulus.

CASE 2

Soft Rock Example

Given: Tertiary sandstone with 32% porosity.

Grain Density	=	2.65 gm/cc.
Acoustic Velocity*	=	8130 ft/sec
	=	123 μsec/ft
Bulk Density*	=	2.17 gm/cc.
Formation Water	=	225,000 ppm TDS.
Reservoir Temperature	=	184° F.
Reservoir Pressure	=	7300 psi.

Estimated net effective stress ≤ 1,000 psi.

*At 100% Water Saturation.

What percent change in velocity could be expected if 75% of the pore space were filled with methane gas (gas gravity 0.8)?

STEP 1

$M = 13.3 \times 10^{10}$ $dynes/cm^2$.

STEP 2

K_s taken from Reference 24 as 37.9×10^{10} $dynes/cm^2$.

STEP 3

Estimated $\overline{C}_b = 3.5 \times 10^{-6}$ psi (??) from experience. Compressibility values on poorly consolidated formations are difficult to obtain. It is believed that many operators are attempting to measure and catalogue such data; however, to this author's knowledge, most of these data are being held confidential.

$\overline{K} = 1/\overline{C}_b = 1.97 \times 10^{10}$ $dynes/cm^2$.

STEP 4

$K_w = 3.71 \times 10^{10}$ dynes/cm^2.

$K_g = 0.186 \times 10^{10}$ dynes/cm^2.

$K_f = 0.244 \times 10^{10}$ dynes/cm^2.

$\rho_g = 0.32$ gm/cc.

$\rho_w = 1.15$ gm/cc.

$\rho_f = 0.528$ gm/cc.

STEP 5

$\overline{M} = 4.59 \times 10^{10}$ dynes/cm^2.

STEP 6

M @ 75% gas saturation = 5.27×10^{10} dynes/cm^2.

STEP 7

ρ_B @ 75% gas saturation = 1.97 gm/cc.

STEP 8

V_p @ 75% gas saturation = 5,370 ft/sec.

F

SUMMARY

The velocity change computed with these assumptions is approximately 34%! However, the computed velocity of 5,370 feet per second, or 186 microseconds per foot, compares reasonably well to observed sonic log values in gas zones having rock and reservoir properties similar to the assumptions made in this calculation.

As a matter of interest if the bulk compressibility of the rock frame is assumed to be 3.0×10^{-5} psi^{-1} (or an increase of nearly an order of magnitude), the computed Gasmann velocity for 75% S_g is 5,000 feet per second. This is an additional 4.5% decrease in velocity from the base, water-bearing case.

- 37 -

ABOUT THE AUTHOR

BRIAN E. AUSBURN is currently a Vice President and Consulting Engineer with J. R. Butler and Company in Houston, Texas. He received his B.S. and M. S. degrees in Geological Engineering from the University of Oklahoma. His past experience includes work as a senior petrophysical engineer with Shell Oil Company, a senior engineer and research project leader with Shell Development Company, and a geological assistant with Amerada Petroleum Corporation. At the Research Institute at University of Oklahoma he served as a research assistant.

SEISMIC MODELING

Stratigraphic modeling: A step beyond bright spot

E. V. Dedman, J. P. Lindsey and **M. W. Schramm, Jr.,** Members, GeoQuest International, Ltd.

15-second summary

The bright spot technique for locating stratigraphically trapped gas is considered a major advance in the field of exploration. And, recent refinements in bright spot technology are providing even better, more useful subsurface information. This article discusses stratigraphic modeling, a procedure which allows detailed and reliable stratigraphic interpretation of porous, commercial sands, using only seismic data. In effect, this new technique provides lithologic answers ahead of the bit.

SEISMIC MODELING is essentially a computational procedure which simulates seismic response that would be generated from an assumed geological cross-section (subsurface definition). This subsurface is normally defined in terms of: (1) interface geometry; (2) interval properties such as velocity, density and attenuation, and (3) other acoustically significant parameters. Further, the subsurface definition should incorporate all pertinent geological and geophysical information and assumptions.

Two major interpretive classes of modeling are structural and stratigraphic. Structural modeling relates to interpretive problems where structural identification is the primary objective—that is, subsurface geometry is a dominant consideration in the model and the trapping mechanism is primarily structural.

Stratigraphic modeling concerns interpretive problems where identification of lithologic changes is the primary objective—that is, the trapping mechanism is primarily stratigraphic.

Obviously, many problems have elements of both types of modeling; however, experience indicates that most modeling efforts can be generally considered in one of these two classes. The reason for making this distinction is that somewhat different formulation, analytical and evaluation techniques are used in the two types of modeling.

Emphasis in this article is on "stratigraphic" modeling.

Stratigraphic modeling dates from the early 1950s when the synthetic seismogram was first introduced. With the advent of sonic logging, a key acoustic property of rocks could be measured and relationships between rock sequence and seismic data studied. Computation of the synthetic seismic response was a one dimensional stratigraphic model. At first, only primary reflections were considered. Then interbed multiples were included and later, effects of attenuation. Today, sophistication of the synthetic seismogram calculation leaves little more to consider. So it is hoped that stratigraphic modeling can be expanded to a more useful tool than the synthetic seismogram.

Two new observations are the bases for reviewing stratigraphic modeling: (1) observed correlation between gas content in offshore Gulf of Mexico sands and seismic event amplitudes, and (2) ability to measure and use the basic propagating seismic wavelet to interpretive advantage. The first observation is more generally known as the "bright spot" phenomenon. From its introduction barely two years ago, bright spot technology progressed from merely recognizing them on seismic records, to mapping areal extent, calculating sand thicknesses and vertical distribution and measuring sand shale ratios in a new way.

Seismic basic wavelet applications are even newer than the bright spot, and will be discussed in more detail later.

Greatest initial use of stratigraphic modeling is with Gulf Coast sand reservoir problems. The reason it

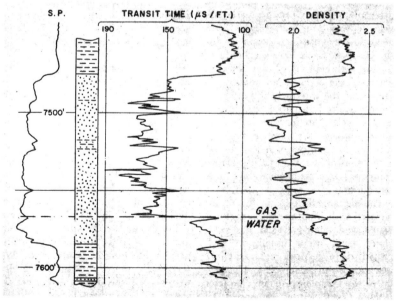

Fig. 1—Acoustic parameters for a gas sand.

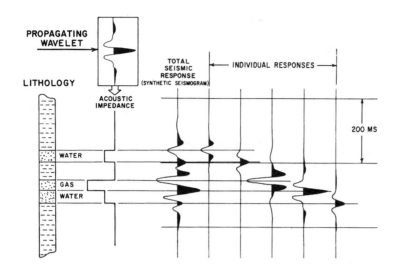
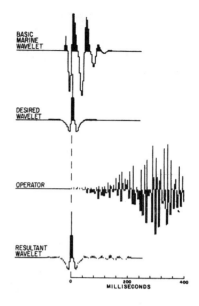

Fig. 2—Relationship between lithology, propagating wavelet and seismic response.

Fig. 3—How wavelet signatures can be processed.

works well and is useful is that gas-water and shale-gas contacts are clearly indicated by both velocity and density changes on logs (Fig. 1). The boundaries produce a 22% reflection coefficient—and cause strong seismic response. The important aspect of this log is that the acoustic anomaly it demonstrates relates to pay thickness.

Vertical lithology and seismic response. Fundamental concepts of stratigraphic modeling are rooted in basic relationships between acoustic lithology and its seismic response. Fig. 2 helps illustrate this. In simplest form, a seismic wavelet (pressure variation, velocity variation, etc.) moves as an acoustic wave through layers of dissimilar rock. Each interface produces a reflected replica of the incident wave. If the acoustic impedance decreases at the interface, the reflected basic wavelet is negative to the incident wavelet. An increase in impedance produces the positive reflected wavelet. Magnitudes of reflections are a function of the acoustic impedance contrast. Large changes produce large reflections.

Individual reflections in Fig. 2, associated with interfaces such as shale-water sand, shale-gas sand, gas sand-water sand, occur usually in such rapid time sequence that their responding wavelets overlap. This produces the complex response character labeled "total seismic response." All elements of total response have some meaning with respect to acoustic lithology of layers it characterizes. In Fig. 2, the meaning is readily apparent, at least for the gas sand. Its top and bottom interface are marked by the two largest legs of the complex. Even the isolated water sand can be seen to have a similar character although weaker in strength.

The propagating wavelet shown in Fig. 2 is ideal. It is symmetrical and relatively broad band. Schoenberger has demonstrated that these two characteristics provide the greatest degree of resolution for seismic responses. But unfortunately, no seismic wavelets are symmetrical and our ability to accurately pick gas sand boundaries in Fig. 2 is gravely impaired.

The basic propagating wavelet. Exact shape and character of a seismic basic wavelet depends on many things: source; receiver; proximity of a boundary, such as the surface; instrumentation; processing it undergoes before we see it, and presumably effects of attenuation as it travels through rock. If this wavelet could be measured, synthetic seismic solutions could be attempted using that shape. This is being done with some success.

But usually we are ahead of the bit, and our goal is to interpret stratigraphic meaning of seismic data. We then might use our measured wavelet to reprocess the data and correct its shape to something more desirable for interpretation purposes.

Wavelet correction. Complexities of the basic wavelet and its associated thin layer responses cause us to wish for a better wavelet. With knowledge of this wavelet shape, it is possible to exchange it for one more desirable (Fig. 3). We would rather it be a sharp spike in character, because that would provide the ultimate in resolution. If this were attempted, results would be disastrous from a stratigraphic interpretation viewpoint. Data would become very noisy, because we have asked for frequency components that did not exist in the original wavelet. The filter that was designed to correct it attempted vainly to build up these components and only managed to magnify noise in these spectral zones.

The most resolved wavelet that we can safely ask for is the "desired wavelet" of Fig. 3. It has the same bandwidth as the original basic wavelet, plus symmetry. By published design techniques, a filter is computed, shown as "operator" in Fig. 4. This is used on the basic wavelet to produce a new wavelet having a very acceptable likeness to our desired wavelet.

What we have accomplished is this: Since the operator will filter a basic wavelet and correct it to the much more desirable wavelet shown, **it will filter the seismic data and convert all interfering original wavelets to the simpler desired wavelet shape.** This produces favorable results for stratigraphic interpretation, as we will show later.

Example model. The following illustration of use of stratigraphic modeling is taken from the authors' more detailed paper on the subject which is referenced in the bibliography. Fig. 4A shows a model response, using a typical marine basic wavelet, to a type of sand deposit. This response suggests faulting because of the typical visual characteristics seen. If so, thickness of the zone involved changes across the fault because the character will not correlate completely.

Fig. 4B shows results of processing our data for wavelet correction. We should be much less willing to call it a fault as a result of our process. In fact, we now can clearly make out two sands, lens-like in cross-section, that pinch out in both directions. We can

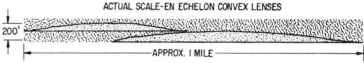

Fig. 4—Model response of typical marine basic wavelet (A), result of processing for wavelet correction (B) and geologic picture of the sand response (C).

even safely speculate that the sands do not communicate since the lower one extends under the upper one, and a shale separation is strongly suggested.

Fig. 4C shows the geological picture of our sand response. Shore-line sands deposited en echelon at the seaward rim of the delta is the answer. These sands have abrupt transitions at top and base on an E-log because they are well sorted and relatively clean.

Needless to say, the correct assessment of the degree of communication between sands that are bright on the seismic has important impact on drilling economics.

Case history. Fig. 5A is a standard seismic section which gives typical indications of a bright spot between shot points 2 and 4 at about 1.4 seconds. Structural closure is evident, and we would expect to have an amplitude processed version of this data for stratigraphic analysis.

Fig. 5B shows the amplitude version, in this case trademarked "Hi-Fi" since data are provided courtesy of Teledyne Exploration, Inc. The prospective zone is indeed bright, and it now becomes important to determine the thickness or thicknesses of gas pay zones involved. Presumably, areal extent of the anomaly can be mapped from the data grid available of which our section is but one line. Our prob-

lem is that the wavelet complex constituting our bright spot is as much a function of the basic wavelet as of the stratigraphy and lithology. Consequently, we need a section processed to correct wavelet shape to something more desirable. By now we know that it is a symmetrical band pass wavelet having the same bandwidth as the original wavelet.

Fig. 5C is the result of this processing on the Hi-Fi section. This technique has been trade-named "Shaper," and reshaping of the event in question is obvious. We now see a simpler complex than in the Hi-Fi version and this simplification, if we could study the data in detail, is a characteristic of all event complexes wherever we look, bright or not. This is the principal badge of reliability for any wavelet correction processing. It is a significant test of having the correct basic wavelet and consequently the correct operator design. It constitutes a necessary but not sufficient condition of valid processing.

Automatic gain control may be applied to the Shaper section (Fig. 5D) to highlight wavelet character at locations other than the bright spot event. Then, low impedance layers that are relatively thin may be interpreted.

The response character to look for is an asymmetrical sequence of two major lobes, the first deflecting to the left (white) and the second de-

About the authors

EMORY V. DEDMAN is a professional geophysical engineering graduate of the Colorado School of Mines. His geophysical experience includes service in both technical and management capacities with Phillips Petroleum Co., Richmond Exploration Co., Venezuela S.A. and The California Co. He has also held positions of vice president and president of Resources Technology Corp. and has worked as an independent consultant.

J. P. LINDSEY received B.S. and M.S. degrees in electrical engineering from Oklahoma State University. His past experience includes work as a research engineer with Phillips Petroleum Co. and as chief research geophysicist and vice president of research and development of Geocom, Inc.

MARTIN W. SCHRAMM, JR., holds a B.S. in petroleum engineering and geology from University of Pittsburgh, an M.S. in geology from the University of Pittsburgh and a Ph.D. from Oklahoma University. He has worked as an independent consultant and in geological and management positions with Gulf Oil Corp., Alex W. McCoy & Associates, Cities Service Oil Co., Cities Service International Co. and White Shield Exploration Corp.

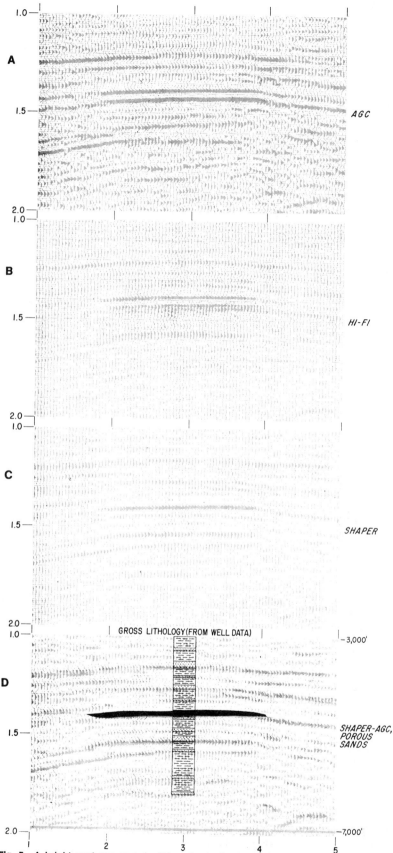

Fig. 5—A bright spot can reveal—(A) a standard seismic section; (B) amplitude version of (A); (C) results from further processing of amplitude version; (D) result when automatic gain control is applied to (C). (D) also shows how sands picked only from processed seismic data compare with lithologic log from well drilled later on the prospect.

flecting to the right (black). These zones turn out to be porous sands for Plio-Pliestocene and upper Miocene-age sediments with porosity in excess of 25% and sand quality suitable for commercial accumulations.

Sand thickness is given by the timing between these major identifying wavelet lobes, with proper allowance for amplitude and tuning effects. Two-way time thickness is converted to feet by use of calibration curves for sand velocities derived from well logs measured in similar rock types. Reasonably good estimates are obtained if only shale velocities are known at depths in question. Amplitudes then are measured for reflection coefficient magnitude and these in turn predict sand velocities relative to those of shale.

Fig. 5D strongly suggests that with amplitudes and wavelet processing, coupled with stratigraphic modeling and the lessons it teaches, one might quite reliably mark the porous sands on the AGC Shaper section. This indeed is possible. Thus, when there are no bright spots and the knowledge of position, quality, thickness and quantity of porous sands is desirable, as in the exploration for oil without a gas cap, we appear to have a new tool.

Fig. 5D also shows the AGC Shaper section with sands between 3,000 feet and 7,000 feet marked, and a lith log from a well. The gas sand is at the high of a sand unit extending down both flanks of the structure. The black lobe marking the base of the gas sand appears depressed in time in the water zone as compared to the gas portion. Close inspection would reveal that the white lobe marking the top of the sand is not time disturbed at the edge of the gas, thus showing time thickness of the water sand to be greater than time thickness of the gas sand. This is not a reasonable result of a uniform sand thickness fully loaded with gas if gas sand interval velocity is assumed lower than water sand interval velocity. And if Fig. 1 is at all representative, this seems to be the case.

The conclusion is that the sand is only gas saturated in the upper portion, and we are seeing a gas-water contact at the black lobe of the gas sand response. The alternative is to assume that the gas saturated portion suddenly thins at the water contact downdip, an unlikely situation.

An E-log from a well drilled on the

prospect confirms our interpretive suspicions. There is 120 gross feet of sand with the top 65 feet gas saturated. If it is big enough areally and not too shallow, and there is a pipe line nearby, it is commercial. Two shale stringers in the sand are not visible to seismic, being below the resolution afforded by the data bandwidth. The lith log (derived from the E-log), superimposed on our sand picks, is converted to time through the use of a suitable time-depth chart.

The comparison between the lith log and sands picked on the seismic is quite satisfying. Although thicknesses are not a perfect match in some places, and the number of individual sands do not always correspond, it is possible to detect—ahead of the drill —major sand units and a good measure of gross total sand in the column. From this data, sand-shale ratio maps of considerable detail can be constructed in unexplored areas with little or no well control.

It is important to note the following points related to judging the match between properly processed seismic data and a log:

• Log correlations over seismic lateral dimensions cannot include all sands seen on every hole.

• The log is inspecting a column with a radius of a few feet, while seismic is responding to a Fresnel zone dimension (300-1,200 feet).

• Sands of commercial interest are those with sufficient lateral dimensions and continuity to constitute adequate hydrocarbon storage volume.

The role of modeling. Interestingly, no modeling was brought to bear on the case history discussion. This does not imply that modeling is not required or is not of valid use. Actually, many of the assertions made about seismic data characteristics were derived from modeling experiences previous to this episode. Also, we haven't addressed the problem of precise prediction of reservoir volume and the variance in this calculation, an exercise which would require careful modeling.

The case history is completed by generating a combination map showing areal extent and an isolith map of the bright spot sand (Fig. 6). The northwest-southeast line crossing the block was the seismic line, and the well location is shown on this line. Neither of the intersecting lines shows

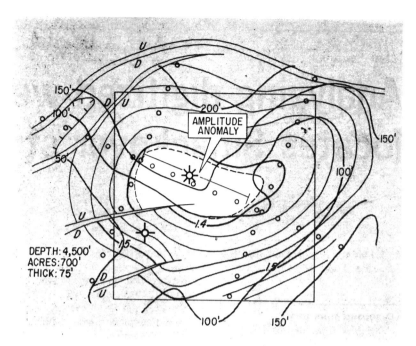

Fig. 6—Combination time-structure map and sand isolith map (color contours) of case history prospect in Fig. 5.

the bright spot. Plotting its extent on the base map overlaying the structural map for that same horizon shows us the likely areal shape. It is made to follow the contour level at the indicated gas-water contact. This produces a picture perfectly commensurate with the failure to show on the intersecting lines and well correlated with structure.

The colored contours of Fig. 6 constitute the isolith map for the bright spot sand. These thicknesses were read from the AGC Shaper data off the bright spot zone and posted on the base map for contouring.

In this case, it would have been possible to contour the gas zone interval had that been desired. To have done so would be another way to illustrate that the gas did not fully fill the sand since the two contour values would have been in conflict.

Conclusions. Based on experience to date, the following conclusions can be drawn about stratigraphic modeling in Tertiary sequences:

1. Modeling of stratigraphic situations of interest is an invaluable tool for helping the geophysicist and geologist understand more fully the phenomena he sees, particularly related to bright spot events.

2. It is necessary to measure and allow for the exact shape of the basic seismic wavelet, either through its use in generating interpretation templates, or more preferably through wavelet correction processing such as the Teledyne Shaper process.

3. Wavelet processed data open the door to stratigraphic interpretation of porous, commercial sands in great detail on the seismic data and with good reliability. This is equivalent to a new sand-shale ratio technique of much greater precision and local reliability than any previous method.

4. Stratigraphic modeling is specifically useful for making careful estimates of pore volume and focusing on the key parameters upon which these estimates depend.

5. The method is applicable to other geophysical provinces than the Gulf Coast and has the general capability of providing lithologic answers ahead of the drill.

BIBLIOGRAPHY

Harms, J. C., and Tackenburg, P., "Seismic Signatures of Sedimentation Models," *Geophysics*, Vol. 37, 1972, p. 45-58.

Sengbush, R. L., Lawrence, P. L. and McDonal, F. J., "Interpretation of Synthetic Seismograms," *Geophysics*, Vol. 26, 1961, p. 138-157.

Tegland, E. R., "Sand-Shale Ratio Determination from Seismic Interval Velocity," 23rd Annual Midwestern Regional Meeting, SEG and AAPG, Dallas, Texas, March 8, 1970.

Domenico, S. N., "Effect of Water Saturation on Seismic Reflectivity of Sand Reservoirs Encased in Shale," *Geophysics*, Vol. 39, 1974, p. 759-769.

Gardner, G. H. F., Gardner, L. W. and Gregory, A. R., "Formation Velocity and Density—The Diagnostic Basis for Stratigraphic Traps," *Geophysics*, Vol. 39, 1974, p. 770-780.

Schoenberger, M., "Resolution Comparison of Minimum-Phase and Zero-Phase Signals," *Geophysics*, Vol. 39, 1974, p. 826-833.

Dedman, E. V., Lindsey, J. P. and Schramm, M. W., Jr., "Stratigraphic Modeling," The Denver Geophysical Society Continuing Education Seminar on New Trends in Seismic Interpretation, Golden, Colo., April 17-19, 1975.

THREE-DIMENSIONAL SEISMIC MODELING†

FRED J. HILTERMAN*

Record sections from three-dimensional acoustic models often contain diffracted events not predictable by classical raypath theory. Several observed and calculated record sections from models of typical geologic structures such as synclines, anticlines, and faults verify this diffraction phenomenon. A careful interpretation of the character and moveout of these diffracted events is required to delineate certain portions of the geologic structures.

A far-field approximation of the retarded potential equation is suitable for direct time-domain evaluation and is used to synthesize the calculated sections. The excellent comparisons between the calculated and observed record sections suggest that the mathematical modeling technique can be a useful tool for enhancing field interpretations.

INTRODUCTION

Seismic record sections indicating faults, synclines, and other geologic structures have been an integral part of the geophysicist's interpretation for many years. The analysis of the seismic events on these record sections can, however, lead to confusing and indecisive interpretations. In this event, the geophysicist has often resorted to mathematical and experimental models.

As early as 1937, Rieber used small-scale seismic models to investigate the diffraction patterns occurring near irregular boundaries. He photographed the acoustic wavefronts diffracted from a fault model to demonstrate the geometric spreading effects of secondary propagation. To some degree, the experimental portion of the present work is an extension of Rieber's three-dimensional model studies.

The purpose of the present study was to develop a mathematical means for predicting the response to energy from a point source impinging on a three-dimensional acoustic model of arbitrary shape. Synthetic record sections displaying not only the geometric moveout but also the character of the pulse were desired. The synthetic record sections were then to be compared with experimental sections in order to verify the mathematical modeling technique.

For the experimental modeling, an eighth-inch condenser microphone and an electric spark simulated the detector array and the energy source. Paper and wood were used as model reflectors. In essence, the modeling was purely geometric, since the boundaries were rigid and shear waves were absent.

Our mathematical model utilizes a far-field approximation of a numerical technique suggested by Mitzner (1967). The development of the mathematical model from the retarded potential equations is given in the Appendix.

EXPERIMENTAL MODEL

Initial investigations of the acoustic pulse shape indicated that a transient with a peak frequency between 30 and 50 khz could readily be generated and propagated in air. The investigations also indicated that attenuation would not significantly alter the assumption of spherical propagation from the point source if measurements were made in the far field. Model and earth prototype units are summarized in Table 1.

Figure 1 illustrates a model and the modeling

---

† Part of thesis T 1297 submitted for Ph.D. degree at the Colorado School of Mines, Golden, Colorado, 1970. Manuscript received by the Editor May 25, 1970; revised manuscript received August 19, 1970.
* Mobil Oil Corporation, Dallas, Texas 75221.
Copyright © 1970 by the Society of Exploration Geophysicists. All rights reserved.

equipment that were constructed using Table 1 as a design criterion. The physical dimensions permitted the recording of a time window in which the reflections and diffractions from the geologic model boundary were uncontaminated with signals from irrelevant boundaries such as the sides of the enclosure. Electronic components are shown in the block diagram of Figure 2.

As illustrated in Figure 1, there were two viewing windows in the front of the enclosure. The right hand window contained a glass panel which reduced air turbulence when data were being collected and also acted as a safety precaution. The left-hand window was open and through this window the geologic models were placed on the movable platform. These geologic models were constructed out of wood or heavy paper; no significant change was observed between the reflected energy from the wood and from the paper.

After the geologic models were placed on the platform, the microphones and source were adjusted vertically so as to simulate the desired depth to the geologic structure. With the positions of the microphones and source fixed, successive shotpoint locations were obtained by translation of the movable platform.

Table 1. Summary of experimental scale factors and units

Time scale factor = $\text{time}_p/\text{time}_m$[1] = 1000
Length scale factor = $\text{length}_p/\text{length}_m$ = 12,000

Quantity	Model	Prototype
Length	1 inch	1000 ft
	1 cm	394 ft
Time	1 μsec	1 msec
Velocity	1120 ft/sec	13440 ft/sec
Peak frequency	40 khz	40 hz

[1] Subscript p represents prototype and subscript m represents model.

As the geologic models were incrementally passed beneath the source and detector, a single trace for each shotpoint was recorded by an oscilloscope camera. The traces were then manually digitized at a two microsecond sampling interval, normalized with respect to the reference microphone, and plotted as a time record section.

EXPERIMENTAL AND MATHEMATICAL RESULTS

A difficulty which arose when discussing the record sections was the ambiguity of the words "diffraction" and "reflection" when applied to the models. It was decided, therefore, to use the definitions set forth by Dix (1952, p. 237). The

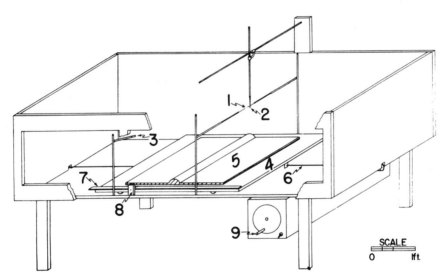

Fig. 1. Drawing of a model and associated equipment. The numbers refer to
1. Spark gap.
2. Eighth-inch condenser microphone (detector).
3. Half-inch condenser microphone (reference).
4. Horizontally movable platform.
5. Model of geologic structure.
6. Platform cable.
7. Meter stick.
8. Horizontal reference point.
9. Platform crank.

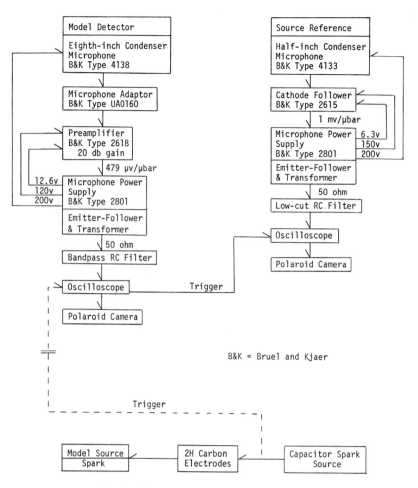

Fig. 2. Block diagram of modeling apparatus.

term reflection denotes an event which yields information concerning the curvature or dip of the boundary. Diffractions include all other events, some of which may yield information regarding the spatial location of an irregularity on the boundary.

An isometric sketch of each model accompanies the record sections. The location midway between the source and detector is shown by a small circle in the sketches and is referred to as the shotpoint location. Coordinate dimensions in the sketches are directly proportional to each other, so that distance can be represented by the shotpoint locations which are one cm apart. Two-way traveltime in microseconds is obtained by adding the appropriate time shift T_0 to the record section time scale.

A small time variation between the calculated and observed sections exists in several places. This small time discrepancy, introduced by experimental error, is not detrimental to the resulting interpretation.

Model A—Syncline

Model A in Figure 3 consists of a concave surface with a center of curvature located below the shotpoints. This type surface configuration with its buried focal line has often plagued the geophysicist. Even after the patterns in Figure 3 are recognized as those from a concave surface, the trace analysis is still not simple. The major problem is picking the exact two-way traveltime for the point where raypath 3 in Figure 4 is normal to the boundary.

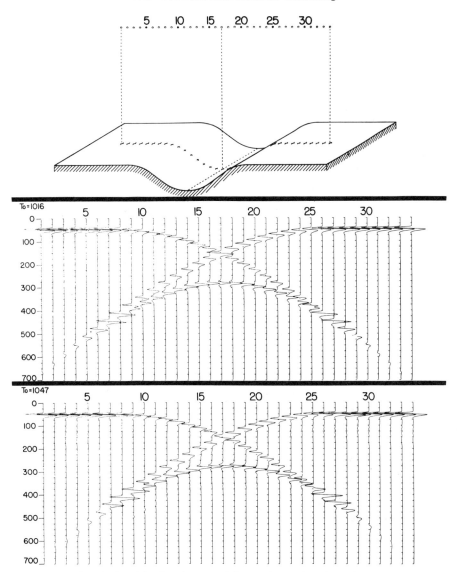

Fig. 3. Calculated (top) and observed (bottom) record sections of Model A.

Two-way traveltimes associated with raypaths 1, 2, and 3 are depicted in the trace analysis of Figure 4. However, this trace analysis was made with the knowledge of the model dimensions and shotpoint location. Also shown by two sets of arrows are the arrival times required for trough or peak reference picks. Using pulse shape alone, we found it difficult to correlate the picks of raypath 3 with those of the other two raypaths.

This dilemma is resolved if the change in pulse shape is considered. According to Dix (1952, p. 365), a reflected harmonic wave exhibits a phase change of 90 degrees every time it passes through a focal line. Fourier analyses of the incident pulse and of the returned energy for raypath 3 verified this phase phenomenon for high-frequency components. The amplitude density spectra were similar except for a constant multiplier.

Whenever a function undergoes a constant phase shift of 90 degrees, it is essentially being convolved in the time domain with the distribution function $(-\pi t)^{-1}$ (White, 1965, p. 11). The convolution of a pulse of finite duration with $(-\pi t)^{-1}$ yields a noncausal function which ex-

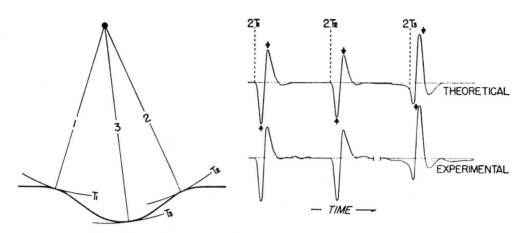

Fig. 4. Raypath geometry and trace analysis for shotpoint 15 of Model A.

tends from minus infinity to plus infinity. This noncausal effect is illustrated in Figure 4 by the nonzero onset energy at time $2T_3$. However, the phase shift for a transient incident wave cannot be exactly 90 degrees because diffracted energy would have to arrive before $2T_1$, the shortest traveltime to the surface.

Applying the reciprocity theorem to the general law that waves in an even number of dimensions leave wakes (Morse, 1948, p. 312), one concludes that the reflection at $2T_1$ has a sharp onset but has a tail that is infinite, though small in magnitude. The same principle holds for the reflection at $2T_2$, except that, due to the wake of the previous

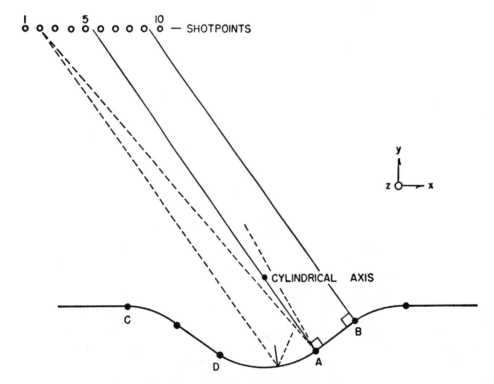

Fig. 5. Raypath limits normal to Model A.

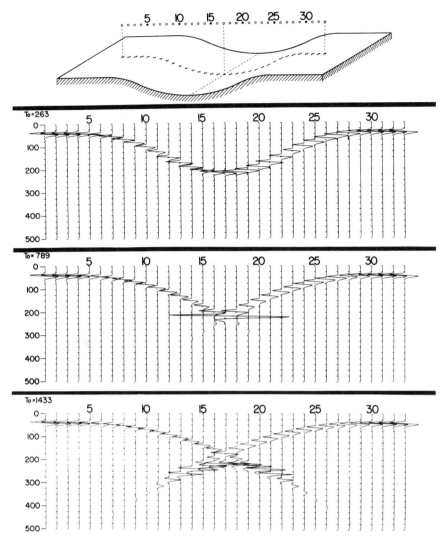

FIG. 6. Calculated record sections of Model B. Radius of curvature of concave-upward surface is 18 cm. Top section has shotpoint surface 10 cm below curvature axis; middle section, 1 cm below; and, bottom section, 10 cm above.

reflection, the reflection does not have a zero-amplitude onset. It is difficult to decide if the onset of the diffraction should be placed at $2T_1$ or at a later time associated with the surface between raypath 1 and 3. The problem is more academic than practical, since in the high-frequency approximation, the diffraction approaches a reflection with an onset time at $2T_3$. In short, this discussion suggests that classical raypath interpretation can be misleading if care is not exercised when concave boundaries are involved.

An additional feature Model A exhibits is an event that apparently transforms from a reflection to a diffraction without an appreciable increase in its time moveout. Figure 3, aided by Figure 5, illustrates this phenomenon. For shotpoints 1 to 5 (and 29 to 34), no raypaths are normal to the far flank of the syncline and hence only diffracted energy can be returned to shotpoints 1 to 5 from the far flank. The term diffraction was chosen, since shotpoints 1 to 5 do not contribute any information regarding the curvature of the anomalous surface. A trace analysis of this event is difficult because the pulse has broadened and does not have a distinct onset time.

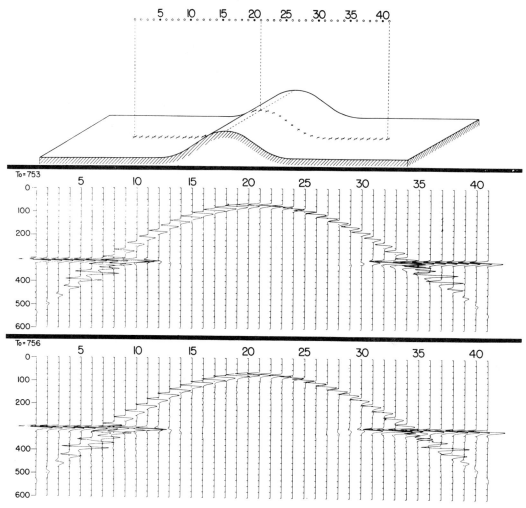

FIG. 7. Calculated (top) and observed (bottom) record sections of Model C.

Model B–Syncline with various depths of burial

Figure 6 illustrates the different calculated wave patterns obtained by varying the depth to the boundary. The same boundary configuration was used in all three of the calculated sections.

The amplitude of the diffraction on the 17th trace in the middle section of Figure 6 is 5 to 6 times that of the reflection amplitude on trace 1. This exact model was not tested experimentally; however, a similar model was tested and a diffraction amplitude $5\frac{1}{2}$ times that of the reflection was observed.

Model C–Large anticline

In Figure 7, the anticline may be considered as the boundary inverse of the syncline shown in Figure 3; that is, the size of the anomaly is the same in both models except that the anomalies are reversed in direction.

The syncline in Model A (Figure 3) was relatively easy to interpret, since the anomaly was several wavelengths high and there were zones in the time section where the diffraction was not masked by the flank reflections. On the remaining models, however, the structures were approximately two wavelengths in relief and the resolution of the events was more difficult.

Model D–Small anticline

The anticline illustrated in Figure 8 has two

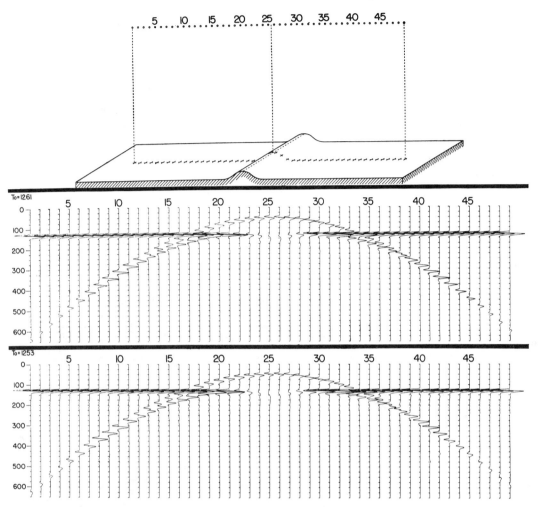

FIG. 8. Calculated (top) and observed (bottom) record sections of Model D.

surfaces that create buried foci; these are the bottom portions of the flanks. This buried foci effect is evident on the record sections if the event in the lower left-hand corner is followed upward toward the right. As the pulse reaches traces 12 and 13, a phasing effect starts. The change in pulse shape is actually a separation of a reflection from the top side of the flank and a later arriving diffraction from the lower portion of the flank. However, the small time difference between the reflection and the diffraction arrival times makes it difficult to interpret the flank of the anticline.

On traces 23 through 28, it appears as if a weak reflection at 125 μsec continues right through the anticline. This event is not a reflection but is a diffraction generated by the curvature of the flanks. In a similar manner, we can show that the energy arriving in the latter portion of traces 1 through 7 does not have a raypath which is normal to the surface. Once again, there is no sharp discontinuity to suggest a diffraction, which the event must be.

Although Model D is the same type of structure as Model C, there are several geometric differences worthy of noting in their patterns. In Model C, the buried focal lines are nearer the shotpoint surface than the buried focal lines in Model D. This geometric difference causes the diffractions from the concave surfaces of Model C to be evident on a fewer number of traces than the diffractions of Model D.

Most of the energy associated with the anomaly

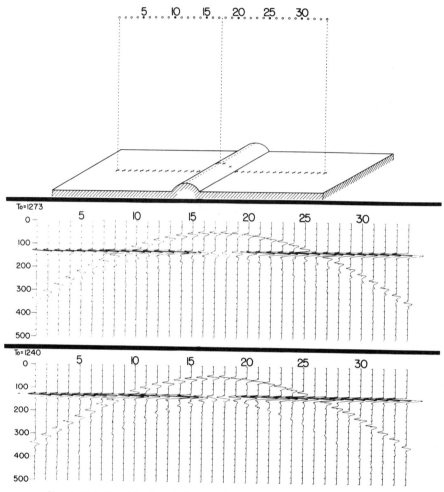

FIG. 9. Calculated (top) and observed (bottom) record sections of Model E.

arrives before the plane reflection in Model C, whereas the converse is true for Model D. However, the moveout of the convex-upward event in both model sections is approximately the same. These last two observations illustrate the importance of migration and the care that must be exercised when interpreting multilayered structures.

Model E–Reef

In some geologic areas it is desirable to outline accurately from seismic data the flanks of small positive anomalies. For example, what event (or events) would distinguish the two different types of flanks shown in Figures 8 and 9?

The boundary of Model E contains two discontinuities in its curvature. The magnitude of the diffracted energy from these discontinuities differs on the calculated and observed record sections. This difference is quite obvious on traces 16 to 19 at 125 μsec and also on traces 1 to 11, where the calculated diffraction from the near edge is larger than the observed diffraction. Both of these discrepancies are mainly due to the mathematical omission of the interacting surface effects. In the mathematical development, only that surface energy which is direct-incident energy is considered to have an effect on the shotpoint. Thus the energy which reflects off the side of the cylinder, strikes the flat surface, and returns to the detector is ignored. Since the boundary is rigid, this reflected-diffracted energy is sufficient

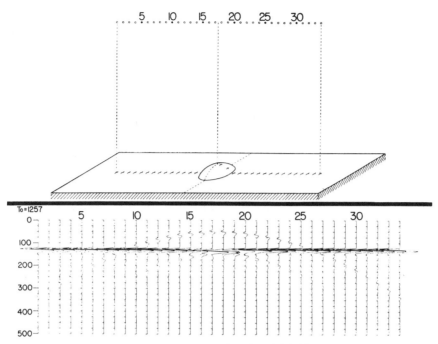

Fig. 10 Observed record section of Model F.

to cause a noticeable pressure variation at the shotpoint. This type surface energy can be incorporated in the theory but the subsequent numerical evaluation is more complicated (see Mitzner, 1967).

Model F–Dome

A complete mathematical record section of Model F in Figure 10 was not computed because a satisfactory algorithm for finding the solid angle had not been developed. A pressure wave reflected from a rigid sphere was calculated and found to compare favorably with the experimental data.

In Model F, the spherical segment has a radius and altitude which are within 0.1 cm of those of the cylindrical segment in Model E. It is not surprising, therefore, that the arrival times of Model F and Model E are approximately equal.

On traces 16 to 19 of Model F, the diffractions have a magnitude approximately equal to that of the reflection on trace 1. This abnormal amplitude is associated with the in-phase reflected-diffracted energy around the base of the spherical segment.

Model G–Vertical fault

Model G in Figure 11 presented unusual problems when a comparison between the observed and calculated pressure behavior was made. One of the problems was resolved when it was discovered that the shotpoint locations for the calculated section were 1.34 cm lower than the experimental shotpoint locations. This accounts for some of the moveout discrepancy between the calculated and observed diffractions in Figure 11.

Reflected-diffracted energy accounts for the amplitude difference between the calculated and the observed diffractions for shotpoints on the downthrown side. However, it is questionable if this observed energy is strongly evident on seismic field data, where the reflection coefficient is usually less than 0.3.

Model H–30-degree fault

For the 30-degree fault in Figure 12, the effect of the reflected-diffracted energy is very minor as shown by the comparison of the calculated and observed responses. A crossover of the diffracted events from the fault edges occurs between shot

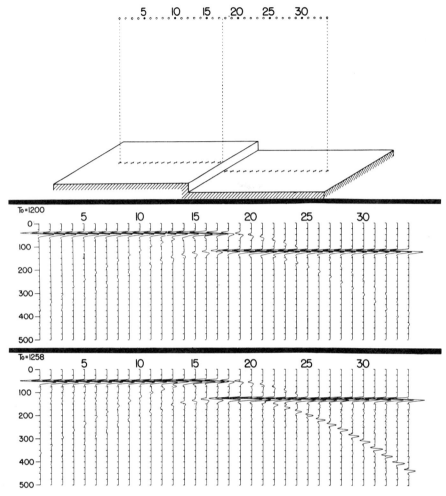

Fig. 11. Calculated (top) and observed (bottom) record sections of Model G.

points 31 and 32. The corresponding energy buildup on these traces is significant and may be used as a criterion for locating the fault plane.

Model I—Vertical fault with drag

The effect of introducing drag on the vertical fault is illustrated in Figure 13 by the two calculated record sections, one with drag and the other without. Angona (1960) has observed similar effects for drag on a two-dimensional model.

CONCLUSIONS

The excellent agreement between the mathematical and experimental model sections demonstrates that the mathematical modeling technique can accurately profile a three-dimensional acoustic model. The mathematical technique is especially suitable for subsequent interpretation, since the boundary segments can be analyzed individually.

Two types of diffractions were encountered. The first had a sharp onset and yielded information regarding spatial points on the boundary, such as the top edge of a fault. The second type associated with concave-upward surfaces had no sharp onset and was not predictable in many cases by raypath theory. Diffractions of this second type are often treated in practice as unwanted noise.

It is realized that neither the mathematical nor the experimental model is an exact representation

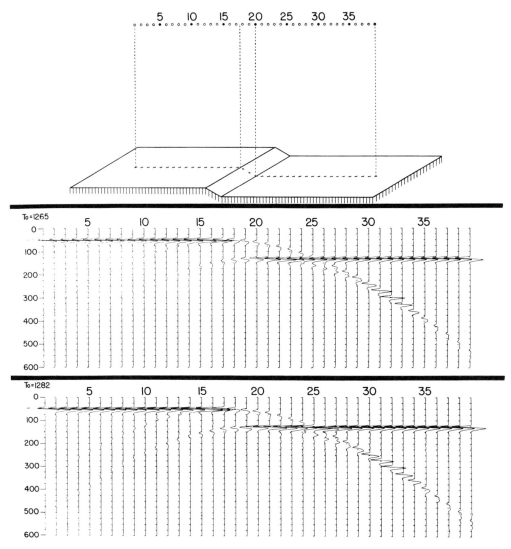

FIG. 12. Calculated (top) and observed (bottom) record sections of Model H.

of the earth prototype, but the insight gained from the reflection-diffraction complexes of the model sections is extremely useful in identifying field events which would otherwise be considered unwanted noise.

ACKNOWLEDGMENT

The author wishes to express his sincere appreciation to Professor John C. Hollister, who acted as thesis advisor at the Colorado School of Mines and who provided valuable suggestions during the model study and preparation of the paper.

Financial support of the work was given by NASA and Mobil Oil Corporation.

The spark-source system and circuitry were furnished by Atlantic Richfield Company and John P. Woods.

APPENDIX

MATHEMATICAL DEVELOPMENT

The pressure field in the fluid medium that results from the interaction of a spherical acoustic wave with a rigid boundary can be formulated several ways. We chose an integral-equation formulation because it is more amenable to sub-

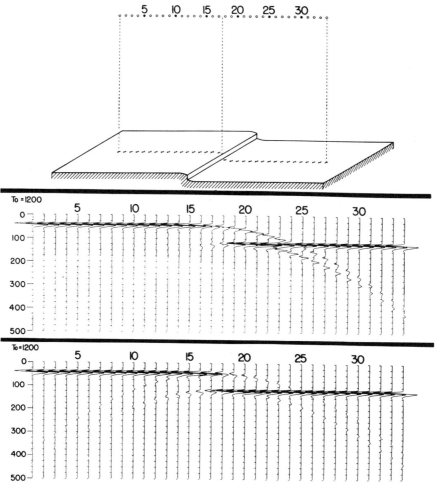

Fig. 13. Calculated (top) record section of Model I and calculated (bottom) record section of Model G. Model I is pictured above the top section; Model G appears at the top of Figure 11.

sequent numerical evaluation than the initial boundary-value method.

It has been found convenient to first express the field variation in terms of a continuous velocity potential $\phi(x, y, z; t)$, which also satisfies the wave equation. The pressure is related to the velocity potential through

$$p(x, y, z; t) = \rho \frac{\partial \phi}{\partial t}(x, y, z; t), \quad (1)$$

where ρ is the equilibrium density of the fluid.

The integral representation of the velocity potential at the field point ϕ_{fp} has been fully developed in terms of an incident potential ϕ_{inc} and a retarded surface potential $[\phi_s]$ by several authors, including Officer (1958, p. 260–265), Båth (1968, p. 192–198), and Drude (1922, p. 159–184). Imposing the rigid-boundary condition of $\partial \phi/\partial n = 0$, we get the integral equation for the velocity potential at the field point as

$$\phi_{fp}(t) = \phi_{\text{inc}}(g; t) + \frac{1}{4\pi} \int_S$$
$$\cdot \left\{ \frac{1}{r^2} \frac{\partial r}{\partial n} \left([\phi_s]_r \right) \quad (2) \right.$$
$$\left. + \frac{r}{c} \left[\frac{\partial \phi_s}{\partial t} \right]_r \right\} dS,$$

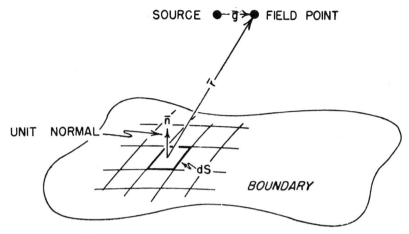

Fig. 14. General configuration of notation used in equation (2).

where Figure 14 along with the following definitions describe the notation used. S = fluid-solid interface (boundary); c = speed of sound in fluid; and $[\phi]_a = \phi(t - a/c)$, a time-retarded function.

Equation 2 is valid only for a point not on the boundary. For a point on S where the curvature is continuous, the potential satisfies the relation

$$\phi_s(\bar{h}; t) = 2\phi_{\text{inc}}(g'; t)$$
$$+ \frac{1}{2\pi} \int_{S'} \left\{ \frac{1}{r'^2} \frac{\partial r'}{\partial n} \left([\phi_{s'}]_{r'}\right) \right. \quad (3)$$
$$\left. + \frac{r'}{c} \left[\frac{\partial \phi_{s'}}{\partial t}\right]_{r'} \right\} dS'$$

(see Figure 15).

If the incident pressure field is continuous, equation (1) can be applied to equations (2) and (3) to yield respectively

$$p_{fp}(t) = p_{\text{inc}}(g; t)$$
$$+ \frac{1}{4\pi} \int_S \left\{ \frac{1}{r^2} \frac{\partial r}{\partial n} \left([p_s]_r\right) \right. \quad (4)$$
$$\left. + \frac{r}{c} \left[\frac{\partial p_s}{\partial t}\right]_r \right\} dS,$$

$$p_s(\bar{h}; t) = 2p_{\text{inc}}(g'; t)$$
$$+ \frac{1}{2\pi} \int_{S'} \left\{ \frac{1}{r'^2} \frac{\partial r'}{\partial n} \left([p_{s'}]_{r'}\right) \right. \quad (5)$$
$$\left. + \frac{r'}{c} \left[\frac{\partial p_{s'}}{\partial t}\right]_{r'} \right\} dS',$$

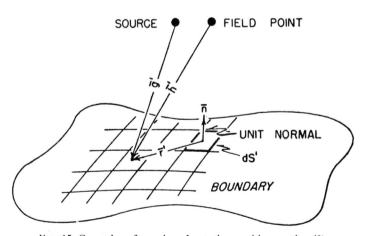

Fig. 15. General configuration of notation used in equation (3).

where the subscript identification has been retained.

At an arbitrary distance from the point source, the incident field $p_{\text{inc}}(a;t)$ can be expressed in terms of a known incident field $p_D(D;t)$ as

$$p_{\text{inc}}(a;t) = \frac{D}{a} p_D(t - a/c + D/c). \quad (6)$$

Mitzner (1967) presented a numerical technique for evaluating equations (4) and (5) in scattering problems. His technique produced excellent results but required a significant amount of computer time. In order to simplify equations (4) and (5), two approximations were introduced. The first approximation was that equation (5) can be written with the incorporation of equation (6) as

$$p_s(\bar{h};t) \simeq 2\frac{D}{g'} p_D(t - g'/c + D/c) \quad (7)$$

without unduly affecting the subsequent evaluation of equation (4).

In the second approximation, the source and field point were relocated at the same spatial point halfway between their original coordinate positions. This new spatial point is the shotpoint location. With this new coordinate definition, the incident pressure in equation (4) was dropped.

The advantage of redefining the source and field point locations is that now $g' = h = r$, so that equation (4) can be expressed in terms of equation (7) as

$$p_{fp}(t) \simeq \frac{D}{2\pi} \int_{\Omega(S)} \left[\frac{1}{r} p_D(t - 2r/c + D/c) \right. \\ \left. + \frac{1}{c} \frac{\partial p_D}{\partial t}(t - 2r/c + D/c) \right] d\Omega, \quad (8)$$

where the variable of integration was transformed to $d\Omega$, the solid-angle differential subtended at the shotpoint.

The integrand in equation (8) is a function of the variable r. The domain of the integrand with respect to r is from $r = R_0$ to $r = \infty$, where R_0 is the shortest distance from the shotpoint to the boundary.

In order to arrange equation (8) in a form suitable for discrete-time analysis, the following substitutions were made: $\tau = r/c$; $g(t) = p_D(t+D/c)$; $h(t) = \partial p_D / \partial t (t+D/c)$.

Inserting the above into equation (8) yields

$$p_{fp}(t) \simeq \frac{D}{2\pi c} \int_{\Omega(S)} \left\{ \frac{1}{\tau} g(t - 2\tau) \right. \\ \left. + h(t - 2\tau) \right\} d\Omega. \quad (9)$$

The continuous functions $g(t)$ and $h(t)$ were of duration T and were specified in discrete time as

$$G(i\Delta) = g(t) \quad (10)$$
$$H(i\Delta) = h(t), \quad (11)$$

where $i = 0, 1, 2, \ldots, N = T/\Delta$, and Δ = sampling rate. The discrete function H was actually expressed in terms of G through a numerical differentiation formula (Hildebrand, 1956, p. 82), since $g(t)$ was a known time-shifted function.

In the numerical evaluation of equation (9), it was desirable first to time-contour the boundary with respect to the traveltime from the shotpoint (see Figure 16). The traveltime to the kth zone was averaged to give

$$T_k = t_0 + (k + 1/2)\Delta, \quad (12)$$

where $t_0 = R_0/c$.

By time-contouring the boundary in the above manner, we made both discrete functions, G and H, constant on a given contour line. We could then replace the integral in equation (9) with a summation summed over the contour intervals. Also the observation time was specified discretely as

$$t = 2t_0 + (m + 1)\Delta. \quad (13)$$

Incorporating equations (10) through (13) into equation (9) yields

$$P(2t_0 + (m+1)\Delta) \\ \simeq \frac{D}{2\pi c} \sum_{k=L}^{M} \Omega(k) \left[\frac{G\{(m-2k)\Delta\}}{T_k} \right. \\ \left. + H\{(m-2k)\Delta\} \right], \quad (14)$$

where $P(i\Delta)$ = sampled approximation of the continuous function $p_{fp}(t)$,

$m = 0, 1, 2, \ldots$, desired number of samples,

$\Omega(k)$ = solid angle subtended at the shotpoint by the kth zone,

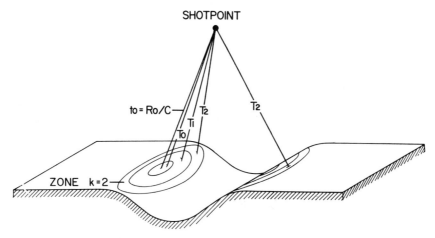

FIG. 16. Time-contouring of boundary with respect to shotpoint.

$$M = \begin{cases} m/2 & \text{for } m \text{ even} \\ (m-1)/2 & \text{for } m \text{ odd, and} \end{cases}$$

$$L = \begin{cases} 0 & \text{for } m \leq N \\ (m-N)/2 & \text{for } m > N \text{ and} \\ & (m-N) \text{ even} \\ (m-N+1)/2 & \text{for } m > N \text{ and} \\ & (m-N) \text{ odd}. \end{cases}$$

Solid-angle calculations

The solid angle Ω subtended at a field point by a nonsingular element is

$$\Omega = \int_S \frac{\partial r}{\partial n} \frac{1}{r^2} dS. \tag{15}$$

In the rectangular Cartesian system equation, (15) becomes

$$\Omega = \int_S \frac{[(u-x)n_x + (v-y)n_y + (w-z)n_z]}{[(u-x)^2 + (v-y)^2 + (w-z)^2]^{3/2}} dS, \tag{16}$$

(see Figure 17).

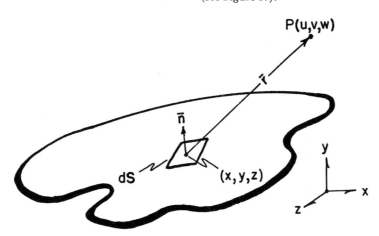

FIG. 17. Solid angle sign convention.

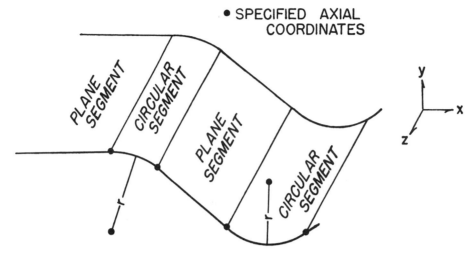

FIG. 18. Method of defining boundary.

Expressing the boundary (surface) as a function of the two independent variables z and x by

$$y = \beta f(x, z) + c, \quad (17)$$

we write the components of the unit normal as

$$(n_x, n_y, n_z) = \frac{\left\{-\beta\left[\dfrac{\partial f(x,z)}{\partial x}\right],\ 1,\ -\beta\left[\dfrac{\partial f(x,z)}{\partial z}\right]\right\}}{\left[1 + \beta^2\left[\dfrac{\partial f(x,z)}{\partial x}\right]^2 + \beta^2\left[\dfrac{\partial f(x,z)}{\partial z}\right]^2\right]^{1/2}}. \quad (18)$$

The boundaries in this study were infinite cylindrical surfaces which we approximated with segments of planes and circular cylinders. As illustrated in Figure 18, the boundary is completely described by specifying the x and y coordinates of the end points of each segment and the radius and axis location of circular segments.

In equation (14), the solid angle $\Omega(k)$ is subtended at the shotpoint by the surface element inscribed by two cones whose vertices are at the shotpoint and whose elements terminate on the boundary. The length of the elements are R_k for the outer cone and R_{k-1} for the inner cone. The subscripted R is the distance from the shotpoint to the middle of the kth contour zone. For $k=0$, the inner cone collapses to a straight line. This process of defining the surface of integration by two cones is similar to time-contouring the surface as discussed previously.

In order to evaluate $\Omega(k)$ associated with the kth zone, the solid angle corresponding to the outer cone was calculated and from this solid angle, the previously-calculated solid angle associated with the inner cone was subtracted. The process was repeated until the upper limit of the summation sign in equation (14) had been satisfied.

Since the boundary was divided into cylindrical segments, the total solid angle for a particular k was found by summing the solid-angle contributions from each segment. Figure 19 illustrates the geometry involved for finding the solid-angle contribution from an arbitrary cylindrical segment. The length of the cone element is R.

If P, the shotpoint location, is in the xy plane, only one-half of the surface of integration needs to be considered because of the symmetry about the xy plane. Applying this principle and the relationships $n_z = 0$, $w = 0$, and $ds = (dx^2 + dy^2)^{1/2} = [1 + (dy/dx)^2]^{1/2} dx$ to equation (16) yields

$$\Omega = 2\int_{x_1}^{x_2}\int_{z_1}^{z_2} \frac{((u-x)n_x + (v-y)n_y)}{((u-x)^2 + (v-y)^2 + z^2)^{3/2}}\,dzds. \quad (19)$$

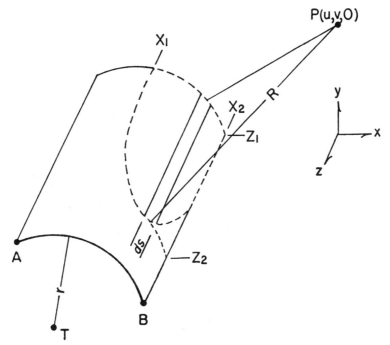

Fig. 19. Geometry and notation used in solid angle contribution from segmented surface.

Integrating equation (19) with respect to z and inserting the limits $z_1=0$ and $z_2=[R^2-(u-x)^2-(v-y)^2]^{1/2}$ gives

$$\Omega = \frac{2}{R}\int_{x_1}^{x_2}\frac{((u-x)n_x + (v-y)n_y)(R^2 - (u-x)^2 - (v-y)^2)^{1/2}}{(u-x)^2 + (v-y)^2}ds. \quad (20)$$

Substituting the appropriate relations for y, n_y, n_x, ds, x_1, and x_2 into equation (20) yields a single integral with respect to x.

REFERENCES

Angona, F. A., 1960, Two-dimensional modeling and its application to seismic problems: Geophysics, v. 25, p. 468–482.

Båth, M., 1968, Mathematical aspects of seismology: Amsterdam, Elsevier.

Dix, C. H., 1952, Seismic prospecting for oil: New York, Harper.

Drude, P., 1922, Theory of optics [trans. by C. R. Mann and R. A. Millikan]: New York, Longmans, Green and Co.

Hildebrand, F. B., 1956, Introduction to numerical analysis: New York, McGraw-Hill.

Mitzner, K. M., 1967, Numerical solution for transient scattering from a hard surface of arbitrary shape—Retarded potential techniques: J. Acoust. Soc. Am., v. 42, p. 391–397.

Morse, P. M., 1948, Vibration and sound: New York, McGraw-Hill.

Officer, C. B., 1958, Introduction to the theory of sound transmission: New York, McGraw-Hill.

Rieber, F., 1937, Complex reflection patterns and their geologic sources: Geophysics, v. 2, p. 132–160.

White, J. E., 1965, Seismic waves: Radiation, transmissions, and attenuation: New York, McGraw-Hill.

Interpretative lessons from three-dimensional modeling

Fred J. Hilterman*

ABSTRACT

Three-dimensional (3-D) seismic modeling has been accomplished by describing geologic surfaces with triangular plates and then computing the seismic response by Kirchhoff wave theory. The resulting time sections illustrate many interesting 3-D phenomena which are useful in interpreting geologic structures.

Three-dimensional resolution studies relate the concept of Fresnel zone reflection to seismic resolution. If high resolution is desired both horizontally and vertically, then not only is a dense field survey required, but also a detailed amplitude study. The dense seismic coverage is required to map the focal line of concave boundary edges, which are difficult to delineate with conventional seismic data.

Additional studies on complex models, such as grabens and 3-D permeability traps, associate interpretational pitfalls to a wandering specular reflection path, that is, "sideswipe." In each geologic model, maximum resolution is obtained on a principal plane line (dip line). If a seismic dip line is not available, the necessity of doing 3-D migration is emphasized, even if it is a migration of the time map.

INTRODUCTION

Seismic modeling is now an integral part of most geophysicists' interpretation. This modeling can involve both forward algorithms, such as one-dimensional (1-D) synthetic seismograms, and inverse algorithms, such as wave theory migration. The present paper will emphasize forward three-dimensional (3-D) modeling and interpretative lessons derived from the modeling. While emphasizing forward modeling, the restrictions and limitations of our current inverse modeling methods will be apparent.

The main theme is to illustrate what interpretative lessons can be exploited from 3-D modeling which might be overlooked with conventional two-dimensional (2-D) modeling. In order to explain some of the differences between 2- and 3-D synthetic seismograms, I decided not to start with a realistic model because the seismograms across even the simplest geologic models are difficult to evaluate. Instead, 2- and 3-D Fresnel zones were modeled, and the output seismic sections were evaluated for both time and amplitude anomalies. These Fresnel zone sections then set the groundwork for realistic geologic models.

THEORY

The seismic pressure response for a zero source-receiver separation was computed for each of our 3-D geologic models according to the Kirchhoff equation described by Trorey (1970). This is a scalar wave-equation solution, so only compressional waves are considered. The normal incidence reflection coefficient is also applied. The present examples were for constant-velocity media, but an elaboration on some of the computational procedures in Kirchhoff theory for multivelocity models was given by Hilterman and Larson (1976).

The approximation generated by using a zero-offset shooting geometry to model a seismic common-depth-point (CDP) stacked section was investigated theoretically by Berryhill (1977) and Trorey (1977, 1978). As pointed out by these authors, a single image point spatially adjusted for the real source-receiver separation can be located; from this image point a zero-offset pressure response can be computed which is a good approximation for the sourcer-receiver offset pressure response. However, for the zero-offset approximation to compare favorably to CDP stacked data, a continuous stacking velocity analysis and application must be conducted during processing. The zero-offset location of the source-receiver will be called the shotpoint.

In Figure 1 the diffraction response for a triangular plate is illustrated. Notice that the theoretical solution is singular and, thus, must be band-limited for future digital analysis. The theoretical solution and a band-limiting procedure are developed in Appendix A. A sampled version of the pressure response

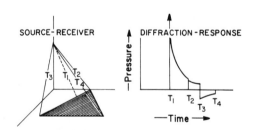

FIG. 1. Diffraction response from a flat plate for a zero-offset source-receiver model. The diffraction response at time T_1 is singular.

Manuscript received by the Editor March 20, 1978; revised manuscript received May 21, 1981.
*Formerly Seismic Acoustics Laboratory, University of Houston, Houston, TX; presently Geophysical Development Corporation.
0016-8033/82/0501—784-808$03.00. © 1982 Society of Exploration Geophysicists. All rights reserved.

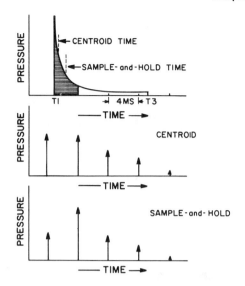

FIG. 2. Comparison of centroid sampling and sample-and-hold sampling of a continuous diffraction response. Times T_1 and T_3 normally do not coincide with discrete sample times.

using a two-point linear interpolator which was based on the centroid time position. No erroneous phase shifts were evident on stratigraphic synthetic seismograms when centroid sampling was applied, while phase shifts were introduced by the sample-and-hold sampling.

In Figure 3, a 3-D geologic boundary is represented by numerous triangular plates. To obtain an accurate representation of the geologic boundary with the smallest number of samples, the author used variable plate sizes rather than a uniform grid. Because the Kirchhoff wave equation is an integration scheme, both the curvature and the area of the boundary must be represented adequately by the sampling technique. Thus, the selection of a nonuniform spatial sampling allows for a denser sampling at tight curvatures, such as the top of the dome, without oversampling the surrounding gentle flanks or flat areas.

Once the boundary sampling is completed, shotpoint locations are easy to define for a series of seismic lines, as shown by the small circles in Figure 3. As indicated before, the pressure at a shotpoint is computed by integrating around the edge of each triangle with respect to time and then summing all the triangle time responses.

FRESNEL REFLECTION ZONES

Spike time sections

In Figure 4, the definition of a Fresnel zone for a transient seismic wavelet is described. Classically, Fresnel zones are related to monofrequency waves (Born and Wolf, 1970), and it seemed a natural extension to select the predominant period of the transient wavelet to describe a Fresnel zone. When the shotpoint is over the middle of a disc, energy reflects from the disc's midpoint, and this response is depicted in the time domain by a spike at the two-way time T_0. Similarly, the coherent diffraction from the edge arrives later in time by the amount T, the wavelet period, and is depicted as an exponential function in the time domain. Classically, the disc (in Figure 4) covers two Fresnel zones, but here we will refer to it as a one-wavelength (1λ) Fresnel zone. The time function, in the upper right, is called the spike time response even though the Fresnel zone was described for a source wavelet with period T. The next step was

from the near edge of the plate in Figure 1 is shown in Figure 2. Only the diffraction response associated with the T1-to-T3 traveltimes portion of the plate is illustrated. Computationally, it is desirable to sample as coarsely as possible; thus, a sampling interval of 4 msec was chosen. Because of the rapid exponential decay at the front end of the diffraction response, it was necessary to use centroid sampling rather than the conventional sample-and-hold sampling. A comparison of the two sampling techniques is shown in Figure 2. As explained in Appendix A, the centroid of the area under the continuous curve between two digital time samples was found, and then this centroid weight was distributed

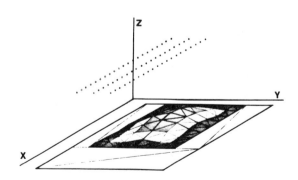

FIG. 3. Boundary description of dome using triangular plates. Zero-offset source-receiver locations are shown by small circles.

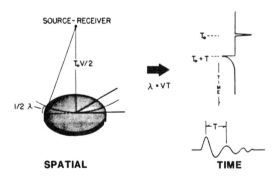

FIG. 4. Description of a one-wavelength Fresnel zone in the spatial and temporal domains.

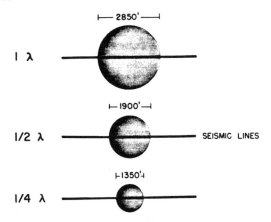

FIG. 5. Model profile lines over three flat Fresnel discs at a depth of 5000 ft. Velocity is 10000 ft/sec and the predominant period of the transient seismic wavelet is 40 msec. Shotpoints are spaced at 200-ft intervals.

to examine the time response for different shotpoints. These are illustrated in Figures 5 and 6.

The one-wavelength Fresnel zone in Figure 5 was based on a 400-ft source wavelength, which amounts to a 40-msec predominant period. The one-half wavelength zone was based on the width of the first lobe of the seismic wavelet, which was 18 msec rather than $1/2\ T$. Similarly, the one-quarter wavelength zone was based on 9 msec. A seismic velocity of 10,000 ft/sec was used throughout the study.

Even though these Fresnel zones were defined for a 400-ft

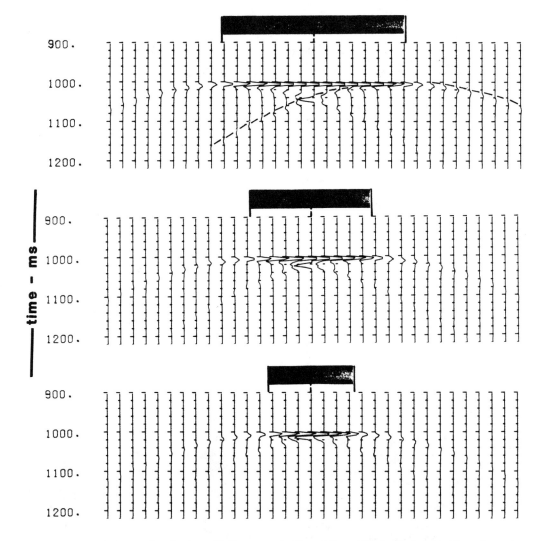

FIG. 6. Spike time sections for the profile lines across the Fresnel discs which are illustrated in Figure 5.

FIG. 7. Model profile lines over 2-D Fresnel plate reflectors which are at a depth of 5000 ft.

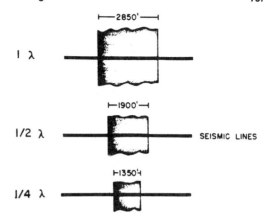

wavelength source pulse, Figure 6 illustrates the seismic response for a spike source pulse. These time sections actually have a two-point source pulse with weights of 0.5 in order to smooth the diffraction response and thus aid the visual presentation. The spike time sections were examined first instead of the wavelet section because the origin of the diffraction energy, and possible tuning effects, are easier to evaluate on the spike time sections.

In Figure 6, the shaded zones represent the spatial location of the discs. Two diffraction tails, which emanate from the right-

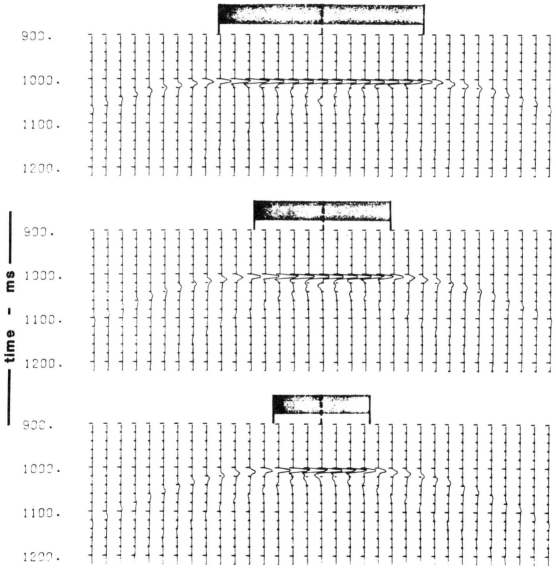

FIG. 8. Spike time sections for the profile lines across the 2-D plates illustrated in Figure 7.

hand edge of the disc, are shown in the upper section. The diffraction tail to the right of the disc has the same polarity as the reflection event, while the diffraction tail under the disc has a polarity opposite to the reflection event. However, the magnitude of the diffraction tail under the disc is larger than the other diffraction. Thus, while the arrival times of these two diffraction branches are symmetric, their magnitudes are not. This unequal distribution of energy on the diffraction branches is easy to visualize because when the shotpoint is over the disc, there is more diffracting edge with approximately the same two-way traveltime than when the shotpoint is symmetrically displaced off the edge. The diffractions come in focus at the midpoint (as discussed by Woods, 1975), and this high amplitude is a significant factor, as will be shown later, in specifying horizontal resolution.

As shown in Appendix B, the midpoint diffraction response is one of the few diffraction responses which can be written easily. At the midpoint, the response due to a spike source wavelet is:

$$p(t) = \frac{1}{vT_0} \left[\delta(t - T_0) - \left(\frac{T_0}{T_1}\right)^2 \delta(t - T_1) \right],$$

where

$p(t)$ = spike diffraction response,
T_0, T_1 = two-way traveltimes from the source to the center and edge of the disc, respectively,
v = velocity, and
$\delta(t)$ = Dirac delta function.

Notice in Figure 6 that the reflection-diffraction amplitude at

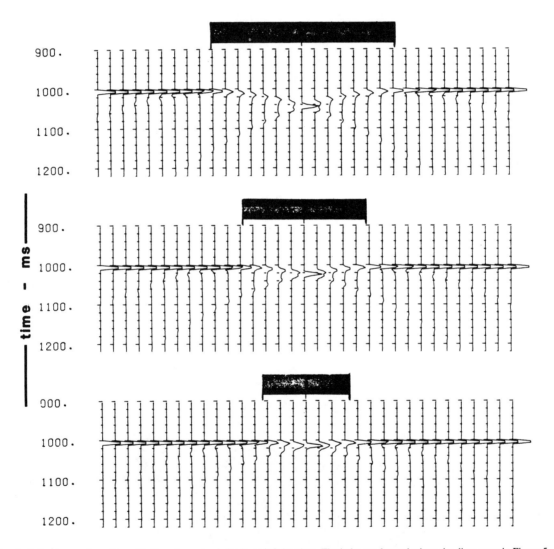

FIG. 9. Spike time sections for profile lines traversing holes in an infinite plate. The holes are located where the discs were in Figure 5.

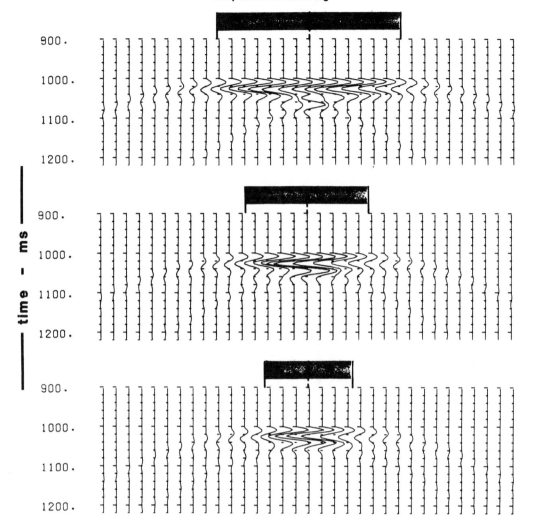

FIG. 10. Wavelet time sections for the profile lines across the disc which are illustrated in Figure 5.

the very edge of each disc is approximately one-half the reflection amplitude at the disc center. Thus, it appears that by marking the half-amplitude point of a horizontal event, the lateral extent of the boundary can be mapped. This concept will be used later, but with a slight variation, to compensate for concave edges. Remember the disc has a continuous convex (circular) edge.

By examining the center traces in the three sections of Figure 6, we notice a dipole response is generated as the disc size becomes smaller. This is similar to a thin bed tuning effect, as described by Widess (1973). Before examining the seismic wavelet sections for these Fresnel zones, the 2-D plate diffractions, which have widths analogous to the Fresnel discs, are shown in Figures 7 and 8.

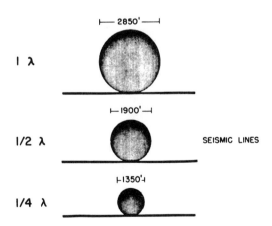

FIG. 11. Model profile lines tangent to three Fresnel reflectors at a depth of 5000 ft.

790   Hilterman

The 2-D Fresnel zone is defined similarly to the Fresnel zone of a circular disc. The most dramatic difference between the time sections in Figures 6 and 8 is that for shotpoints near the middle of the 2-D plate there are no high-amplitude focused diffractions, which do exist for shotpoints centered over the disc. However, the diffractions at the edge and off the 2-D plate are only slightly larger than the corresponding disc diffractions.

If we extrapolate this boundary curvature principle, we would expect diffractions generated by a concave edge to have a greater magnitude than 2-D plate diffractions. This is the phenomenon that Woods (1975) illustrated for a set of holes that correspond in location to the discs shown in Figure 5. The spike time sections for circular holes are given in Figure 9. The diffraction events in Figure 9 have the same magnitude, but opposite polarity, to the diffraction events in Figure 6.

Wavelet sections

The wavelet sections are the result of convolving the 25-Hz source wavelet with the spike time sections. The wavelet sections for the Fresnel discs (Figure 5) are shown in Figure 10.

There are several features worthy of noting in Figure 10. In the upper section there is an apparent frequency broadening under the disc because of the strong diffractions which appear on the section beneath the disc reflection. This effect turns into a Fresnel brightness in the middle section where the amplitude is 1.5 times that of the specular reflection amplitude. For small geologic bodies, the amplitude is a measure of the reflecting area. In the lower section, which is for the one-quarter wave length disc, the time response is approximately the derivative of the seismic wavelet. In summary, disc size is analogous to the more

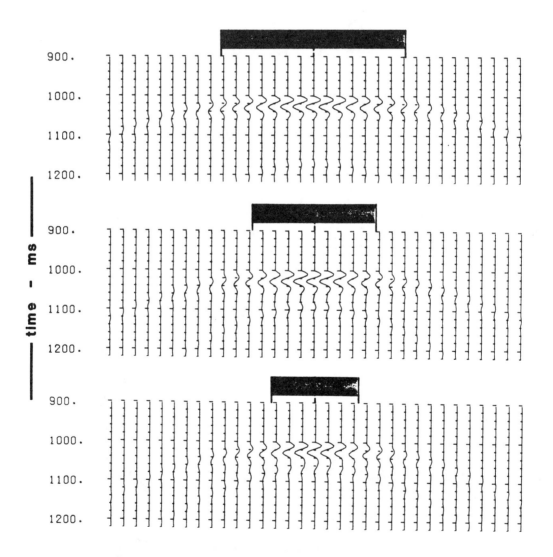

FIG. 12. Wavelet time sections for the profile lines tangent to the Fresnel discs in Figure 11.

familiar effects of bed thickness tuning, as pointed out by Widess (1973). Thin beds are analogous to small spatial Fresnel zones, where both have seismic responses which are the derivative of the source wavelet and the amplitude is a measure of size. However, if a thin bed is also spatially small, then the seismic response takes the form of the second derivative which appears as a high-frequency source wavelet with negative polarity.

Figures 11 and 12 illustrate another set of seismic lines across the three Fresnel zones. The amplitude in the center of each wavelet section is approximately the same. In the middle section, a weak diffraction (1100 msec on center trace) from the far edge gives the appearance of sideswipe from a dome out of the plane of the section. This phenomenon is also observed in the upper and lower sections. However, in the upper section the opposite edge is far enough away that the diffraction is weak and hard to see, while in the lower section the diffraction from the far edge arrives at nearly the same time as the reflection and is thus hard to recognize. The time separation between reflections and diffractions is better illustrated in Figure 6 by examining the traces right under the disc edge on the three time sections. This anomalous domal-appearing sideswipe is also observed for basins.

The wavelet sections across the holes are shown in Figure 13. The ability to resolve the lateral edge of the hole in the lower section is a function of the bandwidth. The buildup of amplitude at the center of the hole is once again at the focal point of the concave edge. Two-dimensional migration of the lower section in Figure 13 will not resolve the horizontal limits of the hole.

In the next example, the amplitude information concerning Fresnel zones will help to interpret the seismic sections across a distributary sand bar.

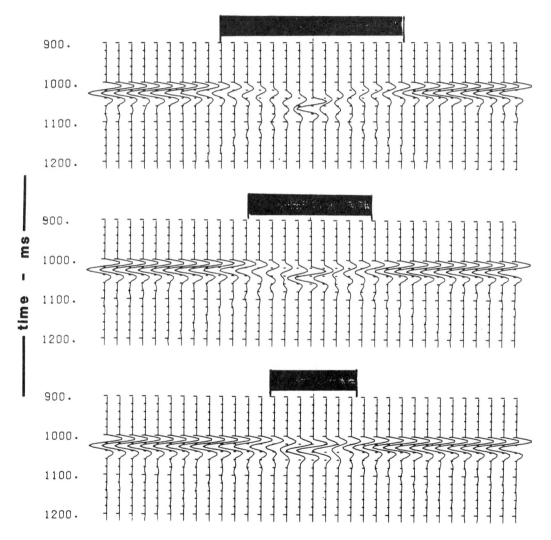

FIG. 13. Wavelet section for profile lines across Fresnel holes. The holes are the same size as the discs shown in Figure 5.

DISTRIBUTARY SAND BAR

The profile lines in Figure 14 were designed on a 1000-ft grid across a hypothetical amorphous sand body. The depth of burial is 5000 ft, and the relief is small enough to be beyond the vertical resolution of the seismic system. From the Fresnel zone experiments, it appears that the edge reflections should be one-half the amplitude of a normal incidence reflection; so we can utilize this criterion to pick the edge of the sand body. This amplitude rule was used on the wavelet sections in Figures 15 and 16.

In Figures 15 and 16 the boxed areas on the time sections represent the true lateral extent of the geologic boundary, while the black line represents the interpreted extent using the half-amplitude rule. On line 2E, the reentrant of the field near the 3N intersection (Figure 14) is not seen, and therefore the field is overestimated. The high amplitude in the misinterpreted zone of line 2E is due to a concave edge, which was previously noted with the Fresnel holes in Figure 13. In a similar fashion, line 4E overestimates the extent of the field. In the middle of line 4E, the erroneous sand bar was predicted because the shotpoints approached the focal line of the concave edge. Unfortunately, the time delay, where the erroneous sand was predicted on line 4E, is approximately 4 msec and would normally be ignored by associating it with either structure or a velocity anomaly over the field. In a similar fashion, the diffraction events off the edge are too small to provide an effective diffraction interpretation. Line 3N (Figure 16) goes through the neck of the field, and applying the half-amplitude rule on the corresponding wavelet section grossly overestimates the field extent.

An areal outline of the estimated field, as predicted by the half-amplitude rule from the sections in Figures 15 and 16, is given in Figure 17. The area enclosed by the dotted line plus the shaded area is the seismic estimate of the total sand body area which is 40 percent too high. All the overestimation is due to concave edges. As will be shown, this can be avoided if a denser seismic grid were used and the concave focusing phenomenon were recognized. In Figures 18 and 19, an amplitude interpretation, using an areal survey, is illustrated. The key to the interpretation is at the intersection of lines 3N and 4E where the high-amplitude nose is plunging southward.

The seismic amplitudes shown in Figure 19 for two of the areal grid lines were normalized with respect to the normal incident reflection amplitude. These normalized amplitudes were contoured as shown in Figure 18. Note that constructive tuning yields amplitudes which are greater than the normalized reflection coefficient. This tuning, of course, was observed on the Fresnel zones due to constructive interference of the diffractions, particularly on the one-half wavelength Fresnel zone. On the amplitude contour map, the concave edge can be recognized by the high-amplitude focal line and the deviation from parallelism of the contours as the focal line is approached. Notice that the convex sides have parallel amplitude contours. This is not the final approach to outlining fields because there were many as-

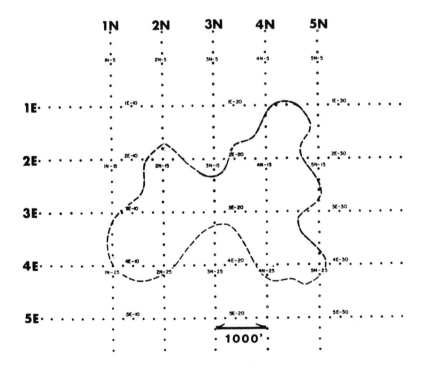

FIG. 14. Map of a hypothetical horizontal sand body buried at 5000 ft. Profile lines are 1000 ft apart.

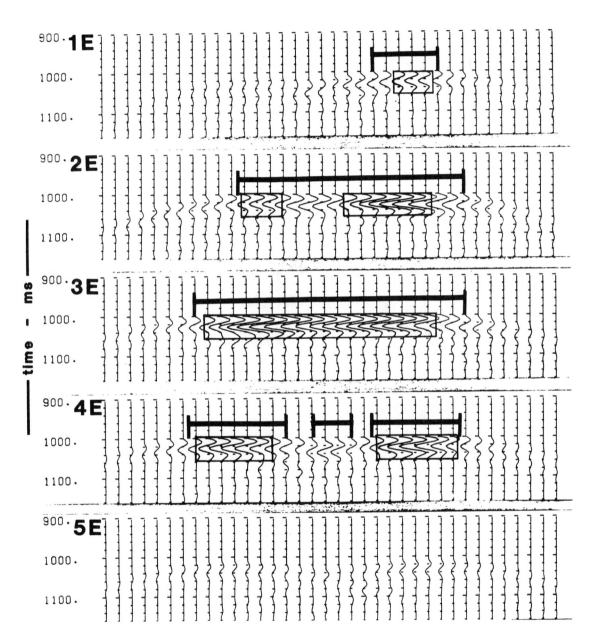

FIG. 15. East-west wavelet time sections across the sand body shown in Figure 14.

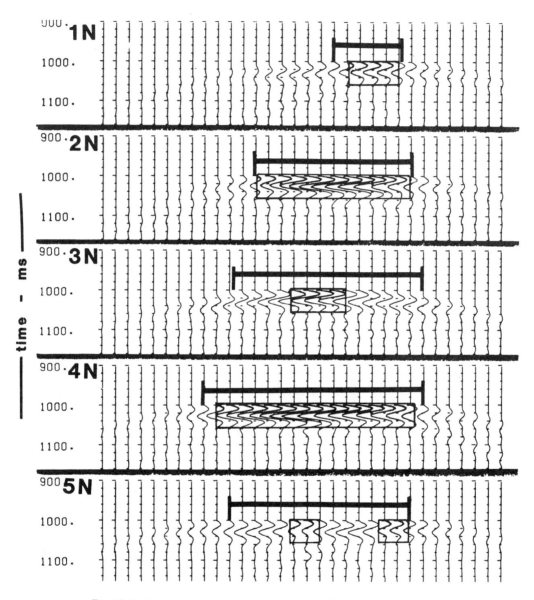

FIG. 16. North-south wavelet time sections across the sand body shown in Figure 14.

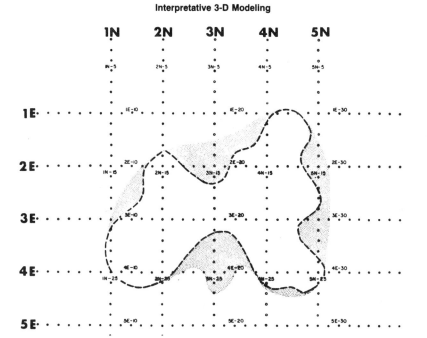

FIG. 17. Base map with seismic interpretation of field outline using one-half amplitude rule on time sections in Figures 15 and 16. The true areal extent is shown by the dashed lines.

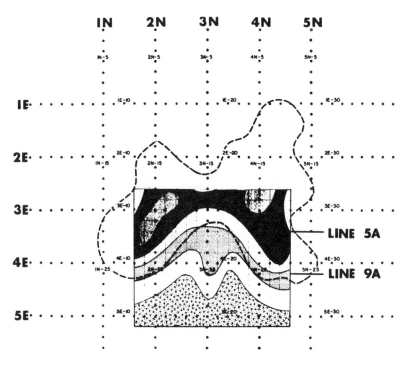

FIG. 18. Amplitude contours based on Figure 19 for 200-ft interval profile lines. Amplitude contours are maximum in the middle of the sand body and decrease continuously outward.

sumptions made, such as constant thickness, constant reflection coefficient, and constant depth. However, with refinement, amplitude studies offer additional information and are potentially an effective method of delineating small fields. Meanwhile, the necessity for a 3-D migration which preserves amplitude has been demonstrated on this model by McDonald et al (1981). Also note that the high-amplitude focal line was not recognized in the northern portion of Figure 17 because the line spacing was too coarse.

In the next section, the obvious difference between 2-D and 3-D modeling, i.e., out-of-plane reflections or sideswipe, will be examined.

SYMMETRIC BASIN AND DOME

A set of profile lines was gathered over a symmetrical dome and basin where the dome and basin had the same relief of 500 ft. The theoretical profile lines were run over the structure at a 5000-ft depth and then for a 10,000-ft depth where the line spacing was the same for both depths. However, only the time sections for the 10,000 ft depth will be shown. Because of the symmetry, there are only three independent lines which are offset at 0, 2000, and 4000 ft from the center of the structures, as shown in Figures 20 and 21.

To illustrate the problems with sideswipe, lines connecting the specular reflection points for the dome at 5000-ft depth were plotted on the base map in Figure 20. Because the specular reflection path is not vertically beneath the profile line but wanders up the dome, 2-D section migration will not properly migrate the events on the corresponding seismic sections. For 2-D geologic bodies oblique to the seismic profile lines, the migration velocity can be adjusted by the angle between the principal dip plane of the geologic structure and the seismic profile plane to yield a fairly accurate migration (French, 1975). However, 2-D geologic symmetry does not exist for this model. In a similar fashion, if a migration using cross lines were attempted, according to the Slotnick (1959) method, it would also fail because the specular reflection paths are not orthogonal at the cross lines. However, a time map migration which accounts for bed curvature is effective (Musgrave, 1961).

In the top section of Figure 22, the classical bow tie response for buried foci is evident. The waveshape for the diffractions from the bottom of the basin are 180 degrees out of phase with the flat plane reflections, as predicted by Dix (1952). This phase reversal occurs because the energy goes through both an inline focus and an out-of-plane focus. That is a 3-D phenomenon.

Of course, the high amplitude that exists on the center trace is due to the source and receiver being at the exact focal point, and this normally would not occur in the field so that this amplitude would not be expected on field data. What is surprising on the center section is how the anomaly heals itself so rapidly with offset. Thus, 2000 ft away from the center there is little evidence of the basin. However, an apparent anticlinal sideswipe, which originates from the far basin flank, is observed at 2.150 sec on the center trace. If the dip at the basin lip were steeper, the amplitude of this sideswipe would be increased and the event would

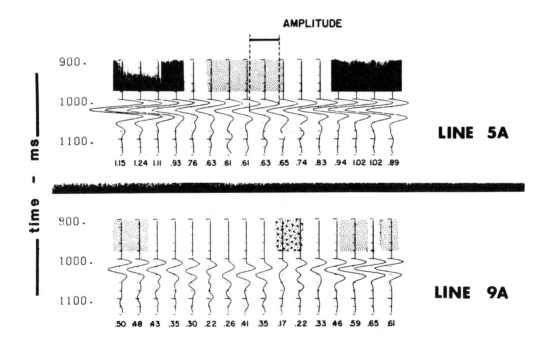

FIG. 19. Wavelet section amplitudes of lines 5A and 9A (Figure 18). The amplitudes are normalized to the amplitude of a normal incidence reflection. The shaded zones correspond to those in Figure 18.

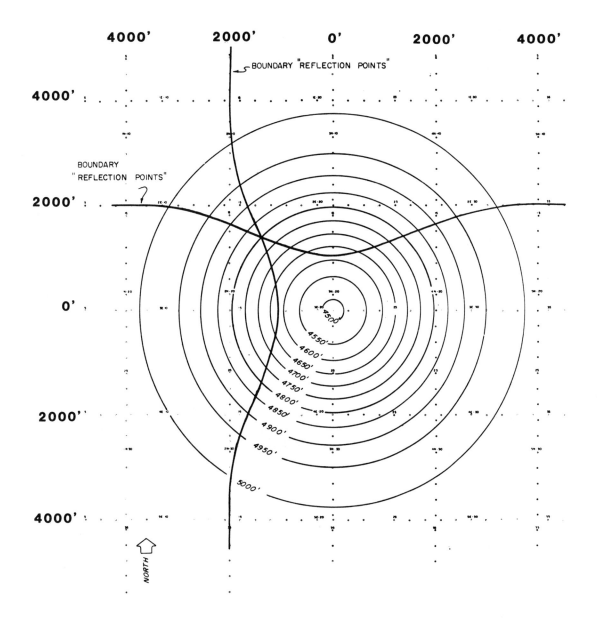

FIG. 20. Base map of dome buried at 5000 ft. Specular reflection paths for lines 2E and 2N are plotted on the base map.

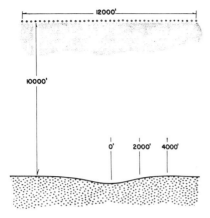

FIG. 21. Cross-section of basin from which wavelet sections in Figure 22 were derived. Inverting the basin yields a dome which corresponds to the sections in Figure 23.

appear as an anticline. In Figure 23, seismic sections for a dome of similar size as the basin are shown.

Looking at the 2000-ft offset section for the dome and the basin, we are thankful that productive structures are domes rather than basins; otherwise, they would be much more difficult to discover. The 2000-ft offset sections suggest that many more domes than basins are mapped. In the bottom section of Figure 23, the bright spot, which is about 2 times larger in amplitude than the rest of the reflection event, is due to sideswipe focusing from the flank. Once again, the need for a 3-D migration is evident. To see how the lack of a 3-D migration can affect an interpretation, a stratigraphic trap is presented in the next section.

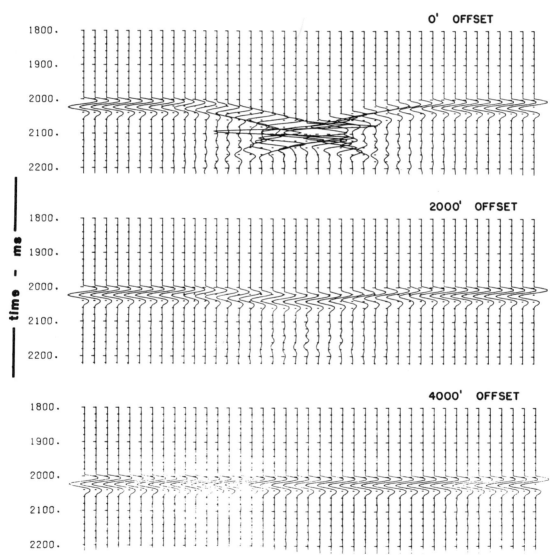

FIG. 22. Time sections across basin as described by depth cross-section in Figure 21.

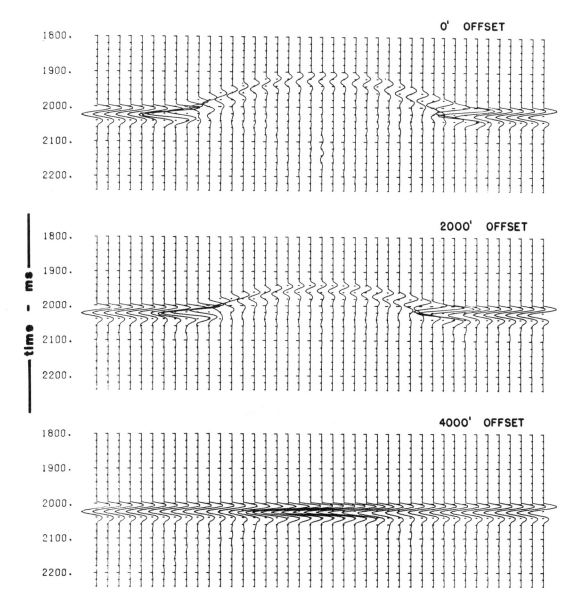

FIG. 23. Time sections across dome as described by the depth cross-section in Figure 21.

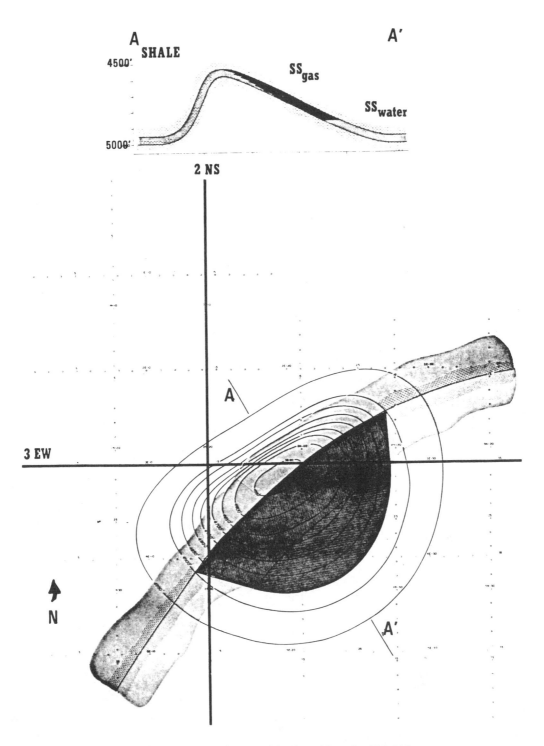

FIG. 24. Stratigraphic trap model with sand formation 50 ft thick.

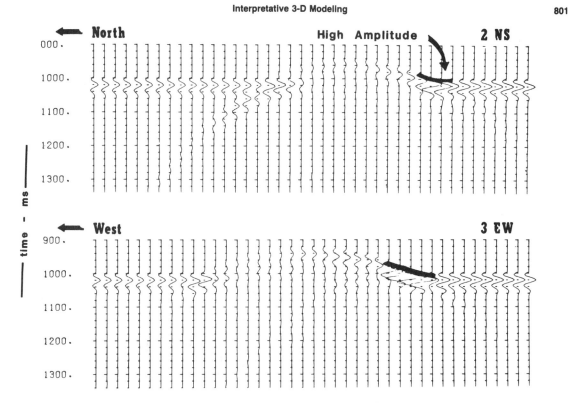

FIG. 25. Time sections across stratigraphic trap described in Figure 24.

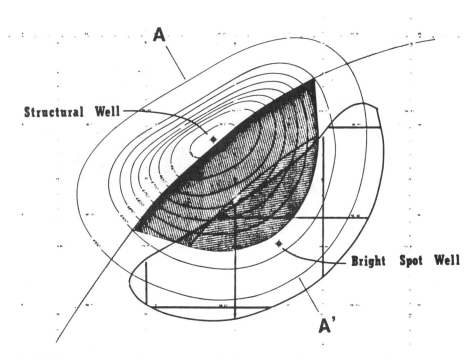

FIG. 26. Bright spot outline of stratigraphic trap. Both structural and bright spot wells are improperly placed.

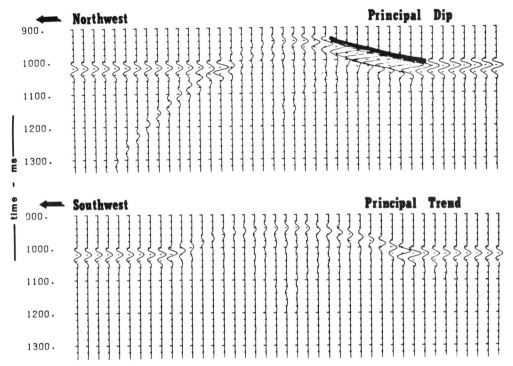

FIG. 27. Principal plane time sections across stratigraphic trap shown in Figure 24.

STRATIGRAPHIC TRAP

As shown in Figure 24, a sand 50 ft in thickness was deposited. The sand shales up toward the northwest to create a premeability barrier, and then uplift created the stratigraphic trap. The profile lines are 4200 ft apart. Because the reflection coefficient of the gas-shale reflector is at least 4 times greater than the other reflection coefficients, a "bright spot" (amplitude) study seemed appropriate. In Figure 25 two time sections, 2NS and 3EW, are depicted.

A high-amplitude bright spot is indicated on the sections. The low-reflection amplitude on the crest is due to the combined effect of curvature and a small reflection coefficient. This small amplitude surely would make it difficult to follow the event across the crest when the signal-to-noise (S/N) ratio is lowered. The diffractions in the top section are due to sideswipe from the steep side of the structure and would not contract totally if a 2-D migration were performed. If the bright spots are picked on all of the seismic sections and mapped, then an unmigrated outline of the stratigraphic field can be drawn on the base map. This was done in Figure 26.

The kidney-shaped outline in the lower right of Figure 26 is the outline of the interpreted bright spot. As is obvious from the figure, if either the top of the structure or the middle of the bright spot were drilled, the results would be disappointing. Two-

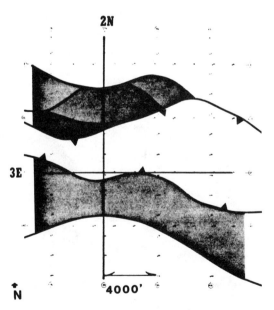

FIG. 28. Graben with 250-ft total throw.

Interpretative 3-D Modeling

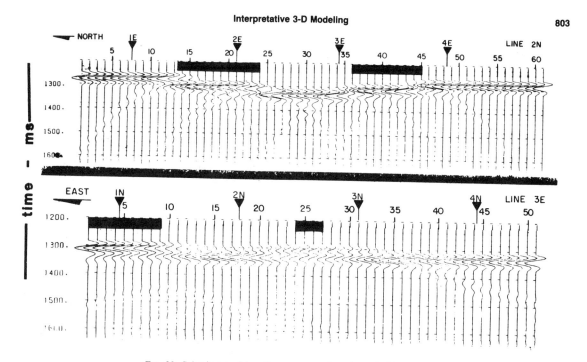

FIG. 29. Seismic wavelet sections across graben shown in Figure 28.

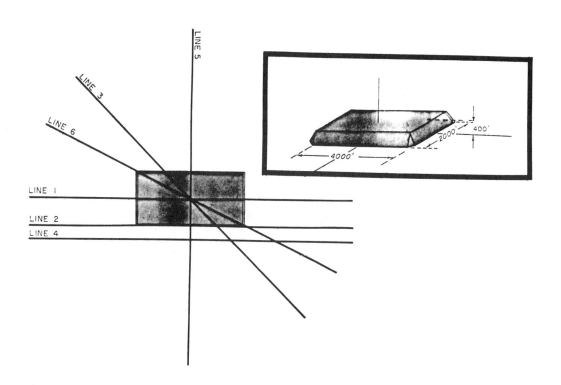

FIG. 30. Reef model with 400-ft vertical relief which has no reflecting interface directly beneath the reef.

dimensional migration of these lines moves the bright spot boundary updip and reduces its size to a more accurate areal size; however, up to 60-ft misties occur at tie lines when 2-D migration is used. Once again, an alternative approach is to migrate the contoured time map. Besides the migration effects, a lesson to learn from this model would be the desirability of having a principal dip line and principal trend line. These are the lines $A - A'$ and the line perpendicular to $A - A'$. In Figure 27 the principal sections are shown.

The advantage the principal dip and trend sections have over oblique sections is that the definition of the structure is increased and not smeared by out-of-plane reflections. That is, the specular reflection path does not wander, but is below the seismic line. Thus, 2-D migration algorithms are effective on principal plane lines. The need for principal plane lines and denser seismic control is also illustrated by the graben example in Figures 28 and 29.

GRABEN

The total vertical displacement for the graben in Figure 28 is 260 ft. There is one reflection surface for each fault block and the faults are vertical. It is recognized that it is important to have several reflectors to correlate across a fault, but the single reflecting surface in this graben model illustrates phenomena which

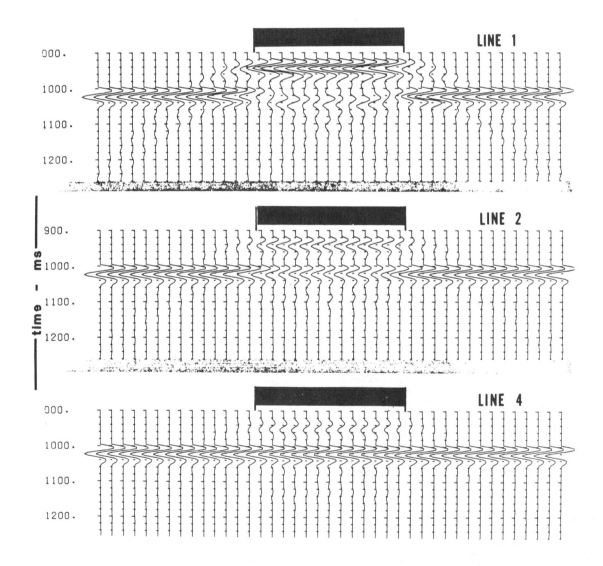

FIG. 31. Seismic sections across reef model shown in Figure 30.

would not have been observed if several surfaces were included.

It is easy to identify the faults on line 2N (Figure 29) as the shaded zones indicate. Both half-amplitude reflection points and diffractions help to pick the fault edges. However, the wormy appearance or leggy character of the reflection event on line 3E is directly related to sideswipe. For instance, on line 3E at the intersection with line 2N the sideswipe diffractions make it almost impossible to observe the bottom of the graben.

The general conclusion interpreters gave after examining these sections is that the seismic grid lines are too coarse and additional reflection events are needed to aid in tying faults on north-south to those of east-west lines. Consider how difficult the interpretation would have been if the fault blocks were dipping.

In the final example, it is shown that it is also possible to underestimate the size of a structure.

REEFS

Six profile lines, as shown in Figure 30, were run across a 4000-ft structure which represents a reef similar to those found in the Michigan basin. The areal extent is 4000 × 2000 ft. One seismic line (line 4) is offset 500 ft from the reef. The seismic sections are shown in Figures 31 and 32.

On line 1 in Figure 31, the half-amplitude rule adequately defines the top edges. Also, the apparent reflection below the reef-

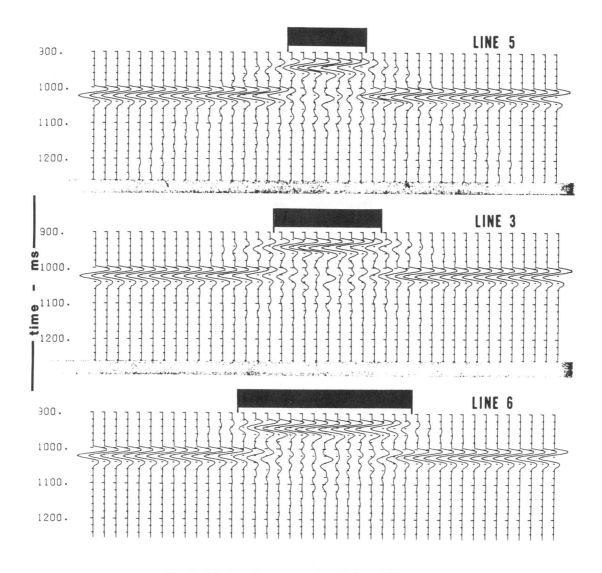

FIG. 32. Seismic sections across reef model shown in Figure 30.

top reflection is due to constructive diffractions which would not exist in field data. On line 2, half-amplitude reflections are obtained from the top and bottom reflector as would be expected since each reflection-diffraction event is from a half-plane. The sideswipe on line 4 is not very strong, but if a small amount of turnover is placed on the top edge, this sideswipe amplitude would greatly increase because then the sideswipe would have a reflection effect as well as a diffraction effect.

In Figure 32, notice that the size of the reef, as would be picked by the half-amplitude rule on line 6, would put the edge three traces inward from the true edge. This apparent shrinkage is due to the fact that line 6 reflects from a quarter-plane as it approaches the edge. Of course, the lower reflection has an amplitude three-quarters that of the full-plane reflection event. Thus, shotpoints on line 6 must be over the structure some distance before the reflection amplitude is one-half that of the normal incidence amplitude. Here is a possibility of underestimating the size of a structure.

CONCLUSIONS

There are four points illustrated by the Fresnel zone models and the five geologic models.

First, the Fresnel zone examples indicate that the spatial size of a reflector affects the seismic response like thin beds do as far as amplitude, timing, and derivative effects are concerned. Thus, a synergistic interpretation of areal extent and vertical thickness must be done.

Second, sideswipe from basins, discs, and domes can appear as anticlinal events. Thus, it appears that many anomalously high structures are mapped. As a corollary, we found that depressions are more difficult to delineate than are structural highs.

Third, the need for dense coverage was emphasized by the overestimation of the areal extent of the gas field. Also, with dense coverage amplitude analysis helps delineate horizontal extent.

Finally, 3-D migration kept suggesting itself as a viable solution to many problem areas that have been areally profiled.

REFERENCES

Berryhill, J. F. 1977, Diffraction response for nonzero separation of source and receiver: Geophysics, v. 42, p. 1158–1176.
Born, M., and Wolf, E., 1970, Principles of optics: Oxford, Pergamon Press, 806 p.
Dix, C. H., 1952, Seismic prospecting for oil: New York, Harper and Brothers, 414 p.
French, W. S., 1975, Computer migration of oblique seismic reflection profiles: Geophysics, v. 40, p. 961–980.
Hilterman, F. J., 1975, Amplitudes of seismic waves—A quick look: Geophysics, v. 40, p. 745–762.
Hilterman, F. J., and Larson, D., 1976, Kirchhoff wave theory for multi-velocity models: Geophysics, in preparation.
McDonald, J. A., Gardner, G. H. F., and Kotcher, J. S., 1981, Areal seismic methods for determining the extent of acoustic discontinuities: Geophysics, v. 46, p. 2–16.
Musgrave, A. W., 1961, Wavefront charts and three-dimensional migration: Geophysics, v. 26, p. 738–753.
Slotnick, M. M., 1959, Lessons in seismic computing: SEG, Tulsa, 268 p.
Trorey, A. W., 1970, A simple theory for seismic diffractions: Geophysics, v. 35, p. 762–784.
——— 1977, Diffractions for arbitrary source-receiver locations: Geophysics, v. 42, p. 1177–1182.
——— 1978, Discussion: diffractions for arbitrary source-receiver locations: Geophysics, v. 43, p. 1259–1273.
Widess, M. B., 1973, How thin is a thin bed?: Geophysics, v. 38, p. 1176–1180.
Woods, J. P., 1975, A seismic model using sound waves in air: Geophysics, v. 40, p. 593–607.

APPENDIX A
COMPUTATIONAL FORM OF KIRCHHOFF'S WAVE EQUATION

Kirchhoff's retarded potential equations were solved by Trorey (1970) for a coincident source-receiver setup where the incident pressure pulse impinges on a flat boundary plate. The reflection coefficient of the plate was assumed to be small.

Referring to Figure A-1, Trorey's diffraction pressure response is given as

$$p(t) = \frac{Z}{\pi(vt)^2} \frac{d\theta}{dt}, \quad (A-1)$$

where

$p(t)$ = diffraction response,
v = velocity, and
t = two-way traveltime.

As indicated by Trorey, $d\theta/dt$ is evaluated around the edge of the boundary. Equation (A-1) is the diffraction impulse response, and normally when doing modeling a source wavelet would be convolved with $p(t)$. Also, the reflection coefficient has been set to unity in equation (A-1).

Notice that as t_x decreases, that is, as the source-receiver location approaches the vertical extent of the diffracting edge, $d\theta/dt$ increases rapidly. This can be seen from Trorey's expressions:

$$\frac{d\theta}{dt} = \frac{t\, t_x U(t - \tau)}{(t^2 + t_x^2 - \tau^2)\sqrt{t^2 - \tau^2}}, \quad (A-2)$$

where $U(t)$ = unit step function.

There are two situations where the expression $d\theta/dt$ requires special attention, that is, when $t = \tau$ and when $t_x = 0$. When $t_x = 0$, $d\theta/dt = (\pi/2)\delta(t - \tau)$, so that the diffraction response of a half-plane is

$$p(t) = \frac{1}{4Z} \delta(t - \tau), \quad (A-3)$$

where $\delta(t)$ = the Dirac function.

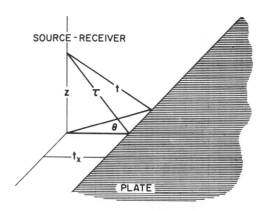

FIG. A-1. General configuration for the notation used in equation (A-1). The variables τ, t, and t_x are the two-way time equivalents for the distances they represent.

The singularity at $t = \tau$ can be avoided computationally by approximating $d\theta/dt$ as

$$\frac{d\theta}{dt} \simeq \frac{\theta(t + \Delta t) - \theta(t)}{\Delta t}$$

which, by referring to Figure A-1, can be written as

$$\frac{d\theta}{dt} \simeq \frac{1}{\Delta t}\left[\tan^{-1}\sqrt{\frac{(t + \Delta t)^2 - \tau^2}{t_x^2}} - \tan^{-1}\sqrt{\frac{t^2 - \tau^2}{t_x^2}}\right], \quad \text{(A–4)}$$

where Δt = sampling interval.

Figures A-2 and A-3 illustrate (with exaggerated exponential tails on the $d\theta/dt$ response) the application of equations (A–1) and (A–4) in computing the response from a triangular plate.

When evaluating $d\theta/dt$ around the edge of the plate, a simple way to remember if the diffraction response is going to be positive or negative is to examine how the incident wavefront approaches an edge. If the wavefront travels across the plate and then hits the edge, it has a negative diffraction response, but if the wavefront strikes the edge of the plate straight on, it is positive.

Sampling of the continuous function $d\theta/dt$

Even with the band-limiting procedure in equation (A–4), a sampling problem became evident when stratigraphic traps were evaluated using a 4 msec sampling interval. Distinct phase shifts on the time sections were evident due to improper sampling. The problem is illustrated in Figure A-4. One conventional method of sampling the continuous functions would be to find the area between the onset time .061 sec and the first integer sample time .064 sec and then linearly interpolate the area between the discrete samples at .060 and .064 sec. In this example, the midpoint time (sample-and-hold time) for the first area would be at .0625 sec; thus a weight of {3/8 area} would be assigned to the sample at .060 sec and a weight of {5/8 area} would be assigned to the .064 sec discrete sample. This linear interpolation scheme would work well for the curve in Figure A-4b because, essentially, it has a constant value over the interval .061 to .064 sec. However, it does not work well for the diffraction response in Figure A-4a. A solution is to find the centroid time and linearly adjust the area on this time rather than the midpoint time. This is done in the following manner:

(a) Compute the area as

$$\text{Area} = \int_{.061}^{.064} \frac{d\theta}{dt}\, dt = \theta(.064) - \theta(.061). \quad \text{(A–5)}$$

(b) Compute θ at centroid time Tc

$$\theta(Tc) = \frac{\text{Area}}{2} + \theta(.061). \quad \text{(A–6)}$$

(c) Referring to Figure A-1 and equation (A–4), we have

$$Tc = \{[t_x \tan \theta(Tc)]^2 + \tau^2\}^{1/2}. \quad \text{(A–7)}$$

The time Tc is used to adjust the area linearly between two discrete samples.

For final display purposes, a spreading loss correction was applied to the synthetic trace. The division by the sampling interval Δt in equation (A–4) will be canceled by the Δt carried in the convolution with the source wavelet. For a specular reflection, the amplitude for a Dirac delta function $\delta(t)$ may be approximated as $1/\Delta t$.

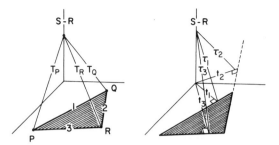

FIG. A-2. Description of triangular plate with respect to source and receiver in terms of two-way traveltimes.

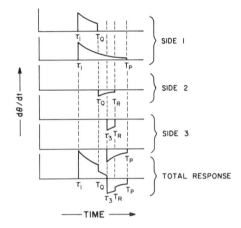

FIG. A-3. Diffraction responses for each edge in Figure A-2, plus the total response. The singular responses have been band-limited for ease in visual presentation.

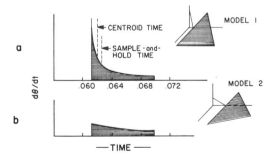

FIG. A-4. $d\theta/dt$ response for one side of a triangular plate for two different offsets t_x.

APPENDIX B
REFLECTION-DIFFRACTION FROM A FRESNEL DISC

Hilterman (1975) expressed the pressure response $p(t)$ at the shotpoint in terms of a convolutional integral. His results are:

$$p(t) = \frac{1}{2\pi v} \left[\left(f * \frac{2}{t} \frac{d\Omega}{dt} \right) + \left(\frac{df}{dt} * \frac{d\Omega}{dt} \right) \right], \quad (B-1)$$

where

- $*$ = convolution operator,
- t = two-way traveltime,
- v = velocity
- $f(t)$ = source pulse, and
- $\Omega(t)$ = solid angle subtended at the shotpoint by the wavefront intersection with the boundary.

For the reflection from a flat plane, the solid angle can be expressed in terms of two-way traveltime as

$$\Omega(t) = 2\pi(1 - T_0/t) U(t - T_0), \quad (B-2)$$

where

- T_0 = two-way vertical traveltime, and
- $U(t)$ = unit step function.

Therefore, if the two-way traveltime to the edge of a Fresnel disc is T_1 when the shotpoint is in the middle, then the solid angle for the disc can be expressed as

$$\Omega(t) = 2\pi(1 - T_0/t)[U(t - T_0) - U(t - T_1)] \\ + 2\pi(1 - T_0/T_1) U(t - T_1). \quad (B-3)$$

Taking the time derivative of the solid angle in equation (B-3) yields

$$\frac{d\Omega}{dt} = \frac{2\pi T_0}{t^2} [U(t - T_0) - U(t - T_1)], \quad (B-4a)$$

and

$$\frac{d^2\Omega}{dt^2} = \frac{2\pi}{T_0} \delta(t - T_0) - \frac{2\pi T_0}{T_1^2} \delta(t - T_1) \cdot \\ \cdot \frac{4\pi T_0}{t^3} [U(t - T_0) - U(t - T_1)]. \quad (B-4b)$$

Equation (B-1) can be rewritten as

$$p(t) = \frac{1}{2\pi v} \left(f * \frac{2}{t} \frac{d\Omega}{dt} + f * \frac{d^2\Omega}{dt^2} \right). \quad (B-5)$$

Inserting equations (B-4a) and (B-4b) into equation (B-5) yields

$$p(t) = \frac{1}{vT_0} \left[f(t - T_0) - \left(\frac{T_0}{T_1} \right)^2 f(t - T_1) \right]. \quad (B-6)$$

Equation (B-6) is the diffraction response due to a disc when the source and receiver are coincident on the vertical axis of the disc.

Stratigraphic Modeling and Interpretation—Geophysical Principles and Techniques[1]

NORMAN S. NEIDELL and ELIO POGGIAGLIOLMI[2]

Abstract A simplistic view of seismic data and their relation to stratigraphy is adopted. Each event or waveform on each data trace is assumed to relate to a sharp acoustic impedance change in the subsurface directly below the trace location. The incorporation of geologic observations and principles thus should permit interpretation of the acoustic parameter changes in regard to lithology and stratigraphy.

Several departures of the real world from the simplistic view can occur, and these seriously complicate the proposed interpretive sequence. Seismic processing methods, however, act to transform the data so that the simplistic view may be adopted. Further, the preservation of seismic amplitudes and the ability to transform seismic waveforms to more beneficial character add a new and quantitative interpretive dimension to the data.

Seismic model studies are considered as a means of establishing precise requirements for making stratigraphic correlations and describing the seismic character of specific exploration objectives. Such studies further enable the development of a quantitative approach to stratigraphic correlation, not only for the thicker lithologic units, but for thin units—under 20 m— as well. Seismic amplitudes and waveform manipulations prove to be the foundation of these analytic procedures.

Quantitative methods of stratigraphic correlation in concert with the newly developed work on interpreting seismic reflection patterns offer significant tools for stratigraphic interpretation. Not only are results of improved resolution and certainty to be anticipated, but the prospect of making stratigraphic correlations in the presence of complex structure draws closer as a routine exploration practice.

RELATING SEISMIC DATA TO STRATIGRAPHY

Stratigraphic interpretation of seismic data requires that the seismic information be expressed in geologic terms. A strictly geologic view of the earth is developed from surface observations, the guiding principles of geologic evolution, and subsurface information from boreholes. Subsurface information includes well-log measurements from a variety of physical sensors. We can correlate seismic data with the geologic view in terms of geometric description and a subset of the log measurements (namely, velocity and density). Such correlation, used in context with geological and geophysical principles, is at the heart of stratigraphic interpretation.

Typical well-log measurements include the spontaneous potential, resistivity, radioactivity parameters, sonic travel time (velocity), and density. Although the vertical detailing of such measurements is excellent, they are limited in their definition of lateral variation in subsurface parameters. Hence, a strong interpretive element which relies on fundamental geologic concepts and principles is required. Still, log measurements provide insight regarding the porosity and fluid content of the rocks and the lithology in general, and thereby enhance stratigraphic conclusions based solely on a more limited field-observation approach.

Seismic measurements of travel time and amplitudes define the subsurface geometry and give estimates of the acoustic impedances related to rock velocities and densities. Vertical detail is limited owing to the lengthy duration of the individual seismic wavelets and the occurrence of overlapping wavelets from closely spaced reflectors. Lateral definition is good although averaged over regions known as "Fresnel zones." Specialized analyses and interpretive methods can make use of indirect seismic information such as waveforms and velocity to provide insights to the nature of porosity, fluid content, and lithology in general. Here again, fundamental principles and strong interpretive elements are applied in concert.

With these backgrounds, we need a specific vehicle by which the geologic view and the seismic information can be correlated. The most elementary form of such a vehicle has been available for some time—the synthetic seismogram (see Peterson et al, 1955; Wuenschel, 1960).

The synthetic seismogram represents the viewpoint of a laterally restricted segment of the earth taken as a horizontally layered medium. A plane wave is assumed to propagate in this segment, and the seismogram is constructed from the partial reflections of the plane wave from each of the planar boundaries.

---

[1]Manuscript received, September 21, 1976; accepted, March 25, 1977.

[2]Consultants, GeoQuest International, Ltd., Houston, Texas 77027.

This paper represents a geophysical distillation of an approach to stratigraphic interpretation developed over a number of years by several individuals at GeoQuest International, Ltd. It is important that these contributions be recognized; most of the discussions document their work. In particular, we cite J. P. Lindsey, E. V. Dedman, M. W. Schramm, Jr., and A. K. Nath.

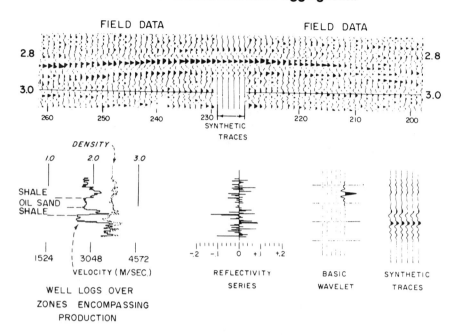

FIG. 1—Correlation of seismic expression of North Sea oil field with geologic data via seismogram synthesis. Synthetic traces inserted into field data at trace 225 which is nearest well location.

Figure 1 illustrates seismic data across an oil field in a North Sea Jurassic sandstone, and also shows a definitive synthetic-seismogram correlation. In this instance, both velocity and density logs were used to calculate reflection coefficients. The particular waveform used in developing the synthetic seismic data matches the propagating waveform of the seismic data. For this comparison, the processed field data were separated at the trace closest to the well location and six repetitions of the synthetic seismic trace were inserted. The agreement in detail is remarkably good. Two important points must be emphasized:

1. Density cannot be neglected as a physical parameter without degrading the effectiveness of our correlation; and
2. The agreement in waveform between the synthetic data and the field data is an essential ingredient of the correlation.

A synthetic-seismogram correlation as in Figure 1 requires:

1. An essentially flat subsurface having lateral continuity and homogeneity over at least a Fresnel zone (the effective subsurface area giving rise to a reflection event),
2. Lithologic boundaries that are well defined rather than transitional, and
3. A common waveform for the seismic data and the synthetic data.

We also make the unstated assumption that each event on the seismic data corresponds to a lithologic boundary, and thus rule out the possibility of noise and other events of no direct geologic significance.

Developing only the geometric viewpoint of the subsurface from a seismic profile necessitates moving beyond the limitations of the synthetic-seismogram model. More realistic models of the subsurface geology are needed for study; such models are accompanied by a variety of wave-propagation phenomena which complicate the interpretive process.

Consider the schematic overthrust-fault model shown in Figure 2 along with its seismic response as computed using wave theory. The hazards of stratigraphic interpretation in such a province are evident. Fully half of the seismic events observed do not arise by reflection but are in fact diffraction contributions. Diffraction events occur from terminating acoustic impedance boundaries or tightly curved convex surfaces and cannot be simply interpreted in the manner of reflection events. The reflection events corresponding to the hydrocarbon-water contact appear only on traces 23

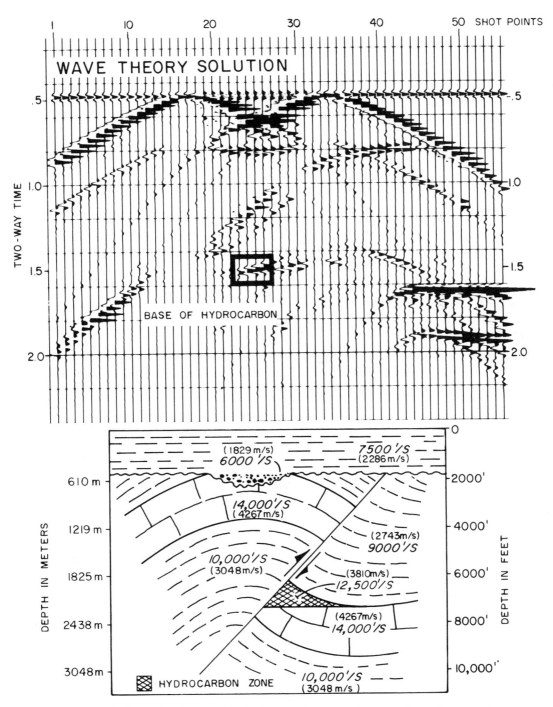

FIG. 2—Overthrust-fault model and computed seismic expression. Diffraction events dominate seismic picture, and bending of seismic rays obscures identification of reflections from hydrocarbon zone.

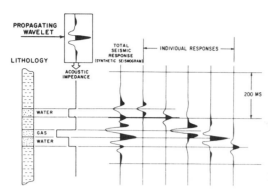

FIG. 3—Idealistic view—relation between lithology, propagating wavelet, and seismic response (after Dedman, et al, 1975).

through 26 as a slightly dipping horizon at 1.5 sec. Displacement of the contact with respect to the eroded crest of the anticline is caused by refraction or the bending of the reflection ray paths in conjunction with plotting seismic traces halfway between the sources and receiver positions (vertical plotting).

Still other difficulties exist which are not illustrated by this figure. At an elementary level, for example, the presence of noise-event alignments or events which have been reflected more than once from shallow horizons (multiple reflections) can also complicate a seismic interpretation.

Seismic data processing, apart from its more routine bookkeeping function, attempts to present the interpreter with data more closely resembling the ideal circumstances which have been assumed, as in the generation of synthetic seismograms. In this light, data processing should be regarded as an extension of the interpretive function.

Distortion of subsurface reflectors by vertical plotting and refraction effects, as well as the presence of diffraction events (as in Fig. 2), has a processing treatment in the form of migration. Migration of seismic sections portrays the subsurface with reflection events corresponding to their points of origin. In migrated sections, anticlines and synclines should appear in their true dimensions, and tight curvatures and bed terminations should be portrayed accurately without "masking" or "smearing" caused by diffractions. Similarly, the elimination of multiply reflected seismic events can be attempted both by stacking and deconvolution procedures (see Peacock and Treitel, 1969, for an account of the latter technique). Since appropriate processing may not always be accomplished, the interpreter frequently is forced to look beyond these obstacles in seeking to define stratigraphy.

The nature of the seismic waveform has been cited as an essential requirement for stratigraphic correlation and is clearly an important factor in controlling seismic resolution.

Figure 3 (see also Schramm et al, this volume) shows a portion of a lithologic log and a corresponding acoustic impedance log. Each contrast in acoustic impedance will be marked by a reflection event having a simple waveform. The polarity or sense of the reflection and its size will indicate the nature of the contrast. Individual reflection events for the model are shown along with their superposition in the resulting seismic trace.

Stratigraphic interpretation would begin with the development from each trace of an acoustic impedance log or, equivalently, a reflectivity series. These results would be correlated from trace to trace to provide the structural considerations, and also would be correlated with available geologic information. By use of geologic principles and insights appropriate to the region, lithologic estimates would be inferred, and, from these estimates and the indicated changes, geometry and depositional patterns, sequences, and history might be interpreted.

Departures from this ideal have been emphasized, and the objective may not always be directly accomplished. Nevertheless, tools such as seismic data processing can assist us in our goal, and the importance of their specific roles can be appreciated.

Good data acquisition and processing are necessary prerequisites for any seismic stratigraphic work. It has been indicated that some ability to capture, manipulate, and generally transform seismic waveforms is needed in order to facilitate the correlations which must be made. Such tools and techniques and their effects on waveform polarities and amplitudes are described in the next section. We also understand that geometric effects, diffractions, and other complicating factors and elements must be included in our interpretations. The role of modeling and, in particular, two- and three-dimensional wave-theory modeling will be discussed specifically in this regard. Finally, the problems of developing principles for making lithologic estimates, improving the vehicles for the correlations, and quantifying the resolution inherent in the seismic expressions must be addressed.

UTILIZATION OF SEISMIC AMPLITUDES AND WAVEFORMS

The rediscovery of the exploration significance of seismic-event amplitudes (documented in the symposium held by the Geophysical Society of

Houston, 1973) is not only a key step in recent progress toward direct detection of hydrocarbons, but is also recognized as one of the cornerstones in the quantitative analysis of thin stratigraphic intervals. In the recent past, difficulties in utilizing seismic amplitudes arose from mechanisms like divergence, reflection loss, and attenuation—which cause the seismogram to become rapidly weaker with increasing reflection time or depth. In the presence of this strong amplitude decay, the interpretive amplitude distinction between a strong reflection and a weak one becomes difficult to discern. Advances in seismic recording technology and new approaches for scaling data prior to display overcame this difficulty.

Attention may now be focused on seismic waveforms. Typical seismic wavelets, particularly as acquired in marine surveys, tend to be lengthy, "leggy," and generally complex in form. The specific waveform for a survey tends to remain fairly consistent from one source discharge to the next, and it propagates in the earth largely unchanged in significant character under most circumstances. This last statement must be qualified by noting that whatever changes are introduced by propagation do not affect interpretive applications when the data are presented to the interpreter at normal plotting scales, where each individual wavelet can have a maximum amplitude no greater than 2 cm.

Recently it has been possible to capture seismic waveforms. Documentation has been accomplished by direct measurements in deep water or by extraction of the waveform from the seismic response of simple lithologic complexes. Appropriate complexes include water-bottom reflections and responses from thick or thin beds having unusually high or low acoustic impedance contrasts—as, for example, Gulf Coast gas sandstones which have low contrasts with respect to the surrounding rocks.

Figure 4 shows first a typical marine wavelet captured by a direct measurement. It is next transformed to a symmetric form approximating a desired form by Wiener filtering. The Wiener filter operation is a well-known signal-processing method which has been used for many years and accomplishes transformations of signals from a given form to a desired form in the best least-squares sense (see Robinson and Treitel [1967] for a discussion of Wiener filters). Application of the Wiener filter to each of the traces from the same seismic survey having the particular propagating wavelet will result in replacement of that wavelet by the symmetric one to a good approximation. Such treatment of data will be called "wavelet signature processing" or simply "wavelet processing."

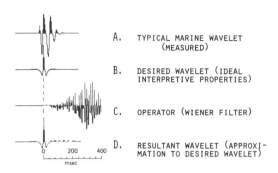

FIG 4.—Wavelet signature-processing transformation of basic marine wavelet to ideal interpretable form by Wiener filtering.

All the advantages of replacing the propagating waveform by one having zero-phase or symmetric properties are not obvious. It is clear from an interpretational viewpoint that each change in acoustic impedance will now be marked by a simple, readily understood waveform, but there are other tangible benefits of the transformation. The shortening of the waveform causes improvement in the effectiveness of standard processing methods for noise and multiple suppression. Similarly, the increased resolution of the shortened waveform permits more highly resolved seismic-velocity estimates to be made.

The resultant wavelet in Figure 4D has precisely the same range of frequency content as the original basic wavelet, and no requirement has been made for an extrapolation of information content. Berkhout (1973, 1974) and Schoenberger (1974) have noted the enhanced resolution and information potential inherent in changing waveforms without broadening the frequency band. Remaining within the well-defined frequency band allows the introduction of improved resolution without the introduction of noise. By contrast, other approaches seeking to increase resolution through broadening of the frequency band (e.g., spiking deconvolution) introduce much noise and generally produce less acceptable results.

A transformation of a Teledyne air-gun waveform is noted in Figure 5. The original waveform (Fig. 5C) has been measured by a direct waterborne arrival to a deeply submerged hydrophone. A surface receiver ghosting operator has been simulated so that this result may be compared to Figure 5A, which is the waveform as extracted from a strong water-bottom reflection in normally recorded data. Despite the effects of the hydrophone array present in Figure 5A, the comparison is most favorable, suggesting the validity of several of the ideas expressed earlier. Note that the

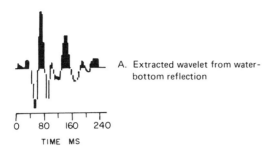

A. Extracted wavelet from water-bottom reflection

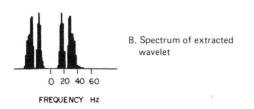

B. Spectrum of extracted wavelet

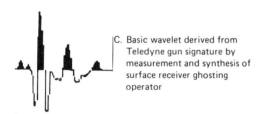

C. Basic wavelet derived from Teledyne gun signature by measurement and synthesis of surface receiver ghosting operator

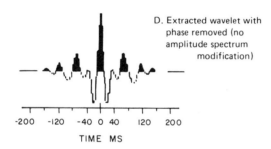

D. Extracted wavelet with phase removed (no amplitude spectrum modification)

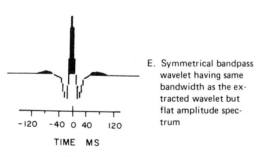

E. Symmetrical bandpass wavelet having same bandwidth as the extracted wavelet but flat amplitude spectrum

FIG. 5—Verification of wavelet consistency and role of amplitude correction in wavelet processing.

amplitude spectrum of the wavelet (Fig. 5B) shows a deep notch at about 25 Hz. Wavelet processing must fill this notch and generally flatten the amplitude spectrum to achieve an ideally interpretable waveform, as is illustrated by contrasting Figures 5D and 5E. In Figure 5D the amplitude spectrum, including the deep notch, remains unchanged, whereas it is smoothed and flattened in Figure 5E.

Examination of representative examples of wavelet-processed data reveals the improved resolution and interpretability. Compare similarly processed seismic time sections over the North Sea oil field considered in Figure 1. In one panel of Figure 6 the original waveform is present, whereas, in the second, a shorter waveform of simple symmetric character has been introduced in place of the original.

The wavelet-processed data have fewer events owing to the removal of multiple reflections. Processing techniques designed to remove multiples perform better where the shorter, more highly resolved wavelet has been introduced. Also, a more definitive interpretation which shows some indication of an oil-water contact appears to be possible.

Hence, the waveforms inherent in seismic data can be manipulated and transformed to simpler waveforms which are more amenable to interpretation. These transformations have been more readily accomplished on marine data, although R. O. Lindseth (1976) has presented results indicating analogous success with land data. In consequence, correlations between seismic data and geologic inputs can be effected with greater rigor to produce more definitive results.

CONTRIBUTIONS OF MODELING AND WAVE-THEORY MODELING

Simulation of exploration seismic data, or seismic modeling, is in fact the successor to the synthetic seismogram, as the vehicle by which proposed seismic correlations with geologic data may be verified, and also as the tool by which geologic hypotheses may be tested. Although publications describing modeling techniques and applications are few in number (see Taner et al, 1970; Shah, 1973), the technology has reached a rather sophisticated level. Neidell (1975) described some of the more advanced considerations and limitations associated with the use of seismic modeling.

Unlike the synthetic seismogram, seismic modeling accepts a description of a subsurface condition in terms of geometry and acoustic parameters, including compressional-wave velocity, density, and an attenuation factor. Two-dimensional modeling systems are designed to view the subsurface condition as having perfect out-of-the-

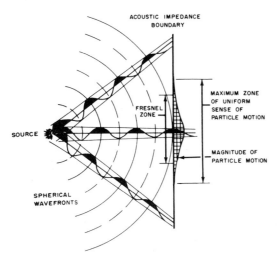

FIG. 7—Spherical wave motion encountering a flat reflector illustrating Fresnel zone region which is principally responsible for reflection event.

plane lateral continuity and homogeneity. A geometry of virtually any complexity may be treated. The seismic parameters can be permitted to vary both horizontally and vertically to represent lithologic transitions and similar subtle stratigraphic effects.

In the discussion regarding a wave-theory modeling illustration for an overthrust fault (Fig. 2), modeling was portrayed as a tool for overcoming vertical plotting and other geometric effects on the seismic section which might complicate the interpretation of stratigraphic objectives, and also for treating diffractions. The following discussion will describe modeling from a more advanced viewpoint. In particular, we can use it to develop insights to the spatial resolution inherent in seismic data and the very origin of reflections from a three-dimensional subsurface representation. Furthermore, model studies are important to the general characterization of specific exploration targets in seismic terms.

Consider the results of a two-dimensional wave-theory modeling study seeking to determine how features comparable in extent with a Fresnel zone will appear on a seismic section. The specific calculations illustrated are developed from the basic theories and concepts proposed and developed by Hilterman (1970).

Figure 7 illustrates the Fresnel zone. Spherically radiating wave motion encounters a flat acoustic impedance boundary. The maximum-sized region of this boundary, over which the particle motion is all in the same direction, is indicated. A Fresnel zone, although directly related to the size of this region, has a somewhat smaller value since some allowance must also be made for the respective magnitudes of particle motion within this region.

If the source is at a distance R from the reflecting boundary where R represents travel distance at a velocity V over a two-way time of t seconds, and the dominant frequency content of the seismic data is f_c, then a Fresnel zone radius r_f is approximated by

$$r_f \simeq \frac{V}{4}\sqrt{\frac{t}{f_c}}$$

This mathematical relation is derived using basic geometry and an empirical weighting. For the values $V = 3,000$ m/sec, $t = 1$ sec, and $f_c = 25$ Hz, $r_f = 150$ m.

Seismic illumination of the subsurface may then be considered analogous to a "searchlight beam," and the subsurface area illuminated by this beam is considered to be the Fresnel zone. With such an intuitive point of view, we can consider fundamental questions such as how small in extent a subsurface feature must be as compared to a Fresnel zone in order to be detectable, and what the nature of its seismic expression must be in order to facilitate detection.

Note in Figure 8 that, as a sandstone body approaches the lateral dimension of a Fresnel zone and becomes still smaller, the seismic signature loses all reflection character and appears in the diffraction form corresponding to a reflecting point. The lessons are clear: (1) thin beds of small spatial extent have a seismic expression, but not dominantly of reflective nature; and (2) although the form of the signatures are alike for sandstone bodies smaller than the Fresnel zone, their size can be gauged by noting the differences in the strengths of the seismic events. The smaller extent bodies produce weaker events by occupying only portions of the full Fresnel zone.

Amplitude processing is concerned with the preservation of seismic strengths. With the mechanism just observed, it was seen that, *where relative amplitudes have significance, they may be related to the identification of small lithologic inhomogeneities.*

As a practical matter, the size of the Fresnel zone is initially related to the quarter wavelength of the dominant frequency component of the seismic waveform and the near-surface velocity. For the expression for r_f, the size of the Fresnel zone grows in proportion to the stacking velocity, which may be determined readily from seismic data. The case study examined by Figure 8 is a

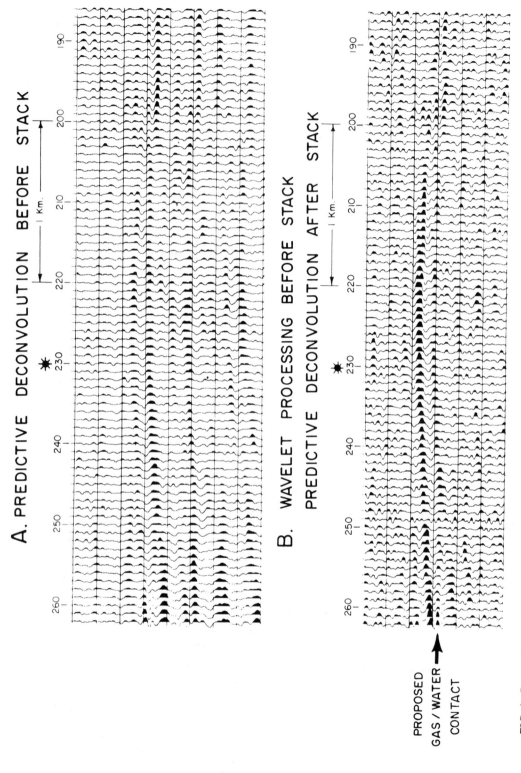

FIG. 6—Processed seismic expressions of North Sea oil field with original waveform and wavelet processed results showing enhanced interpretability and resolution of the latter.

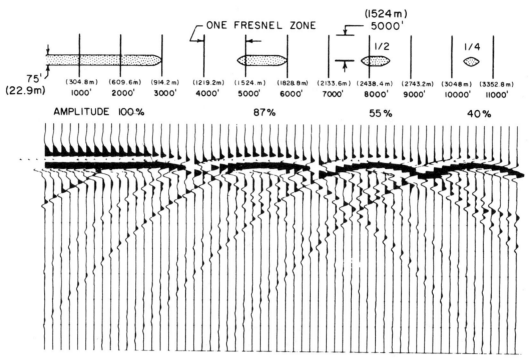

FIG. 8—Wave-theory model response for sandstone bodies of varying lateral extents illustrating significance of Fresnel zone size.

typical one for intermediate-depth Gulf Coast sandstones.

Next we can investigate the role of the Fresnel zone from the point of view of a three-dimensional wave-theory model. Figure 9 shows a reeflike box structure over which two seismic profiles are calculated. Line 1 crosses the structure through its long axis; line 4 parallels line 1 but is 152.4 m from the edge of the structure. The structure is 1,524 m below the surface.

Line 1 shows the structure clearly, along with all diffraction events at its termination. The superposition of the diffraction events from the horizon at the base of the reef almost suggests the continuity of this horizon under the structure. Observe that reflection amplitudes over the reef are comparable in magnitude with those of the reflective horizon on which it was developed.

Line 4 suggests a structure above and below (though more subtle here) the reference horizon. The reflections above the reference horizon are from the flat-topped portion of the structure (Fig. 9). The Fresnel zone which gives rise to reflection events lies partially on the reef top and partially off of it. In fact, the reduced-amplitude event seen for the reef top is proportional to that part of the Fresnel zone area which is on top of the reef.

The reference-horizon reflection amplitude seen below the reef-top reflection is also reduced since it, too, represents only a part of the Fresnel zone (although this effect is more difficult to see). Below the reference horizon the weak events are formed by re-enforcement of diffractions from the base edges of the reef acting out of the profile plane.

Three-dimensional model studies show that all seismic interpretations must be visualized in a three-dimensional environment. Such model studies are probably the most powerful means presently available for teaching the principles of seismic interpretation.

Another Fresnel zone-related model study of stratigraphic nature provides insight to the circumstances in which the synthetic seismogram may be reliably employed as a correlation tool. The particular case study is called the "sand-shale interfingering model," and the applicable illustrations are Figures 10 through 15 (see Lindsey et al, 1976).

Subsurface lithologic boundaries may range from sharp contacts to gradational ones. The lack of a well-defined boundary will understandably influence the seismic reflections. Additionally, there will be a lateral averaging over the Fresnel

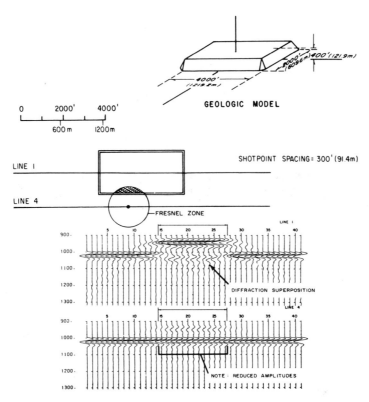

FIG. 9—Three-dimensional wave-theory modeling over "box car" reef illustrating significance of Fresnel zone in producing reflections.

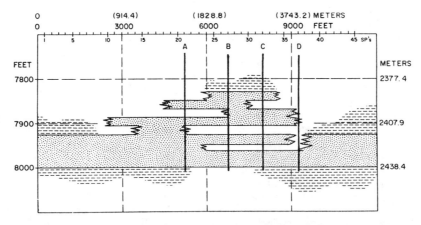

FIG. 10—Sandstone-shale interfingering model illustrating lateral lithologic transition.

Stratigraphic Modeling and Interpretation

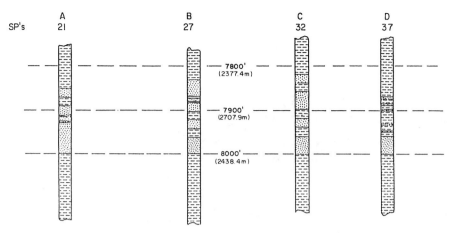

FIG. 11—Lithologic logs for four wells (Fig. 10) of sandstone-shale seismic section.

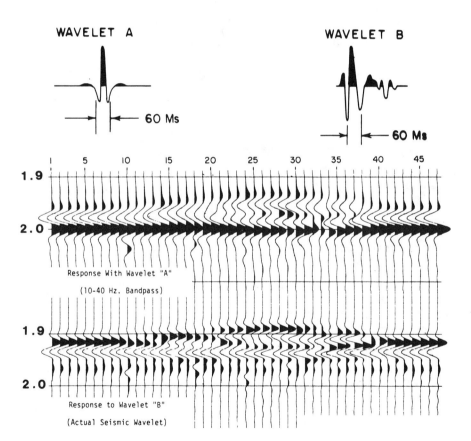

FIG. 12—Sandstone-shale interfingering-model seismic response.

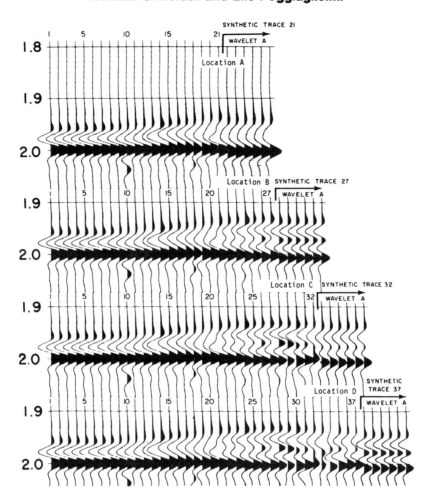

FIG. 13—Comparison of seismic response with wavelet A to synthetic seismogram using wavelet A.

zone. Where a diffuse boundary is involved, there is some weakening of the reflection response; there is also weakening in lateral continuity. Further, the seismic response may differ from a response predicted by the synthetic seismogram developed from well-log measurements even if a common seismic waveform is used. The well log measures the vertical lithologic sequence within a few inches of the borehole; seismic waves average several hundred feet laterally and thus may differ from the well log in describing the boundary character by introducing spatial variations.

The sandstone-shale interfingering model attempts to show some of the effects of a diffuse boundary. A sandstone of approximately 30.5 m thickness is embedded in a shale. The sandstone base is uniform over a large area, but the sandstone top is locally interfingered with the overlying shale. Four holes are shown at locations A through D (Fig. 10). The lithologic logs for these holes are shown in Figure 11.

A seismic wave-theory response to the complete sandstone unit is shown in Figure 12. The upper version uses a seismic wavelet that is symmetrical and polarized such that a positive reflection (transition from a "soft" or low-acoustic-impedance rock to a "hard" rock) will be displayed as a black or right-hand deflection. This is an ideal wavelet for visual detection of detailed stratigraphy and acoustic lithology. The lower seismic response is the same model using a wavelet actually present in some marine seismic data; the wavelet is not symmetrical or regular in any sense. Figure 13 through 15 compare both wave-

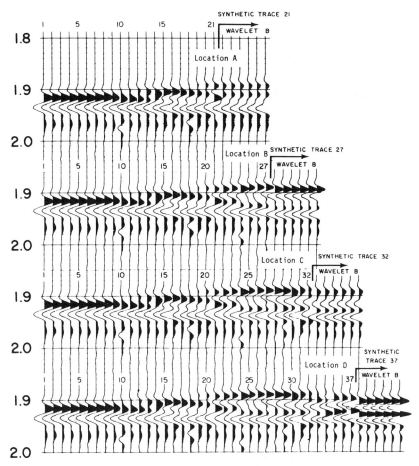

FIG. 14—Comparison of seismic response with wavelet B to synthetic seismogram using wavelet B.

lets A and B for the wave-theory seismic response to the synthetic seismogram derived from the logs using these same wavelets. The synthetic trace repeated six times is shown adjacent to the well location in each case. Contrast the good agreement at locations A and C with location D (indicated on Fig. 13 and 14). Note also the complete and obvious lack of correlation resulting from the use of differing waveforms as in Figure 15.

This study offers important guidelines for using simple synthetic seismograms. First, it is essential that the seismic waveform be common to the synthetic and the actual data. Then, we understand that the synthetic seismogram and the actual data should correlate only to the extent that the sequence of the log extrapolates laterally for at least a Fresnel zone distance (about 304.8 m for this case; Fig. 10). It is interesting to see how the application of more advanced tools does not rule out the use of the more elementary ones, but rather instructs us in their proper and appropriate use.

The well-defined base of the sandstone section shows a consistency in amplitude and arrival time that sets it apart from the transitional upper boundary. The effects of the transition appear as a generally reduced amplitude or, more prominently, as a waveform distortion.

The search for reefs in Michigan involved elements present in a model study executed by Nath (1975). The remarkable improvement in exploration success for Silurian reefs in Michigan was the result of seismic data-processing techniques which first treat the severe statics problem that had previously degraded the data quality (see especially McClintock, 1977). Next, the preservation of seismic-amplitude information for land data constituted a significant further step; and

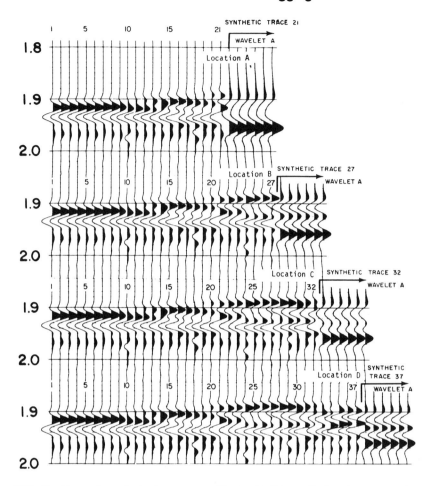

FIG. 15—Comparison of seismic response with wavelet B to synthetic seismogram using wavelet A.

the seismic characterization of the reef objective through modeling completed the interpretive sequence.

As Nath's summary concludes, the reef signature is identified as a break in continuity of certain reflections. After consideration of lithologic differences in acoustic terms, we note from the ray-trace display of Figure 16 that geometrical effects of the tight curvature disperse the reef reflection response over a broad ground surface area, causing weak responses. Hence, geometrical considerations prevail, making consideration of the acoustic differences between the anhydrite, evaporite, and carbonates of secondary importance. The reef, which is by definition a stratigraphic trap, has a seismic indication which seems stratigraphic in appearance, yet is structural in origin. The indicated seismic extent of the reef has also been much exaggerated by the same ray-spreading mechanism.

Figure 17 illustrates a model primary-reflection time section developed from a North Sea location. A strong indication of a stratigraphic prospect, a pinchout, is suggested at 3.3 sec. The particular study is based on a schematic representation of a North Sea horst block, which is shown in Figure 18.

The seismic expression of the model with a simple waveform as shown in Figure 19 is readily interpreted in structural terms. The typical, longer waveform section (Fig. 17), however, shows a strong indication of a pinchout near the center of the horst block. In this case, the superposition of the lengthy waveforms and the slightly discordant

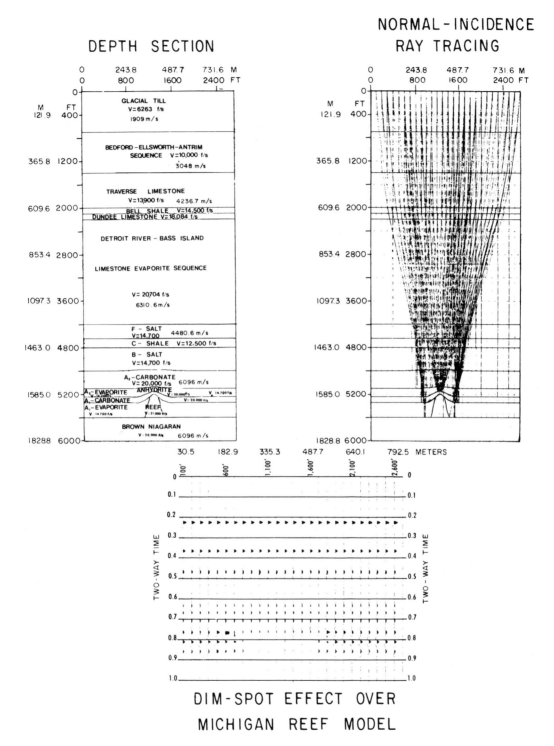

FIG. 16—Model study delineating seismic expression of pinnacle reef (after Nath, 1975).

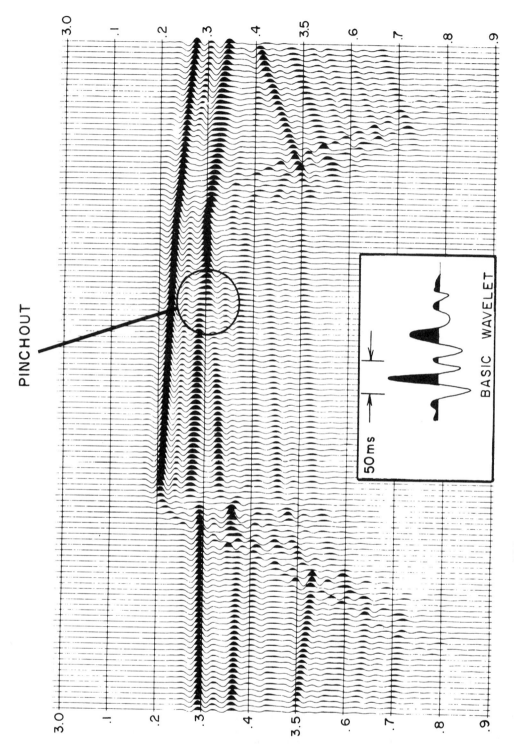

FIG. 17—North Sea horst-fault model, wave-theory solution (primaries only).

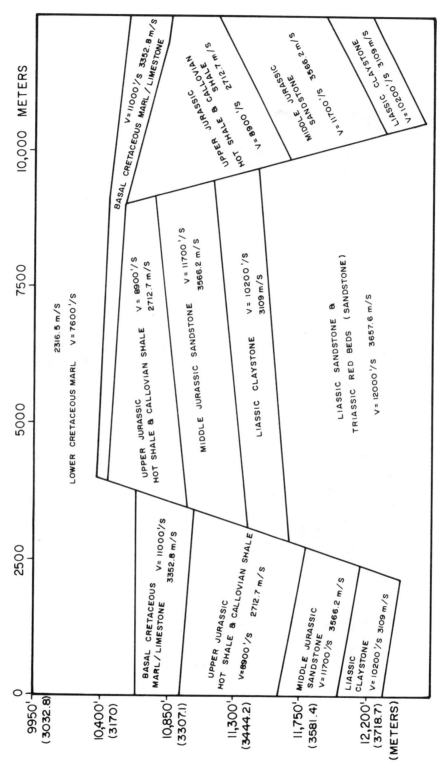

FIG. 18—North Sea model, geology and seismic parameters.

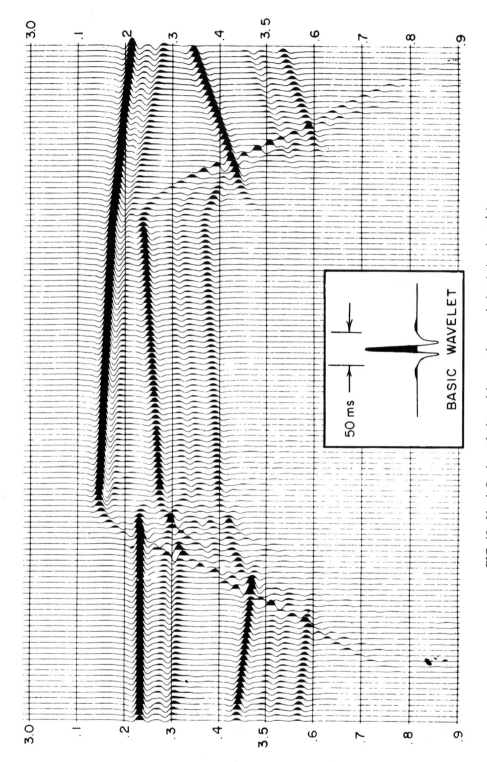

FIG. 19—North Sea horst-fault model, wave-theory solution (primaries only).

Stratigraphic Modeling and Interpretation

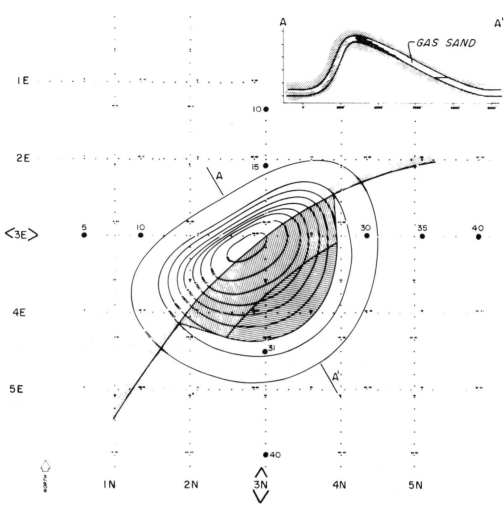

FIG. 20—Three-dimensional model-C depth map.

geometry have conspired to suggest a stratigraphic element which does not exist.

In spite of conventional processing methods, the indication of a stratigraphic prospect exists, even when yet another typical but lengthy waveform is introduced. Only the introduction of the symmetric shortened wavelet clarified the seismic expression. Clearly, then, we must be cautious of some stratigraphic-prospect indications on the grounds that the prospect itself is a product of the seismic expression and does not actually exist geologically.

The final study is a three-dimensional wave-theory model having both structural and stratigraphic elements. The structural contours and an indication of the lithology are posted on a base map (Fig. 20), which shows also the trace locations for the computed seismic profiles. The structure consists of a sandstone body which has a facies change to shale short of the crest. The sandstone is partially gas-filled.

Figure 21 shows north-south and east-west seismic profiles across the structure. Since the principal axes of the feature are rotated by about 45° from these directions, the picture is far from straight forward. Still, the most pronounced features of the prospect are the amplitude anomalies to one side of the crest. The profiles of the principal axes are shown in Figure 22. They help considerably in clarifying the interpretation, but they usually are not available in exploration studies.

One of the principal lessons of this study is obvious. If we honor the structural interpreter's maxim of "drilling on the highs," the pay zone

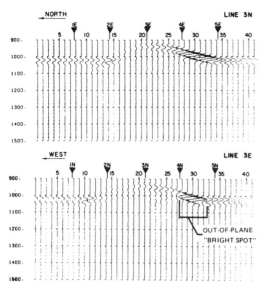

FIG. 21—Three-dimensional model-C seismic profiles.

will be missed. Instead, we must be guided by stratigraphic precepts; and these must be knowledgeably applied. The drilling of the "bright spot" on profile 3E of Figure 21 will result in a dry hole. In this case we are looking at reflections from out-of-the-plane of the profile.

Experience with seismic modeling has shown that it is unlikely for significantly differing hypotheses of the subsurface geology to lead to similar seismic expressions, particularly where we are willing to model with separated source and receiver positions, thus more closely simulating the information content of stacked seismic data. Most of the ambiguity in the model studies is concerned with subtle geologic elements rather than the gross conceptualization of the subsurface condition. These subtle elements are, of course, quite important in consideration of stratigraphic situations; hence, case studies help to suggest the full potential for stratigraphic interpretation based on available tools and techniques.

Only our ingenuity and understanding of geological and geophysical principles truly limits exploration based on the technology which has been developed.

QUALITATIVE AND QUANTITATIVE STRATIGRAPHIC CORRELATIONS

Variations in the characteristics of the stratigraphic targets which we can identify from seismic data require a refined use of those data and the available interpretation tools. The studies previously described in some sense document this viewpoint. In such refined approaches, information may often be forthcoming which bears significantly on still other hydrocarbon-related disciplines. We can demonstrate such a circumstance by considering in another context Figures 1 and 6.

The wavelet-processing discussion used two data panels (Fig. 6) taken from a North Sea oil field. Panel A shows a typical processed section which was subjected to predictive deconvolution before stack. In panel B, wavelet-processed results have been similarly deconvolved. At the left-hand edge of the wavelet-processed panel B, there is an event having geologic discontinuity (see arrow) which strongly suggests an oil-water contact. In this environment—a Jurassic sandstone under the Kimmeridgian unconformity—an unusual sequence of fluid mechanisms is required to develop sufficient acoustic contrast between the oil- and water-filled portions so that a seismic event becomes visible.

The top of the sandstone is clearly visible as a trough, but, to the right of trace number 205, some change in lithology is taking place because we lose sight of this trough. Beyond the portion of data shown and continuing to the right, the sandstone top becomes well defined once again with yet another possible indication of an oil-water interface. If the implied lithologic change is indicative of a change in porosity, then serious consequences of a reservoir-engineering nature may be implied.

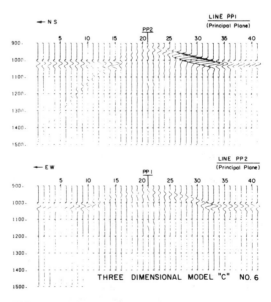

FIG. 22—Three-dimensional model-C seismic profiles.

The ability to define reservoirs with precision as indicated and even to suggest porosity changes from seismic data is well beyond the scope of usual stratigraphic interpretations. Correlation of the synthetic seismogram with seismic data for the North Sea as shown in Figure 1 corresponds also to this oil field. The Jurassic sandstone is clearly visible as a low-velocity low-density zone on the well logs and is correspondingly denoted by a trough-and-peak sequence in the seismic data.

Where thick lithologic units exist in the subsurface, quantitative approaches to stratigraphic correlations using the tools and methods described can be readily conceived. Thick units are those whose boundaries are marked by reflection events having sufficient separation in time to be clearly resolvable. In such a case, the necessary calibrations between arrival time and depth, and between arrival-time difference and thickness, are established using velocity information from nearby well-log measurements. If the lithologic unit is sufficiently well defined and thick, it may even be possible to develop the velocity information from the seismic data alone (see Taner and Koehler, 1969, for a basic discussion of seismic-velocity determination).

For normal seismic data, thick lithologic units typically encompass 20 m or more depending on depth of burial, regional velocity variation with depth, and specific characteristics of the effective seismic wavelet. This value is, in fact, very much in line with our usual view of the thickness resolution inherent in our seismic data (see Sheriff, 1976). Because many exploration situations are concerned with beds having less than 20-m thickness, the matter of their resolution in quantitative terms is far from academic. Hence, we must consider in analytical terms the thin-bed stratigraphic resolution potential inherent in seismic data.

The exposition of wavelet-processing concepts and practices gave an indication that the resolution of seismic data, at least in qualitative terms, was related to waveform (see Fig. 6). Later, a model study relating to Fresnel zones indicated that seismic amplitudes held the key to seismic resolution in the spatial dimension (see Fig. 8). We may thus anticipate that these same factors will play analogous, but analytical roles in establishing seismic resolution in the travel-time sense.

Ricker (1953), in an early work on resolution, set the pattern for most subsequent thinking. According to Ricker, the breadth of a seismic waveform entirely controlled its resolution potential. Widess (1973) carried this philosophy to a logical conclusion. Widess cited this limit of resolution as being 1/8th of the wavelength of the waveform central frequency. He noted from a synthetic-seis-

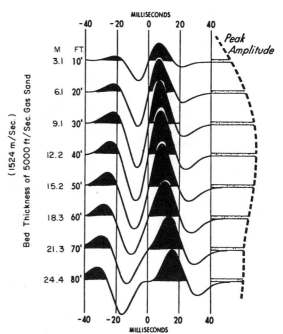

FIG. 23—Response of low-velocity layer—synthetic-seismogram study for 25-Hz Ricker wavelet.

mogram study that, as the thickness of a high-velocity bed dropped below 1/8th of the dominant wavelength, the peak-to-trough separation of the reflection signature became invariant.

Figure 23 shows computed synthetic seismograms for a single low-velocity bed for a variety of thicknesses. A 25-Hz Ricker wavelet is taken as the seismic waveform, and invariance is observed below 12.2-m thickness, in accord with results of Widess. Peak-to-trough amplitudes are now also indicated. At the thickness where peak-to-trough separation becomes invariant, "tuning" occurs and maximum reflection amplitude through event reinforcement is seen. For bed thicknesses smaller than the tuning thickness, all resolution information appears to become encoded in the event amplitudes. Widess' results exhibited the same phenomenon, which went unrecognized.

J. P. Lindsey (Geophysical Society of Houston, 1973) explicitly appreciated and utilized the principle of amplitude encoding of thin beds. Of course, in our earlier discussion of the Fresnel zones, we encountered a similar phenomenon in that the amplitudes of seismic signatures for point reflectors related to the spatial extent of small subsurface reflectors.

Hence, wavelet-processed data, taken in conjunction with data in which reflection strengths

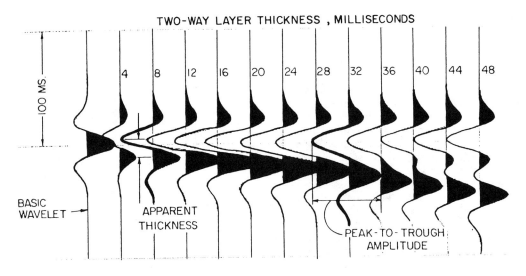

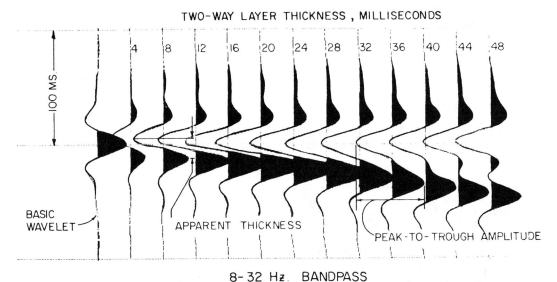

FIG. 24—Synthetic-seismogram studies of thin beds.

are controlled so as to have significance, offer a powerful stratigraphic approach to the resolution of thin beds. The practical bounds of the method and specific mechanisms remain to be explained. Still, we must not overlook the profound economic impact on exploration for stratigraphic objectives of this new viewpoint toward seismic resolution.

It is important to stress again that our new viewpoint toward resolution rests in good measure on our knowledge of the propagating waveform and the simplicity of its structure. Where we have no such reference, our insights have far less effect, as is illustrated by the work of Meissner and Meixner (1969). Although thin-bed effects were studied, intuitions received little guidance owing to use of a rather realistic, complex, lengthy seismic source pulse. Clearly, then, we shall want to apply this quantitative technology only to wavelet-processed results in which our requirements for a waveform reference are satisfied.

In Figure 24 some synthetic-seismogram studies for thin beds are repeated, this time using two differing symmetric waveforms—a 20-Hz Ricker wavelet and an 8 to 32-Hz Butterworth bandpass wavelet. The thicknesses are now presented in two-way travel-time units. For every such example a characteristic amplitude tuning thickness is noted, whereas, for lesser thicknesses, all thickness information is amplitude encoded.

From synthetic-seismogram studies of this type, we can develop sets of calibration curves as are illustrated by Figure 25. The vertical scale presents actual bed thickness in milliseconds, and the two horizontal scales, in turn, give apparent thickness measured or peak-to-trough time separation and the measured peak-to-trough amplitude in arbitrary scale units. The specific curves shown are taken directly from the synthetic-seismogram study using the 20-Hz Ricker wavelet. If the actual thickness of the bed and the apparent thickness, as determined by peak-to-trough separation, were the same value, the time-resolution calibration curve would follow the diagonal dashed line. For the thicker lithologic units or beds, this dashed line is in fact followed, showing that the resolution in travel time can be accurately accomplished.

For the particular 20-Hz Ricker wavelet under consideration, the smallest possible peak-to-trough separation which can be observed approaches an asymptote of about 17.3 msec (see Time Resolution Limit on Fig. 25). Also, at a bed thickness of 19 msec, the indicated seismic strength or amplitude tunes to a value about 40% greater than the normal amplitude level observed for bed thickness of 45 msec or more. It follows that, for seismic data processed to have a 20-Hz Ricker wavelet as the propagating wavelet, thickness estimates for beds having apparent thickness (from peak-to-trough observations) of less than 25 msec are best determined by using calibrated amplitude values.

Recall that the technology for transforming and manipulating seismic waveforms gives us the ability to impress any waveform we choose into the data so long as it does not include frequency components not present in the originally propagating waveform. Calibration curves for amplitude and travel-time measurements as shown may

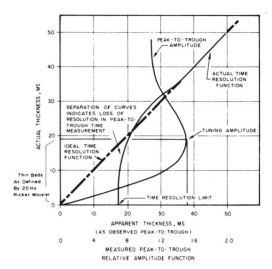

FIG. 25—Resolution calibration curves for 20-Hz Ricker wavelet developed from synthetic-seismogram study.

be developed for *any* specific symmetric wavelet used for wavelet processing. Further, an analytic means now exists for distinguishing between thick and thin beds and resolving thin beds as a function of a known propagating wavelet.

Several practical applications are apparent. First, it is clear that a thinning bed, or one which is pinching out, will have its clearest seismic expression when the tuning thickness is reached. The interpretive significance of this observation is profound for several reasons. At a most elementary level, we see that a thin gas sandstone will have the highest amplitude reflection not necessarily at its thickest portion or where there is most gas, but, rather, where the tuning thickness is attained. This is usually near the edges of the gas-filled zone. Of more consequence to this discussion is the fact that the observation of tuning amplitude levels on the processed seismic section gives us one of two calibration values needed to relate amplitudes on the seismic section to the arbitrary scale of amplitudes on our calibration-curve plot. A second calibration value is normally determined from observation of the normal amplitude level for a thick lithologic unit. The need for preservation of the significance of seismic amplitudes through processing is clearly indicated.

Next, we consider a wave-theory response from a schematic model after Lindsey et al (1976) which shows a thin shale stringer in a sandstone body of uniform thickness (Fig. 26). The 15.2 m of the gross body character leads to an invariance of waveform. Net shale content can be estimated directly by noting the peak-to-trough amplitudes.

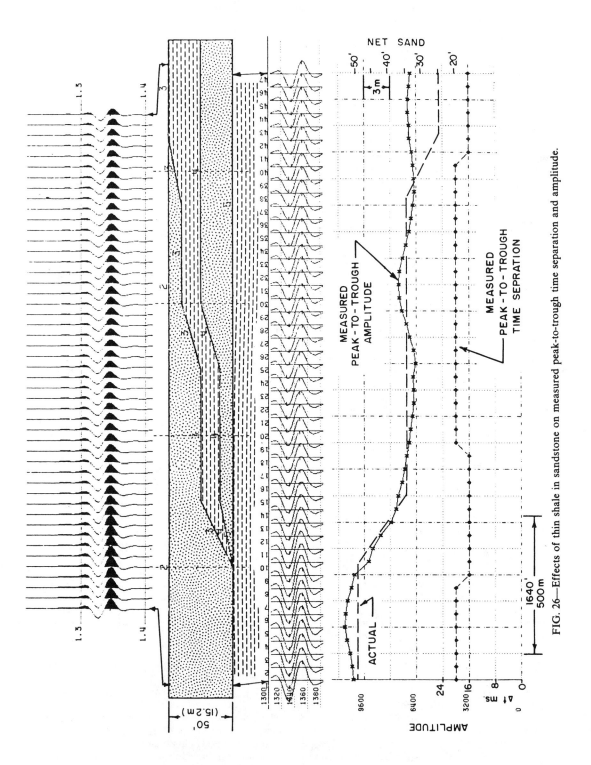

FIG. 26—Effects of thin shale in sandstone on measured peak-to-trough time separation and amplitude.

Stratigraphic Modeling and Interpretation

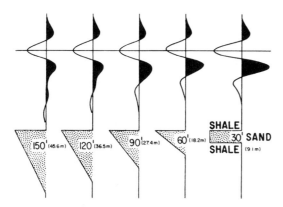

FIG. 27—Seismic response for vertical transition contacts of variable thickness.

At the bottom of the figure, an output from a computer analysis monitors the measured peak-to-trough time separation and peak-to-trough amplitude on a trace-by-trace basis. Excellent agreement is noted between the net sandstone estimates and the measured amplitudes. The peak-to-trough time separation in this case tells us nothing because we are dealing with a thin bed. Note that the most serious departures of the measured net sandstone curve from the actual curve correspond to places in the model where beds terminate. These terminations give rise to diffraction events and contaminate the reflections, causing departures from the theoretical considerations which were presented, as the latter were based on the more simplified theory of synthetic seismograms.

The model study of Figure 26 presents both a limitation on the method proposed and one of a series of cautions which must accompany its use. Although a net shale estimate may be derived, the position or distribution of that shale is not indicated in the seismic response. In an ideal case we might be able to identify 5 m in a 15.2-m unit, but we could not determine its position or distribution in the unit (i.e., five 0.5-m stringers would appear precisely as one 2.5-m stringer, etc.). We further see that diffractions which were not included in our considerations can disturb the desired quantitative correlations.

Other cautions can be derived from elements of the model. In developing Figures 23 and 24, for example, the low-velocity unit is encased between similar materials and waveform calibrations are computed accordingly. For the circumstance where the encasing lithology above and below differ significantly in acoustic properties, a correspondingly adjusted amplitude calibration would be needed. Here geologic information developed principally from nearby well logs would be necessary to develop such amplitude calibrations in anticipation of the drill.

Further, transitional contacts over depth have not been included in the model considered. We may readily appreciate the effect of such contacts by synthetic-seismogram studies as shown in Figures 27 and 28. Figure 27 shows that the principal effect of an intermediate-thickness transition zone is a decrease in amplitude, the black peak reflection being somewhat smaller than the preceding trough. A similar effect was observed for spatially derived transitions as indicated by the study of Figures 10 through 14. The ability to resolve specific transitional effects as suggested by Figure 28 is quite limited. Hence, if these effects are to be included in a quantitative analysis, a geologic basis for their incorporation must first be developed.

By way of summary, we consider next another model by Lindsey et al (1976) encompassing a transitional lithology against a structural background similar in concept to the three-dimensional model study of Figures 20 through 22. A fault edge has been introduced in Figure 25. We put into practice the computer analysis of peak-to-trough measurements and determine gross lithology, but now in the presence of a structural component. Note that amplitude level of the water sandstone has been interpreted and that all amplitude variation exhibited is of lithologic origin and not related to thinning or thickening of the particular unit.

The quantitative techniques described have been in use now for some time in a variety of practical contexts. In all their applications, however, the geophysical tools and principles described here play prominent roles. These methods are significant, but there is still more to the seismic study of stratigraphy than the delineation of

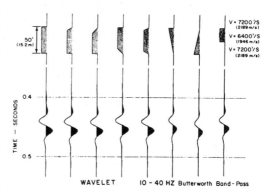

FIG. 28—Seismic response to abruptness of vertical contacts (after Meckel and Nath, this volume).

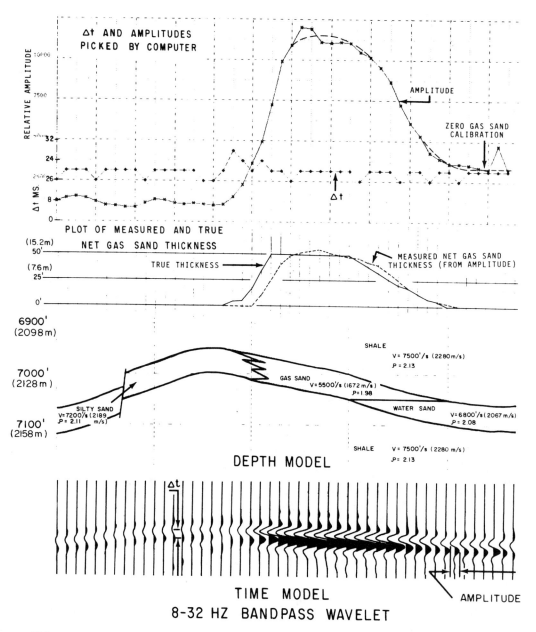

FIG. 29—Updip-permeability-barrier model study and peak-to-trough time separation and amplitude resolution analysis.

thin beds. We must cite, in particular, the pioneering work of Harms and Tackenberg (1972), Sangree and Widmier (1974), and Vail et al (1976), who set out to achieve stratigraphic objectives by identification of sedimentary processes from seismic patterns. Spatial and temporal resolution in concert with these approaches offers a comprehensive and most powerful technology for stratigraphic interpretation in a more complete sense.

SUMMARY

The relation of seismic data to stratigraphy can be understood conceptually starting from a rather naive point of view. Taking a single seismic trace, we may assume that each reflection event consists of a simple symmetric wavelet and denotes a change in lithology corresponding to a change in acoustic impedance. The polarity and size of the reflection events allow the development of an underlying reflectivity series and acoustic impedance log. These types of seismic-derived information, when interpreted from trace to trace and in light of geologic principles and available subsurface information, permit accurate correlation of seismic data with the subsurface information where these coincide, and permit extrapolations elsewhere.

The difficulties cited represent departures of real seismic data from the ideal data of the naive approach. For each such problem, some technique of seismic data processing is available as a mechanism by which the real data can be transformed or improved to have characteristics approaching the ideal.

Seismic modeling is understood to be the successor to the synthetic-seismogram calculation. Illustrations of modeling in two and three dimensions have brought forth a variety of considerations of important interpretational significance. Lessons from the model studies warn the interpreter to account constantly for the three-dimensional nature of the subsurface in both structural and stratigraphic terms.

Finally, the stratigraphic resolution potential of seismic data can be expressed in quantitative terms. The computer-based procedure which was developed uses both time measurements and amplitudes in the amplitude-processed, wavelet-processed environment to achieve high-resolution lithologic delineation.

The future for stratigraphic interpretation and finding hydrocarbons in stratigraphic environments seems particularly bright at this time in regard to the available tools and methods. Newer tools and procedures may improve our capabilities further. The practical applications of these concepts and techniques as they pertain to actual exploratory and development projects are discussed and illustrated in an accompanying paper by Schramm et al.

REFERENCES CITED

Berkhout, A. J., 1973, On the minimum-length of one-sided signals: Geophysics, v. 38, p. 657-672.
―――― 1974, Related properties of minimum-phase and zero-phase time functions: Geophys. Prosp. (Netherlands), v. 22, p. 683-703.
Dedman, E. V., J. P. Lindsey, and M. W. Schramm, Jr., 1975, Stratigraphic modeling: a step beyond bright spot: World Oil, v. 180, no. 6, p. 61-65.
Harms, J. C., and P. Tackenberg, 1972, Seismic signatures of sedimentation models: Geophysics, v. 37, p. 45-58.
Hilterman, F. J., 1970, Three-dimensional seismic modeling: Geophysics, v. 35, p. 1020-1037.
Geophysical Society of Houston, 1973, A symposium: lithology and direct detection of hydrocarbons using geophysical methods: Houston.
Lindseth, R. O., 1976, Mapping stratigraphic traps with seislog (abs.): 46th Ann. Mtg. Soc. Exploration Geophys., Houston.
Lindsey, J. P., M. W. Schramm, Jr., and L. K. Nemeth, 1976, New seismic technology can guide field development: World Oil, v. 183, no. 7, p. 59-63.
McClintock, P. L., 1977, Seismic data-processing techniques for northern Michigan reefs: AAPG Stud. Geol. No. 5 (in press).
Meckel, L. D., and A. K. Nath, 1977, Geologic considerations for stratigraphic modeling and interpretation: this volume.
Meissner, R., and E. Meixner, 1969, Deformation of seismic wavelets by thin layers and layered boundaries: Geophys. Prosp. (Netherlands), v. 17, p. 1-27.
Nath, A. K., 1975, Reflection amplitude, modeling can help locate Michigan reefs: Oil and Gas Jour., v. 73, no. 11, p. 180-182.
Neidell, N. S., 1975, What are the limits in specifying seismic models?: Oil and Gas Jour., v. 73, no. 7, 144-147.
Peacock, K. L., and S. Treitel, 1969, Predictive deconvolution theory and practice: Geophysics, v. 34, p. 155-169.
Peterson, R. A., W. R. Fillipone, and F. B. Coker, 1955, The synthesis of seismograms from well log data: Geophysics, v. 20, p. 516-538.
Ricker, N., 1953, Wavelet contraction, wavelet expansion and the control of seismic resolution: Geophysics, v. 18, p. 769-792.
Robinson, E. A., and S. Treitel, 1967, Principles of digital Wiener filtering: Geophys. Prosp. (Netherlands), v. 15, p. 311-333.
Sangree, J. B., and J. M. Widmier, 1974, Interpretation of depositional facies from seismic data: Continuing Education Symposium, Geophysical Society of Houston.
Schoenberger, M., 1974, Resolution comparison of minimum-phase and zero-phase signals: Geophysics, v. 39, p. 826-833.
Schramm, M. W., Jr., E. V. Dedman, and J. P. Lindsey,

1977, Practical stratigraphic modeling and interpretation: this volume.

Shah, P. M. 1973, Ray tracing in three-dimensions: Geophysics, v. 38, p. 600-604.

Sheriff, R. E., 1976, Inferring stratigraphy from seismic data: AAPG Bull., v. 60, p. 528-542.

Taner, M. T., and F. Koehler, 1969, Velocity spectra—digital computer derivation and applications of velocity functions: Geophysics, v. 34, p. 859-881.

——— E. E. Cook, and N. S. Neidell, 1970, Limitations of the reflection seismic method; lessons from computer simulations: Geophysics, v. 35, p. 551-573.

Vail, P. R., et al, 1976, Interpretation of seismic sequences from reflection patterns: Preprint, 29th Ann. Mtg., Midwest Sect. Soc. Exploration Geophysicists, Dallas.

Widess, M. B., 1973, How thin is a thin bed?: Geophysics, v. 38, p. 1176-1180. Wuenschel, P. E., 1960, Seismogram synthesis including multiples and transmission coefficients: Geophysics, v. 25, p. 106-129.

Seismic Attributes: Amplitude, Frequency, Phase, Velocity

Synthetic seismic sections of typical petroleum traps

BRUCE T. MAY* AND FRANTA HRON‡

Interpretation of seismic reflections remains a key problem in seismic exploration because reflections have complex behavior, especially near geologic structures. One method to gain an understanding of this complex behavior is to study synthetic seismic sections of models of typical petroleum traps as computed by zero-offset ray tracing for primary P-waves only. These synthetic sections have features of significant interpretative value to the practicing geophysicist, such as variations in reflection amplitudes and complexities in reflection-time geometries. Asymptotic ray theory was applied to calculate reflection amplitudes, accounting for mode conversion and three-dimensional geometric divergence of ray tubes in the presence of curvilinear interfaces. This suite of synthetic seismic sections illustrates the difficulties in making correct seismic interpretations of geologic structures and suggests three conclusions: (1) The customary assumption that seismic sections are simple images of geologic cross-sections is an oversimplification and can lead to erroneous interpretations. (2) Variations in overlying strata produce marked disturbances in reflection amplitude and traveltime. (3)' Critical reflections that provide key structural information are often difficult to recognize and are apt to be ignored, or misidentified, especially if trace processing has been cursory. A fundamental principle of interpretation is underscored by this study: CDP stacked seismic sections must be interpreted so as to be consistent with structural depth models because stacked sections are not simple images of geologic structures.

INTRODUCTION

Reflection identification is one of the foremost problems in the interpretation of exploration seismic data. This reflection interpretation problem is not trivial and can reach severe proportions, often in regions of keen economic interest. The seismic data customarily are CDP (common-depth-point) stacked seismic traces (Mayne, 1962) and interpretation effort is most often concentrated in and around geologic structures such as faults or salt domes. As pointed out by Taner et al (1970), there are several problem areas with "potentially serious consequences" if interpretation and computational procedures are not carefully carried out. Perhaps the most fundamental problem area, or at least the one to be studied first, is reflection identification on CDP stacked sections because this is frequently the first processing output where detailed interpretation is concentrated.

CDP trace stacking, as well as many other trace processes, has been geared to enhance reflection events which simulate zero-offset P-wave reflections that can be easily interpreted as simple images of geologic structures. However, several factors combine to disrupt and complicate reflection continuity so that interpretation can become very difficult. Because of the horizontal displacement that reflections suffer (cf., "inverse migration," Hagedoorn, 1954, p. 92), seismic events can be broadly dislocated in the vicinity of even rather simple geologic structures. As shown in the experiment below, reflection dislocation is often accompanied by multivalued primary reflections (cf., "non-uniqueness of primary reflection travel paths," Taner et al, 1970, p. 551), broken reflections, and crossing of shallow reflections over deeper ones. Apparent reflection continuity is further interrupted because of rapid variations of reflection

---

Presented at the 45th Annual International SEG Meeting, October 15, 1975 in Denver. Manuscript received by the Editor January 24, 1977; revised manuscript received January 5, 1978.
*Amoco Production Co., P. O. Box 591, Tulsa, OK 74102.
‡Formerly Amoco Production Co., Tulsa; currently University of Alberta, Physics Dept., Edmonton, Alta., Canada T6G 2E1.
0016-8033/78/1001-1119$03.00. © 1978 Society of Exploration Geophysicists. All rights reserved.

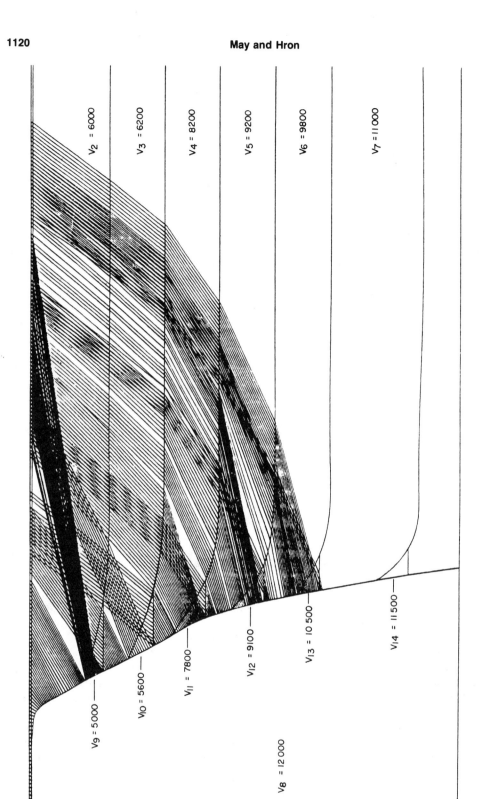

FIG. 1. Normal incidence ray plot for a salt dome flank model. (For corresponding time section see Figure 23.) Note the effects of ray focusing and pronounced nonuniform sampling (spacing of reflection points) along the salt dome flank.

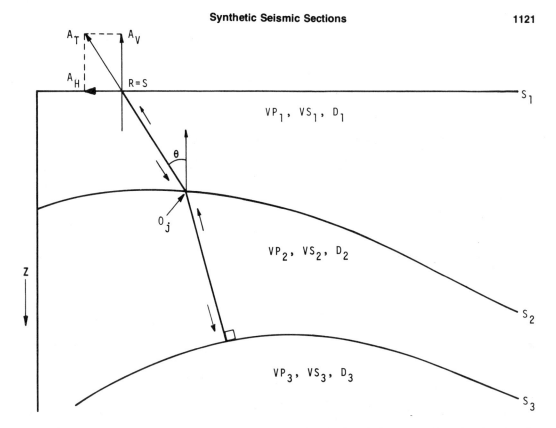

FIG. 2. Schematic ray representation for zero offset tracing, i.e., for identical source-and-receiver location, in homogeneous layers separated by curvilinear interfaces. The total amplitude A_T of the ground motion at the receiver R is decomposed into its vertical and horizontal components with the amplitudes A_V and A_H, respectively.

amplitude. Reflector curvature, for example, causes rapid changes in geometric spreading and rapid changes in wave mode partitioning, resulting in significant variation of reflection amplitude. In addition, stacking velocities (Taner and Koehler, 1969) change rapidly due to raypath configuration changes, producing erroneous reflection amplitude variations on stacked sections if stacking velocity functions are not carefully handled. Such conditions as these are often unaccounted for in conventional data processing, and final seismic sections can have important events attenuated beyond recognition. These are some of the reasons why interpretation of stacked sections becomes difficult.

Generally, CDP trace processing is accomplished in a two-stage transformation of nonzero offset trace data: first, to simulate zero-offset recording at all trace distances, normal moveout is removed; second, all traces within a CDP trace gather are stacked to obtain statistical powering (see, e.g., Robinson, 1970, p. 442; Mayne, 1962). Such processing is done at the exclusion of information about range effects on amplitude and phase, multiple reflections, and converted waves. Historically, range effects have been neglected, and all wave modes except first-arrival P-waves have been regarded as undesirable noise.

With this in mind, computation of synthetic seismograms in this paper involves once-reflected P-waves (primaries) with normal incidence on the reflector. The zero angle of incidence on the reflecting interface requires coincidence of source and receiver locations. With the exception of plane parallel layers, nonzero refraction angles will generally occur even for zero-offset rays and result in energy partitioning due to wave mode conversion. Synthetic seismograms based on this type of ray tracing are "noise-free" because they contain no additional information, such as S-waves, multiple reflections, point source diffrac-

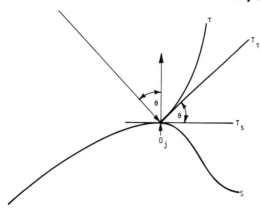

FIG. 3. Within asymptotic ray theory both the wavefront τ and the curved interface S are replaced by their tangential planes T_τ and T_s constructed at the point of incidence 0_j.

tions, and random real noise. These synthetic sections are therefore ideal for studies of interpretation problems.

If a structural model is represented by homogeneous layers separated by curved boundaries, individual rays consist of straight line segments and zero-offset ray tracing can be executed easily. Figure 1 illustrates a set of zero-offset rays reflected from a single steeply dipping boundary. Note the effects of ray focusing and pronounced nonuniform sampling in the subsurface. Ray spacing at the surface is uniform, the apparent gaps being created by refraction (bending) of raypaths crossing layer boundaries near the salt dome flank.

We designed our modeling experiment with the objective of showing the relationships between models and synthetic seismic sections for typical petroleum trapping structures. Commonly, interpretations of CDP-stacked seismic sections are made as though stacked time sections were simple images of geologic cross-sections. Through comparisons of these models and their synthetic sections, it will be demonstrated that, in general, it is erroneous to follow this oversimplified interpretative procedure.

THEORETICAL BACKGROUND

Parameterization of the model

Structural models generated for the ray tracing experiment in this report represent three-dimensional geological structures with a vertical plane of symmetry. They consist of L homogeneous, isotropic, ideally elastic layers overlaying a half-space and separated by $L + 1$ curvilinear boundaries. Under these conditions, individual rays are plane curves confined to the vertical plane of incidence and three-dimensional models can be reduced essentially to two-dimensional ones. Thus all interfaces can be represented by a set of $L + 1$ curves

$$z = s_\ell(x) \quad \ell = 1, \ldots, L + 1, \quad (1)$$

with continuous first and second derivatives. In many cases, the functional dependence of horizon depth z on horizontal distance x is obtained for each horizon with the help of cubic spline functions applied to a limited number of known data points.

Elastic properties of each layer are characterized by P-wave velocity (VP_ℓ), S-wave velocity (VS_ℓ), and a volume density (δ_ℓ), $\ell = 1, \ldots, L + 1$. [A half-space is formally considered as the $(L + 1)$-th layer in our notation.] Although these three elastic parameters are independent of one another for any homogeneous isotropic medium, simple empirical relations among them are often used to decrease the number of input data. In practice, S-wave velocity (VS) and density (ρ) are often related to the P-wave velocity (VP) by simple relations

$$VS_\ell = VP_\ell/\sqrt{3} \quad \text{(for Poisson's ratio } \sigma = 0.25\text{)},$$

and

(2)

$$\rho_\ell = 0.23\, VP_\ell^{1/4} \quad \text{(Gardner et al, 1968).}$$

($\ell = 1, 2, \ldots, L + 1$ is the layer index number.)

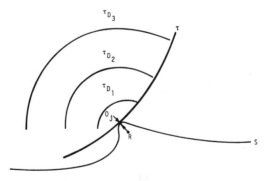

FIG. 4. The case when the radius of curvature R of the reflector at the incidence point 0_j is comparable to a wavelength. The seismic energy of the incident wave (wavefront τ) is transformed into the energy of a diffracted wave (wavefronts τ_{D_1}, τ_{D_2}, τ_{D_3}) emanating from that part of interface S.

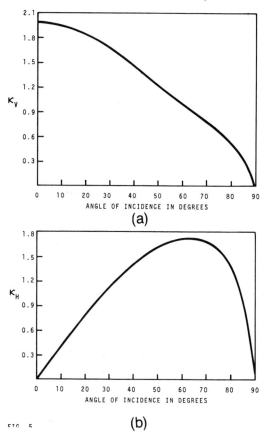

FIG. 5. Moduli of free surface conversion coefficients for vertical (a) and horizontal (b) components, if $VS_1/VP_1 = 0.576$.

Ray amplitude computation

Asymptotic ray theory (Hron and Kanasewich, 1971; Červený and Ravindra, 1971) has been used for amplitude computation because of its versatility and conceptual clarity. It is based on the assumption that the total wave field can be decomposed into an infinite number of contributions corresponding to all possible raypaths between the source and receiver. The displacement vector **W** of each individual ray is then expressed in the form of a ray series in inverse powers of predominant frequency ω_0 of the source wavelet $f_0(t)$ as

$$\mathbf{W}(\mathbf{r},t) = \sum_{\nu=0}^{\infty} \frac{\mathbf{A}_\nu(\mathbf{r})}{(i\omega_0)^\nu} f_\nu[t - \tau(\mathbf{r})]. \quad (3)$$

Here $A_\nu(\mathbf{r})$ and $\tau(\mathbf{r})$ stand for the amplitude of the νth term and the arrival time of the ray as functions of position vector $\mathbf{r}$, respectively, and $i = \sqrt{-1}$.

Although asymptotic ray theory gives us only a high-frequency approximation to the exact solution, it is the only theory that can be applied reliably and efficiently for such complicated media that consist of anisotropic layers (Daley and Hron, 1977, 1978) or inhomogeneous layers separated by curved boundaries (Pšenčík, 1974). Unfortunately, asymptotic ray theory cannot be used under some conditions (e.g., in the vicinity of caustics) when alternative methods have to be used (Hron et al, 1977). Even though alternative methods are known and have been combined with asymptotic ray theory successfully (Hron, 1973), they were not used for the production of results in this report. Thus, ray amplitudes in the vicinity of caustics (where asymptotic ray theory yields an infinite amplitude) are not computed, and the corresponding wavelets are missing on the synthetic seismic sections.

The actual evaluation of ray amplitudes in the computation of synthetic sections presented in this report has been done with the help of zero-order approximation of asymptotic ray theory when only the first term ($\nu = 0$) is used in the ray series. In this approximation, the total amplitude A_T of the ray incident on the surface is given (see Hron and Kanasewich, 1971 or Červený and Ravindra, 1971) as

$$A_T = \frac{1}{P'' \cdot P'} \prod_{j=1}^{n-1} K_j, \quad (4)$$

where n is the number of segments in the ray, $K_j (j = 1, \ldots, n - 1)$ are coefficients of reflection and refraction related to the seismic plane waves, and $P = P'' \cdot P'$ stands for the geometrical spreading of the ray tube. (In our notation, P'' represents the spreading in the vertical plane containing the raypaths or "in-plane" spreading, whereas P' is the spreading in the direction perpendicular to the plane of incidence or "out-of-plane" spreading.)

In case of homogeneous layers, the quantities P' and P'' can be written as (Alekseev and Gel'chinski, 1959)

$$P'' = \sqrt{\ell_1 + \sum_{j=2}^{n} \ell_j \prod_{\nu=1}^{j-1} \Delta_\nu},$$

and

$$P' = \sqrt{\frac{1}{V_1} \cdot \sum_{j=1}^{n} \ell_j V_j}, \quad (5)$$

where

$$\Delta_j = \frac{V_{j+1}}{V_j} \cdot \frac{\cos^2 \theta_j}{\cos^2 \chi_j} + \frac{r_j}{R_j} \cdot \frac{1}{\cos^2 \chi_j} \cdot \left(\frac{V_{j+1}}{V_j} \cdot \cos \theta_j \pm \cos \chi_j\right), \quad (6)$$

and

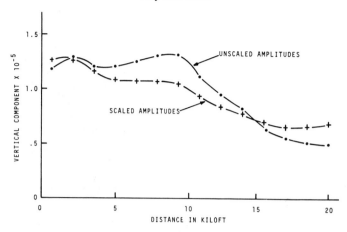

FIG. 6. Illustration of the undesirable effect of amplitude scaling necessary for section plotting. Scaled amplitudes are taken from reflection "c" in Figure 20.

$$r_{j+1} = \frac{r_j}{\Delta_j} + \ell_{j+1}.$$

Here,

- j is an ordinal number of the ray segment ($j = 1$ corresponds to the segment radiated by the source, while $j = n$ is related to the last segment of the ray that is registered by the receiver);
- V_j is a velocity of wave propagation along the jth segment;
- ℓ_j is a length of the jth segment;
- θ_j is the angle of incidence of the jth segment;
- χ_j is the corresponding angle of reflection or refraction;
- r_j is the radius of curvature in the plane of incidence of the impinging wavefront at the jth point of incidence; and
- R_j is the "in-plane" radius of curvature in the plane of incidence of the interface at the vicinity of the jth point of incidence.

According to our sign convention, R_j will be assigned a positive ($R_j = |R_j|$) or a negative ($R_j = -|R_j|$) value depending on whether the vicinity of the jth point of incidence is convex or concave for the wave incident at that point. If this vicinity is approximated by a straight line, R is set to a certain value, say, $R_j = 10^{60}$. In equation (6), the plus sign is taken if the ray is reflected, whereas the minus sign is taken if the ray is refracted.

As P'' in equation (4) becomes an imaginary number for rays whose arrivals are on the reversed branches of traveltime curves, the amplitude A_T may also be a complex number expressed by its modulus and phase. If A_T does not have a real value for a given ray, distortion of the original wavelet takes place after A_T is convolved with the source function $f_0(t)$. (This is clearly visible, for example, in Figure 12 at c'' corresponding to one of the flank reflections.) The correct choice of the argument of P'' depends on the form of Fourier transform used for the computation of the spectrum $S(\omega)$ of the source wavelet. If the direct Fourier transform were chosen as

$$S(\omega) = \int_{-\infty}^{\infty} f_0(t)\, e^{-i\omega t}\, dt, \qquad (7)$$

P'' has to be written as

$$P'' = e^{-i\pi/2} \sqrt{\left| \ell_1 + \sum_{j=2}^{n} \ell_j \prod_{\nu=1}^{j-1} \Delta_\nu \right|}. \qquad (5a)$$

If, however, the complex conjugate to equation (7) were selected as a direct Fourier transform, the complex conjugate to equation (5a) must be used instead.

Some basic features of the zeroth approximation in asymptotic ray theory can be inferred easily from equation (4) without becoming involved in mathematical foundations of the theory. One can see that with the increasing value of the cross-section $P = P' \cdot P''$ of the ray tube, the modulus of ray amplitude A_T decreases, resulting in a smaller energy flux. In other words, it is assumed that energy associated with seismic wave propagation travels along the ray tube without passing through its walls.

The product of coefficients of reflection and refraction of seismic plane waves in equation (4) character-

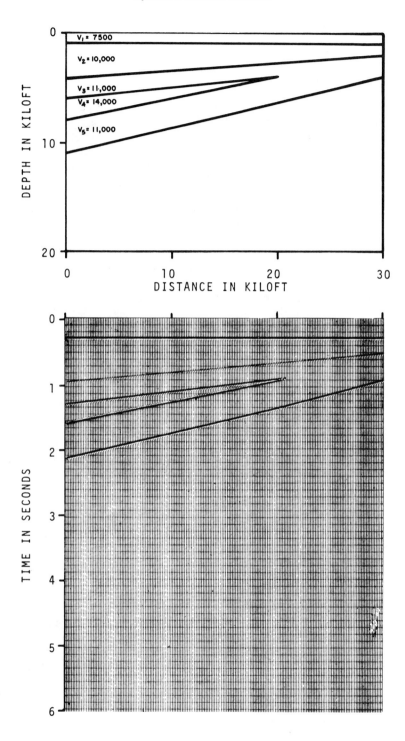

FIG. 7. Wedge model and corresponding normal incidence seismic section. This structural model simulates a sandstone wedge surrounded by lower velocity shale.

izes the partition of seismic energy occurring during the encounters of the seismic wave with individual interfaces. The use of coefficients related to plane waves also indicates that, within asymptotic ray theory, both the curved wavefront and the curvilinear interfaces are replaced by tangential planes at the point of incidence O_j (Figure 3). Then partition of seismic energy, which takes place because of the existence of two reflected (P and S) and two refracted waves (also P and S), can be evaluated in the same way as if a plane wave impinged on a plane interface at the same angle θ_j. This approximation is valid only for smoothly curved interfaces and is not justified in the vicinities of sharp corners, where the values of the radii of curvature R_j are of the same order of magnitude as the wavelengths. Then, as a result, the absolute value of P'' in (5) increases while the total amplitude A_T decreases, indicating that the seismic energy of the incident wave is transformed into the diffracted wave emanating from that point (Figure 4).

Finally, equation (4) cannot be used in the vicinity of points where the geometrical spreading of the ray tube $P = P'' \cdot P'$ approaches zero. Such points are called caustics, and on the surface they are characterized by the cusps on traveltime curves.

Wave interference at the datum surface

When the seismic wave impinges on the surface, two additional reflected waves (P and S) are generated and interfere with the incident wave. The reflection from the surface can be combined with the decomposition of the ground motion into horizontal and vertical components with the help of so-called coefficients of surface conversion (Cervený and Ravindra, 1971 or Daley and Hron, 1977, among others). Typical values of surface conversion coefficients κ_H and κ_V for incident P-wave and isotropic medium are given in Figure 5. More numerical results, including the values of surface conversion coefficients for anisotropic media, are presented by Daley and Hron (1978), together with appropriate theoretical formulas.

Thus, the vertical A_V and horizontal A_H amplitude components may be obtained from the total amplitude A_T of the incident wave [see equation (4)] as

$$A_V = |A_V| e^{i\phi_V} = \kappa_V A_T, \qquad (8)$$

and

$$A_H = |A_H| e^{i\phi_H} = \kappa_H A_T.$$

Here $|A_V|$ and $|A_H|$ stand for the moduli while ϕ_V and ϕ_H represent the phases of these two components which become complex numbers for reverse branches of traveltime curves.

The surface conversion coefficients are not needed if the receiver is removed from the surface as it is in offshore situations. Then the vectorial summation of displacement vectors of all interfering waves must be carried out instead.

Figure 5 illustrates the effect of the free surface on the reflection amplitudes. Figure 5a shows that the vertical component of ground motion starts with the value $\kappa_V = 2$ and decays to zero as a function of incidence angle. The horizontal component (Figure 5b) increases to a maximum and then declines to zero. It is apparent from Figure 5 that the omission of the surface influence caused by neglecting the surface conversion coefficients in equation (4) introduces a substantial error in the evaluation of reflection amplitudes.

Within the framework of asymptotic ray theory, both vertical and horizontal components in equation (8) are frequency independent. They can be used for the computation of the vertical and horizontal components of ground motion for any high-frequency source whose source function $f_0(t)$ is known by using the Fourier transform in the form given by equation (7).

The zero-order approximation of asymptotic ray theory is well suited as a tool in amplitude computation due to its conceptual clarity (being identical with the geometric optics approximation) and the relative ease with which it can be applied. The amplitude computation and its incorporation into existing ray tracing procedures is a quite inexpensive venture increasing CPU time generally by less than 15 percent.

Time-variant amplitude scaling

Reflection amplitudes computed at zero offset generally decrease with increasing traveltime because of divergence, or expansion, of the ray tube with increasing travel path. In order to overcome this amplitude decrease to observe all reflections on one seismogram, it is necessary to apply an inverse time-variant amplitude scaling function. Such a function creates trace-to-trace amplitude distortion even though care is taken when choosing the scaling function. This scaling distortion arises because reflections transversing non-horizontal layers have rather complicated amplitude-time and amplitude-distance characteristics.

For studies involving amplitude computations, scaling should be avoided or be compensated for. However, to display all reflections on a single synthetic seismogram, reflection amplitudes on each seismic section in this paper have been subjected to exponential, time-variant scaling of the form

$$A = A_0 e^{bT}.$$

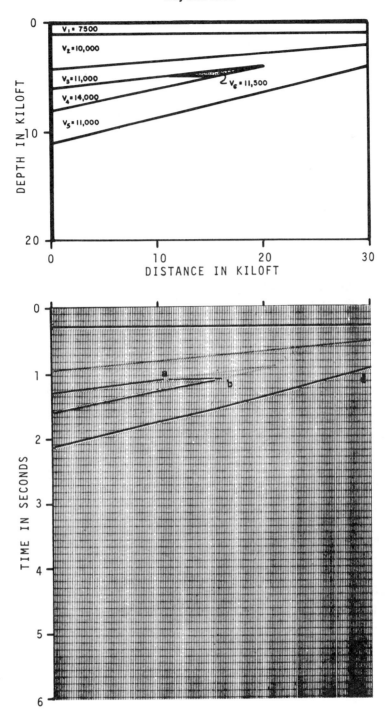

FIG. 8. Wedge model with reservoir. The same model as in Figure 7 except for insertion of a low-velocity reservoir at the wedge apex. Note the likelihood of wedge shape misinterpretation on the time section.

A_0 and b are determined in a least-mean-squares sense from the amplitudes on a specified reference trace. As seen in Figure 6, substantial amplitude distortion can occur by application of even such a simple scaling function.

GEOLOGIC DEPTH MODELS AND SEISMIC TIME SECTIONS

Simplified models of typical petroleum traps are illustrated in this section. The principal objective of obtaining the seismic response of these models is to provide data for studying the relationship between model geometries and seismic section geometries. The second objective is to determine whether the reflection amplitude response of a specific model geometry could affect the interpretation of the corresponding seismic section. It will be shown that in many cases seismic section geometries and reflection amplitudes cannot be anticipated without numerical modeling, even for rather simple models.

In the following examples, horizontal scales for the depth models and synthetic seismic sections are the same. Therefore, direct comparisons can be made from models to sections, keeping in mind that models are displayed in the depth domain and sections are displayed in the time domain. In these comparisons, it is noteworthy that reflectors have been sampled at different subsurface increments because of raypath geometry, namely, focusing and defocusing (see Figure 1). Because we are attempting to gain a rather complete understanding of the geometric relationship between models and their normal incidence seismic response, each portion of any given reflection, such as a multivalued time branch, potentially has the same importance as any other portion of that reflection, and it should be interpreted as such. That is, on a stacked section, no portion of any reflection should be underestimated or ignored in processing or interpretation solely because it is not readily identified or understood. It will be shown below that certain reflections that are poorly understood may contain the most important information.

Models are presented in order of increasing complexity with and without horizontal "reservoir fluid interfaces" for comparative purposes. There are no diffractions, multiple reflections, and P-S converted waves shown in the following synthetic sections.

Stratigraphic wedge

Figure 7 is a model simulating a sandstone wedge with a P-wave velocity of 14,000 ft/sec surrounded by a lower velocity shale ($V_P = 11,000$ ft/sec), and the lower diagram is the synthetic section for that model. Figure 8 illustrates the same model as Figure 7 after filling the upper portion of the wedge with a low velocity ($V_P = 11,500$ ft/sec) material simulating a gas zone in an otherwise brine-saturated sand. Note the marked amplitude decrease in Figure 8 at the upper end of the sand-shale wedge and the possibility of making an erroneous interpretation of the wedge shape because of the amplitude decrease. Note also that dipping events are migrating farther than the flat "contact event" from the fluid interface (points a and b), creating a gap at a and an overlap at b. An additional interpretation difficulty would be encountered if diffraction effects were present. Diffractions at a and b would tend to "heal" the reflection gaps and further complicate the seismogram.

Figure 9 shows how the reflections from the wedge would appear if the sandstone wedge had a velocity ($V_P = 11,500$ ft/sec) not very different from the shale. The reservoir velocity has been decreased by the same amount as the reservoir velocity decrease in Figure 8. The entire upper wedge has a velocity significantly lower than the surrounding shale and therefore all reservoir boundaries have high-amplitude response. Due to the reservoir velocity contrast, there is a 180 degree phase shift of the reflection from the reservoir upper boundary.

Unconformity

Figure 10 illustrates a simple unconformity. On the time section, reflection amplitudes change at the unconformity because of velocity variations in layers beneath the unconformity. Also, note that although the reflectors are linear (in depth), they have an apparent structure and appear to be concave downward in time. This model has layers of constant interval velocity, so the effect creating apparent structure (curved reflections) likely would be more pronounced on the real seismic trace data because of the usual increase of velocity with depth-of-burial.

Figure 11 is of the same unconformity model as Figure 10 except for the introduction of a low-velocity zone simulating a hydrocarbon bearing zone beneath the unconformity. Note the effect of this introduction: (a) apparent abnormal location of the base of the low-velocity zone reflection (the "contact event") in time; (b) amplitude decrease of the unconformity at the top of the reservoir; (c) amplitude change of the reflections from the layer containing the low-velocity zone; and (d) distortion and "broken" appearance of the reflection below the layer containing the low-velocity zone.

The disturbed character of reflections beneath dis-

(Text continued on p. 1140)

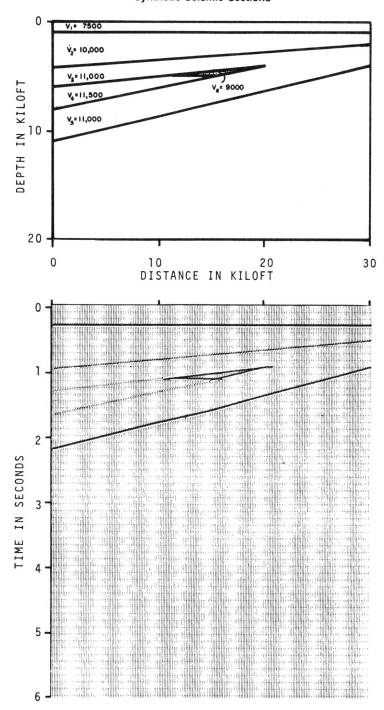

FIG. 9. Wedge model with low-velocity reservoir. The same model as in Figure 7 except the sandstone wedge velocity is not very different from the surrounding shale velocity. A corresponding reservoir velocity decrease gives rise to high-amplitude definition of the reservoir. (cf., Figure 8.)

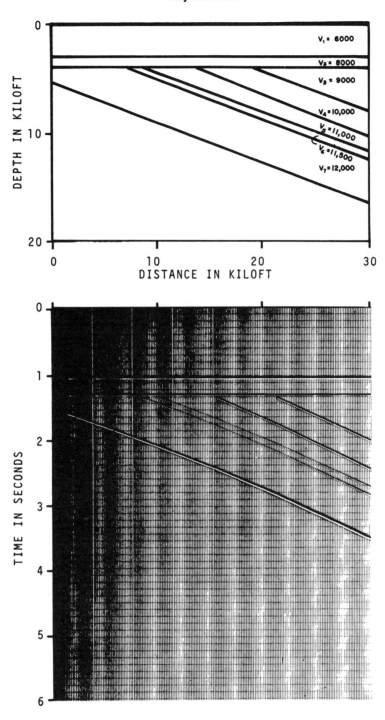

FIG. 10. Unconformity model.

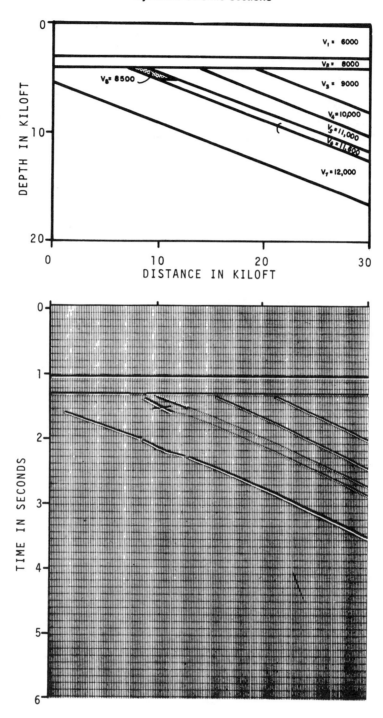

FIG. 11: Unconformity model with reservoir. On the time section, note: low amplitude of unconformity reflection above reservoir, high amplitude of side and bottom of reservoir reflections, overlapping of reservoir bottom reflection with one side, and broken appearance of reflection beneath reservoir.

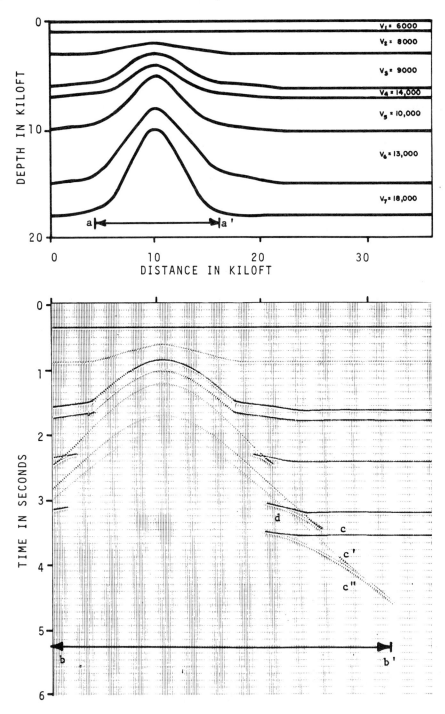

FIG. 12. Antiline model. Note: the apparent size difference between the structural model and time section because of horizontal displacement of reflections; low amplitudes of anticline crest reflections; and multi-valued time-distance flank reflections of substantial amplitude.

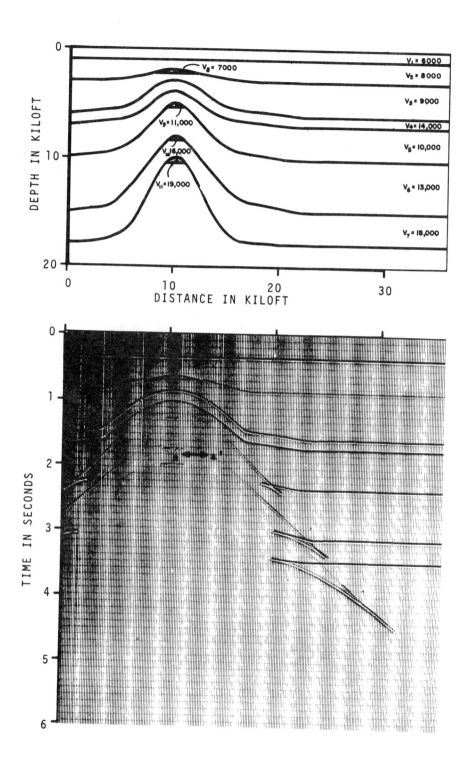

Fig. 13. Anticline model with reservoirs. Note separation of reservoir crest and reservoir base reflections $(a\text{-}a')$ due to horizontal displacement of reflections.

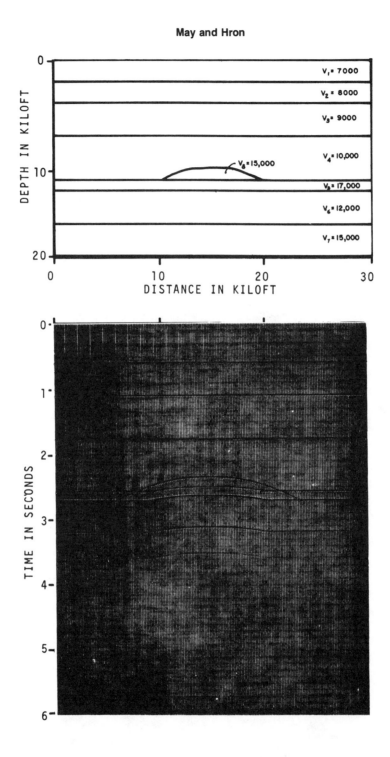

FIG. 14. Reef model. Note the overlap of the reef crest reflection with deeper reflections and the pull-up and amplitude change of reflections beneath reef.

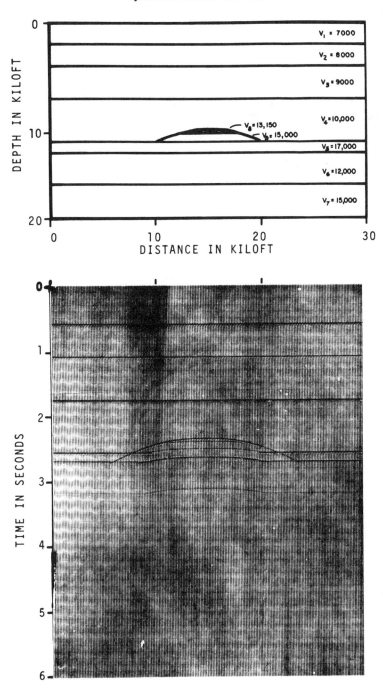

FIG. 15. Reef model with reservoir.

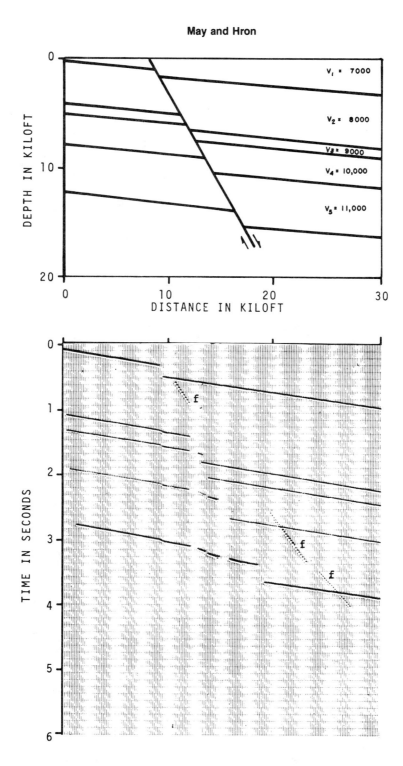

FIG. 16. Normal fault model. Note the amplitude and position of fault plane reflections labeled "f". Note further nonlinear reflections from linear reflectors beneath the fault.

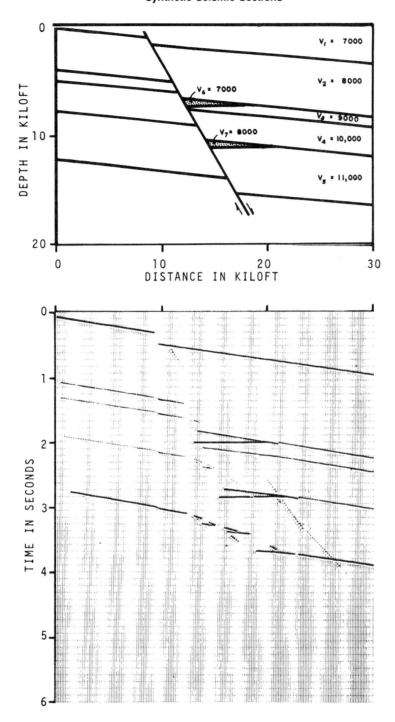

FIG. 17. Normal fault model with reservoirs. Note the shape difference of the reservoir outlined by reflections compared to the actual structural shape. Note the increased broken character of reflections beneath the fault due to presence of discordant reflectors defining the reservoirs. (cf., Figure 16.)

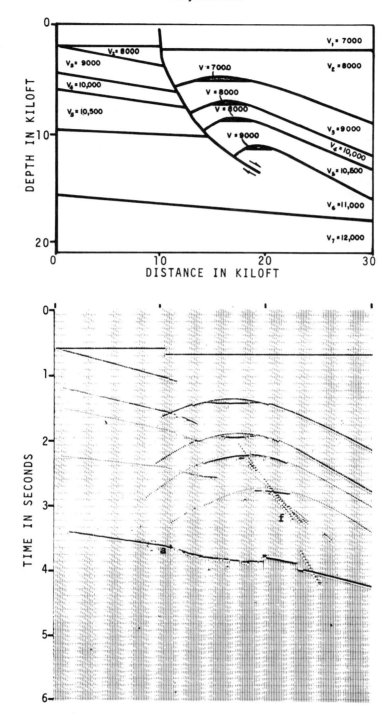

FIG. 18. Growth fault model. There is a major shape difference between the structural model and time section, especially in the deeper reflectors. Note the position and amplitude of the fault plane reflections, "f".

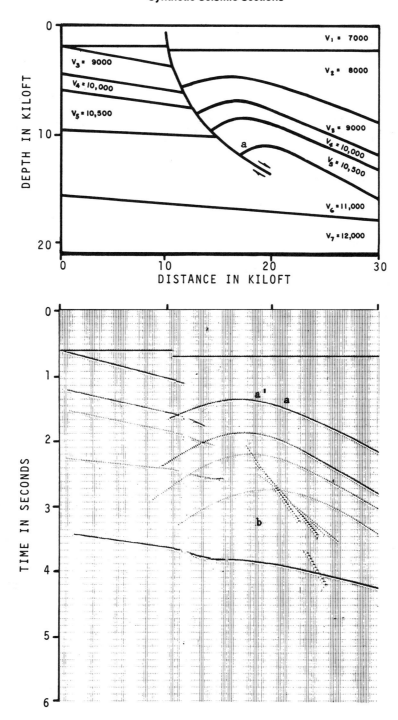

FIG. 19. Growth fault model with reservoirs. Note the migration distance separating the reservoir crests and reservoir bases (a-a').

(Text continued from p. 1127)
cordant reservoir reflectors such as illustrated in Figure 11 is a general result. If diffractions were included in the synthetic sections presented here, the seismic response would be further complicated. Reflection discontinuities shown here would not be abrupt and the gaps in the reflections would "heal". Similar reflection characteristics can be anticipated at any location where the presence of a reservoir effectively introduces a refracting horizon at some angle discordant to regional dip. This effect is prevalent in the following models.

Anticline

Figure 12 is an illustration of a simple anticline. First, compare the extent of the structure a-a' to the structural extent b-b' on the time section, a result of reflection migration. Also, note that on the time section some reflections are multivalued; that is, they arrive at more than one time at the same CDP location; for example, at points c, c', and c''. Likewise, shallow reflections cross other, deeper reflections as at d.

This model illustrates another general result: crossing of shallow reflections over deeper ones and the multivaluedness of reflections show that time sections are not simple images of structural depth models. Customary interpretation of this synthetic section could lead to misinterpretation of the section by insertion of either faults or diffraction patterns, neither of which exist on this structural model.

The correct, complete interpretation of primary P-wave reflections on the time section requires identification of all reflection segments on the section—anything less would represent an incomplete interpretation. It is seen, then, that through modeling, reflections and reflectors can be correlated, although the configurations of each are dissimilar. Therefore, time sections must be interpreted so that they are consistent with structural depth models even though the reflections in themselves may appear to be geologically contradictory.

On the synthetic section in Figure 12 there is a reflection amplitude increase due to focusing of rays in the concave upward regions and, perhaps more importantly, a reflection amplitude decrease at the anticline crest. Reflection amplitudes from the flanks observed at large (horizontal) migration distances have magnitudes comparable to amplitudes of off-structure reflections. Thus, reflections from steeply dipping beds such as anticline flanks are likely to be observable in real data, even though the reflections may be displaced miles from the subsurface reflection point.

Note that reversed time-distance branches have a 90 degree phase shift and also that the amplitude increase such as at point c' in Figure 12 is a type of focusing due to time-distance multiplicity and reflection wavelet superposition.

Figure 13 is the same anticline model as in Figure 12 except for the introduction of four low-velocity hydrocarbon zones at the anticline crest. The important feature of this diagram is the appearance of the flat "contact events" of the low-velocity zones on the time section; very substantial separation now exists between the deeper flat events and corresponding reservoir crest reflections (see a-a').

Reef

Figure 14 contains an illustration of a reef model. Of interest are reflections a from the top surface of the reef that cross two reflections b and c immediately below the reef. There is an apparent velocity pull-up and amplitude change for all reflections below the reef. Also note the multivalued nature of reflections c and d below the reef edges produced by raypath refraction at the steeply dipping reef flanks.

Figure 15 contains the same reef model as in Figure 14 except for the introduction of a low-velocity (hydrocarbon) zone in the reef crest. Note the reduction in apparent pull-up of the deeper reflections in Figure 15.

Normal fault

Figure 16 is an illustration of a normal fault. Reflectors beneath the fault are in a "shadow zone"; that is, the reflections coming from beneath the fault are disturbed, an effect that increases with increasing reflector depth. The migrated position and relatively strong amplitude of the fault plane reflection f should be noted.

Figure 17 shows the effect of changing the velocity of two zones in the fault model of Figure 16, simulating fault-trapped gas reservoirs. Note the resulting reflection phase changes at the downdip edge of the reservoirs, the increase in disturbance of the deeper reflections in the "shadow zone", and the high-amplitude reflections from the base of the low-velocity zones. There are also subtle amplitude changes along the reflections that originate under the fault.

Growth fault

Figure 18 illustrates a growth fault such as those frequently found in southern Texas. The following important points arise primarily because of horizontal displacement of reflections: (a) fault plane reflection f has a significant amplitude and consists of segments (with large traveltimes and large migration distances) that intersect anticlinal reflections; (b) flank reflections from the left side of the anticline cross regional dip reflections (there are no diffractions shown on this section); and (c) substantial structural shape change is apparent on the time section which is particularly noticeable in the deeper reflections (compare reflector at a to reflection a').

Interpretation of time sections can become difficult at this level of model complexity. Complete interpretation of zero offset data is impossible without having a depth model in mind.

Figure 19 illustrates the result of adding low-velocity zones (hydrocarbon reservoirs) at the anticlinal crests of the growth fault model in Figure 18. Note the effect of these zones on the crest reflections. Due to migration, the size of the low-velocity zone, as measured by the reflection capping the zone, appears to have a very large (horizontal and vertical) extent. The base of the low-velocity reflections now has a time-structure which is not only of lesser extent than the cap reflections (cf., a and a'), but also has a concave upward shape created by the geometry of the overlying layers. This effect increases with depth. Note also the increased distortion of the deeper layers at b due to the introduction of the low-velocity zones.

Thrust fault

A model of a thrust fault is shown in Figure 20. Note the location, amplitude and discontinuous character of the fault plane reflection f. Note also the multivalued reflections a below the fault plane and the zone b of "no data" below the fault.

Figure 21 shows the effect of adding two low-velocity zones at possible hydrocarbon trap locations. The time section of this model illustrates the difficulty that would occur in attempting to locate exploration targets beneath the fault plane such as point a on the depth model which now appears at point a' in the time section. Not only has the reservoir base reflection migrated to the right, but it now exhibits strong "apparent" dip. Note the distorting influence of the crest low-velocity zone on the shape of reflections beneath the low-velocity zone (compare time sections in Figures 20 and 21).

Salt dome flank

Perhaps the most intriguing of the results obtained from this direct modeling experiment are associated with the more complicated structural models, as exemplified by the salt dome flank in Figure 22. The most prospective targets are upturned reflectors flanking the dome. Reflections from these interfaces are multivalued: the deepest reflector locations defined by points a-b transform in time to the reflection points a'-b'. Cursory CDP trace processing of real data in situations similar to this model would result in attenuation of the multivalued reflections if stacking velocities appropriate to regional dip only were employed for normal moveout corrections.

Note the high-amplitude and large horizontal displacement of the reflection c from the salt dome face. Note also that beyond one second of traveltime, the slope of the reflection from the salt dome face decreases as reflector dip increases; that is, reflections migrate farther for reflectors of steeper dip. Figure 23 illustrates the result of introducing low-velocity zones in prospective trap regions flanking the dome face.

From this model, the following observations are important in salt dome flank exploration: (1) A reflection, if observable, from a reservoir base may appear to be located in the salt dome interior on the time section. (2) A reflection from the base of a reservoir, although horizontal on the structural model, may dip into the salt. (3) Considerable migration of some events has occurred; for example, points a-b on the structural section migrate to a'-b' on the time section. (4) The reflection from the salt dome face can be a high-amplitude event. With the introduction of the low-velocity reservoirs, the reflection from the salt face becomes more discontinuous and has significant amplitude changes along it. For example, c-d has migrated to c'-d' on the time section in Figure 23.

Overhanging salt dome

A severe problem exists when interpreting real trace data recorded over an overhanging salt dome like that illustrated in Figure 24. Two important questions are: (1) Is there overhang? (2) Can the low velocity zone be detected? It seems improbable that sufficient energy could reach the steep reflectors under the salt edge because of mode conversion at steeply dipping interfaces; note that even though there is a large "umbrella" reflection a' from the salt face a, very little information is being received from the dipping reflectors under the overhang. A small amplitude reflection b' is being received from the horizontal "contact event" at b. Interior reflections from the inner salt face cause the crossover pattern at c'.

CONCLUSIONS

Three factors that affect the interpretation of seismic sections are apparent from these results.

(Text continued on p. 1147)

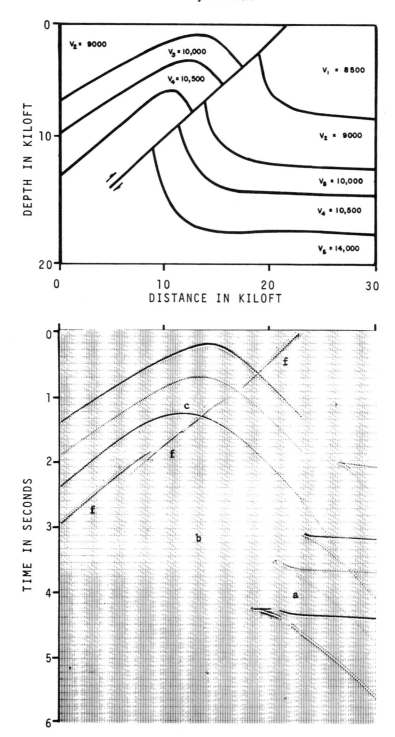

FIG. 20. Thrust fault model. Note the shape of deep reflections (*a*) beneath the fault compared to the actual structural shape. Note the apparent reflection data void (*b*) beneath the fault.

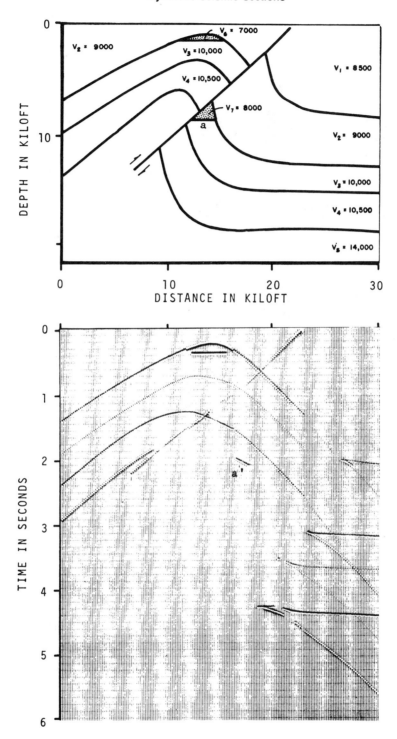

FIG. 21. Thrust fault model with reservoirs. Compare to Figure 20. Note the spatial location of reflection a' and note its dip to the right.

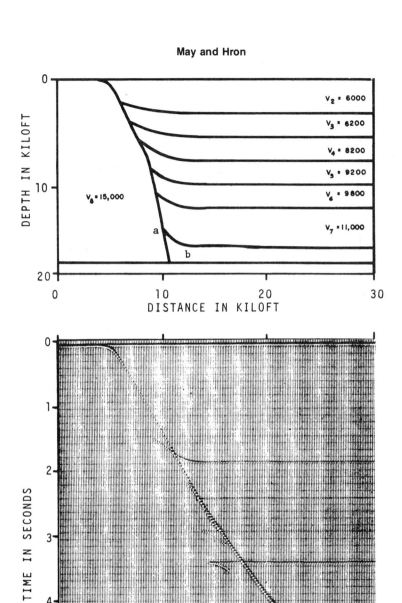

FIG. 22. Salt dome flank model. Note the position and amplitude of the salt dome face reflection (*c*). Note the reflection attitude of the updip portions of flanking horizons, in particular compare *a-b* to *a'-b'*.

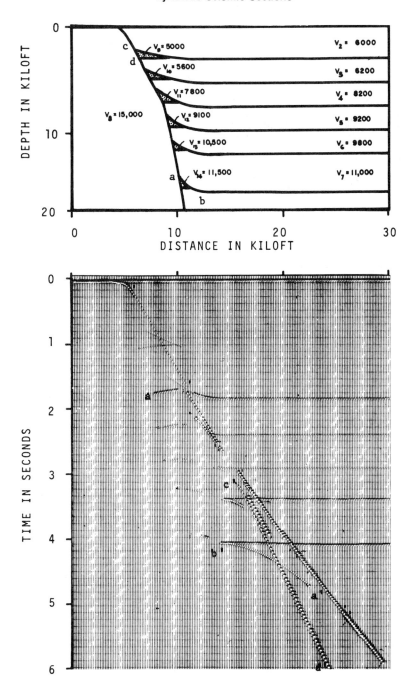

FIG. 23. Salt dome flank model with reservoirs. Note the attitude of reservoir base reflection (*a*). Compare *a-b* to *a'-b'* and *c-d* to *c'-d'* locations.

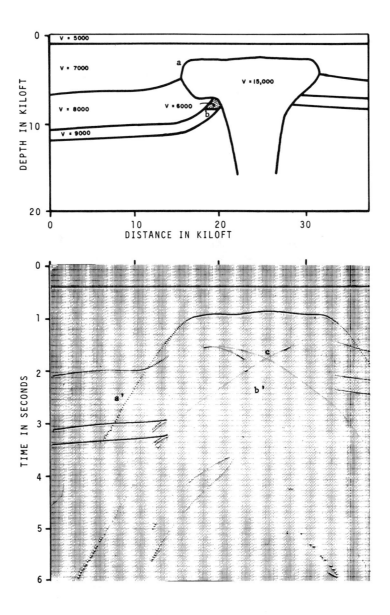

FIG. 24. Overhanging salt dome model. Note the shape change between the structural model and time section, particularly the upper salt face. Note the horizontal location of prospective reflecting horizons beneath the salt overhang and their positions on the time section. The crossing reflections at c are from the internal salt faces. Note the position and attitude of reflector b at b'.

(Text continued from p. 1141)

1) Complicated relationships exist between geologic structures and corresponding seismic sections. Seismic sections are not simple images of structural geometry. Therefore, a fundamental principle of interpretation which should be adhered to is: *composited (i.e., CDP stacked) time sections must be interpreted so that they are consistent with structural depth models.* This requires identification and correct interpretation of all reflection events, including multivalued and broken reflections. The resulting geometry as sketched on a time section may appear to be geologically unacceptable (e.g., crossing reflections). But when viewed in conjunction with a structural depth model, such an interpretation will be recognized as being both geologically and geophysically consistent.

2) From a more detailed point of view, considerable variation in amplitude or traveltime can occur along any given reflection because of the disturbing influence of overlying layers. Reflections may or may not be observable, depending on changes in reflection or transmission coefficients (compare Figure 8 to Figure 9). Naturally, the angular dependence of these coefficients complicates this problem. Changes in bed curvature alone (Figure 12) may produce significant amplitude variations that bear no relationship to changes in the lithologic properties of the reflecting horizon. Similarly, variations in interval velocity or structure of overlying strata can produce significant anomalies on deeper reflections (Figure 17).

3) Results from this model study imply that cursory trace processing of seismic data can attenuate important reflections beyond recognition. The complex nature of reflections (e.g., multivalued and crossing reflections) emanating from most structures shows that great care must be taken during trace processing to properly preserve all reflections.

These three conclusions combine to produce an undesirable situation: reflections that could provide the most important structural information are frequently the most difficult to identify and understand.

ACKNOWLEDGMENTS

The authors would like to express their gratitude to the several persons who assisted in the preparation, and offered suggestions for, the material in this paper. We wish to acknowledge especially the assistance of P. K. Link and A. E. Dowdy, and express our appreciation to Amoco Production Co. for permission to release this information.

REFERENCES

Alekseev, A. S., and Gel'chinski, B. Y., 1959. On the ray method of computation of wave field for inhomogeneous media with curved interfaces, *in* Problems in the dynamic theory of seismic wave propagation, v. 3: G. I. Petrashen, editor, Leningrad Univ. Press, Leningrad (in Russian), p. 107–159.

Červený, V., and Ravindra, R., 1971, The theory of seismic head waves: University of Toronto Press.

Daley, P. F., and Hron, F., 1977, Reflection and transmission coefficients for transversely isotropic media: Bull. SSA, v. 67, p. 661–676.

—— 1978, Reflection and transmission coefficients for seismic waves in ellipsoidally anisotropic media: Submitted for publication in Geophysics.

Gardner, G. H. F., Gardner, L. W., and Gregory, A. R., 1968, Formation velocity and density—The diagnostic basis for stratigraphic traps: Presented at the 38th Annual International SEG Meeting, October 3 in Denver.

Hagedoorn, J. G., 1954, A process of seismic reflection interpretation: Geophys. Prosp., v. 2, p. 85–127.

Hron, F., 1973, A numerical ray generation and its application to the computation of synthetic seismograms for complex layered media: Geophys. J. Roy. Astr. Soc., v. 35, p. 345–349.

Hron, F., and Kanasewich, E. R., 1971, Synthetic seismograms for deep seismic sounding studies using asymptotic ray theory: Bull. SSA, v. 61, p. 1169–1200.

Hron, F., Daley, P. F., and Marks, L. W., 1977, Numerical modelling of seismic body waves in oil exploration and crustal seismology: Proc. A.S.M.E. Sympos. on Computing Methods in Geophysical Mechanics, Atlanta, p. 21–42.

Hron, F., Kanasewich, E. R., and Alpaslan, T., 1974, Partial ray expansion required to suitably approximate the exact wave solution: Geophys. J. Roy. Astr. Soc., v. 36, p. 607–625.

Mayne, W. H., 1962, Common reflection point horizontal stacking techniques: Geophysics, v. 27, p. 927–938.

Pšenčík, I., 1974, Ray amplitudes of waves propagating in inhomogeneous media with curved interfaces: Proc. XIV. E.S.C. Assembly, Trieste, p. 171–175.

Robinson, J. C., 1970, Statistically optimal stacking of seismic data: Geophysics, v. 35, p. 436–446.

Taner, M. T., and Koehler, F., 1969, Velocity spectra–digital computer derivation and application of velocity functions: Geophysics, v. 34, p. 859–881.

Taner, M. T., Cook, E. E., and Neidell, N. S., 1970, Limitations of the reflection seismic method; Lessons from computer simulations: Geophysics, v. 35, p. 551–573.

Application of Amplitude, Frequency, and Other Attributes to Stratigraphic and Hydrocarbon Determination[1]

M. T. TANER and R. E. SHERIFF[2]

Abstract Improvements in seismic data acquisition and processing techniques make it possible to observe geologically significant information in seismic records which has not been evident in the past. New types of measurements help in locating and analyzing geologic features, including some hydrocarbon accumulations. Analysis of a seismic trace as a component of an analytic signal permits the transformation to polar coordinates and the measurement of quantities called "reflection strength" and "instantaneous phase." These, plus several other quantities derived from them, are called *attribute measurements* and can be coded by color on seismic sections. Such color displays permit an interpreter to associate measurements and changes in measurements with structural and other features in the seismic data. They thus facilitate identification of interrelations among measurements. A series of examples shows how such analysis and display helps in locating and understanding faults, unconformities, pinchouts, prograding deposition, seismic sequence boundaries, hydrocarbon accumulations, and stratigraphic and other variations which might be misinterpreted as hydrocarbon accumulations.

INTRODUCTION

The concept of correlation based on seismic "character" has been used by geophysicists since the beginning of reflection exploration. Interpreters have learned that certain reflecting sequences or depositional situations are characterized by a distinctive waveform and have used this for jump correlation, correlation across faults, and verification that reflections have been picked correctly. They have sometimes noted changes in the waveform and have correlated these with changes in the thickness of intervals, in facies, in the number and thickness of beds, etc. Character interpretation has been an art—a geophysicist with long experience in one particular area has learned the significance of certain pattern changes, whereas an equally good geophysicist without local experience would not (a) recognize the changes or, (b) be able to explain their significance even if recognized.

Today geophysicists more easily recognize changes attributed to subsurface conditions due to three factors:

1. Better recording and processing techniques, so that data are less distorted and less obscured by noise. Data must be recorded and processed so as to minimize nongeologic variations. Amplitude information and a wide band of frequencies must be preserved, the seismic waveform shortened and stabilized, multiple energy and other noise removed, and the data repositioned (migrated) without amplitude or waveform distortion. Variations which remain on a seismic record must be associated with variations in the subsurface to derive stratigraphic meaning.

2. Availability of more types of measurements made on seismic data, providing more ways of looking at the data. The arrival time of events is the most common measurement made on seismic data, followed by measurements of the amount of normal moveout (variation of arrival time with source-to-geophone distance) of amplitudes and, to a lesser extent, of the dominant frequency (or dominant period). An objective of this paper is to show how additional meaningful measurements can be made.

3. Better displays for communicating measurements to an interpreter so that he can see significant interrelations. Commonly we have too many measurements (position along the seismic line, arrival time, normal moveout, amplitude, dominant frequency, etc.) to permit an intelligible display on two-dimensional graphs (record-sections). Multiple displays are then used, such as record sections at different amplitudes, or sections using different stacking velocities, or those using different filters. Separate displays are also made of velocity analyses, spectral plots to show frequency content, and other quantities. The multiple display technique often makes it very difficult to relate measurements from one display to another. A second objective of this paper is to show how color displays help in this communication problem.

SEISMIC WAVES AS ANALYTIC SIGNALS

Seismic waves which we ordinarily detect and record can be thought of as an analytic signal with both real and imaginary parts, of which only the real part is detected and displayed. Another way of expressing this point of view is to call it a "time-dependent phasor" (Bracewell, 1965). This

---

[1]Manuscript received, October 1, 1976; accepted, January 19, 1977.

[2]Seiscom Delta Inc., Houston, Texas 77036.

Many contributed to this paper and to the work on which it is based. Special acknowledgment must go to N. A. Anstey, F. Koehler, and Seiscom Delta management.

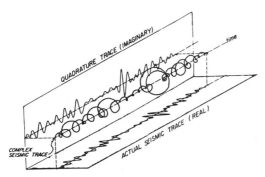

FIG. 1—Complex seismic trace as generated by a vector whose length varies with time, rotating as a function of time. The actual seismic trace is the projection of this vector onto the real plane and the quadrature trace is the projection onto the imaginery plane..

point of view looks on the observed seismic trace $g(t)$ as expressed by:

$$g(t) = R(t) \cos \theta(t).$$

The quantity $R(t)$ is the envelope of the seismic trace and $\theta(t)$ is the phase.

One can imagine a vector perpendicular to the time axis (Fig. 1) whose length varies as a function of time; this vector also rotates about the time axis as a function of time. The projection of the head of this rotating vector on the real plane gives the conventional seismic trace, $g(t)$. The head of the vector can be projected onto the imaginary plane to give the *quadrature trace*. The quadrature trace, $h(t)$, is expressed by:

$$h(t) = R(t) \sin \theta(t).$$

Hilbert transform techniques (Bracewell, 1965) permit us to generate the quadrature trace from the observed real trace so that both portions are available for analysis.

The seismic trace is usually a measure of the velocity of motion (with geophones) or of pressure variations (with hydrophones) which result from the passage of seismic waves. A seismic wave involves the moving of particles of matter from their equilibrium positions and thus involves kinetic energy. Hence the conventional seismic trace, $g(t)$, may be thought of as a measure of kinetic energy. The particle motion is resisted by an elastic restoring force so that energy becomes stored as potential energy. As a particle moves in response to the passage of a seismic wave, the energy transfers back and forth between kinetic and potential forms. The quadrature trace, $h(t)$, may be thought of as a measure of potential energy. Ordinarily, geophone output is proportional to the velocity of particle motion, which means that the kinetic energy is proportional to the square of the amplitude of the quadrature trace. Observed in this way, the quantity $R(t)$ in the above equations can be thought of as being proportional to the square root of the total energy of the seismic wave at any given moment.

REFLECTION STRENGTH AND INSTANTANEOUS PHASE

We can solve the foregoing equations separately for $R(t)$ and $\theta(t)$. We call $R(t)$ the "reflection strength" and $\theta(t)$ the "instantaneous phase:"

$$R(t) = [g(t)^2 + h(t)^2]^{1/2}$$
$$\theta(t) = \tan^{-1}[h(t)/g(t)]$$

These equations are solved for every sample point so that $R(t)$ and $\theta(t)$ have independent values at each point rather than being averages over a number of samples.

On arrival of the seismic reflection, the reflection strength first increases then decreases, thus a reflection is evidenced by a local maximum in the reflection strength. Polarity at the time of the maximum may be either positive or negative, depending on whether the reflection coefficient is positive or negative, how the interference of successive reflections affect the waveform, what conventions have been assumed in the recording-processing-display procedures, and what phase shifts have been introduced by recording and processing. If the magnitude of the reflection strength maximum is displayed in relation to the phase at the time of the maximum, polarity as well as the amplitude of the envelope will be illustrated. In subsequent figures (8, 19, 20), reflection strength is color-coded and superimposed on the real trace to accomplish this purpose.

Reflection strength may have its maximum at phase points other than peaks or troughs, especially where the reflection is the interference composite of several subreflections. Thus maximum reflection strength associated with an event is more meaningful than merely the amplitude of the largest peak or trough and reflection strength measurement differs from conventional amplitude measurement in a fundamental way.

Observing where (within an event) the maximum reflection strength occurs gives additional information. The color-coded reflection strength display provides a measure of reflection character. It is sometimes an aid, for example, in distinguishing reflections from massive reflectors and those which are interference composites. Reflections from massive interfaces tend to remain constant over a large region. Such reflections provide the best reference for smoothing or flattening data or for measuring time-thickness variations which might indicate differential compaction, local or regional thinning, facies changes, velocity variations, etc. Reflections which result from the interference of several separate reflections tend to vary along a seismic line as the thickness or contrast of the individual component reflector changes. Variations which are systematic with structure may indicate growth during deposition. Unconformities often show changes in reflection strength character as the subcropping beds change. This may be the indicator for unconformities which are otherwise difficult to detect. The quantitative aspect of reflection strength measurement may aid in the lithologic identification of subcropping beds if it can be assumed that deposition is constant above the unconformity (so all the reflection coefficient changes can be attributed to the subcropping bed). Seismic sequence boundaries tend to have fairly large reflection strength.

The instantaneous phase is a quantity independent of reflection strength. Phase emphasizes the continuity of events; in phase displays (shown later in Figs. 9, 13, 16, 18), every peak, every trough, every zero-crossing has been picked and assigned the same color so that any phase angle can be followed from trace to trace. Weak coherent events thus are brought out. Phase color displays use the color wheel such that $+180°$ and $-180°$ are the same color because they are the same phase angle. Such phase displays are especially effective in showing pinchouts, angularities, and the interference of events with different dip attitudes.

FREQUENCY

The time derivative of the instantaneous phase is called *instantaneous frequency*. Like instantaneous phase, it is a value appropriate to a point rather than being an average over some interval. The instantaneous frequency can vary abruptly, which is sometimes an advantage because abrupt changes do not get lost in an averaging process. It is also sometimes a disadvantage because there may be so many changes that the interpreter cannot comprehend them. (Instantaneous frequency is displayed in Figs. 10, 15).

It is useful to smooth frequency measurements; smoothing can be done in many ways, such as using "time windows" of varying shape and length in time. One particularly useful scheme is to use a weighting according to reflection strength, which produces "averaged weighted frequency." (This quantity is displayed in Figs. 11, 17, 21).

The above methods of determining frequency—both instantaneous frequency and weighted frequency—are different from the more familiar Fourier-transform methods in which the data over an appreciable length of trace are fitted with sine wave curves of different frequencies, amplitudes, and phase shifts. The amplitudes of the different frequency components are thus averages over the entire part of the trace being fitted, rather than values appropriate to a single instant in time.

In some areas frequency has been a good indicator of condensate reservoirs; these are associated with a characteristic low-frequency anomaly directly underneath them. Use of weighted frequency as an indicator of a condensate reservoir is an empirical relation based on a number of observations. The mechanism by which such zones attenuate high frequencies is not known.

POLARITY

The sign of the seismic trace (whether positive or negative) when the reflection strength has its maximum value is determined and called *polarity*. A magenta or blue color is assigned to suggest that the reflection coefficient (if the event indicates a single dominant interface) is positive or negative. The intensity of the color is varied according to the magnitude of the reflection strength (polarity displays shown, Figs. 12, 14).

USE OF COLOR SECTIONS FOR DISPLAY

An interpreter usually faces a major problem in assimilating large masses of data. The ability to extract more from the data compounds this problem by requiring more data for optimum comprehension and interpretation.

Color displays help show the significance of measurements and interrelations. Color effectively adds another dimension to comprehension; color-coded quantities can be superimposed on a conventional record-section plot so that both the conventional data and the color-coded quantity can be seen simultaneously, thus making interrelations easier to see.

Color has become increasingly important, though the literature on the subject is limited. Balch (1971) discussed the use of color seismic sections as an interpretation aid, and occasional advertisements in *Geophysics®* have illustrated limited use of color. Most uses of color on seismic sections have been either simple, using few colors, or difficult to reproduce. Seis-Chrome® displays of seismic attribute measurements have been used for several years but little has been published. The Seis-Chrome® process color-codes data retaining the fidelity of digital processing; a number is associated with each location, according to the attribute being displayed, and a color is assigned to each number or range of numbers. Usually a one-to-one correspondence is established so that each number is represented by a different color. A color key is commonly provided to show the numerical value which a color represents, and thus to permit quantitative interpretation of color changes. (The set of color keys usually used is shown in Fig. 7, and a diagramatic representation of the encoding is shown in Fig. 22.) Different or additional colors could be used; whatever color code is defined produces exactly the same color whenever the same number occurs.

The seismic measurements most commonly color-coded are: (1) reflection strength, (2) phase, (3) frequency, either instantaneous or weighted, (4) polarity, and (5) velocity. However, quantities displayed in color are not restricted to this particular assortment; other quantities such as dip, rate of change of dip, or cross-dip have also been displayed.

EXAMPLES AND GEOLOGIC MEANING

Geologic interpretation of color-coded displays of attribute measurements is illustrated with examples of five seismic lines. Interpretation of some features is shown on the conventional black and white seismic sections in Figures 2 through 6, where letters locate features referred to subsequently (color displays of these lines are shown in Figs. 8-22). All of the color sections were made by the Seis-Chrome® process and are copyrighted by Seiscom Delta Inc.

Ideally, displays of various attributes are available for each line because different displays bring out different features; there is useful information where the same feature is indicated, and also where different features are indicated. However, only selected displays are shown here because the intent is to illustrate their use rather than to make a complete interpretation.

Figure 7A illustrates the color code often used for reflection strength. The color steps indicate "dB" less than the maximum reflection strength on the section (or in the area, if a number of sections are being processed to allow meaningful line to line comparisons); thus "O dB" indicates the maximum amplitude. Since attention usually is focused on the strongest reflections, these are assigned the red end of the spectrum, although the color assignment is arbitrary and a different color assignment could be made.

Figure 7B shows the code commonly used for phase. The colors represent a color wheel, that is, $-180°$ is the same color as $+180°$, so that colors are continuous.

Figure 7C shows frequency in 2-Hz steps. Frequencies lower than 6 Hz, including occasional negative frequencies (when the phasor temporarily reverses its sense of direction), are left uncolored.

The bar graph of Figure 7D is used for polarity displays; it shows whether the phase is positive or negative at the points where reflection strength has a local maximum. The color hues are divided into five ranges according to the reflection strength. Where the sense of polarity is known, magenta is used to indicate a positive reflection coefficient and blue shows a negative one.

OFFSHORE LOUISIANA EXAMPLE

Figures 8 and 12 show data for the offshore Louisiana line shown in Figure 2; this line is 10 mi (16 km) long. These (and all of the following sections) are "squash plots," that is the data have been compressed horizontally so as to include greater length of line. The vertical exaggeration which results is often helpful in stratigraphic interpretation because many miles of data can be seen at a glance. The squash plot does, however, distort structure. For example, faults appear to be steeper than they are. A black and white gain-equalized plot (Fig. 2) forms the background of the color sections except for phase; a phase plot in black and white is sometimes used as a background.

Figure 8 shows reflection strength; several gas and oil accumulations are known to occur along this line. Gas accumulations often show as "bright spots," high-amplitude reflections, indicated by red and yellow colors (events A, B, C in Fig. 2). The shallow gas accumulations (A, B) are noncommercial and the bright spot at 1.35 sec, (C), corresponds to a commercial gas field. Below this gas field there is condensate production at 2.1 sec, (D). The amplitude anomaly associated with this condensate zone is not especially obvious although a flat spot attributed to the water level surface can be seen at about 2.2 sec.

The phase display (Fig. 9) emphasizes the continuity of events; it is especially useful where the signal to noise ratio is poor, although this benefit is not well-illustrated in this example. Discontinuities, faults, angular unconformities, pinchouts, zones of thickening and thinning offlap, onlap, interfering events, and diffractions stand out

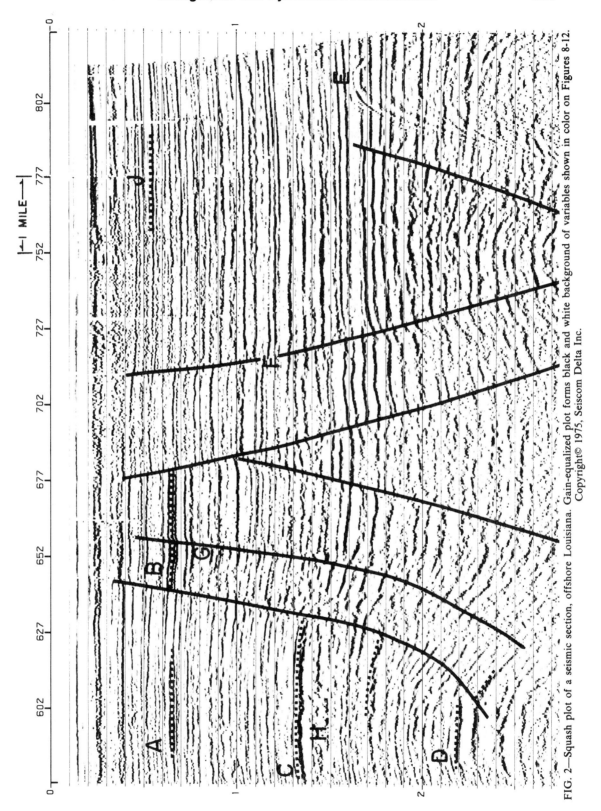

FIG. 2—Squash plot of a seismic section, offshore Louisiana. Gain-equalized plot forms black and white background of variables shown in color on Figures 8-12. Copyright© 1975, Seiscom Delta Inc.

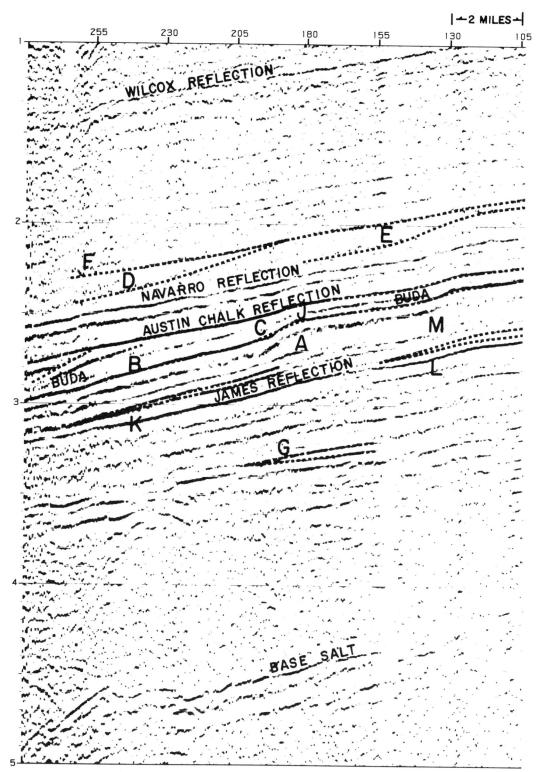

FIG. 3—Seismic section of north-south dip line, East Texas. Black and white section matches color Figures 13-15. Copyright© 1975, Seiscom Delta Inc.

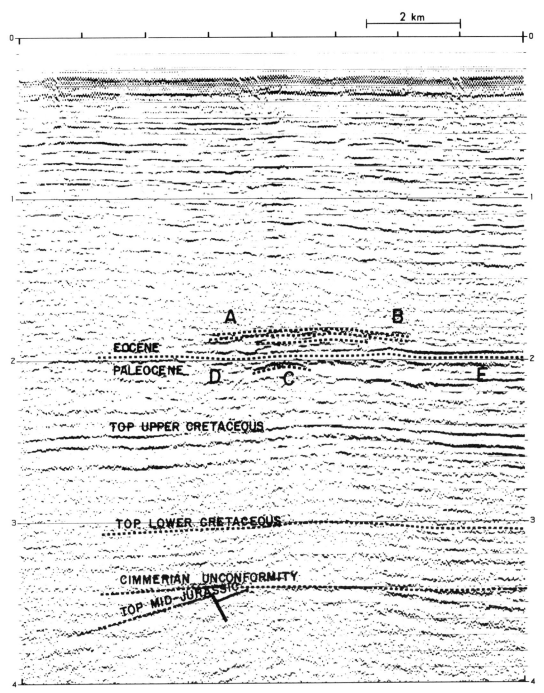

FIG. 4—Seismic section of line in North Sea. Corresponds to color Figures 16-17. Copyright© 1976, Seiscom Delta Inc.

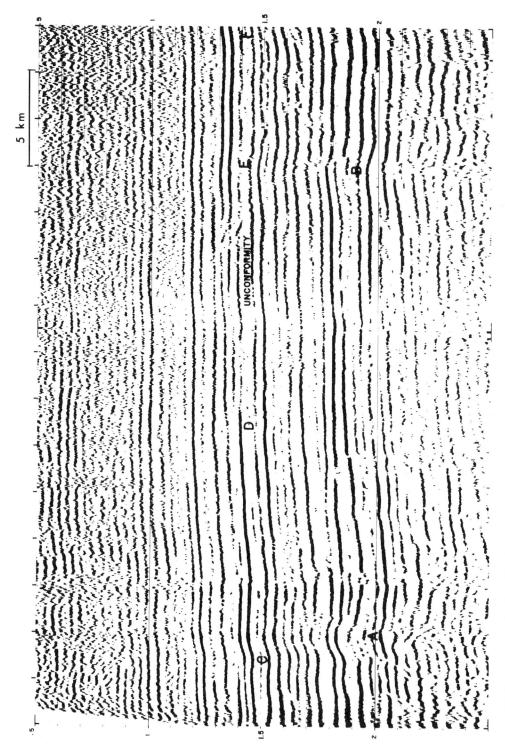

FIG. 5—Seismic section from line in Western Canada. Black and white corresponds to color Figures 18-19. Copyright© 1975, Seiscom Delta Inc.

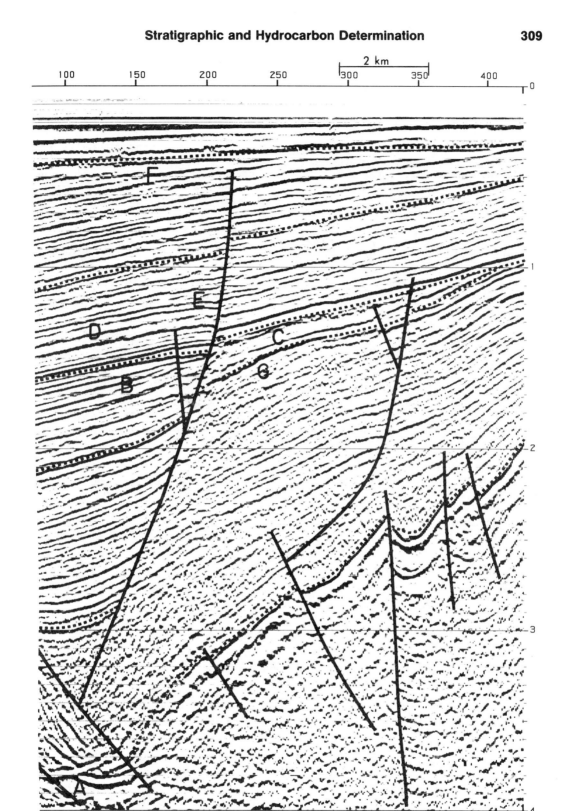

FIG. 6—Seismic section of line offshore Alaska. Interpretation is speculative due to lack of well control. Figure corresponds to color Figures 20-21. Copyright© 1975, Seiscom Delta Inc.

clearly in the phase display. The diffractions cresting at 1.65 sec at the right of the section (E) come from off (to the side of) the line and so do not indicate anything about the geology along this line.

Figure 10 shows instantaneous frequency. The color patterns help in correlating across faults (e. g., F). Lateral changes in the color pattern indicate that something has changed about the reflection. Such changes focus the interpreter's attention on the places where a change occurs even though the nature of the change may not be indicated.

Figure 11 shows the frequency weighted by reflection strength and then smoothed. The rapid variation of frequency which is sometimes distracting on instantaneous frequency displays is thus averaged out. Sharp shifts toward lower frequencies underneath the gas sandstones (G, H) indicate loss of high frequency in the gas zones. The condensate accumulation (D) shows as a distinctive low-frequency anomaly; such anomalies often characterize accumulations in this area.

Polarity for this line is shown in Figure 12. This display makes clear that the magenta bright spot at about 0.55 sec (J) is of a different kind than the blue bright spots associated with gas accumulations (A, B, C). The magenta bright spot indicates a positive reflection coefficient, an increase in acoustic impedance (perhaps an increase in calcium carbonate content), whereas gas accumulations usually have lowered acoustic impedance because of the lower velocity and density of the gas sandstones. The flat spot at 2.2 sec at the left of the section (D) appears to be associated with a positive reflection coefficient.

EAST TEXAS EXAMPLE

Figures 13, 14, and 15 show a north-south dip line in East Texas; this section is shown in black and white in Figure 3. Figure 13 shows phase. The outstanding feature on this line is the nonporous Edwards reef, the left edge of which is just to the right of the center at 2.6 sec (A in Fig. 3). The event which extends from 2.75 sec at the left edge to 2.3 sec at the right is the Austin Chalk reflection. Below the Austin Chalk to the left of the Edwards reef, the Woodbine Sandstones pinch out (B, C). The fore-reef Woodbine Sandstone produces gas in this area. The phase display clearly shows the prograding depositional pattern of the Woodbine.

The prograding pattern of the Midway section (D, E) is seen for about 0.2 sec above the Navarro reflection. Regions of onlap and offlap show so nicely in the phase display that it is helpful in picking seismic sequence boundaries. Such sequence boundaries include F and the Navarro, the top and base of the prograding Woodbine; the James, evidenced by the onlapping pattern above it as seen at K and L; and the onlap pattern at G.

Figure 14 is a polarity display. Note the change in polarity (change from blue to magenta) of the Austin Chalk reflection above the Edwards reef (J). To the right, the reflection coefficient is positive, associated with limy, high-velocity rocks, whereas to the left it is negative because of the lower velocity sandstone-shale Woodbine sequence. The changes in polarity as determined from the phase at peak reflection strength generally agree with the results from interval velocity calculations based on normal moveout along this line. Strong reflectors at the bottom of the section are interpreted as the base of the Louann Salt.

Mapping the time between reflectors (isotime maps) is an important interpretive tool to delineate geologic features. Its reliability deteriorates if the waveshape of the reflections being timed also varies. The lateral constancy of the color pattern of the polarity, reflection strength, or other measurements is helpful in selecting good reference reflections (where there is a choice) for such isotimes. For example, the lateral change of pattern of the James reflection makes it clear that isotime maps referenced to it will include the effect of variations which involve the James as well as variations elsewhere.

Figure 15 shows instantaneous frequency along this line. Note the distinctive pattern in the Edwards reef zone (A to M).

NORTH SEA EXAMPLE

Figure 16 shows phase for a line in the North Sea. A black and white copy of this section is shown on Figure 4. Production is derived from turbidite sandstones of Paleocene and Eocene age which piled up as mounds. The strong reflector at 1.95 sec, approximately the Eocene–Paleocene boundary, shows little structural relief. Turbidite sandstones (A, B on Fig. 4) appear to have been deposited on top of this interface although irregularities just under it (C) may indicate Paleocene turbidites. The turbidites have distinctive character in the phase display.

Figure 17 is the weighted frequency display of the same line. The zones of low frequency just beneath the turbidites are attributed to loss of higher frequencies in the gas accumulations. Thus the boundaries of this low frequency zone (D, E) may indicate the limits of the production. The low frequency anomaly is not uniform across the field, probably indicating variabilities in the accu-

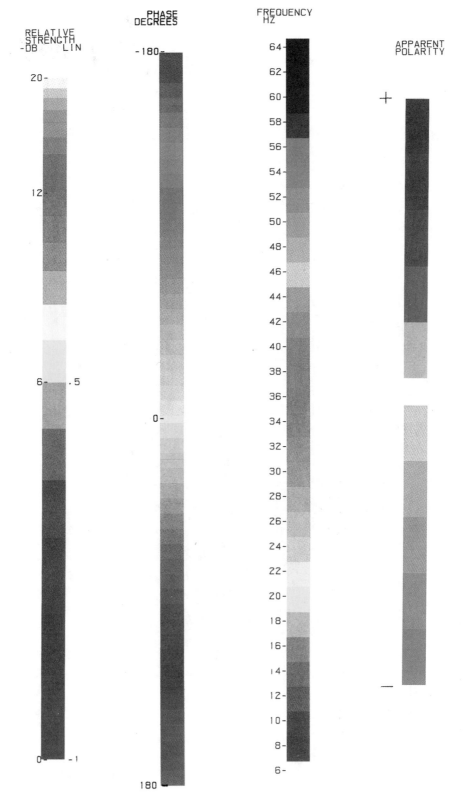

Figure 7 - Color Keys

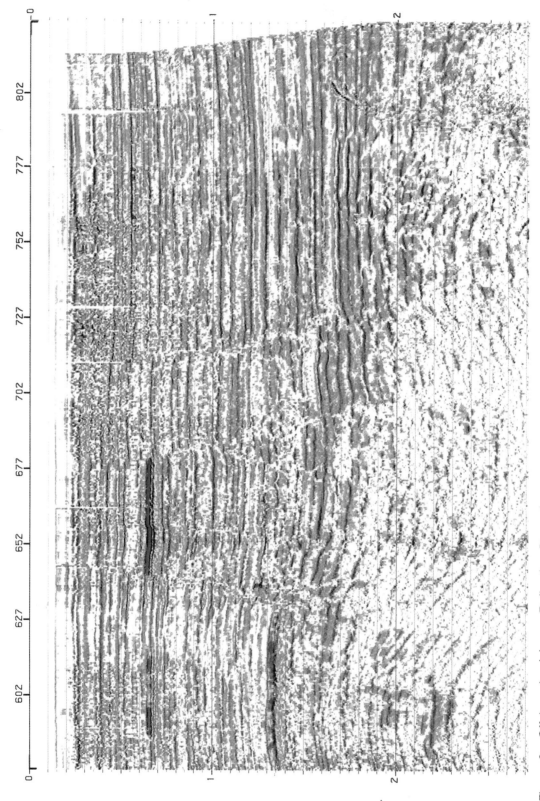

Figure 8 - Offshore Louisiana - Reflection Strength

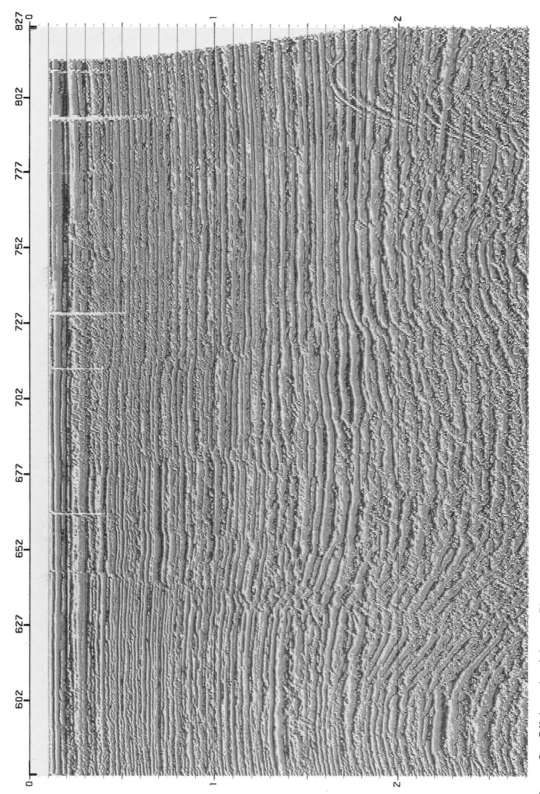

Figure 9 - Offshore Louisiana - Phase

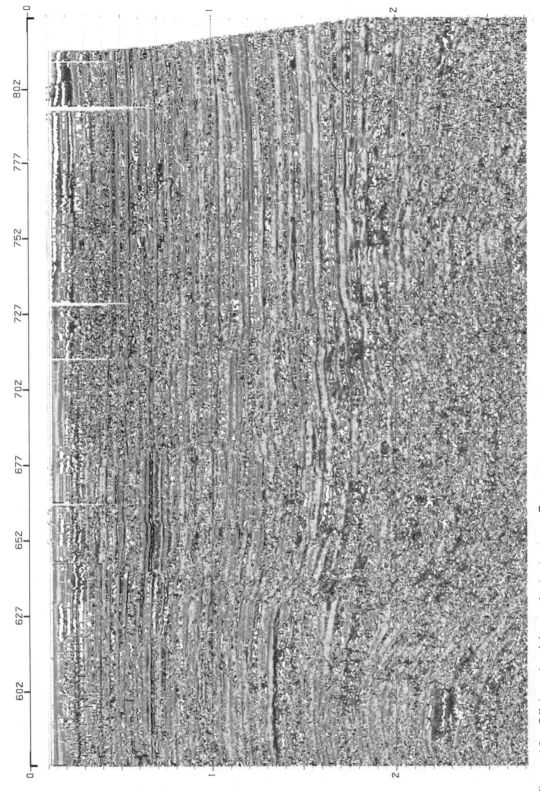

Figure 10 - Offshore Louisiana - Instantaneous Frequency

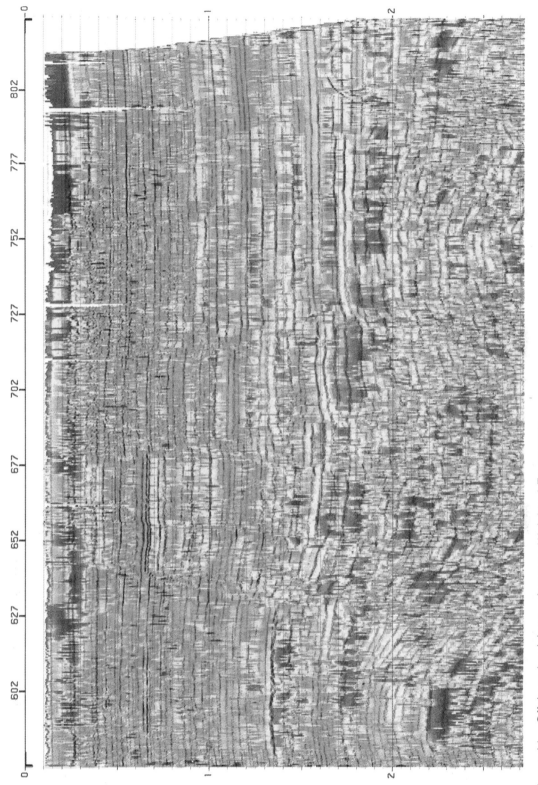

Figure 11 - Offshore Louisiana - Average Weighted Frequency

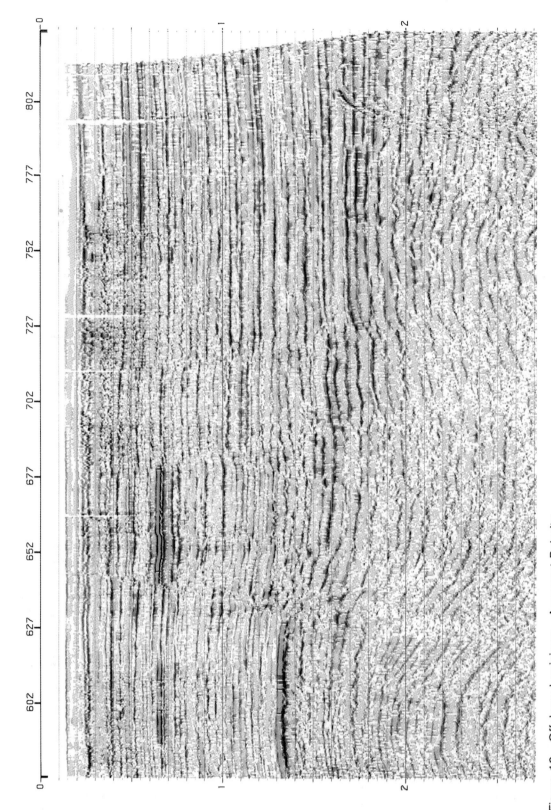

Figure 12 - Offshore Louisiana - Apparent Polarity

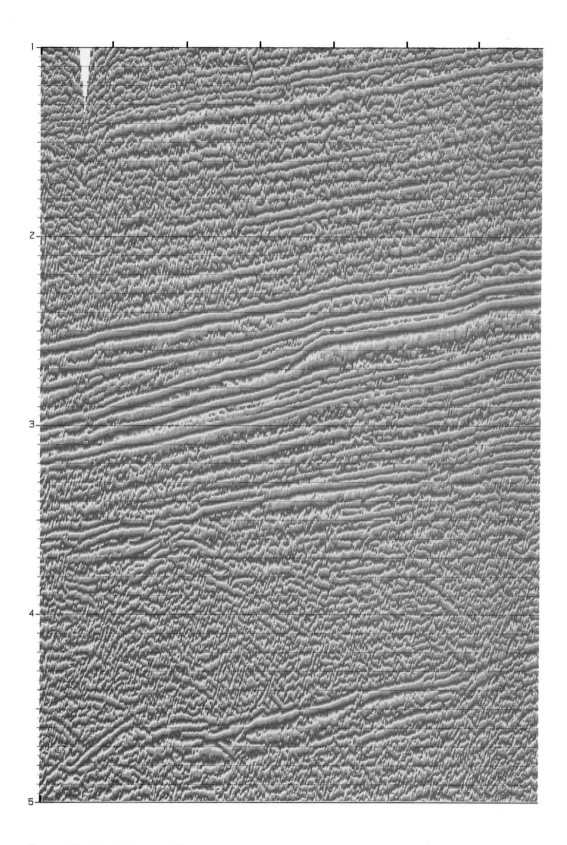

Figure 13 - East Texas - Phase

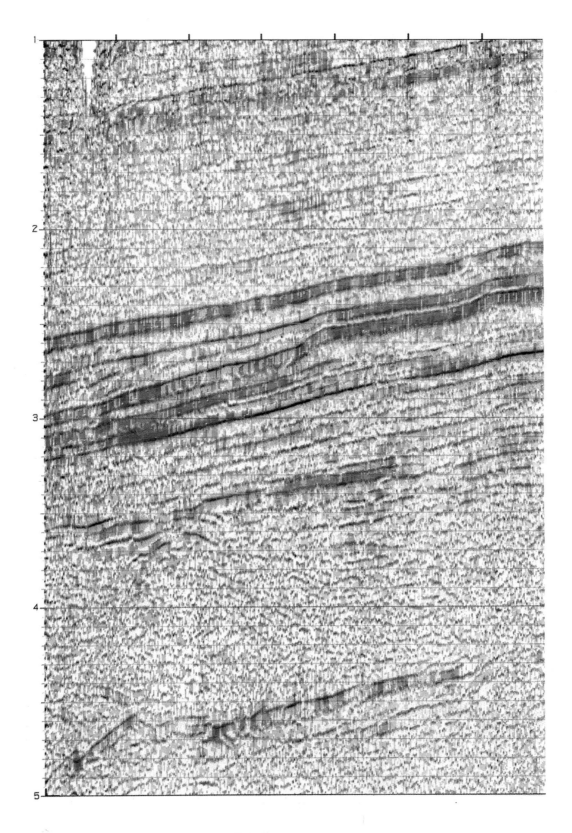

Figure 14 - East Texas - Apparent Polarity

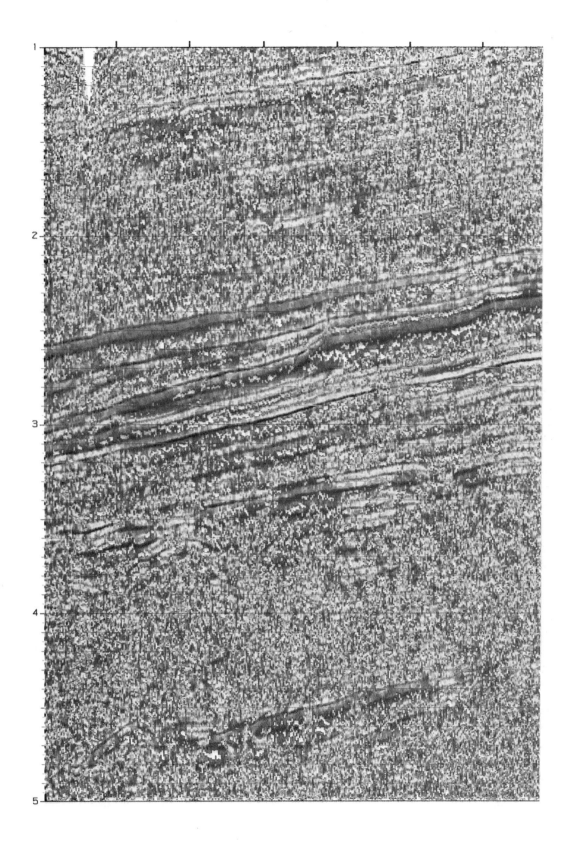

Figure 15 - East Texas - Instantaneous Frequency

Figure 16 - North Sea - Phase

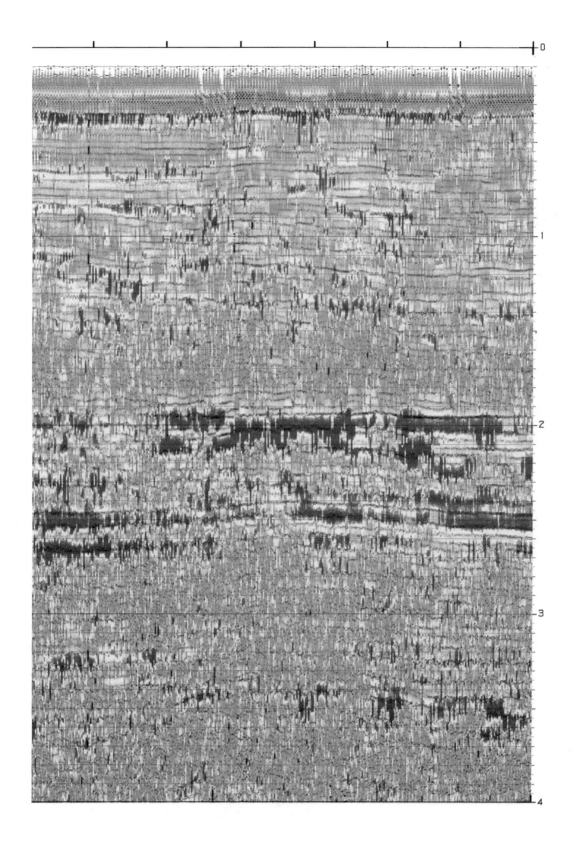

Figure 17 - North Sea - Average Weighted Frequency

Figure 18 - Western Canada - Phase

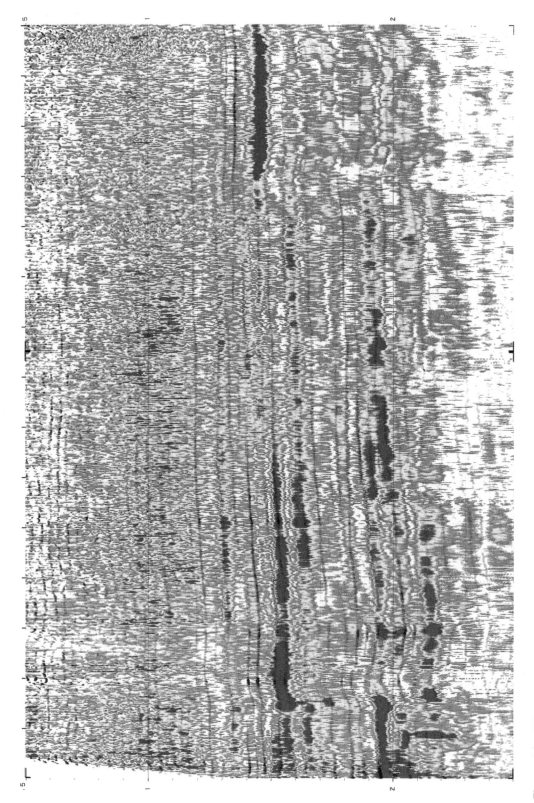

Figure 19 - Western Canada - Reflection Strength

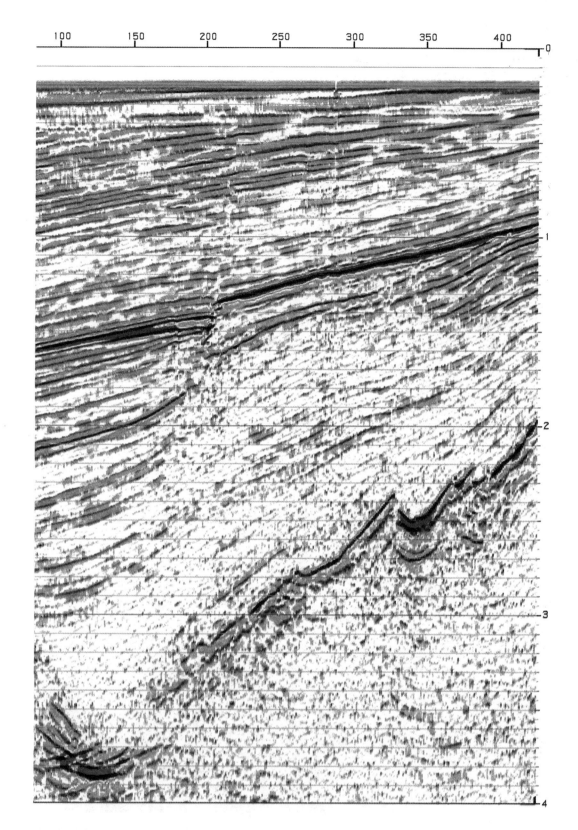

Figure 20 - Offshore Alaska - Reflection Strength

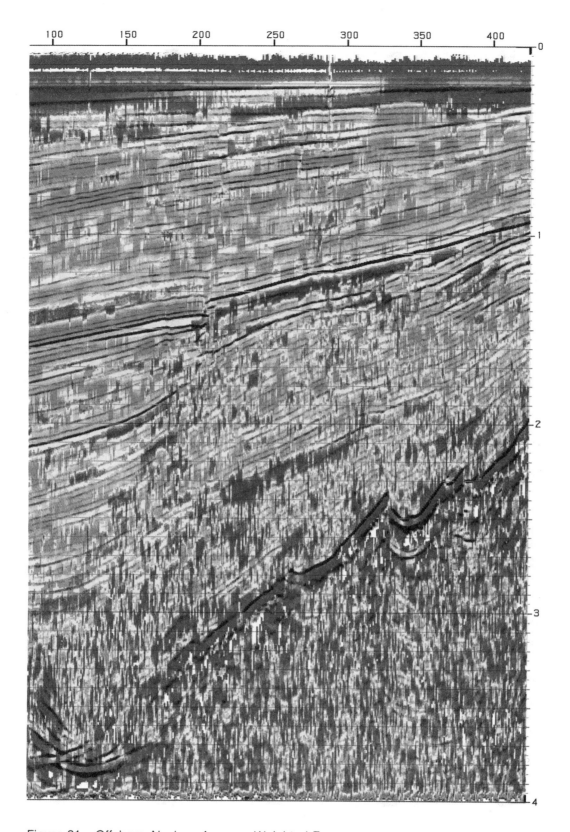

Figure 21 - Offshore Alaska - Average Weighted Frequency

Figure 22 - Attributes for Same Portion of One Trace

mulation. This line shows over 50 msec of structure at the top of the Cretaceous at about 2.4 sec.

WESTERN CANADA EXAMPLE

Figures 18 and 19 show phase and reflection strength for a seismic line across reefs in western Canada; this section is also shown on Figure 5. There is a reef just below 2 sec toward the left side of the section (A). Differential compaction over and off the reef has produced drape in the sediments for an appreciable distance above the reef. The reef is a weak reflector; it shows mainly as an interruption in the lateral continuity. Another smaller, less obvious reef is interpreted toward the right side of the section (B), similarly indicated by changes in lateral continuity of reflection strength. Reflection strength appears to be greater at the reef level on the seaward (left) side of the reefs.

An unconformity lies at about 1.55 sec at the left side of the section and 1.45 sec at the right side. A basal sandstone lying on this unconformity is absent in the regions where the reflection strength is particularly large (C to D and E to F). The bright spots indicating the absence of the basal sandstone appear to terminate abruptly above reefs (C, E), suggesting that differential compaction over the reef produced some relief on the unconformity surface at the time of deposition of these basal sandstones so that they are absent for some distance in the landward direction. The abrupt seaward termination of these bright spots thus provides additional evidence for the underlying reefs.

This bright spot interpretation could not be made without considerable well control in the area. Much more meaning can be obtained from seismic sections when well information can be combined with the seismic interpretation. Color-attribute sections extend well information to the surrounding region, such as that of mapping the distribution of this basal sandstone. (Likewise, noting the lateral extent of the color pattern measures changes from what is shown in the well.)

OFFSHORE ALASKA EXAMPLE

Figures 20 and 21 show a line offshore Alaska. A black and white representation of this line is shown on Figure 6, but this interpretation is speculative because there is no well control. These data have been migrated by wave equation migration which preserves amplitudes and reflection character. Migration has sharpened features and resolved buried foci and conflicting dips; however, there is appreciable cross-dip which has not been allowed for in the migration (which assumes that the data are two-dimensional). The interfering events at (A) differ in cross-dip.

Figure 20 shows reflection strength. The strong reflecting event extending from 1.6 sec at the left to 0.9 sec at the right is an angular unconformity. Change in polarity of the reflection across the fault is indicated by the red color being superimposed on a trough between the black lines of the background gain-equalized section (B) to the left of the fault, and on a peak (C) to the right of the fault.

Figure 21 shows averaged weighted frequency. Commonly entirely different features stand out in one display compared to another. Thus the strong unconformity downthrown (B) does not stand out whereas a shallower reflection (D) does. The lack of an upthrown counterpart to (D) in the frequency-character is attributed to rapid growth of the fault at the time of deposition. Confirming this interpretation, note the considerable downthrown rollover into the fault here and just above this event (E), which is nearly gone in the shallower beds, although the fault clearly extends much higher in the section. Several seismic sequence boundaries have visible effects on the frequency patterns (as at F, G), as well as the unconformity cited above (B, C). Sequence boundaries are most evident in phase displays (the phase display for this line is not included) but they also show effects on reflection strength, frequency, and other displays. More information is obtainable by using a set of displays of different attributes synergetically than by interpreting them individually.

APPENDIX: INTERRELATION OF ATTRIBUTES

An enlarged portion of a seismic trace (actually part of one of the traces shown on Figs. 3, 13, 14, 15) is shown in Figure 22 along with the color representations of the reflection strength, phase, instantaneous and weighted frequency, and polarity.

REFERENCES CITED

Bracewell, R. N., 1965, The Fourier transform and its applications: New York, McGraw-Hill, p. 268-271.

Balch, A. H., 1971, Color sonagrams—a new dimension in seismic data interpretation: Geophysics, v. 36, no. 6, p. 1074-1098.

Reilly, M. D., and P. L. Greene, 1976, Wave equation migration: 46th Ann. Mtg., Soc. Exploration Geophys., Houston.

Sheriff, R. E., 1976, Inferring stratigraphy from seismic data: AAPG Bull., v. 60, p. 528-542.

Taner, M. T., et al, 1976, Extraction and interpretation of the complex seismic trace: 46th Ann. Mtg., Soc. Exploration Geophys., Houston.

REFLECTIONS ON AMPLITUDES*

BY

R. F. O'DOHERTY** and N. A. ANSTEY**

Abstract

O'Doherty, R. F., N. A. Anstey, 1971, Reflections on Amplitudes, Geophysical Prospecting, 19, 430-458.

Modern seismic recording instruments allow precise measurements of the amplitude of reflected signals. Intuitively we would expect that this amplitude information could be used to increase our knowledge of the physical properties of the reflecting earth.

The relevant factors defining the amplitude of a reflection signal are: spherical divergence, absorption, the reflection coefficient of the reflecting interface, the cumulative transmission loss at all interfaces above this, and the effect of multiple reflections.

Of these factors, three—spherical divergence, the reflection coefficient and the transmission loss—are reasonably clear concepts (though the estimation of transmission loss from acoustic logs caused some difficulties in the hey-day of synthetic seismograms). Absorption still presents considerable problems of detail, but our understanding has increased significantly in recent years.

The factor least well understood is undoubtedly the effect of multiple reflections. Multiple paths having an even number of bounces can have the effect of delaying, shaping and magnifying the pulse transmitted through a layered sequence. Simple demonstations of this phenomenon can be made using elementary thin plates, and these can be presented for various synthetic and real sequences of layers. Such demonstrations lead one to explore the relation between the spectrum of the transmitted pulse and the spectrum of the reflection coefficient series.

If it were possible to isolate the amplitude and shape variations imposed by absorption within a layer, there would be a chance that this measure of absorption would be useful as a correlatable or diagnostic indication of rock properties. If it were possible to isolate the amplitude and shape variations imposed by multiple reflections, there would be a chance that this measure would be useful as an indication of cyclic sedimentation and of the dominant durations of the sedimentary cycles. However, the separation of these two effects constitutes a formidable challenge. The very difficulty of this separation suggests that it may be opportune to review the quantitative estimates of absorption made by field experiments.

Introduction

Why is it that in some areas we need 100 kg of explosive, while in others we need little more than a cap?

---

* Presented at the 32nd meeting of the European Association of Exploration Geophysicists, Edinburgh, May 1970.
** Seiscom Limited, Sevenoaks, Kent, England.

Part of the answer lies, of course, in the noise background. But there must be more to it than that; in the early part of the record, long before the amplitude of the reflection signal dies away to the noise level, we observe one amplitude decay rate in one area and a grossly different one in another. Why?

And why is it that no routine quantitative use is made of seismic amplitudes? Surely the amplitudes must be related to the geology in some meaningful way?

Indeed, was not this one of the principal considerations which led us to adopt binary-gain recording with such enthusiasm? What happened?

We cannot answer these questions fully. Nevertheless it seems opportune to study seismic amplitudes in some detail, and to take note of any features of the amplitude decay which might possibly be indicative of the geology.

We start with a review of the factors which determine the amplitude variations of the seismic reflection signal.

Factors Affecting Reflection Amplitudes

In this study we are concerned primarily with the variations in reflection amplitudes imposed by the subsurface geology. Thus we exclude amplitude considerations which merely define the *scale*—such factors as instrument sensitivity, source energy, and the geophone-ground coupling—and we assume a broad-band instrumental response. This allows us to define five major factors affecting the variations of amplitude: spherical divergence, interface reflection coefficients, absorption, interface transmission losses, and multiple reflection effects. These we discuss in turn.

1. *Spherical divergence*

The familiar law of conservation of energy, when applied to a spherical wavefront emanating from a point source in a uniform lossless material, tells us that the intensity diminishes as the inverse square of the radius of the wavefront (figure 1a). Translated into the type of measurements made in seismic work, this says that the pressure amplitude of the seismic wave is inversely proportional to the distance travelled. As always, we are grateful when nature produces a simple relationship.

But nature is just mocking us. The earth is not uniform, and in the presence of an increase of seismic velocity with depth the wavefronts are generally not spherical. Therefore the amplitude decay is subject to an additional effect associated with refraction (figure 1b).

For a representative case, the decay of amplitude due to spherical divergence is illustrated in figure 2. The overall decay of 50 dB is referred to a very early reflection at 0.1 s, and thus is appropriate only to a geophone close to the source; more distant geophones would record much less decay. In both cases we may well find that spherical divergence accounts for the majority of the

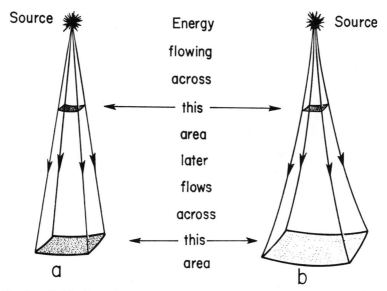

Fig. 1. The inevitable decay in amplitude associated with geometrical divergence (a) in a uniform material (b) in a material whose velocity increases with depth.

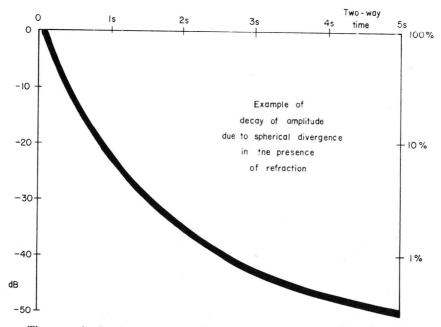

Fig. 2. The magnitude of the geometrical decay, in a typical case. The 0 dB level represents the amplitude of a supposed "first" reflection at 0.1s.

total decay observed on our records; we shall see later what are the circumstances under which this is so.

Spherical divergence in itself conveys no geological information, and so we hasten to compensate it. In the past, this has often been done by multiplying each sample by a factor proportional to the depth (or even the raw time) to which it corresponds; in the following material, however, we are assuming that the compensation also takes full account of refraction.

Even so, the compensation cannot be all we would wish. For example:

—In the usual case when primary reflections and long-period multiples may arrive simultaneously, but with different effective velocities, it is not possible to provide exact compensation for both. Proper compensation for the primaries ordinarily leaves the multiples too large.
—Similarly, compensation on the basis of horizontal layering ordinarily yields an excess of amplitude for reflectors showing strong dip.
—The law assumes a point source. Much seismic work nowadays is done with a source array; such arrays appear as a point source at low frequencies, but may be appreciably directional (showing a less-than-spherical loss) at higher frequencies. Properly, therefore, divergence compensation for such sources should include a frequency-dependent term.

The complications represented by these three items should not worry us too much. Basically, spherical divergence is an effect which is highly predictable and simply understood.

2. *Interface reflection coefficients*

As we know, a seismic reflection is generated at every geological interface across which there is a contrast of acoustic impedance. For present purposes, the acoustic impedance is represented by the product of density and velocity, or ρV. Then at normal incidence (and we must note this limitation) the pressure-amplitude reflection coefficient is given by the difference of the ρV values divided by the sum of the ρV values.

The reflection coefficient, we remember, is not a measure of the physical or geological properties of a layer, but only of the contrast of properties between two layers.

Within most sedimentary sequences, a reflection coefficient of $\pm$ 0.2 would be regarded as large. Values higher than this are observed, but (except near the surface) these values are unusual. Values of $\pm$ 0.1 are found in abundance, and lower values in profusion.

For seismic purposes, the geologic column is represented by a *reflection coefficient series* (or reflection coefficient log), identifying and quantifying the interface contrasts (figure 3). This is, of course, the conceptual germ of the synthetic seismogram.

If it were possible to isolate from the seismic reflection record the reflection coefficient series (with all the magnitudes correct), and if by an independent measurement we could establish the product of density and velocity in the first two layers, then in principle the acoustic impedance in every other layer could be computed. If interval velocities are known by other means, then layer densities could be deduced.

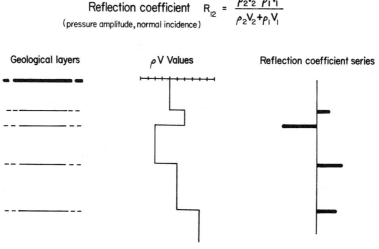

Fig. 3. The physical reality of a layer sequence (left) may be depicted in terms of its ρV log (centre) or its corresponding reflection coefficient series (right).

Therefore one of the long-term objectives of the seismic method must be the determination of the reflection coefficient series with all its magnitudes correct. The determination of the detailed *shape* of the reflection coefficient series is, of course, the objective of a spiking deconvolution process; now we ask also that all the reflection coefficient *magnitudes* should be correct. We can see immediately that we are unlikely to be successful (since, for one thing, our objective would require the complete removal of multiple reflections), and we can see also that use of it to obtain densities is subject to cumulative errors. However, it remains an objective.

3. *Absorption*

We have seen that spherical divergence, while it acts to diminish seismic amplitudes at distance, does not involve any loss of seismic energy—merely a spreading of it over a greater area of wavefront. Again, the processes of reflection and transmission at interfaces do not involve any loss of energy—merely a redistribution of it in the forward and backward directions. Absorption, however, is different; it diminishes seismic amplitudes, as a function of the distance travelled, by an irreversible conversion into heat.

This loss is known to be frequency-selective. A seismic pulse, representing a spectrum of frequencies, loses amplitude by a progressively greater absorption of its higher frequencies. In this sense, we note, the decay of amplitude introduced by absorption cannot properly be divorced from the change of spectrum.

Nowadays we accept that absorption in dry earth materials is related to a power of frequency very close to the first. This can be made eminently reasonable if we consider the seismic wave emanating from a sinusoidal source into a large homogeneous expanse of rock material. If we "freeze" the pattern of particle displacements at a certain instant, we see a succession of alternate compressions and rarefactions. The distance between successive compressions is a wavelength, at the frequency of the source and the velocity of the material. Then, because of absorption, we see a decay in the pressure amplitude from one compression to the next. So if we accept an absorption coefficient proportional to the first power of frequency, we accept, substantially, that this decay in acoustic pressure over each wavelength is a constant (which we might expect to be characteristic of the rock in its given environment). The proportional loss, over one wavelength, is substantially independent of frequency; it is therefore usually expressed in decibels per wavelength.

In deference to the theoretical workers, we should pause a moment to note that this simplified view ignores certain mathematical difficulties; nevertheless it seems to be sufficiently close to reality to warrant our using it for the present.

Let us illustrate the implications for a rock material having an absorption characteristic of 0.2 dB/wavelength and a velocity of 3 000 m/s, and let us consider a path length of 300 m in this material. At a frequency of 100 Hz this distance represents ten wavelengths. We expect each of the ten compressions to be 0.2 dB less in amplitude than the one before; thus the amplitude of the second compression is about 98% of that of the first, the third 98% of the second, and so on. At any other frequency the decay is likewise exponential. So, over the 300 m distance (corresponding to ten wavelengths at 100 Hz, or to one at 10 Hz) the loss is 0.2 dB at 10 Hz, 1 dB at 50 Hz, and 2dB at 100 Hz. Thus it is a simple matter to draw the effect of absorption on the spectrum of the propagating pulse; this is done in figure 4.

So we accept that decay of any sinusoidal component is exponential (in any one material), but what can we say about the amplitude of the composite pulse? Alas, very little; it all depends on the characteristics of the pulse which constituted the "input" to the absorbing earth—on the characteristics of the seismic source.

Under such circumstances our usual approach is to see what would happen to a pure spike input, and to reason onwards from that. So, adopting a spike input, we find that there are immediately two approaches we must consider. The first is appropriate to any circumstances in which we can fairly accept that we observe a single pulse in isolation (so that we can actually measure the amplitude of a selected peak of the pulse); the second applies when we see only a complex of overlapping pulses.

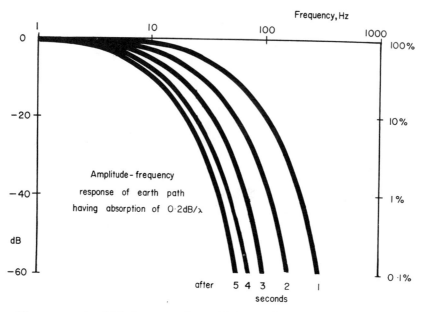

Fig. 4. The progressive high-frequency loss with increasing travel-time, illustrated for a uniform absorption of 0.2 dB/wavelength.

For a single pulse in isolation, two features affect the peak amplitude of the pulse: the peak amplitude decays as the higher frequencies are absorbed, and the peak amplitude decays as the pulse is lengthened by dispersion. The second effect occurs because the velocity of propagation of the sinusoidal components is slightly dependent on frequency, so that components which are in phase at the peak of the pulse early in its history are no longer exactly in phase at later times. To quantify the effect of dispersion on the peak amplitude of the pulse we must make some assumption about its magnitude; the assumption which is both convenient and physically reasonable is the minimum-phase assumption. (For a rudimentary account of minimum-phase behaviour, dispersion, absorption and other matters related to the present discussion, see section 2.3.12 and chapter 3.1 of volume 1 of Evenden, Stone and Anstey, 1970. For a more advanced account, including some practical evidence, see O'Brien,

1969). On this basis, we can display the pulses which correspond to the amplitude spectra of figure 4, and see the effect of absorption on their peak amplitudes; this is done in figure 5.

As we can see, the decay is very rapid at early times; the flat spectrum of the spike means that a considerable proportion of the spike energy is carried at the high frequencies, and this is quickly lost.

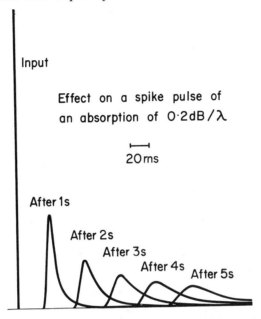

Fig. 5. The absorptive effect of fig. 4 translated into the time domain, on the assumption of minimum phase.

So we must accept the relevance of the source characteristics—if the source does not emit high frequencies, this rapid collapse of amplitude does not occur. Thus when we said earlier that in some practical cases spherical divergence accounts for the majority of the observed decay, we can guess that this indicates a *low-frequency narrow-band source*. There is nothing we can say about the decay of peak amplitude with time, until we know the characteristics of the source pulse (Gurvich and Yanovskii, 1968).

When we are not concerned with a specific pulse observed in isolation, but with a complex of overlapping pulses, one aspect of the problem changes. Manifestly, the broadening of the pulse by absorption and dispersion must increase the chances of overlap; manifestly, also, the resultant amplitudes may be increased or decreased by the overlap, according to the reflector signs, the reflector spacing and the pulse shape. To obtain a useful generalization we must go all the way to a reflection coefficient series having close but random reflector spacing (so that all reflected pulses overlap several or many times), and then look at *average* conditions within a window. Under these circumstances, clearly, dispersion loses its significance; it yields a broadening which is additional to that produced by absorption, but which, unlike the latter, does not involve a loss of energy. Looking at a window containing a random superimposition of pulses, we see dispersion merely as a phase effect which modifies the shape of the composite waveform without changing the energy evident in the window.

We have said that a knowledge of the source characteristics is necessary before we can calculate the amplitude decay due to absorption. The source characteristics, of course, are certain to be different from one source to another, and may also be different from one shot to another. Although this is really all that can be safely said, one would feel guilty about abdicating the discussion on such an unsatisfactory note. So let us consider at least one specimen case;

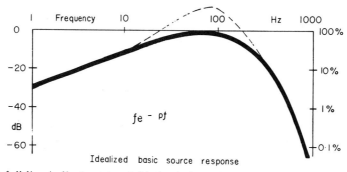

Fig. 6. The full line indicates a tractable basic form for the spectrum of the pulse generated by a small explosive charge in a competent material; normally a resonant peak (such as that shown dashed) is superimposed on this basic form.

for example, let us consider some reasonable form of source spectrum such as that illustrated by the heavy line in figure 6. Then we can show that this basic curve dictates an average amplitude decay, in the presence of absorption, which follows slightly less than a $-3/2$ power of travel time.

In practice most sources exhibit a response more peaked than this (such as the one shown dashed in figure 6); obviously the effect of the peaked response superimposed on the basic curve is to modify the decay, to a greater or less extent.

We can summarize our preparatory discussion of absorption thus:
— It is reasonable to accept, at the present stage of our knowledge, that absorption varies very nearly with the first power of frequency.
— This means that the loss in decibels over a fixed distance in a single medium is proportional to frequency, or that at a single frequency the loss in decibels is proportional to travel time.

—The effect of this loss on the amplitude of the propagating pulse cannot be established until the source characteristics are known.
—As a generalization of the last item, the decay appearing as a result of any frequency-selective effect which is progressive with travel-time must depend also on the constant frequency-selective effects (source, detectors, instruments, filtering) along the path of the signal.
—It is probably reasonable to expect that the absorption mechanism exhibits minimum-phase behaviour.
—The decay of amplitude of a single pulse observed in isolation is slightly greater than that of the average amplitude of a profusion of overlapping pulses.
—Only by chance could the pulse amplitude decay conform strictly to the popular exponential. (We note in passing that this does not exclude the use of exponential compensations for particular purposes. Before dereverberation, for example, we may be forced to use them; after dereverberation we may remove their effect and then apply a better correction if we know one.)

4. *Interface transmission losses*

Again invoking conservation of energy, we know that energy reflected from an interface is not available to be transmitted through it. Clearly, the larger the reflection coefficient, the greater is the transmission loss. We shall need to employ this again later, so let us take particular note of it: More up, less down.

The relationships between the reflection and transmission coefficients are depicted in figure 7. Clearly the transmission loss is unaffected by the sign of the reflection coefficient. Setting aside spherical divergence and absorption for the moment, we can see that the amplitude of a seismic reflection is the product of its own reflection coefficient with the product of all the two-way transmission coefficients of the interfaces above it.

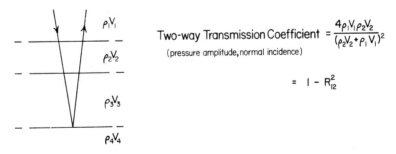

Amplitude of reflection from third interface = $R_{34}(1-R_{12}^2)(1-R_{23}^2)$

Fig. 7. The relationship between the transmission and reflection coefficients.

Are transmission losses a major effect, or a minor one? Intuition is not a very good guide on this question, and it helps to have before us some illustrative values. Figure 8 gives the two-way transmission loss as a function of the number of interfaces for reflection coefficients of ± 0.05, ± 0.1 and ± 0.2.

Our first conclusion is that the transmission loss associated with a single reflector—even a strong one—is virtually insignificant. We would expect a handsome reflection from a reflection coefficient of 0.2, and we might feel that there would be a major diminution of reflections below it; we see, however, that such reflections are diminished by only 0.4 dB, or 4%. The corresponding diminution introduced by a reflection coefficient of 0.05 is 0.02 dB (0.2%) which

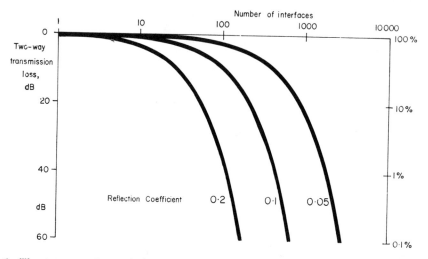

Fig. 8. The two-way transmission loss through a number of interfaces, for a range of reflection coefficients.

is in itself negligible. So if we think of the crust of the earth in terms of a few major layers (representing the coarsest division into geological epochs), we are led to expect quite small transmission losses.

The second conclusion, however, is that the cascaded transmission loss through a great number of interfaces is certainly not negligible. In particular, a large number of individually insignificant interfaces can have at least as great an effect as a few major ones. And, since we know from our first glance at almost any outcrop that the earth's stratification can be very fine, we realize that we have to start thinking rather carefully about geology before we can assess the true significance of transmission losses.

In particular, we find we have to make an immediate distinction between two extreme types of stratification, and the sedimentary processes which give rise to them. We illustrate these in figure 9, using artificial acoustic logs. Both logs show the same systematic increase of acoustic impedance with depth; however, the upper one implies a profusion of thin layers tending to alternate in their ρV values, while the lower one implies slow and progressive variations of ρV value. As geophysicists we would describe the upper log as high-frequency, the lower one as low-frequency. The more fundamental description, however, must be the geological one: we think of the upper log as representing thin layers laid down by a *cyclic pattern of sedimentation* which systematically tended to interleave high-velocity and low-velocity materials, while we think of the lower log as representing a *transitional pattern of sedimentation* which systematically tended to make steady gradations of velocity within basically thick layers.

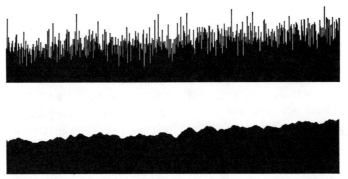

Fig. 9. Artificial acoustic logs prepared to illustrate the geophysical significance of cyclic layering (upper) and transitional layering (lower).

We shall use the terms *cyclic* and *transitional* to describe these two types of layering; in doing so, however, we note that our present concern is with the *acoustical* properties of the layering, and that cyclic sedimentation in this sense need not correspond exactly with cyclic sedimentation in the geological sense.

The relevance of the distinction between cyclic and transitional layering, in the present context, is brought home to us when we compare the transmission loss for the two cases of figure 9; for the lower log it is quite insignificant, while for the upper log it is more than a thousand times as great, and certainly significant.

This happens, of course, because the reflection coefficients in the lower case are smaller, so that the transmission loss is also smaller. But, within the constraints on velocity and density known to exist in the real earth, this is inevitable for a transitional log; large reflection coefficients and large transmission losses can be maintained only if the large reflection coefficients tend to alternate in sign.

Thus we see several clear situations. We see the situation of a massive layer which contains no significant reflectors and therefore contributes no significant transmission loss. We see the situation of a single interface having a large reflection coefficient and a minor transmission loss. We see the situation where this interface is transitional instead of abrupt; in this case the progressive transition makes the transmission loss even less significant—just as an acoustic horn matches between low and high impedance, and so reduces the loss. And we see the situation where thin layers tend to alternate between high and low ρV values, and so provide (if there are many of them) the possibility of large transmission losses.

So, in hopes that we may be led to a technique for distinguishing between transitional and cyclic geology, we ask: How can we assess the magnitude of transmission losses in the real earth?

This proves an unexpectedly knotty problem. It is indisputable that, over a given up-and-down path in the earth, there must be a definite and altogether real transmission loss. But the obvious way to obtain a measure of it—from a velocity log—proves to be full of difficulties.

First, we observe that a velocity log taken with a 1 m receiver spacing is much more active than one taken with a 2 m spacing. This reminds us that a velocity log does not identify (except in a blurred sense) layers having a thickness significantly less than the receiver spacing; the transmission loss computed from a cyclic log at 1 m spacing is greater than that obtained with a 2 m spacing. How far does the effect go? Geologically, we feel that, although very fine layering obviously exists, reflection coefficients between very thin layers are likely to be small. But the potential number of very thin layers is enormous.

Second, we know that not every wiggle on a velocity log represents a corresponding formational change; errors associated with the borehole are inevitable, and the difficulties of compensating these increase as the receiver spacing is reduced.

Third, if we attempt our transmission-loss evaluation digitally, we must be careful to ensure a proper sampling interval. Obviously much of the early implementation of synthetic seismograms was in violation of this; probably this did not matter too much for those applications, but it is essential in any attempt to evaluate transmission losses.

Fourth, we have to think about the nature of geological "interfaces". Sometimes, for sure, they are real discontinuities, properly represented by the simple equation; others may be very smooth gradations (caused, for example, by variations in porosity with progressive changes of grain size) within which the transmission loss is virtually zero.

Finally we have to ponder the acoustics of the situation. The cited equation for transmission loss applies to plane waves, and we usually side-step this by restricting ourselves to the far field, where our spherical waves are almost plane.

But if we are concerned with very thin layers this may not be defensible (see, for example, Hagedoorn, 1954).

So we have to admit that, although the concept of a transmission loss in the real earth is a clear one, we are not well placed to assess the magnitude of the loss.

If we do the best we can, with the logs available today, we emerge with what appears to be a ridiculous result. As we have said earlier, a record from a narrow-band low-frequency source, properly compensated for spherical divergence, shows very little decay attributable to transmission loss—at most a few decibels per second. But values of transmission loss computed from velocity logs often work out at 40-50 dB/s—a figure which, if real, would mean that the seismic reflection method could not possibly work as it does.

So something is wrong. Probably part of the reconciliation lies in the considerations set out above, but part—the greater part—must involve multiple reflections. Just as it is physically ridiculous to think of reflection without multiple reflection, so it is improper to consider transmission without multiple reflection.

5. *Multiple reflection effects*

For many of us, the first intimation that multiple reflections affect the amplitude of "primary" reflections came with the introduction of digitally-generated synthetic seismograms. In those days it was customary to calculate at least three synthetic traces: primaries without transmission losses, primaries with transmission losses, and primaries with all multiples and with transmission losses. Just as we have seen above, the second of these—primaries with transmission losses—usually decayed away to nothing so quickly as to be useless. The most obvious effect of including the multiples was to increase the amplitude of the *primaries*, sufficiently to offset most of the transmission losses.

The explanation—that primary paths are systematically reinforced by very-short-delay "peg-leg" multiple paths—was given and developed by Anstey (1960), Trorey (1962), d'Erceville and Kunetz (1963), Bois and Hemon (1963), Bois, Hemon and Mareschal (1965), Delas and Tariel (1965), Mikhailova, Pariiskii and Saks (1966), and Berzon (1967).

In retrospect, we can see that the explanation was always implicit in the classical acoustic exercise, "the case of the thin plate." In figure 10a we see the basic situation: the direct transmitted signal is followed after a short delay by a 2-bounce multiple reflection whose amplitude, referred to the direct transmitted signal, is just the product of the upper and lower reflection coefficients. The most important feature of this situation is that the sign of the multiple reflection is *always the same as that of the direct signal*; the reflection coefficients are opposite in sign when viewed from above, and one of them is in fact viewed from below—so the product is always positive.

Our first reaction is to see whether we are looking at a major effect or not. Quickly we note that if both reflection coefficients have a magnitude even as high as $\frac{1}{2}$, the multiple reflection has an amplitude of only $\frac{1}{4}$ of that of the direct signal. For more realistic magnitudes, the multiple is very small. However, before we discard the effect as insignificant, we should consider the case of Figure 10b, where we postulate that, somewhere along its path, the direct signal

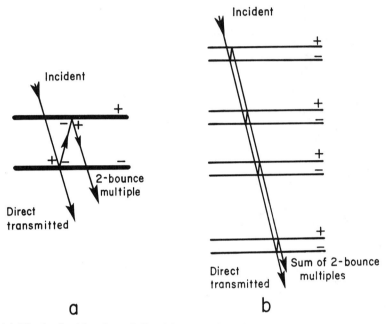

Fig. 10. (a) The basic thin plate, defined between interfaces having reflection coefficients of opposite sign.
(b) The cumulative effect of the multiple reflections from four such thin plates.

encounters four such thin plates. Then if the reflection coefficients are again $\frac{1}{2}$, we can see that the composite multiple reflection is now *equal to the direct signal*. When we include the return path through the same sequence, the composite multiple reflection has double the amplitude of the direct signal.

If the reflection coefficients are of magnitude 0.1, then it takes 50 thin plates before the multiply-reflected signal becomes equal to the two-way direct signal, but the conclusion is eventually the same: *the multiply-reflected signal in a series of thin plates bounded by interfaces of opposite polarity is always of the same sign as the direct transmitted signal, and tends to overtake it in amplitude*. At this stage we can see, in a gross sense, that in a cyclic sequence of layering—a succession of thin plates—there is an inbuilt multiple-reflection mechanism which acts to compensate the large transmission loss that would otherwise occur. Clearly, we are now concerned to determine the *degree* of this compensation: will it be so great that we lose all hope of distinguishing between cyclic and transitional layering by their effect on amplitudes?

Basically, this is a matter of geology; so, let us return for a moment to the geological matters broached in the last section. In particular, let us return to the question: Do geological considerations lead us to expect any connection between the thickness of a layer and the reflection coefficients at its boundaries?

The thick layers, we have said, often represent geological epochs. In the nature of things, there can be only a few of them. The interfaces bounding them may well have large reflection coefficients; the chances are that both are positive.

At the other end of the scale, we have guessed that the very thin layers are likely to be bounded by very small reflection coefficients—we can scarcely conceive large reflection coefficients separated by a matter of centimetres. If the reflection coefficients are small, we can allow the possibility of sequences of transitional layering, while still keeping the overall variations of acoustic impedance within observed limits. We as geophysicists cannot say whether transitional or cyclic layering is the more likely; that is a question for the geologists.

Between the very thin and the very thick layers there must be some range of thicknesses where appreciable reflection coefficients first become possible. Over and beyond this range, the observed limits on acoustic impedance mean that *sustained cyclic layering is more likely than sustained transitional layering*.

So these considerations lead us to expect a middle range of layer thicknesses, bounded by significant reflection coefficients tending to be opposite in sign.

If these guesses have any foundation, we are immediately at odds with one of the favourite assumptions in the theory of seismic processing. Often, in processing, we invoke the assumption that the reflection coefficient series is a train of spikes of random amplitude, spacing and polarity; now we are saying that the *earth's stratification is the result of natural laws, that these provide some predictable constraints, and consequently that the outcome is not completely random.*

Is there some simple check we can make, to resolve the issue? Yes, there is; the assumption of randomness usually expresses itself as an assumption that the auto-correlation function of the reflection coefficient series is a simple spike (so that the auto-correlation function of the ideal seismic record is the auto-correlation function of the seismic pulse)—we can check whether this is true.

Let us try it first on the two synthetic examples of figure 9. Figure 11b is the auto-correlation function of the reflection coefficient series corresponding to the cyclic log of figure 9a; figure 11a is that corresponding to the transitional log of figure 9b. Clearly *neither* is a simple spike. In particular, we note that the first few values of the transitional auto-correlation function are all positive, whereas the second value of the cyclic auto-correlation function swings strongly negative. We shall see the significance of this in a moment.

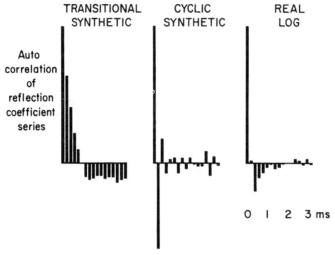

Fig. 11. The auto-correlation functions corresponding to (left) an artificial transitional log, (centre) an artificial cyclic log, and (right) the real log of fig. 12.

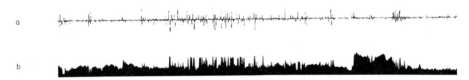

Fig. 12. The reflection coefficient series (a) corresponding to a segment of a real acoustic log (b).

In figure 11c we see the auto-correlation function of a "real" reflection coefficient series. In fact, the reflection coefficient series is that shown in figure 12a, and the real velocity log from which it was derived is shown in figure 12b. The derivation was made on the usual assumption that density variations can be neglected; any case we can demonstrate under this restriction is likely to remain essentially sound when the restriction is removed. The log represents about 1400 m of depth, and 0.343 s of one-way time. The original sampling, before conversion to time, was at 0.765 m of depth, chosen to be less than the receiver spacing of the logging tool. To the eye, the reflection coefficient series could well be random.

But the auto-correlation function (figure 11c) says no. We can see the spike at zero lag, of course, and the values for lags beyond about 2ms are compatible with randomness, but between zero and 1 or 2 ms the values have a different message. They are telling us that, for layer thicknesses up to about 10 m, the stratification shows systematic deviations from randomness.

If we recall the actual operations used in the construction of the auto-correlation function—shifting, multiplying, and adding—we can see very easily that the auto-correlation function of a reflection coefficient series has a direct physical significance. The value at the first lag represents the sum total of all the 2-bounce multiple reflections occurring in layers of one unit of thickness. The value at the second lag represents the sum total of all the 2-bounce multiple reflections occurring in layers of two units of thickness. Similarly, values at higher lags represent 2-bounce multiples of longer periods. If the value is positive, it means that the sum total of all the corresponding 2-bounce multiples is of opposite polarity to the direct transmitted signal; if it is negative, the multiples reinforce the direct signal.

So the auto-correlation in figure 11c is confirming our geological guesses about the likely relation between the thickness of a layer and the magnitude and signs of the interfaces bounding it.

It is true that the equivalence is not exact; our earlier thinking regarded a layer in the sense of a geological entity, whereas figure 11c regards a layer as defined by *any* pair of interfaces. Further, the physical interpretation of the auto-correlation function in terms of 2-bounce multiples ignores the transmission loss in all intervening interfaces. Nevertheless the auto-correlation function suggests two conclusions for the particular log under study:

—Layers of very small thickness (typically about 1 m or less) show a weak tendency to be of transitional type, being bounded by interfaces of like sign. The small magnitude of the first lag value could be due either to a fortuitous offsetting of transitional and cyclic effects, or—and this seems more likely—it could be confirming our guess that very thin layers are likely to be bounded by small reflection coefficients.
—Layers in the thickness range 1-10 m tend to represent cyclic changes, being bounded by interfaces of opposite sign. This is a clear and positive effect.

So the evidence, at this stage, allows us to say that the earth may contain:

—A very large number of very thin layers, whose boundaries have small reflection coefficients and introduce small transmission losses; these losses may be increased somewhat (if the layers are transitional) by multiple reflection effects.
—A smaller (but still large) number of less thin layers, whose boundaries tend to have appreciable reflection coefficients but to be of opposite sign; the

very large transmission losses to be expected in this case tend to be offset by multiple reflection effects.

—A small number of thicker or much thicker layers, whose boundaries tend to have large reflection coefficients; these interfaces give rise to the reflections we see on normal records, but their comparatively small number means that the transmission losses introduced by them are minor.

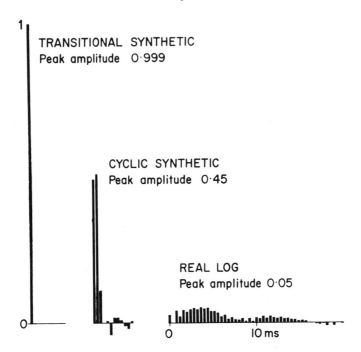

Fig. 13. The transmission response (that is, the impulse response of the two-way transmission path) for (left) the artificial transitional log, (centre) the artificial cyclic log, and (right) the real log of fig. 12.

Before we leave the auto-correlation function, we should note this interesting fact: We can manipulate the positive-lag auto-correlation function to approximate the actual time form of the 2-bounce multiple reflections by removing the zero-lag value, by reversing the sign of all other values, and by doubling the time scale to yield two-way multiple reflection times.

Although this gives us an easy way of assessing whether a particular reflection coefficient series will systematically reduce its transmission loss by its 2-bounce multiple reflections, it leaves us wondering what is the effect of the 4-bounce and higher-order multiples. For this, and to eliminate the approximation involved in neglecting transmission loss within the multiple part of the path, we must go through a complete ray-tracing process to find the form of the complete transmitted signal. It makes good sense to do this for two-way transmission; then we say that we will inject a spike pulse at the top of the layered sequence represented by the log, and calculate the form of a pure isolated reflection as it would be after two-way transmission down and back. We do this in figure 13a for the synthetic cyclic log of figure 9a, in figure 13b for the synthetic transitional log of figure 9b, and in figure 13c for the real log of figure 12. (The techniques for this type of calculation have been given by several previous workers; see, for example, Baranov and Kunetz, 1960; Trorey, 1962.)

As we expect, the spike pulse is scarcely changed, either in amplitude or in form, by transmission through the transitional sequence (figure 13a). In fact there is a small tail added (Bortfeld, 1960), but for a log of the limited extent involved here it is too small to be significant. The change of amplitude due to transmission loss is from 1 to 0.999.

The cyclic sequence, however, produces a much more marked effect (figure 13b). The transmission loss (that is, the diminution of the first point of the transmitted signal) is from 1 to 0.44. A significant positive tail is added, extending to three points. The sum of the amplitude values for the first three points is 0.994. This, obviously, is very interesting; it is telling us that the decrease of amplitude caused by transmission is at least partially compensated by multiple reflection—at the expense of a smearing-out over time. We begin to sense that there will be great difficulty in distinguishing between transitional and cyclic layering by studies on amplitudes alone.

Our greatest concern, of course, attaches to the real log, for which the two-way transmitted pulse is given in figure 13c. The change of amplitude due to transmission loss is from 1 to 0.027, which obviously cannot be the effective value. A very significant positive tail, extending to some 16 ms, is added by the multiple reflections; the sum of the amplitude values over this systematically reinforcing tail is 0.874. Clearly, the effect of the very-short-delay multiples dominates the directly transmitted signal.

And this, we remember, is with a two-way travel time of only 0.686 seconds. If we visualize a deep earth section having the same layering characteristics as are evident in our short piece of log, we can auto-convolve figure 13c sufficient times to represent the two-way travel path to any desired depth. Figure 14a is a repeat of figure 13c; figure 14b represents the approximate form of the output after a transmission corresponding to two-way travel time of 1.372 s, figure 14c that corresponding to 2.744 s, and figure 14d that corresponding to 5.488 s.

Of the several conclusions implicit in these diagrams, let us first stress the one concerned with the very first point—the direct arrival—which obviously becomes quite negligible in all of them.

At reflection times of usual interest, it would make no significant difference if the direct "primary" reflection path did not exist; the useful seismic information is carried by the very-short-delay multiple reflections.

Now let us look at the *form* of the transmission responses of figure 14. Clearly one effect of the lengthening path is to broaden the output pulse; in a

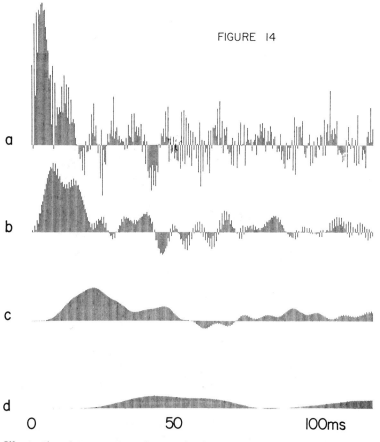

Fig. 14. Illustration (*a*) repeats and extends the two-way transmission response of the real log in fig. 13. Illustrations (*b*), (*c*) and (*d*) are successive auto-convolutions of (*a*), representing the effect of longer path lengths in a statistically similar sequence of layers.

coarse sense, the broadening is similar to that produced by absorption (although the mechanism is entirely different). And, just as the broadening produced by absorption is associated with a high-frequency cut, so the broadening produced by very-short-delay multiples implies a high-frequency cut.

In fact, this conclusion was always present in our simple argument "More up, less down". For one look at the reflection coefficient series of figure 12*a* tells us that the total signal reflected back to the surface must have a low-frequency cut; the high-frequency appearance—the cyclic nature of the sedimentation—virtually guarantees this. Then we can apply our simple argument to spectra just as convincingly as to amplitudes, and conclude that if the reflected signal has a low-frequency cut the transmitted signal must have a high-frequency cut.

Of course, we should be able to do better than just "More up, less down". In the Appendix is set out the derivation of an approximate relationship between

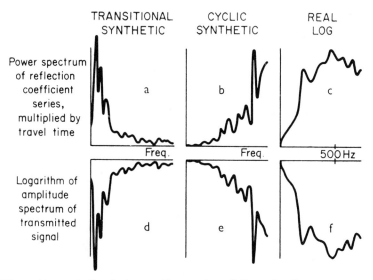

Fig. 15. The anti-correlation between the spectra of the reflecting sequence and of the signal transmitted through it. Frequencies which are selectively reflected are poorly transmitted. More up, less down.

the amplitude spectrum $T(\omega)$ of the transmitted pulse and the power spectrum $R(\omega)$ of the reflection coefficient series:

$$T(\omega) = e^{-R(\omega)t}.$$

This relationship is simply checked by comparing the power spectrum of a given length of the reflection coefficient series with the logarithm of the amplitude spectrum of the pulse transmitted through it. In figure 15 we do this for our two synthetic logs and for the real log. The reflection coefficient series for the illustrative transitional log has the expected low-frequency spectrum (figure 15a), and the signal transmitted through it has a correspondingly inverse spectrum (figure 15d). The reflection coefficient series for the illustrative cyclic log has the expected low-frequency cut (figure 15b), while the signal transmitted through it has an inverse high-frequency cut (figure 15e).

Both examples fit well with the approximate relation above. Particularly satisfying is the real log itself; the power spectrum of the reflection coefficient series (figure 15c) is accurately mirrored by the amplitude spectrum of the transmitted pulse (figure 15f).

This excursion into the frequency domain gives us an alternative and interesting way of looking at the combined effects of transmission loss and very-short-delay multiple reflections. For if we imagine a reflection coefficient series whose spectral structure is such as to have no content at a particular frequency, then there is, in effect, no overall loss at that frequency. However, since the transmitted signal is minimum-phase (and since, therefore, the phase at that frequency depends on the phase at all other frequencies), that frequency can experience a delay even though there is no loss (d'Erceville and Kunetz, 1963; Sherwood and Trorey, 1965).

Further, just as we were beginning to lose all hope of distinguishing between transitional and cyclic sequences by their effect on amplitudes, we see now that there may be additional help available in the frequency domain; specifically, the pulse transmitted through a transitional sequence has a low-frequency cut, while that transmitted through a cyclic sequence has a high-frequency cut.

The formidable difficulty in deriving benefit from this, of course, is that of distinguishing between the effect of the high-frequency cut due to cyclic layering and that due to absorption. Both are progressive, both involve a loss of amplitude and a broadening of the transmitted signal. And, fortuitously, the degree of high-frequency cut associated with cyclic layering may look very much like a constant dB/wavelength effect (at least over a restricted frequency band). Indeed, if we draw a smooth curve to approximate the spectrum of figure 15f over the first 100 Hz, we emerge with a high-frequency cut of about 0.3 dB/wavelength.

The magnitude of this figure immediately throws all our thoughts into disarray. For it raises the clear possibility that the loss of high frequencies due to passage through a cyclic sequence of layers may be greater than the loss of high frequencies due to absorption—the multiple-reflection effect may dominate the absorption effect. This, in turn, says that our records might look much the same if absorption did not exist. Further, in turn, we are led to question the magnitudes that have been quoted for absorption—is it possible that the experimenters have been ascribing to one mechanism an effect which actually owes much to another?

We do not know. What seems most likely is that the two effects co-exist, both contributing a high-frequency cut and one sometimes dominating the other. The seismic pulse returned from any discrete reflector is therefore the interaction or convolution of the pulse shape contributed by the source, the pulse shape contributed by absorption, and the pulse shape contributed by the very-short-delay multiple reflections.

In figure 16 we see along the top line three possible breadths for the pulse representing the combined effects of absorption and the source. In the second line we see these reproduced at 0.027 of the amplitude, to illustrate what would be the loss in amplitude caused by two-way transmission through the log of figure 12 in the (unreal) absence of the very-short-delay multiple reflections. In the third line we see the pulses of the first line convolved with the two-way transmission response of figure 14a, incorporating the very-short-delay multiple reflections. This convolution provides the means for recombining, in effect, the amplitude contributions smeared out over time by the multiple reflection process. The amplitude actually obtained depends, clearly, on the relative breadths of the absorption pulse and the transmission pulse, and on the nature of the high-frequency and low-frequency pulse-shaping effects near

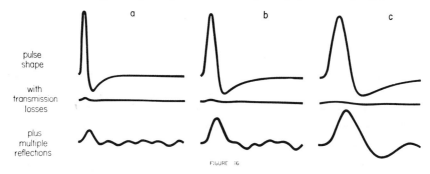

Fig. 16. The effect of convolving three seismic pulse shapes with the pulse formed by transmission through the log of fig. 12, without and with the effect of the very-short-delay multiple reflections.

the source. In general terms, however, we can see that pulses of likely shapes can be transmitted through a strongly stratified earth without amplitude losses of more than a few decibels per second—the observed values.

Summary and Conclusions

The loss of amplitude associated with the geometrical divergence of a wavefront is physically a clear and simple effect. It can be compensated with reasonable safety and reasonable accuracy. When this is done, little decay remains on records using low-frequency narrow-band sources.

After correction for divergence, the amplitude of a reflection depends on its own reflection coefficient and on all the losses incurred above it. If the reflection is discrete the process of reflection itself is not frequency-selective (though, of course, "tuning" effects occur if the reflector is part of a complex). The losses occurring above the reflector are all frequency-selective; the main ones are absorption and the combined effects of transmission and multiple reflection at interfaces. Before we can achieve the desired measurement of reflection coefficient we must be able to quantify these losses; this must require the study of the amplitude loss and the spectral change in combination. The explosives manufacturers can draw some consolation from the observation that these effects are best measured on records from a wide-band source.

The expected effect of absorption can be presented fairly simply; however, the mechanism of absorption—the cause rather than the effect—is not very clear. Further, a case can be made that the magnitude of the effect is in question, since some of the early experimental work may not have taken sufficient account of very-short-delay multiple reflections.

The effects of very-short-delay multiple reflections on "primary" reflection amplitudes and spectra are critically dependent on the nature of the stratigraphy. The distinction between transitional and cyclic sedimentation seems to be a basic and a helpful one, though much more work is required to formulate the connection between the mechanism of sedimentation and the consequent constraints on the reflection coefficient series. This could prove an important and rewarding subject for academic research; although the synthetic seismogram is no longer fashionable, its usefulness is certainly not exhausted.

Transitional sequences, if they are to have realistic upper and lower bounds of acoustic impedance, can show only a small transmission loss; this loss is increased slightly by the effect of very-short-delay multiple reflections. The multiple-reflection effect for a transitional sequence has a low-frequency cut.

Cyclic sequences, still within the same bounds, can have enormous transmission losses; these losses are largely offset, at low frequencies, by the effect of very-short-delay multiple reflections. The transmission process for a cyclic sequence therefore appears to have a high-frequency cut.

The log tested in the present work showed predominantly cyclic stratification in the range of layer thicknesses from 1 to 10 m.

There is an anti-correlation between the power spectrum $R(\omega)$ of the reflection coefficient series and the amplitude spectrum $T(\omega)$ of the pulse transmitted through it; the simple relationship $T(\omega) = e^{-R(\omega)t}$ seems to be a satisfactory approximation.

All of this amounts to a hope—but not a promise—that the combined study of amplitude decay and spectral change on seismic records can lead to some definition of the statistics of the reflection coefficient series, and that this in turn will be interpretable in terms of the type of geological sedimentation.

References

ANSTEY, N. A., 1960, "Attacking the problems of the synthetic seismogram", Geophysical Prospecting 8, 242-259.
BARANOV, V. and KUNETZ, G., 1960, "Film synthétique avec reflexions multiples", Geophysical Prospecting 8, 315-325.

Berzon, I. S., 1967, "Analysis of the spectral characteristics of a thin-bedded sequence", in "Seismic Wave Propagation in Real Media", Consultants Bureau, 1969, New York and London.

Bois, P. and Hemon, C., 1963, "Etude statistique de la contribution des multiples aux sismogrammes synthétiques et réels", Geophysical Prospecting 11, 313-349.

Bois, P., Hemon, C. and Mareschal, N., 1965, "Influence de la largeur du pas d'échantillonnage du carottage continu de vitesses sur les sismogrammes synthétiques à multiples", Geophysical Prospecting 13, 66-104.

Bortfeld, R., 1960, "Seismic waves in transition layers", Geophysical Prospecting 8, 178-217.

Delas, C., and Tariel, P., 1965, "Calcul d'un film synthétique a partir d'un très grand nombre de couches", Geophysical Prospecting 13, 460-474.

d'Erceville, I. and Kunetz, G., 1963, "Sur l'influence d'un empilement de couches minces en sismique", Geophysical Prospecting 11, 115-121.

Evenden, B. S., Stone, D. R. and Anstey, N. A., 1970, "Seismic Prospecting Instruments": volume 1, "Signal Characteristics and Instrument Specifications"; Geoexploration Monographs, Series 1 No. 3; Gebrüder Borntraeger, Berlin and Stuttgart.

Gurvich, I. I. and Yanovskij, A. K., 1968, "Seismic impulses from an explosion in a homogeneous absorbing medium", Izvestiya, Academy of Sciences USSR, Physics of the Solid Earth, English edition (AGU), 634-639.

Hagedoorn, J. G., 1954, "A process of seismic reflection interpretation", Geophysical Prospecting 2, 85-127.

Mikhailova, N. G., Pariiskii, B. S. and Saks, M. V., 1966, "The spectral characteristics of multiple transition layers", Izvestiya, Academy of Sciences USSR, Physics of the Solid Earth, English Edition (AGU) 6-12.

O'Brien, P. N. S., 1969, "Some experiments concerning the primary seismic pulse", Geophysical Prospecting 17, 511-547.

Sherwood, J. W. C. and Trorey, A. W., 1965, "Minimum-phase and related properties of the response of a horizontally stratified absorptive earth to plane acoustic waves", Geophysics 30, 191-197.

Trorey, A. W., 1962, "Theoretical seismograms with frequency and depth dependent absorption", Geophysics 27, 766-785.

Appendix

The transmission response of a set of thin layers

Consider a section whose acoustic response for normally incident energy can be completely described by a set of N reflection coefficients $r(j)$ equally spaced in time. We shall try to show that a frequency-domain relationship exists between the series r and its transmission response T.

Strictly, we deal with the early part of T only. We assume that the effective length of the transmitted pulse is sufficiently short that the differential transmission loss of its components can be neglected. This is not unrealistic for sedimentary layers. However, we do need a more restrictive assumption. If we define

$$a(l) = \sum_{j=1}^{N} r(j)\, r(j+l), \qquad (1)$$

where l represents the delay of a multiple relative to the direct arrival, the expected value of $r(j)\, r(j+l)$ is taken as $a(l)/N$ (that is, the series is assumed stationary). Values of $r(j)$ outside the section are to be read as zeros for the purpose of expressions such as (1).

An impulse in layer N gives rise to a response in layer 0 which consists of a direct arrival and a set of multiply-reflected trains $_k s$ characterized by $2k$ internal reflections. If the direct arrival has unit amplitude, the first multiple is

$$_1s(l) = -\sum_{j=1}^{N} r(j)\,r(j+l), \quad l > 0 \tag{2}$$

where only the differential transmission loss factor, of the form $\prod_{w=j+1}^{l-1}(1 - r_w^2)$, has been neglected. It is convenient to consider this as equivalent to a more general function $m(l)$ defined by

$$\begin{aligned} m(l) &= -a(l) & l &> 0 \\ m(l) &= 0 & l &\leq 0. \end{aligned} \tag{3}$$

To evaluate the second multiple, we consider the expression

$$c(l_1, l_2) = \sum_{j_1=1}^{N} r(j_1)\,r(j_1 + l_1) \sum_{j_2=1}^{j_1+l_1-1} r(j_2)\,r(j_2 + l_2) \tag{4}$$

which represents a contribution to $_2s$ at a delay of $l_1 + l_2$. The suffixes on j and l now indicate the order in which the multiple reflections occur. We can approximate equation (4) by

$$\begin{aligned} c(l_1, l_2) &\backsim \sum_{j_1=1}^{N} r(j_1)\,r(j_1 + l_1) \sum_{j_2=1}^{j_1} m(l_2)/N \\ &\backsim \sum_{j_1=1}^{N} \{m(l_1)/N\}\{j_1\,m(l_2)/N\} \\ &\backsim \tfrac{1}{2}\,m(l_1)\,m(l_2). \end{aligned} \tag{5}$$

For an estimate of the total arrival on $_2s$ at a delay l we sum this, writing $l - l_1$ for l_2:

$$_2s(l) = \sum_{l_1=1}^{l-1} \tfrac{1}{2}\,m(l - l_1)\,m(l_1). \tag{6}$$

This is a convolution and states that, if the multiples are normalized by the direct arrival, the second multiple is half the autoconvolution of the first; physically this makes sense if we think of m as the basic multiple-generating filter. The average amplitude of the first multiple within the section is half its final value; we would expect it to grow linearly. If we repeat this reasoning for the higher-order multiples it is found that a simple recursive relationship exists which can be written

$$_k s = 1/k \;_{k-1}s * m \tag{7}$$

We can sum all these series in the frequency domain, defining

$$M(\omega) = \sum_{l=1}^{N} m(l)\, e^{-i\omega l\tau} \tag{8}$$

where τ is the two-way transmission time within a layer. The transform of the pulse is now given by

$$T'(\omega) = 1 + M(\omega) + \frac{M^2(\omega)}{2!} + \frac{M^3(\omega)}{3!} \ldots = e^{M(\omega)}. \tag{9}$$

For the seismic problem we are interested in two-way transmission. Happily we do not have to go through the argument again, as the direction of propagation affects only the way we number the reflection coefficients. The shape of the two-way transmitted pulse is obtained by squaring T':

$$T''(\omega) = e^{2M(\omega)}. \tag{10}$$

This still describes a pulse whose first arrival is unity. If the original impulse is unity, we have to multiply T'' by $\prod_{j=1}^{N} \{1 - r(j)^2\}$. We can approximate this by considering the relationship

$$\lim_{x \to \infty} \left(1 - \frac{v}{x}\right)^x = e^{-v}, \tag{11}$$

so that, for large N, we would expect $e^{-a(0)}$ to be a satisfactory estimate of the direct arrival. This gives our final pulse spectrum

$$T(\omega) = e^{-a(0) + 2M(\omega)}. \tag{12}$$

To derive useful information from this expression we could note that the amplitude spectrum of T is defined by the real part of the exponent, which is identical to the transform of $-a$. If we define the power spectrum $R(\omega)$ of the reflection coefficient series as the transform of a normalized by the travel-time $t = N\tau$, we can write this as

$$|T(\omega)| = e^{-R(\omega)t}. \tag{13}$$

Equation (12) defines T as a minimum-phase function, so that it has at least one property in common with the exact response (Sherwood and Trorey 1965).

As to the interpretation of the layer thickness τ, any analysis that uses the concept of a reflection coefficient is necessarily discrete, but it is reassuring to note that the transform of a reflection coefficient series can have an interpretation in terms of acoustic impedance q, since

$$\sum_{j=1}^{t/\tau} r(j)\, e^{-i\omega j\tau}$$

converges to

$$\int_0^t \frac{d(\log q)}{2\,du} e^{-i\omega u}\,du.$$

(Therefore the filtering of a q log is a legitimate procedure, but the filtering of $\log q$ is better.)

These considerations lead us to suppose that the power spectrum of a reflection coefficient series is a meaningful concept, which does not depend on the choice of τ.

VELOCITY SPECTRA AND THEIR USE IN STRATIGRAPHIC AND LITHOLOGIC DIFFERENTIATION[*]

BY

ERNEST E. COOK[**] and M. TURHAN TANER[**]

Abstract

Velocity Spectra which were originally developed for the optimum stacking of seismic data have been found to give considerable information concerning lithologic and stratigraphic changes in the geologic section. In the Gulf of Mexico shale sections and sand bodies have been recognized on the Velocity Spectra display, and in the Caribbean last year a first attempt was made to utilize Velocity Spectra information for the determination and mapping of lithology. Since that time, a new program has been developed which takes dip into account when computing the interval velocity. This program has been applied to a seismic section in the North Sea which has resulted in a geologic model derived from interval velocities which were found to be quite consistent. Such a model can be of great value in geological interpretation.

Introduction

Computations of seismic velocities have been commonly made since the beginning of seismic exploration. Dix (1955), Slotnick (1959), Musgrave (1962), and others devised basic manual procedures for application to single coverage records. Utilization of longer geophone spreads in Common Reflection Point shooting (Mayne 1962, 1967), however, required more accurate determination of dynamic corrections for optimum stacking. In most instances, these corrections were estimated using nearby well velocity surveys and simple straight or curved ray path methods; resulting misalignments were corrected by a residual moveout. Manual T^2/X^2 analyses using common depth point trace gathers provided the most useful velocities. Later, conversion of these methods to digital processing has been described by several authors; in particular, Schneider and Backus (1968), Garotta and Michon (1967), and Taner and Koehler (1969), and has since been the subject of various as yet unpublished papers.

The Velocity Spectra

The Velocity Spectra program, in common with other velocity analyses, was initially derived to determine the optimum stacking velocities. However,

---

[*] Presented at the 30th Annual Meeting of the European Association of Exploration Geophysicists, Salzburg, June, 1968.
[**] Seismic Computing Corp., Houston, Texas.

during its use over the past two years it was noted that the spectra display itself could be used for correlation of reflections, and that the interpreted primary reflection velocities could be used to determine the depth and dip of the reflecting interfaces as well as the interval velocities between reflectors. The Velocity Spectra display consists of a number of individual Velocity Spectra computed at 24 millisecond intervals between given minimum and maximum apparent RMS velocities.

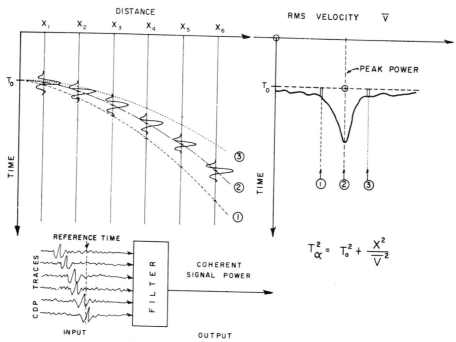

Fig. 1. Derivation of Velocity Spectrum.

Common depth point traces gathered and displayed in order of their distance from the energy source will show a common reflecting horizon as a hyperbola. Therefore, each unique set of reflections having this character will correspond to an apparent RMS velocity and a normal incidence time. Velocity Spectral analysis computes and displays the coherent power among a set of common depth point traces according to various hyperbolic curves as given by the equation $T_x^2 = T_0^2 + X^2/\overline{V}^2$. A Velocity Spectrum is computed at a given normal incidence time by hyperbolic searches covering a gate of 50 milliseconds made in velocity increments of 100 feet per second. This is then repeated down the record at 24 millisecond intervals. The amplitude of the resulting trace, which is indicative of the coherent power in the Spectrum, is measured by a multichannel filter which is schematically shown in Figure 1. This figure shows three hyperbolic curves and their corresponding power on the Spectrum.

Figure 2 shows an actual example taken from the Gulf of Mexico in an area where reflections are plentiful and mostly primaries. Drilling results have indicated that each of these primary reflections corresponds to a sand body. Shales, of course, fill the intervening section. In Figure 3, loss of primary reflecting energy at about 2.3 seconds indicates a stratigraphic change. In this particular case, a deep well has confirmed this interval to be a thick, soft shale.

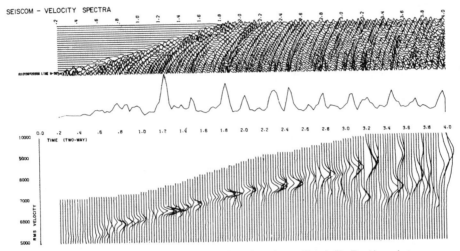

Fig. 2. Typical Velocity Spectra Display in a Good Reflection Area.

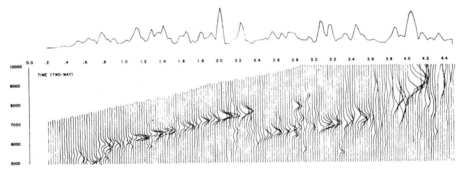

Fig. 3. Velocity Spectra Display showing Stratigraphic Change.

This method transforms all reflections from the time-distance domain, as they would appear on the common depth point trace gathers, to normal incidence time-apparent RMS velocity domain. As explained, this is accomplished by analysing the power of reflected energy arriving according to various hyperbolic paths. It is obvious that the Velocity Spectra display will contain all types of reflections and all other energy coherent in a hyperbolic path within the chosen time and velocity limits. Therefore, this display may be utilized:

1. To determine the velocity function for optimum stacking by interpreting primary reflections.
2. To check the final section against any procedural, interpretational or computer errors.
3. To determine the presence and the effect of multiple reflection interference.
4. To determine normal incidence time and RMS velocities of major reflectors for comparison with the final section for use in more accurate depth, dip and interval velocity computations.
5. To obtain stratigraphic and structural information by observing changes in reflection character appearing on the Velocity Spectra display and lateral changes in interval velocities derived from these curves. Initial attempts along these lines were described by Cook (1967).

Where dips are flat, accurate interval velocities can be derived directly from a single Velocity Spectra display. However, when dipping beds are involved, it is necessary to use a different approach that takes dip into account and computes both interval velocity and dip (Figure 4).

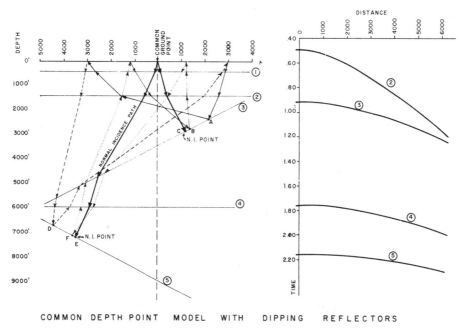

Fig. 4. Common Depth Point Model with Dipping Reflectors.

Our interval velocity-dip computation program is basically an iterative model generating procedure. This program uses interpreted normal incidence time and apparent RMS velocities from two adjacent or very close Velocity Spectra displays to synthesize a series of two-dimensional seismic models in which computed RMS velocities and normal incidence times converge iteratively to the observed values. A similar procedure was described by Sattlegger (1965). The program output consists of the reflection time, RMS, average and interval velocities, dip rate and angle, vertical depth to the reflector, layer thicknesses at the ground point, the depth to the reflector at normal incidence point and its horizontal displacement.

Geologic Application of the Velocity Spectra

A. Preliminary Observations

An example taken from the Netherlands North Sea will illustrate how Velocity Spectra can be used to indicate lithologic and stratigraphic changes. The section chosen was in some respects a problem section, in that it was quite difficult to determine the exact location of the base Zechstein reflection simply by inspection. In fact, even a detailed study made with only the dip information will still leave the location of the base Zechstein in considerable doubt over most of the section.

A total of 20 Velocity Spectra displays, including 9 pairs on which the interval velocity-dip computation program was used, were run along this line. Locations are shown on the section (Figure 5).

Let us now consider a few of these Velocity Spectra displays more closely. Velocity Spectra display 2B (Figure 6) was one upon which our interval velocity model generating program was run. We have used various kinds of shading to indicate variation in interval velocity down the section. Within the Tertiary section, velocities between 5,000 and 6,000 feet per second are distinguished from those between 6,000 and 7,500 feet per second. In the underlying Cretaceous chalk section, velocity intervals between 7,500 and 10,000 feet per second, 10,000 and 14,000 feet per second, and greater than 14,000 feet per second are indicated.

Below the Cretaceous chalk we have made no attempt to distinguish the various geologic ages which may possibly range all the way from lower Cretaceous through the lower Triassic. However, velocities less than 11,000 feet per second are distinguished from those lying between 11,000 and 13,000 feet per second, and those greater than 13,000 feet per second. On this specific example there are no velocities lying between 11,000 and 13,000 feet per second. Velocities over 14,000 feet per second, where we believe they are from Permian rocks, are also shaded distinctly.

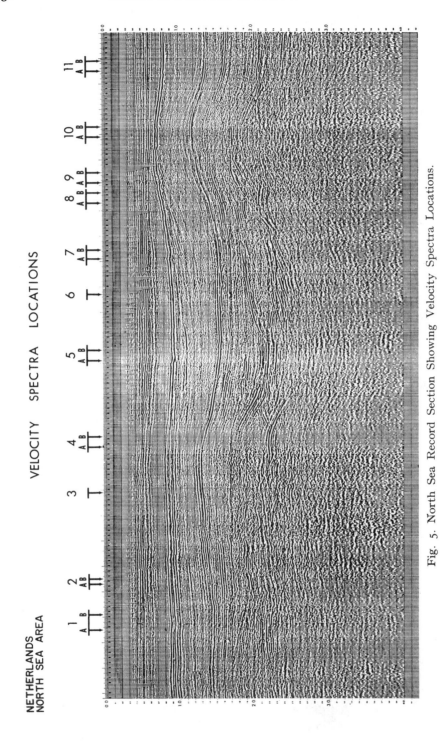

Fig. 5. North Sea Record Section Showing Velocity Spectra Locations.

On this particular spectra display (Figure 6), it is interesting to observe that, of the band of three reflections seen at the base of the Tertiary, it is in fact the center one which represents the actual increase in velocity which occurs at the top of the Cretaceous chalk. It will also be noticed that immediately above the base of the Tertiary there is a low velocity layer having velocities lower than 6,000 feet per second.

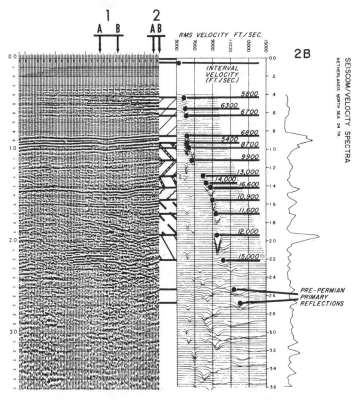

Fig. 6. Velocity Spectra Display 2B Showing Interval Velocities.

The base of the Cretaceous chalk is also very well defined on this spectra display. There is a large increase in the interval velocity of the chalk from a value of 9,900 feet per second in the upper layers to a value of 16,600 feet per second in the basal chalk. We make no attempt here to place any geologic significance on this very high velocity at the base of the chalk, but we would make the comment that our Velocity Spectra display studies in the Netherlands North Sea have shown this kind of velocity variation within the chalk is not infrequent.

Immediately underlying the chalk we see lower velocity material ranging from 10,900 feet per second up to 12,000 feet per second. Below this lies an interval which has a velocity of 15,000 feet per second that we suggest would be very close to that expected for the Permian. Also noteworthy on this spectra display is the presence of two pre-Permian reflections which have RMS velocities lying clearly on the primary curve, and which can thus be identified with confidence.

The next Velocity Spectra display at location 3 (Figures 7 and 8), was not used for interval velocity computations. It illustrates the manner in which our multiple attenuation capability was used to remove two strong multiple reflections, thereby showing clearly the existence of primaries arriving in the same time interval. This further allowed a determination of the velocity change occurring at the top of the Permian which made identification of this formation in this particular location more certain.

Velocity Spectra display 5A (Figure 9) is one of the pairs from which interval velocities were computed. Here the whole Tertiary section has velocities greater than 6,000 feet per second, although the thin layer at its base still shows a velocity reversal. Within the Cretaceous chalk a change in the velocity of the chalk from top to bottom is seen again. Below the chalk, velocities ranging from 13,300 to 14,200 feet per second are found, which observation is consistent with what might be expected in the Triassic section. The Permian section appears to be relatively thin and of higher than usual velocity suggesting that most of the salt has been squeezed out of it leaving the higher velocity limestones, dolomites and anhydrites.

Velocity Spectra display 7A (Figure 10), is another of the pairs from which interval velocities have been derived. Here again the whole Tertiary section appears to have velocities greater than 6,000 feet per second, except for the low velocity layer which again stands out quite clearly at the base. Consistency with the last spectra display is readily apparent. The Cretaceous chalk also shows velocity variations quite similar to those noted in the last set of spectra. By contrast with the previous figure (Figure 9), the material underlying the chalk actually has higher velocities in its upper layers than in its lower. Those familiar with the North Sea area may well identify this circumstance as a geological condition sometimes occurring within the Triassic. Strong reflections are clearly seen at the top and base of the Permian, and contain between them an interval velocity of 15,100 feet per second, characteristic of the salt beds which are its major constituent. Noteworthy on this Velocity Spectra display are two primary arrivals from steeply dipping beds below the base Zechstein.

Velocity Spectra display 10B (Figures 11 and 12), illustrates a further example of the use of our multiple attenuation techniques in assisting or helping to identify the Permian section. Again, the top Permian reflection is obscured by the arrival of strong multiples on the original Velocity Spectra display.

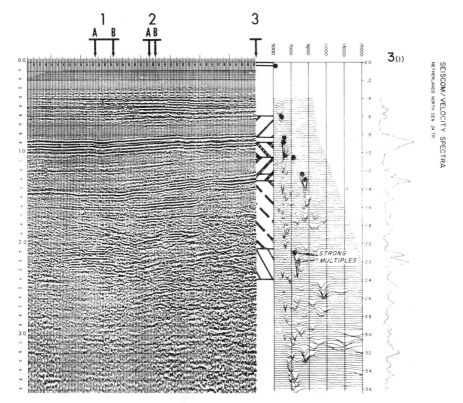

Fig. 7. Velocity Spectra Display 3 with Strong Multiple Reflections.

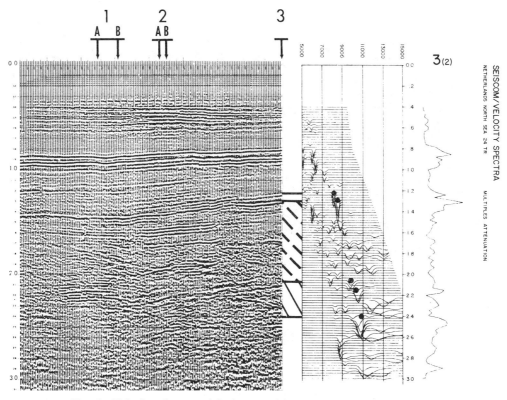

Fig. 8. Velocity Spectra Display 3 with Strong Multiples removed.

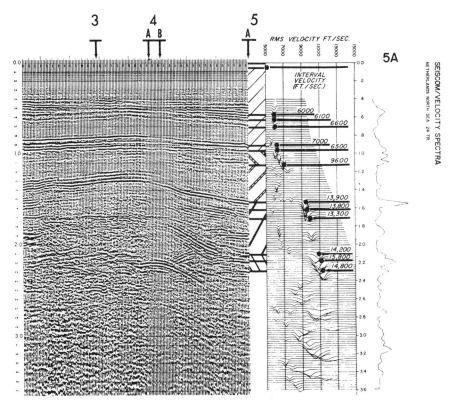

Fig. 9. Velocity Spectra Display 5A showing Interval Velocities.

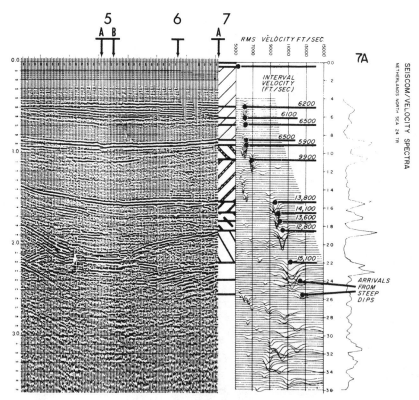

Fig. 10. Velocity Spectra Display 7A showing Interval Velocities.

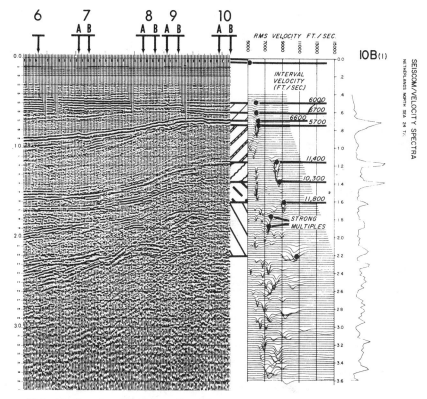

Fig. 11. Velocity Spectra Display 10B with Strong Multiples showing Shallow Interval Velocities.

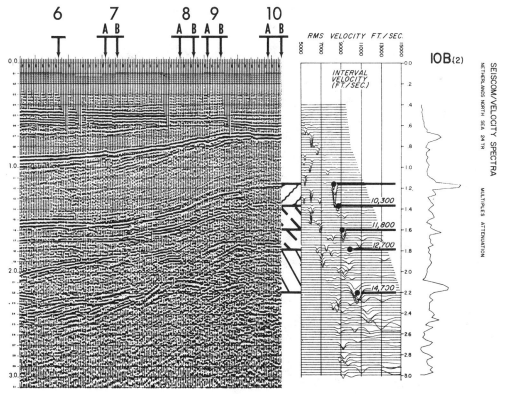

Fig. 12. Velocity Spectra Display 10B with Strong Multiples removed showing Deep Interval Velocities.

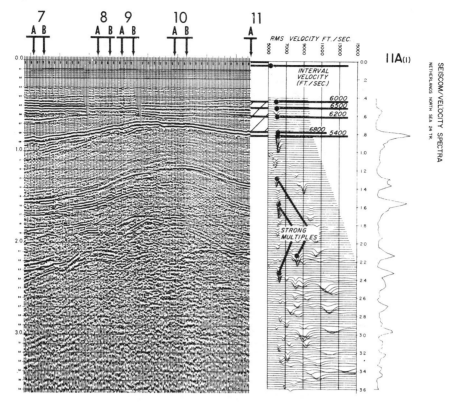

Fig. 13. Velocity Spectra Display 11A with Strong Multiples showing Shallow Interval Velocities.

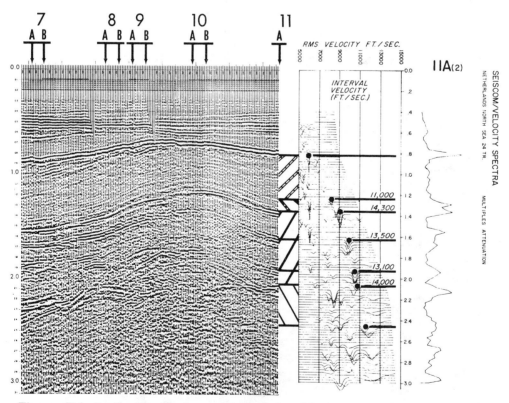

Fig. 14. Velocity Spectra Display 11A with Strong Multiples removed showing Deep Interval Velocities.

On the multiple attenuated Velocity Spectra display, a pick can be made which gives a velocity of 14,700 feet per second which can probably be associated with Permian. Once again the low velocity layer can be seen at the base of the Tertiary. It will also be noticed that on this set of spectra much lower velocities appear between the chalk and the top of the Permian. This could have considerable geological significance.

Finally, Velocity Spectra display 11A (Figures 13 and 14), again makes evident the basal low velocity zone in the Tertiary. Much of the section below the Tertiary was disturbed by strong multiples, and the multiple attenuation technique was used to bring out the primary reflections which showed considerable variation in velocity between the top and bottom of the Cretaceous chalk. The lower velocities reach a value of 14,300 feet per second.

Below the chalk and above the Permian, velocities are quite high when compared with those seen on the last Velocity Spectra display. Although in this particular example a base Zechstein reflection can be picked at about 2.45 seconds, it is not possible to compute an accurate interval velocity for this zone, because no comparable event appeared at this time on the companion set of spectra, 11B. However, the change in slope of the curve at 2.08 seconds indicates that the velocity of this material is considerably higher than 14,000 feet per second, so we are able to identify it as Permian.

B. The Composite Picture

Returning to the original section and placing on it shadings corresponding to the various correlations (Figure 15), shows they fit together fairly well. Those geologists and geophysicists experienced in the North Sea area will appreciate that this constitutes considerable information regarding the stratigraphic and lithologic changes occurring across this section. For example, there are slight but consistent changes in the velocity of the upper Tertiary which must have some stratigraphic significance. The ubiquitous low velocity layer at the base of the Tertiary perhaps indicates a basal conglomerate. Although there are places where this velocity exceeds 6,000 feet per second, it is nevertheless distinctly lower than the the velocities directly above it.

Within the Cretaceous chalk, we see considerable changes in velocity. Abrupt changes appear on the left side of the section while changes are gradual and consistent on the right side. Further study of the character of these changes would probably reveal valuable information about the tectonic movements occurring during Cretaceous deposition.

In the section immediately below the chalk, considerable velocity changes occur which must again correspond with differences in lithology. To those experienced in this area, they could surely be made to tell an illuminating tale. As for the Permian, where before much of our interpretation would

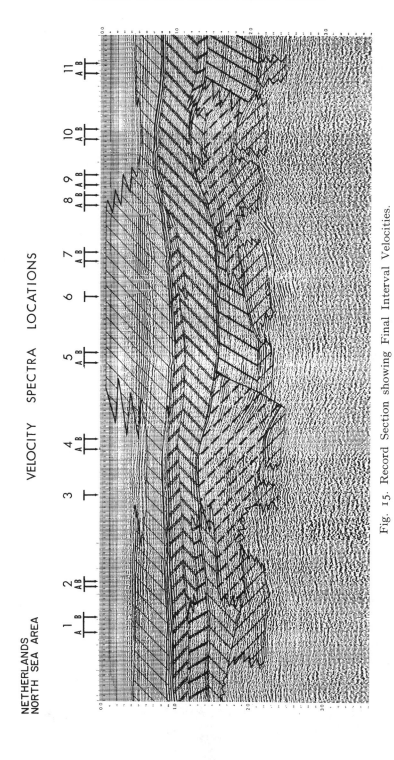

Fig. 15. Record Section showing Final Interval Velocities.

have been guess work, we now have a number of places where we can determine Permian velocities. This, in conjunction with the dip evidence observed on the section, gives some confidence to the prediction of the base Zechstein.

Conclusions and Comments

In closing, it is appropriate to mention the accuracy involved. Obviously the error involved in any interval velocity determination made with these methods will depend on many factors. The four most important will be:

1. *The RMS velocity search grid.* If the normal incidence time intervals were reduced from 24 milliseconds to perhaps 4 milliseconds, or if the velocity search interval were reduced from 100 feet per second to say 10 feet per second, then obviously greater accuracy could be obtained.
2. *The quality of the reflections.* Clearly the better the reflections, the sharper will be the coherency peaks on the Velocity Spectrum.
3. *Changing dip.* We have noticed, and not unexpectedly, that rapidly changing dips down the section lessen accuracy. This arises because the model assumes straight line interfaces. Future models will use curved interfaces.
4. *Anisotropy.* The model assumes an isotropic homogeneous medium between each interface. If anisotropy factors are known, these may be included in the computation.

The most important and most significant conclusion which can be drawn from this study is that by the use of Velocity Spectra displays, even with this simple model, it is possible to obtain interval velocities along a section which are consistent and which show changes in a regular manner. These interval velocities and their changes must relate to the stratigraphy and lithology of the particular section and are of great value in geological interpretation.

We would like to take this opportunity to thank many of the personnel of Seiscom for preparing and processing data presented in this paper. We would also like to thank The Louisiana Land and Exploration Company for making their data available for presentation in this paper.

References

Cook, E. E., 1967, Geophysical Reconnaissance in the Northwestern Caribbean, Presented at the 37th Annual Meeting of S.E.G., Oklahoma City.
Dix, C. H., 1955, Seismic Velocities From Surface Measurements, Geophysics 20, 68-86.
Garotta, R. and Michon, D., 1967, Continuous Analysis of the Velocity Function and of the Move Out Corrections, Geophysical Prospecting 15, 584-597.
Mayne, W. H., 1962, Common Reflection Point Horizontal Stacking Techniques, Geophysics 27, 927-938.
Mayne, W. H., 1967, Practical Considerations in the Use of Common Reflection Point Techniques, Geophysics 32, 225-229.

Musgrave, A. W., 1962, Applications of the Expanding Reflection Spread, Geophysics 27, 981-993.
Sattlegger, J., 1965, A method of computing interval velocities from expanding spread data in the case of arbitrary long spreads and arbitrarily dipping interfaces, Geophysical Prospecting 13, 306-318.
Schneider, W. A. and Backus, M. M., 1966, Dynamic Correlation Analysis, Geophysics 33, 105-126.
Slotnick, M. M., 1959, Lessons in Seismic Computing, S.E.G. Publication, p. 194.
Taner, M. T. and Koehler, F., 1969, Velocity Spectra—Digital Computer Derivation and applications of Velocity Functions, Geophysics 34, 859-881.

OUTLINING OF SHALE MASSES BY GEOPHYSICAL METHODS†

A. W. MUSGRAVE* AND W. G. HICKS**

Shale masses are here defined as large bodies of shale at least several hundred feet in thickness. These may be formed either as diapiric masses or as depositional masses. The shale masses are like salt masses and the two are many times combined to form domal masses; they both may form the updip seal for stratigraphic accumulation of oil.

The shale masses exhibit the following properties by comparison to the normal section: (1) low velocities—in the range of 6,500 to 8,500 ft/sec with very little increase of velocity with depth, (2) low densities—estimated to be in the range 2.1 gm/cm³ to 2.3 gm/cm³, (3) low resistivities—approximately 0.5 ohm-m, and (4) high fluid pressures—about 0.9 overburden pressure. These properties all seem to be caused by the high porosity and low permeability of these large shale masses.

Maps and cross sections of an example area block 113, Ship Shoal Area are shown. The low shale velocities were measured by acoustic logs and verified by refraction shooting. The low densities were deduced from gravity maps. The low resistivities are shown on electric logs, and high pressure is evidenced by the drilling difficulties with heaving shales.

These physical properties allow the outlining of the shale mass by one or more of the following ways: the gravity method is used to outline the low density material, the seismic reflection method is used to outline the lack of reflection contrast and in some cases map the velocity configuration, the seismic refraction method is used to indicate the velocity of the anomalous mass, thereby differentiating between shale and salt.

INTRODUCTION

The first purpose of this paper is to define what is meant by shale masses as used in this discussion. The second object is to describe the properties of these masses that are measured and to show the measurements of the properties by including them in a discussion of an example area. The third purpose is to cover the geophysical techniques used in mapping these physical properties of shale masses. The geophysical methods employed are gravity, seismic refraction, and seismic reflection surveying.

Definition of a shale mass

A shale mass as used in this paper is defined as a large body of shale of approximately 500 ft in the smallest dimension. Discussion of such bodies of shale is not new. These quantities of shale have been called mud lumps, mud volcanoes, mud flows, shale gouge or sheath, brecciated shale, and other terms. A study of these terms indicates that the shale masses described are probably diapiric; that is, they pierce the overlying sediments. However, the above definition does not necessarily make diapirism a condition. It is possible that a shale body can be depositional and have the dimensions indicated. One additional definitive requirement of the mass, whether it is depositional or diapiric, is that it has no continuous porous zones connected to the outside formations. It must include no aquifer such as a sand layer connecting to outside porous areas beyond the shale mass.

Properties of a Shale Mass

A shale mass as defined has distinctive properties which make it an interesting geological feature to study by geophysical measurements. In the Gulf Coast area where most of the study has been made, there are distinctive differences between shale masses and the surrounding normal sand-shale sequences. The properties of the masses which have been identified are as follows:

1) The velocity of shale masses is particularly low and ranges approximately from 6,500 to 8,500 ft/sec, changing only slowly with depth and with area. The velocity measurements were made by refraction methods through large masses of

---

† Presented at the 33rd Annual International SEG Meeting in New Orleans, October, 1963. Manuscript received by the Editor January 23, 1966. Revised manuscript received April 20, 1966.

* Mobil Oil Company, Geophysical Services Center, Dallas, Texas.

** Editor's Note: Mr. Hicks, formerly with the Field Research Laboratory, Socony Mobil Oil Company, Dallas, Texas, did a great deal of work on this project before his untimely death in 1961. The senior author has chosen to recognize the important contributions of his former colleague by honoring him as co-author.

shale and also by continuous velocity logs through large bodies of the shale mass material.

2) The shale masses have low resistivities of the order of one-half ohm-m or less, which is roughly half what is expected in a normal shale. These resistivities were measured from the electric log, and correspond roughly to 2,000 millimhos or greater as measured by conductivity logs (Wallace, 1965).

3) The shale masses have low densities compared with normal shale sections. Density measurements are difficult to make and are not as exact as the other type measurements. They are indicated by minimum anomalies observed from gravity meter surveys. One reference is made in Russian literature (Tsimel'zon, 1959) which indicates that they find gravity minima over shale masses which they refer to as mud volcanoes. The density is estimated to vary between 2.1 and 2.3 gm/cm^3 in the shale mass.

4) Another property of shale masses is that they exhibit very high fluid pressures. In other words, when a shale mass is penetrated, it has a tendency to heave or to flow into the hole and squeeze the drill stem. In addition, heavy mud is required to prevent the blowout of small pockets of gas. Often in the drilling of this material, gas collects during the period when a trip is made to change bits and has gained the name of "trip gas." The pressure is as high as 0.9 of the overburden pressure. In other words, the fluids are supporting substantially all of the weight of the overburden rather than the pressure being on the rock matrix itself.

GEOPHYSICAL METHODS USED FOR OUTLINING SHALE MASSES

The reflection seismic method can be used to map shale masses, and this is normally done by mapping from a cutout of reflections. If there is present a homogeneous mass, there will not be any reflecting interfaces, and therefore no reflections will occur. Of course, there is a possible ambiguity since the mass can be some other type of homogeneous mass such as salt or a combination of shale and salt which is referred to as the domal mass (Atwater, 1959). The refraction seismic method can be used to differentiate between salt and shale of a reflection cutout which has been mapped. This is because salt has a velocity of about 15,000 ft/sec, higher than the normal section; and the shale has a lower velocity than the normal section. The normal section in the Gulf Coast has a velocity between 8,000 and 12,000 ft/sec. The gravity method can also be used because of the low density involved; however, again ambiguity exists as to whether shale or salt is the anomalous material. The shape ambiguity that exists in gravity mapping is now further complicated by not knowing the type of material that is causing the anomaly.

EXAMPLE AREA: BLOCK 113, SHIP SHOAL, LOUISIANA OFFSHORE

The location of this example area relative to the State of Louisiana and the City of New Orleans is indicated on Figure 1.

Velocity

Figure 2 is a map of the domal mass, actually made of two separate salt masses and a shale mass. Profile AA' (Figure 3) crosses the larger salt stock and shows the shale mass lying on the south side of it. When the A-1 well on this section was drilled, a refraction line was shot to the southwest for velocity information and an observed wavefront chart was prepared (Figure 4). Later when the D-1 well of Block 117 was drilled, a refraction profile was shot to the southeast from this well in the same manner and again the wavefronts were plotted on the distance versus depth plot. The difference in the two plots indicates that a large shale mass exists and that this is not a simple shale sheath around the salt dome. This gives some measure of the difference in velocity between the two materials, that is, between the normal section and the shale mass. Also in the D-1 well, a continuous velocity log (CVL) was taken and a form of this is shown in Figure 5. The right-hand portion of the figure shows the interval velocity plotted with averaged increments over each of 100 ft of depth, and a dashed line representing an average increase of velocity with depth is indicated. In the left-hand portion of the figure this line is plotted vertical and the variation from this line plotted as a normalized velocity. This is then compared with a percent sand indicated from an interpretation of the electric log. This percentage is also averaged over the same 100-ft increments. It is noted that a reasonably good correlation exists between the normalized velocity and the percent sand. Profile BB' (Figure 6) indicates that the shale stands well above the lesser salt stock at this location, thus indicating that the

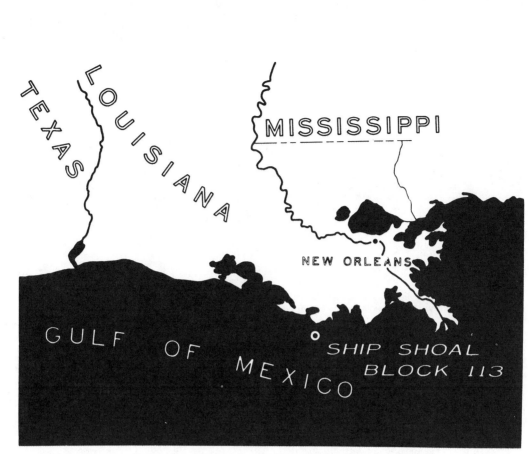

Fig. 1. Location map showing location of the example area.

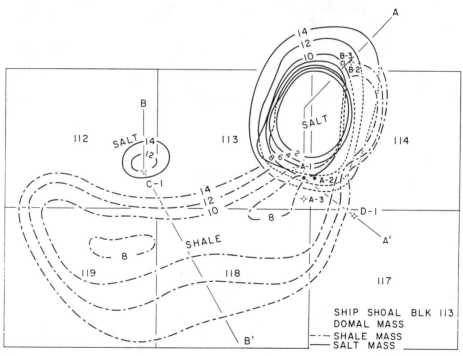

Fig. 2. A detailed map showing the configuration of the salt masses and the shale mass on the example area in block 113, Ship Shoal area.

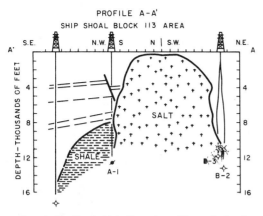

Fig. 3. Profile A-A' indicated on Figure 2 showing shallow salt stock and shale mass laying along side of salt.

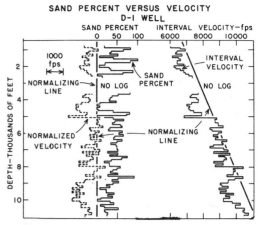

Fig. 5. Percentage of sand versus velocity in the D-1 well, block 117, Ship Shoal area. The curve on the right-hand side shows an interval velocity plot; in the central part of the figure the percentage sand as taken from the electric log is plotted and to the left-hand side is the normalized version of the velocity plot showing the comparison between the amount of shale and the low velocity.

shale may have flowed independently of the salt. The C-1 well (Figure 7), presented in the same manner as the D-1 well, indicates similar conditions down to the shale mass which was found by the drill 500 ft above the top of the salt. Here the shale exhibits some of its extraordinary characteristics. First, the velocity is extremely low compared with that of sands or even with interbedded shales. Secondly, the velocity decreases over a zone several hundred feet in width. Also, note that the resistivity plotted on this same figure from the electric log is below one-half ohm-m. In Figure 8, the A-3 well is a directional hole drilled off of the A platform to the south. This well would have gone outside of the shale had it been a shale sheath. However, it still drilled into a large anomalous shale mass in the depth range from 8,700 to 11,500 ft. Again note its variation from

the normalizing line on the velocity scale making it appear extremely out of position. Also note the low resistivity through the shale. Figure 9 shows a compilation of sand velocities measured from several wells around this dome and indicates that the velocity of sand increases from about 6,000 ft/sec at the surface to about 11,500 ft/sec at 12,000 ft depth with a reasonable scatter of velocity about that point. Figure 10 shows a similar plot for shale velocities from the same group of wells. Here it indicates that interbedded shales, that is, thin beds of shale between sand

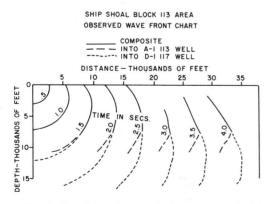

Fig. 4. Wavefront chart made from a composite of refraction information from the wells in the A-1 block 113, and D-1 block 117 in the Ship Shoal area.

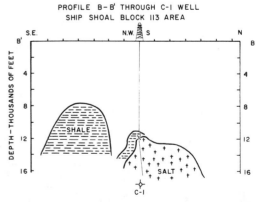

Fig. 6. Profile B-B' through the C-1 well, block 112, Ship Shoal area indicated on Figure 2. This shows a deep salt stock and a shallower shale mass in the adjacent area.

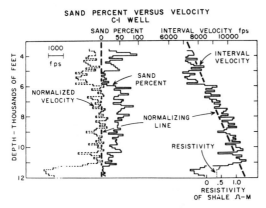

Fig. 7. Sand percentage versus velocity in the C-1 well block 112, Ship Shoal area. Again this shows the same information as illustrated in Figure 5, but in this case it also shows low velocity shale in the bottom of the hole. Also note the resistivity of this shale mass.

layers generally increase in velocity from 6,000 ft/sec at the surface to 9,500 ft/sec at 12,000-ft depth. However, note that the shale mass as indicated by the crossed symbols has a lower velocity below approximately 8,000 ft where the shale mass was encountered. From other information it

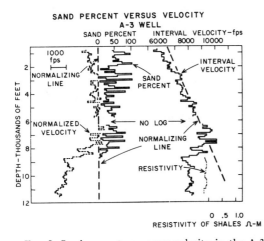

Fig. 8. Sand percentage versus velocity in the A-3 well, block 113, Ship Shoal area. Again this shows the same information as indicated in Figure 7. However, note the phenomenal low velocity mass in the all-shale section between 9,000 and 12,000 ft depth.

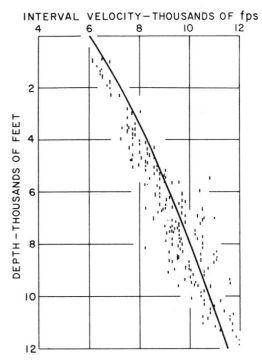

Fig. 9: Plot of sand velocity versus depth from a composite of several wells in the example area. Velocities of sands from several wells were plotted on a composite plot to develop an average interval sand velocity curve for the area.

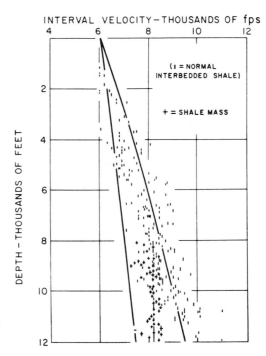

Fig. 10. Plot of shale velocity versus depth from a composite of wells in the example area. The different symbols used for the shale mass in the lower parts of the holes show the predominantly lower velocity existing in deep shale masses.

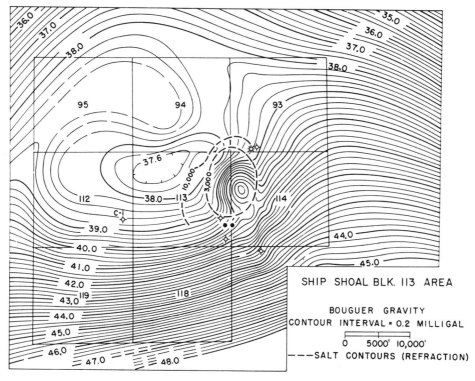

Fig. 11. Bouguer gravity map of the example block 113 Ship Shoal area. The large minimum anomaly and the small maximum anomaly in the eastern central portion of this minimum are shown.

appears that the second line increasing from about 6,000 to about 7,500 ft/sec at 12,000 ft is more representative of what might be expected for the shale mass. Here the shale mass velocities are somewhat higher than this line, but the line does serve as a lower limit of that information.

Density

Figure 11 shows a Bouguer gravity map of the area with a large minimum indicated and a positive due to the shallow salt mass where the large stock penetrates near the surface. From this map, a residual was prepared as shown in Figure 12 which indicates that a large minimum exists due to the salt mass and the shale mass combined to form the domal mass. Note that a sharp positive occurs due to the salt spire at the end of the salt mass. An attempt was made to determine what portion of this anomaly is due to the shale and to the salt that has been mapped by refraction down to 15,000 ft by making the density assumptions indicated in Figure 13. These assumptions are that salt has a constant density of about 2.15 from the surface down, that the density of the normal section increases from about 1.95 to about 2.45 at 15,000 ft, and that the shale mass increases in density but at a lesser rate from about 2.10 to about 2.25 at 15,000 ft. From these density assumptions and the map in Figure 2, a calculation was made of the gravity effect due to this structure above 15,000 ft. The gravity effect from this one possible solution is shown in Figure 14. The difference between this and the residual map shown in Figure 12 is due to at least two things. First of all, the assumptions may be incorrect and secondly; and most important, we have neglected the effect of the material that exists between 15,000 ft and the base of the salt at a depth of 40,000 ft according to some estimates. It can be seen that the configuration is right qualitatively. The calculated magnitude of 2.0 mgal is over one-third of the 5.6 mgal anomaly measured from the map. Possibly the balance of the anomaly could be accounted for by a similar calculation if the salt and shale configuration were known between 15,000 ft and 40,000 ft.

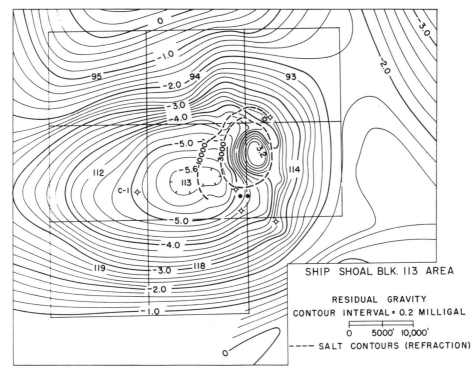

Fig. 12. The residual gravity map of the example block 113 Ship Shoal area. The large minimum anomaly and the small maximum anomaly in the eastern central portion of this minimum are shown.

Pressure in shale masses

Normal pressure measured for the fluids within sediments in the Gulf of Mexico is 0.465 times that of the overburden pressure. This ratio (λ) of fluid pressure to overburden pressure for the average shale or mudstone (Figure 15) at 12,000 ft is 0.9 if its porosity is 20 percent; conversely, if the fluid pressure is 0.9 of the overburden pressure at this depth, it indicates that the porosity of the shale is 20 percent. Of course, this porosity is not necessarily related to permeability, but it still may be the reason why the low densities are measured. Figure 16 is a diagrammatic representation of the problems which occur in connection with a shale mass. The lithology is indicated along the right-hand side of the figure. The next line to the left indicates the velocity configuration which might exist, indicating a relatively smooth increase in velocity through the normal interbedded sands and shales, a sloping dropback in velocity at the top of a hypothetical shale mass at 10,000 ft, and a relatively low interval velocity within the shale mass. This is actually accounted for by the next curve to the left which indicates that the geostatic overburden pressure increases at the rate of one pound per square inch per foot of depth. This is the normal increase with depth of burial. As stated above, the normal hydrostatic

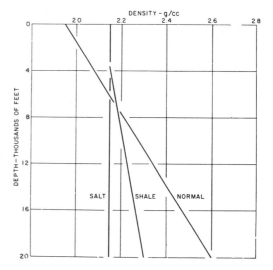

Fig. 13. Assumed density contrast for salt, shale, and normal section in the example area.

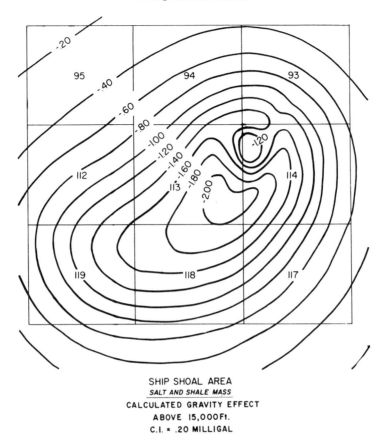

Fig. 14. Calculated gravity effect for sediments above 15,000 ft in the example area. Minimum effect is due to domal mass including salt and shale.

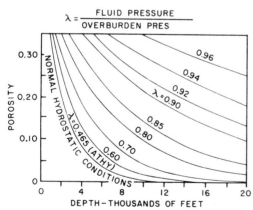

Fig. 15. The lambda values for an average shale or mudstone where lambda equals fluid pressure divided by overburden pressure. This is a plot of depth versus porosity for various values of lambda (After Rubey and Hubbert, 1959).

pressure increases at 0.465 psi/ft of depth, and, therefore, down to 10,000 ft we have an increasing effective pressure between these two lines. This largely accounts for the increase in velocity. Then as we move into the shale mass the hydrostatic pressure increases to 0.9 of psi/ft. Then the differential or effective pressure reduces considerably, and this causes the velocity to drop back to a smaller value (Hicks and Berry, 1956).

Seismic refraction

The next discussion concerns the mapping of a shale mass by refraction methods and the difficulties encountered in trying to map a low velocity material. Figure 17 represents a time versus instantaneous velocity plot representative of a velocity configuration for which a wavefront chart was drawn as indicated in the lower portion of Figure 18. It can be seen that raypaths from 0.020 to 0.100 pass through the upper two layers and

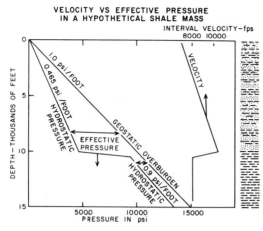

Fig. 16. Velocity versus effective pressure in a hypothetical shale mass. The right-hand side shows a stratigraphic column in which interbedded shales overlie a large shale mass. The next curve shows the velocity distribution for this type of section, showing the gradual increase in velocity down to the shale mass, a rapid but not instantaneous decrease in velocity, and then a slight increase with depth of the shale. The left-hand portion of the figure shows the normal geostatic overburden pressure of one psi/ft the normal hydrostatic pressure of .465 psi/ft and the effective pressure being the difference between these. However, within the shale

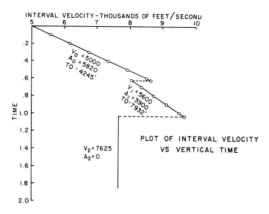

Fig. 17. Represents a plot of interval velocity versus vertical time, showing the beginning velocity and the acceleration constants for the equation of instantaneous velocity, $v = v_0 + at$ for linear increase in velocity with time.

mass the hydrostatic pressure increases to nearly 0.9 psi/ft and the effective pressure decreases very rapidly and remains very small.

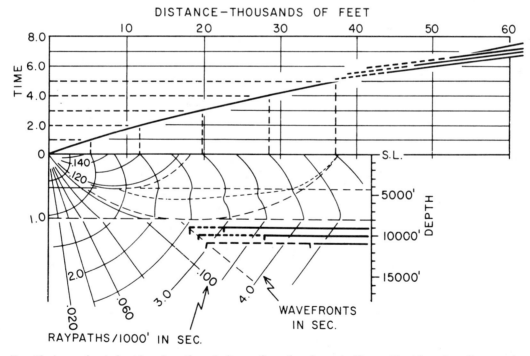

Fig. 18. A wavefront chart based on the velocity configuration shown in Figure 17 with a time distance plot showing the surface refraction that would be observed for various salt configurations and depths within the shale.

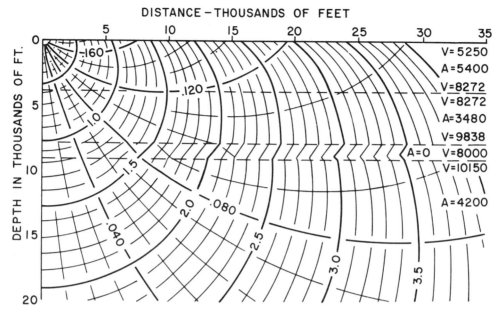

Fig. 19. A wavefront chart developed with the constants shown along the right-hand side and used as the basis for the hypothetical plots of shale masses down in Figure 20 and 21.

into the shale mass. They have no means of returning and are lost forever unless some high velocity material is encountered. Even so the low velocity shale acts as a shield to prevent early first break refractions from being obtained. For instance, it is shown that if a salt dome exists at 9,000 ft depth, 1,000 ft below the top of the shale, the first 5,000 ft of salt will occur as a secondary break and that only when the salt extends longer than 5,000 ft at this level will it occur as a source of first break energy. Likewise, it is shown that if it is buried 2,000 ft within the shale, it is necessary for the stock to be more than 10,000 ft long in order for it to show as a first break; and if it is 3,000 ft within the shale it will be required to be 15,000 ft long before it will show on a first break, thus indicating the shielding power of the low velocity material. This amply indicates that refraction methods cannot be used to map a low velocity material such as a shale mass when it has low relief. Such mapping is only possible if the shale mass has vertical or near vertical sides. However, if the receiver and the source can be placed on opposite sides of the interface, a good map can be made by refraction methods. In order to demonstrate this, some basic information is taken from an earlier paper on Musgrave, Woolley and Gray (1959). The wavefront chart in Figure 19 was used to develop the information shown in Figure 20. In this figure a well drilled into a shale mass from 10,000 ft to 18,000 ft was used to map the shale by methods of refraction loci as described in the paper on salt domes. Note that the loci curves fall within the low velocity material which in this case is the shale rather than the normal section. The tangent to these loci represents the map of the shale. Also note that the shortest time path passes through a minimum amount of the shale material; therefore a flat lying shale bed can only be mapped a short distance from the well. Figure 21 indicates the same conditions except that the shale mass has a steeper flank. In this case, it is possible to map almost as deep as the total depth of the well and determine the limits of the shale.

Reflection

The next problem to be discussed is the error in reflection mapping due to velocity variations caused by the shale mass. Figure 22 shows a shale mass which is either depositional or diapiric. Here it is shown as a depositional mass. Assuming that the velocity is measured along either flank in the normal section with a normal increase in velocity with depth down to 14,000 ft and that the shale mass has a velocity of 8,000 ft/sec, a depth error

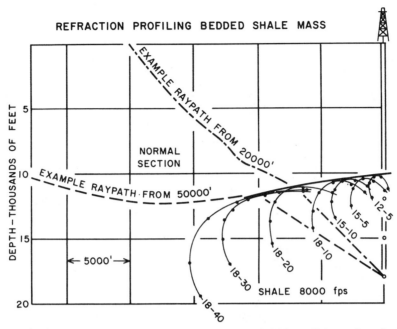

Fig. 20. A refraction profile of a bedded or depositional shale mass in which a well shown along the right-hand side is drilled 8,000 ft into the shale mass and recording points along the surface extending to the left are used to record the information needed to draw the refraction or least time loci outlining the shale mass. Each loci is tagged by two numbers; the first is the receiver depth in thousands of feet and the second is the horizontal distance to shot in thousands of feet.

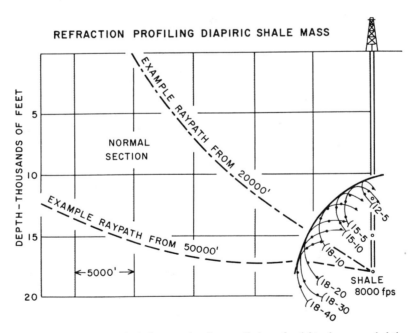

Fig. 21. A refraction profile of a diapiric shale mass showing a well along the right where recorded times are used in conjunction with the wavefront chart to map the diapiric shale outline.

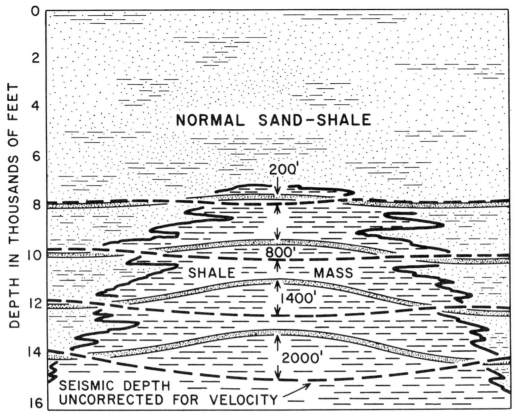

Fig. 22. Depth errors from assumed velocity in the presence of a shale mass. Dashed lines indicate the map position of the sand if allowance is not made for the change in velocity within the shale mass, thus showing that depth errors up to 2,000 ft can occur within reasonable drilling depths simply because of low velocity shale mass which is not accounted for in seismic reflection mapping.

would be encountered if a sand member could be mapped through at any of the shown depths. Thus if an 8,000-ft sand exists, a 200-ft depth error is shown; if a 10,000-ft sand exists, it is mapped 800 ft too low. At 12,000 ft it is mapped 1,400 ft too low and at 14,000 ft it is mapped 2,000 ft too low. Thus an actual anticline of 1,000 ft in magnitude is mapped as syncline 1,000 ft in magnitude just because the velocity is known to have changed in this amount due to a massive shale existing above the mapped sand. Thus areas that look unprospective and not worth the measuring of velocities may well be important when the velocities are known.

Another problem in reflection mapping is that a cutout may exist on the section. This may mean that a homogeneous mass exists. This may be either salt or shale. From reflections it is impossible to tell the difference unless it is possible to measure the velocity through the material by reflections from the under side. If the mass is buried a few thousand feet, multiples may exist within the mass which makes it very difficult to observe an actual coutout as shown in Figure 23. The shale mass, however, may sometimes be determined by a measure of the reflection amplitude as indicated in Figure 24. The upper part of the figure is a variable area section showing random lineups within a diapiric mass and the lower portion is the measure of the relative amplitude over this diapir. The amplitude trace, which is plotted relative to a normalized line, shows that higher amplitudes exist where the reflections occur, and lower amplitudes exist where the multiples occur in the center portion of the line. This idea is developed a little further diagrammatically in Figure 25. This may be either a depositional or a diapiric mass, but in this case again assume that it

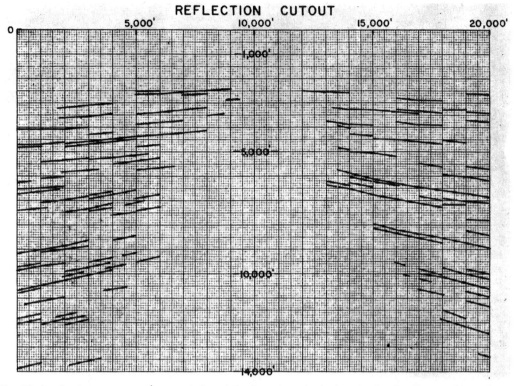

Fig. 23. A reflection cutout on the normal plotted depth section of seismic reflection depths caused by a homogeneous mass of salt or shale.

is a depositional mass with sand layers terminating against the shale flank. Reflections normally occur at these interfaces within the normal sand shale sequence, and multiples exist within the shale mass. Since there is a structural anomaly at the top of the shale mass, multiples within it show increase in structure with depth; and if phantom mapping is used, it is possible that a group of maps could be drawn that would indicate increase in structure within the shale mass and possibly a well might be drilled on these multiples. This could all be avoided if proper precautions were taken to measure the velocity and identify multiples. The shale interface could be mapped and the well drilled into the potential reservoirs on the flank of the shale mass.

SUMMARY

It appears that large shale masses occur which have particular properties; these are probably due to lack of permeability and to the fact that the anomalous shales entrap water that would be forced out if permeability permitted. This causes them to have such properties as high fluid pressures, low velocities, low densities, and low resistivities. These physical properties allow the outlining of the shale mass by one or more of the following ways: the gravity method is used to outline the low density material, the seismic reflection method is used to outline the lack of reflection contrast and in some cases map the velocity configuration, the seismic refraction method is used to indicate the velocity of the anomalous mass thereby differentiating between shale and salt.

ACKNOWLEDGMENTS

We wish to express thanks to Mobil Oil Company for allowing this material to be published.

REFERENCES

Atwater, Gordon I., and Forman, McClain J., 1959, Nature of growth of South Louisiana salt domes and its effect on petroleum accumulation: Bull. A.A.P.G., v. 43, p. 2592–2622.

Hicks, W. G., and Berry, J. E., 1956, Application of continuous velocity logs to determination of fluid saturation of reservoir rocks: Geophysics, v. 21, p. 739–754.

Hubbert, M. King, and Rubey, W. W., 1959, I. Me-

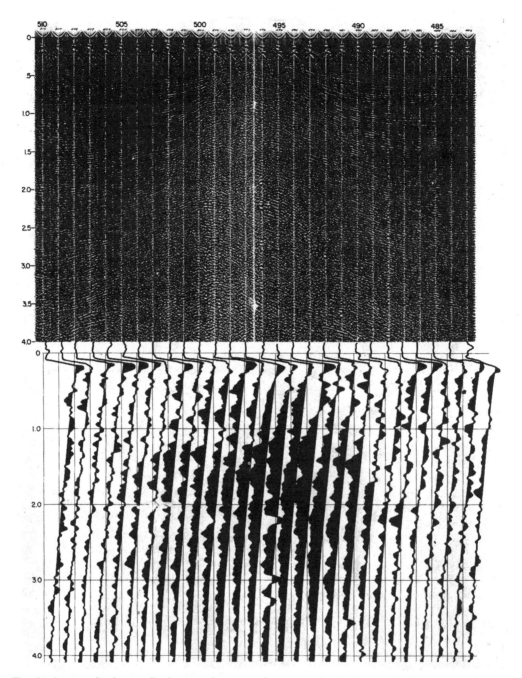

Fig. 24. A true reflection amplitude comparison over a large anomaly of salt or shale. The lower portion of the figure shows the low amplitude of the energy in the vicinity of the anomaly. The curves represent the logarithmic plot of the amplitude of the signal versus time. High amplitude is shown to the right and low amplitude is shown to the left of the normalizing curve.

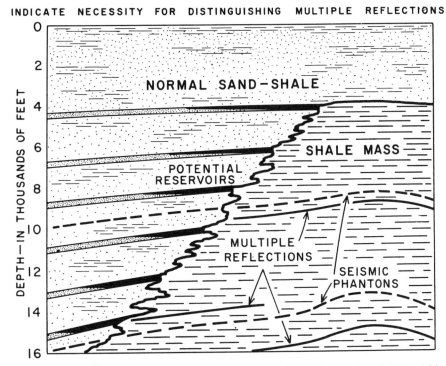

Fig. 25. Hydrocarbons may be trapped by shale mass. The figure shows the necessity of distinguishing between multiple reflections, which may occur as the predominant lineups within a shale mass, and the true reflections in a normal sand shale sequence.

chanics of fluid filled porous solids and its application to overthrust faulting: Bull. G.S.A., v. 70, p. 115–166.

Musgrave, A. W., Woolley, W. C., and Gray, Helen, 1959, Outlining of salt masses by refraction methods: Geophysics, v. 25, p. 141–167.

Rubey, W. W. and Hubbert, M. King, 1959, II. Overthrust belt in geosynclinal area of western Wyoming in light of fluid-pressure hypothesis: Bull. G.S.A., v. 70, p. 167–206.

Tsimel'zon, I. O., 1959, The relationships between local gravitational anomalies and the tectonics of gas and oil bearing regions: Izvestiya Vysshikh Uchebnykh Zavedenii. Neft'i Gaz., no. 8, p. 13–16.

Wallace, W. E., 1965, Abnormal subsurface pressures measured from conductivity or resistivity logs: Log Analyst, v. 5, p. 26–38.

GEOPHYSICS

Three-dimensional seismic monitoring of an enhanced oil recovery process

Robert J. Greaves* and Terrance J. Fulp*

ABSTRACT

Seismic reflection data were used to monitor the progress of an in-situ combustion, enhanced oil recovery process. Three sets of three-dimensional (3-D) data were collected during a one-year period in order to map the extent and directions of propagation in time. Acquisition and processing parameters were identical for each survey so that direct one-to-one comparison of traces could be made. Seismic attributes were calculated for each common-depth-point data set, and in a unique application of seismic reflection data, the preburn attributes were subtracted from the midburn and postburn attributes. The resulting "difference volumes" of 3-D seismic data showed anomalies which were the basis for the interpretation shown in this case study.

Profiles and horizon slices from the data sets clearly show the initiation and development of a bright spot in the reflection from the top of the reservoir and a dim spot in the reflection from a limestone below it. Interpretation of these anomalies is supported by information from postburn coring. The bright spot was caused by increased gas saturation along the top-of-reservoir boundary. From postburn core data, a map of burn volume distribution was made. In comparison, the bright spot covered a greater area, and it was concluded that combustion and injection gases had propagated ahead of the actual combustion zone. The dim spot anomaly shows good correlation with the burn volume in distribution and direction. Evidence from postburn logs supports the conclusion that the burn substantially decreased seismic velocity and increased seismic attenuation in the reservoir. Net burn thicknesses measured in the cores were used to calibrate the dim-spot amplitude. With this calibration, the dim-spot amplitude at each common depth point was inverted to net burn thickness and a map of estimated burn thickness was made from the seismic data.

INTRODUCTION

Improving the efficiency of reservoir production can increase proven reserves. The final stages in the production of a field are enhanced oil recovery (EOR) processes. Effective management of EOR processes requires detailed reservoir description and observations of the volume of the reservoir being swept by the process. High-resolution 3-D reflection seismic surveying can be an effective tool in obtaining reservoir description, and, as demonstrated by this case study, can in some cases actually map the EOR process as it proceeds.

In this case study, 3-D seismic reflection data were used to monitor the propagation of a pilot in-situ combustion (fireflood) process. Three identical 3-D seismic surveys were recorded over the pilot site at preburn, midburn, and postburn times. In this way, the combustion propagation was monitored over (calendar) time.

Acquisition and computer processing of the data were iden-

Manuscript received by the Editor August 18, 1986; revised manuscript received January 26, 1987.
*ARCO Oil and Gas Company, Research and Technical Services, 2300 West Plano Parkway, Plano, TX 75075.
© 1987 Society of Exploration Geophysicists. All rights reserved.

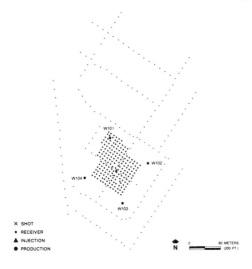

FIG. 1. 3-D seismic survey shot and receiver geometry with locations of production and injection wells.

tical for each set of survey data, so that a direct comparison of the individual data sets could be made. To facilitate interpretation, the attributes of the seismic traces were calculated using Hilbert transform techniques as described by Taner and Sheriff (1977). Reflection strength, in this paper referred to as "envelope amplitude," was then used in the analysis of the reflection seismic data. In a unique application of reflection seismic data, the envelope amplitude traces from the preburn data volume were subtracted from their counterpart traces in the midburn and postburn data volumes, generating "difference volumes."

The combustion process substantially increased in-situ temperature and gas saturation in those sections of the reservoir affected by the burn. Both seismic velocity and density of the reservoir were changed. Zones with altered properties were detected by anomalous amplitude responses in the reflection from the top of the reservoir and from a limestone formation directly below the reservoir. The direction of the combustion propagation and estimates of its volume were based on the interpretation of these anomalies as observed in the difference volumes. The interpretation was supported by data available from monitor wells and postburn coring.

BACKGROUND

Three 3-D seismic surveys were shot over a period of 15 months. The first (preburn) survey was recorded several months previous to ignition of the combustion process. The second (midburn) survey was recorded four months after ignition, and the final (postburn) survey was shot ten months after ignition.

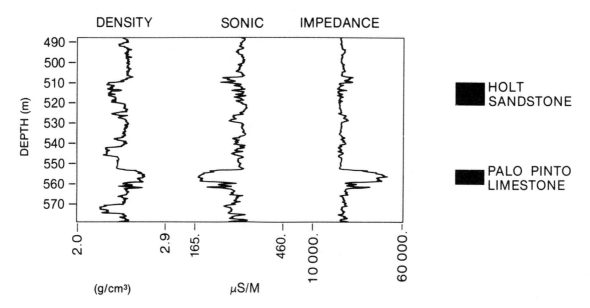

FIG. 2. Example of sonic and density logs with calculated impedance for the stratigraphic section including the Holt sandstone (reservoir) and the Palo Pinto limestone.

The objectives of the seismic program were to

(1) detect a change in seismic reflection character attributable to the combustion process,
(2) determine the direction of burnfront propagation, and
(3) determine the volume of reservoir swept by the combustion process.

The basic premise was that an increase in gas saturation in the reservoir formation would produce measurable changes in reflection amplitude. Bright spots and dim spots, caused by anomalous gas concentrations, are well-known phenomena in exploration seismology. Increased gas saturation in the parts of the reservoir reached by the combustion process was expected to create bright spots and dim spots in the shadow zone (Sheriff, 1980). The 3-D data would be used to map that progression in time.

The EOR program consisted of a five-well pilot test covering a very small portion of the Holt Field in north-central Texas. The test consisted of four production wells separated by 90 m (300 ft) with a central injection well (Figure 1). The engineering objective was to propagate the combustion process from the injection well radially outward, creating and flushing an increased oil saturation zone, the oil bank, toward the production wells. Although the concept is simple, the implementation is quite difficult and is very sensitive to the details of the reservoir geology.

The reservoir is the Holt sand, a 12 m (40 ft) thick sandstone capped by a 2.5 m (8 ft) thick limestone encased as a unit in thick shale (Figure 2.) The sand is silty and laced with shale stringers and some calcite cementation zones. In this

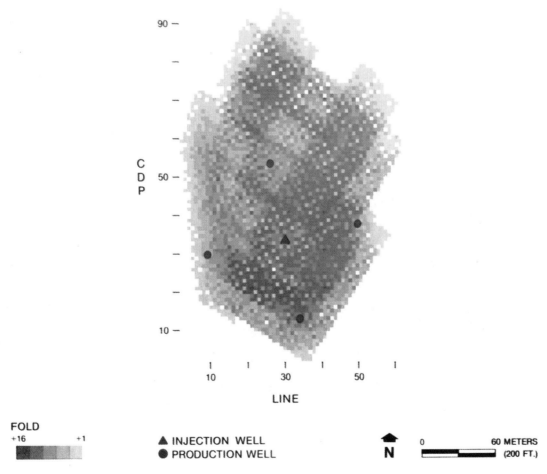

FIG. 3. CDP fold distribution of the seismic surveys. Each CDP bin covers a 3 × 3 m (10 × 10 ft) area.

part of the field, the sand occurs at about 500 m (1650 ft) and dips to the north at 10 degrees. A thin limestone occurs about 45 m (150 ft) below the reservoir and is identified as the Palo Pinto limestone. From extensive core analysis, the horizontal permeability of the sand was found to be several times greater than the vertical permeability. Numerous fractures were observed, and an average orientation of N27E was measured. Although a detailed model of the reservoir and the burn process was not constructed, some effects of the process were anticipated. The combustion process would primarily propagate updip (to the south) due to the differing fluid densities. Propagation would primarily occur laterally within the reservoir from the initiation points. Vertical propagation would be limited to fractures or other natural permeability pathways. Finally, propagation might be further guided to the southwest along fracture-induced permeability pathways.

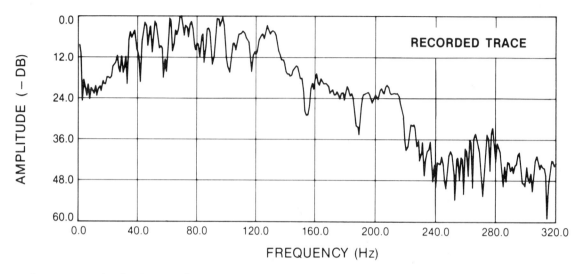

a. Pre-processed seismic trace from a shot record.

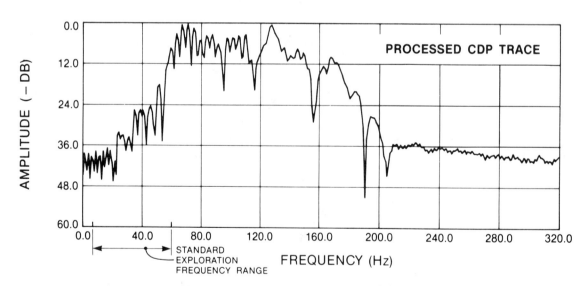

FIG. 4. An example of the power spectra of the trace data before and after processing. The spectra are for the window from 0.350–0.750 s which includes the reflection sequence from the Holt sandstone, to and including the Palo Pinto limestone.

ACQUISITION AND PROCESSING

Several factors guided the choices of acquisition parameters:

(1) the target area was very small (90 m × 90 m);
(2) the target was relatively shallow (500 m);
(3) the data collection was to be repeated as identically as possible; and
(4) the amplitudes and spatial extents of the seismic anomalies would probably be quite small.

Simple seismic modeling based on well logs and the anticipated effect of increased gas saturation indicated that *detection* of the burnfront would be straightforward. However, very high-resolution seismic data would be required to map the lateral extent of the process and determine the net burn volume. A calculation of the resolution limit was made based on Widess (1973), and the center frequency required to resolve 7.5 m (25 ft) vertically was determined to be 100 Hz. The necessary resolution was felt to be achievable, given the shallow depth of the reservoir, if proper acquisition and processing techniques were applied.

The data collection array consisted of a modified 3-D patch geometry. Figure 1 shows the positions of shotpoints and receiver stations within the test area. This patch style of survey allowed the shot pattern to be arranged as necessary, such that the CDP data were collected with high fold over the area of primary interest (Figure 3) even though surface access was limited by buildings, wells, pipelines, etc. Furthermore, the geophones could be permanently installed at each receiver location to guarantee that the receiver array would be duplicated in each survey. A modification of the simple patch geometry was made to account for the migration of reflection points updip, by extending shot and receiver locations downdip (to the north).

The receiver group spacing was 6 m (20 ft) with a single high-frequency (40 Hz) marsh geophone comprising each group. Each of the 182 receivers was buried 6 m (20 ft) below the surface. The 165 shotpoints were distributed along crossed lines with a 12 m (40 ft) spacing between individual shots. Each shot consisted of a 2.5 kg (3 lb) dynamite charge buried at 23 m (75 ft). The recording system used was a 192-channel GUS-BUS with sampling interval of 1 ms and band-pass recording filters set at 50 Hz low-cut and 320 Hz antialias. The burial of both shots and receivers improved the signal-to-noise (S/N) ratio of the recorded data by eliminating air-wave noise and substantially reducing the amplitude of the surface-wave noise. The 50 Hz low-cut filter was chosen at this high level to eliminate surface-wave noise further and to preserve the dynamic range of the recording system for digitization of the desired high-frequency signal. For recording shallow reflection seismic data, it is especially important to eliminate surface-wave noise that can seriously degrade the quality of the shallow data window. The small, deeply buried shots, the high-frequency phones and the low-cut filter all combined to eliminate this problem. The resulting frequency range of the recorded data was substantially higher than the range of standard exploration seismic, as shown in Figure 4a, and yet the range retained the minimum two-octave bandwidth considered necessary for high resolution.

The computer processing of the 3-D data sets used a standard sequence designed for 3-D CDP data. Throughout the processing, extra care was taken to retain true relative amplitude and the maximum usable frequency range. The traces were gathered into 3 × 3 m (10 × 10 ft) CDP bins. The statics and normal-moveout (NMO) corrections were quite small due to the simple geologic structure and small area of the test. For more structurally complex geology, it would have been more difficult to make proper velocity adjustments because the patch geometry has the disadvantages of uneven fold and offset distribution. Three-dimensional surface-consistent statics were computed and were found to be on the order of 2 to 3 ms. Normal-moveout corrections were applied using a datumed root-mean-square (rms) velocity function derived from the well control. Standard spiking deconvolution was applied before stack. A phaseless deconvolution technique was applied to balance further the usable spectrum of each stacked trace. In the final processing step, the data were migrated using an f-k migration algorithm and the velocity function derived from the sonic logs. This approach to migration was deemed adequate, given the localized area of interest and the simple velocity structure. The frequency spectrum of a fully processed trace, windowed in the reflection zone of interest, indicates that the 40–180 Hz bandwidth of the recorded data was enhanced during processing and the center frequency of 100 Hz was obtained (Figure 4b).

As postprocessing steps, the data were properly phase-corrected using well control, and the seismic attributes were calculated for each data set. To remove the geologic structure from the reflectors of interest, static adjustments were made, thereby allowing horizon views to be sliced from the 3-D data volume. Finally, in a unique step, the preburn horizon envelope amplitude at each level of interest was subtracted from the corresponding values in the midburn and postburn data volumes. The preburn data were used as the baseline seismic expression relative to which change was observed. Anomalies in the difference volumes were then interpreted directly.

OBSERVED ANOMALIES

Bright spots

Comparison of the envelope amplitudes of the reflection event at the top of the Holt sand reservoir revealed an increase in amplitude, a "bright spot," which developed after the combustion process was initiated. In Figure 5, a 2-D north-south section, line 14, is shown as it appeared at preburn, midburn, and postburn times. The reflection from the top of the Holt sand is identified as a trough occurring at about 385 ms. At this horizon, the envelope amplitudes at preburn time compared to midburn time show a zone of increased amplitude near well W104, with maximum change between CDP 16 and CDP 30. By postburn time, the bright spot had increased in lateral extent from CDP 16 to CDP 36, but it had not increased in maximum amplitude relative to midburn time.

Horizon slices at the top of the Holt sand, from the envelope amplitude difference volumes, are displayed in Figures 6a and 6b. The midburn difference shows a positive amplitude anomaly in the southwestern side of the data. This corresponds to the bright-spot development observed in line 14,

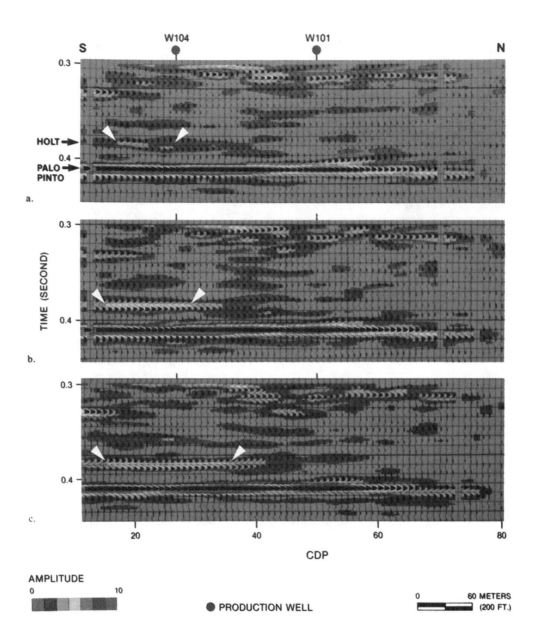

FIG. 5. Line 14, from (a) preburn, (b) midburn, and (c) postburn 3-D seismic data volumes. The reflection wiggle traces are overlain by a color scale of the calculated envelope amplitude. Dip was removed by static shifts before display. A bright spot was created (see arrows) at the top of the Holt sandstone by midburn time (b), and it increases in extent by postburn time (c). A dim spot in the reflection from the Palo Pinto limestone formed just below the peak of the bright spot.

Figure 5, at midburn time. Another, smaller bright spot is located to the southeast of the injection well at line 43, CDP 21. The difference amplitude at postburn time, Figure 6b, shows that the bright spot has grown to cover most of the area within the production wells, the midburn peak to the southwest has shifted downdip toward well W104, and the maximum amplitude of the difference anomaly has increased by about 10 percent.

Dim spots

The strong reflection centered at 410 ms in line 14 (Figure 5) is identified as the Palo Pinto limestone. In line 14, a slight decrease in envelope amplitude occurs in the shadow of the bright spot centered around CDP 22. At midburn time, the decrease in amplitude is about 10 percent, but by postburn time, the decrease is nearly 25 percent, as marked by the change from deep orange and red shading to yellow.

A similar display of another north-south section, line 33 in Figure 7, shows a more substantial dim spot. This anomaly does not coincide with any bright spot at the Holt level at midburn time and only a modest Holt bright spot at postburn time. The dim-spot anomaly, pointed out by the arrows within the figure, is also stronger at midburn time than at postburn time. This lack of spatial coincidence (between bright and dim spots) is important in the interpretation of the results as described below.

The difference slice at the Palo Pinto reflection (Figure 8) clearly shows this anomaly. The dim spot at midburn time, Figure 8a, covers much of the pilot area with two negative-amplitude anomalies. One peak is located at the injection well, but the stronger peak lies about 30 m (100 ft) south-southwest

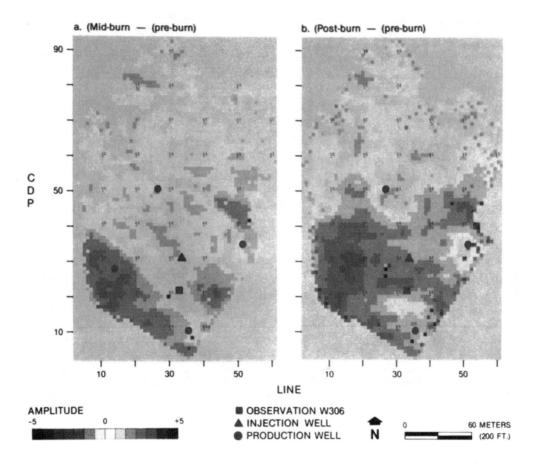

FIG. 6. The difference in envelope amplitude at the top of the Holt sandstone (.385 s) displayed in horizontal time-slice form. Bright spots occur as positive anomalies. Well locations are marked for position in the subsurface at the top of the Holt sandstone.

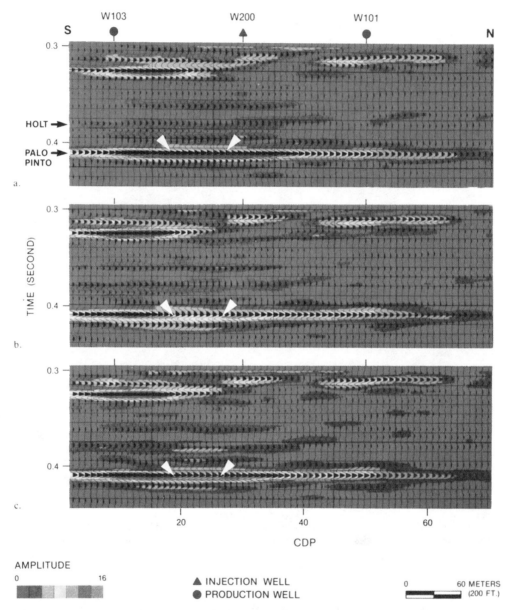

FIG. 7. Line 33, from (a) preburn, (b) midburn, and (c) postburn 3-D seismic data volumes. The reflection wiggle traces are overlain by a color scale of the calculated envelope amplitude. Dip was removed by static shifts before display. A dim spot was created (see arrows) in the Palo Pinto reflection (.410 s) by midburn (b), but it decreases somewhat by postburn (c).

of the injection well. The anomalies do not coincide with bright spots in the Holt reflection. A lower amplitude lobe of the dim spot extending to the southwest edge of the data does correlate with the maximum bright spot observed at midburn time. The dim spot observed at postburn time is lower in amplitude and extends over a significantly smaller portion of the pilot area. The two peaks of the midburn anomaly have merged into a ridge extending approximately southwest-northeast across the injection well with the larger area and peak of the anomaly to the southwest of the injection well.

INTERPRETATION

Combustion model

A simple model of the in-situ combustion process, based on combustion tube experiments, was described by Tadema (1959). The combustion process within the reservoir can be divided into various zones, with each zone defined by its relative temperature and fluid saturations. The "combustion zone" propagates through the reservoir and is defined by maximum oxidation of the heaviest, or immobile, hydrocarbons. In its wake is left the "clean-burnt sand," a hot reservoir matrix with high gas saturation. Ahead of the combustion zone are several zones at lower temperatures and with distinctive percentages of oil, water, and gases until at some distance the original reservoir temperature and fluid mixture are encountered. Of particular interest are (1) that the clean-burnt sand zone has been subjected to very high temperatures, and (2) that combustion gas, as well as some injection gas, are forced ahead of the combustion zone. If this model were expanded into three dimensions, it would consist of a series of concentric rings which propagate radially from the injection well. The model is quite simple and does not at all account for geologic complexities, but it is useful as a starting point for the

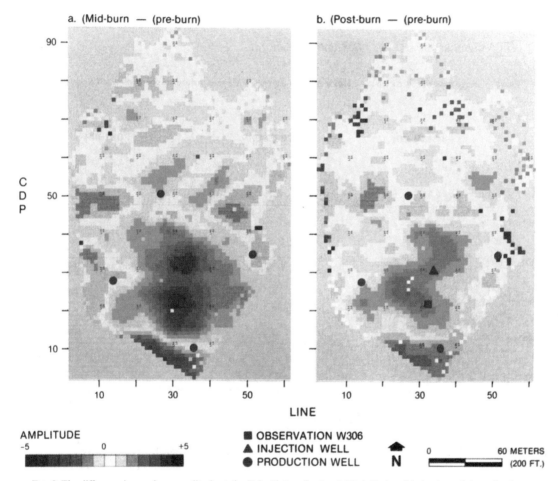

FIG. 8. The difference in envelope amplitude at the Palo Pinto reflection (.410 s) displayed in horizontal time-slice form. Dim spots occur as negative anomalies. Well locations are marked for position in the subsurface at the top of the Holt sandstone.

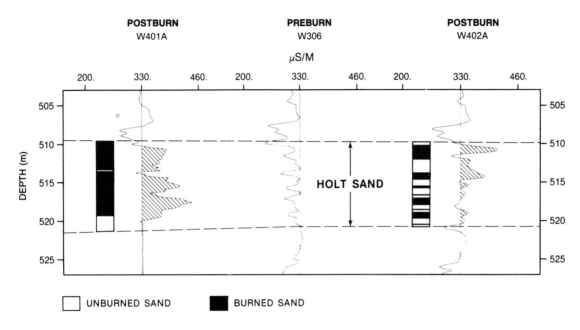

FIG. 9a. Comparison of sonic traveltime logs from preburn well W306 and postburn core wells W401A and W402A.

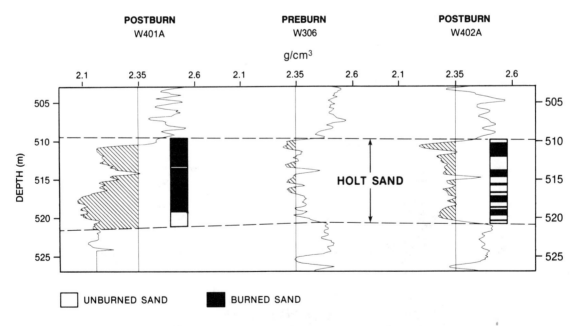

FIG. 9b. Comparison of density logs from preburn well W306 and postburn core wells W401A and W402A.

interpretation of the seismic anomalies in terms of the physical process of in-situ combustion.

Generalizing this model to the Holt sand reservoir, it is important to point out that the Holt sand reservoir, at the initiation of the burn, had little to no gas saturation.

Detection

As stated, the first objective of the seismic program was to detect a change in reflection character attributable to the combustion process. The bright spots and dim spots are considered true combustion-caused anomalies for the following reasons. First, the changes do occur in the reflection from the reservoir and the reflection just below it, as expected. Searches through the difference data volumes substantially above and below the reservoir reflection showed no extended, coherent anomalies. Second, the background noise level of the difference data volumes is substantially lower than the observed anomalies. Figures 6 and 8 show this by the amplitudes to the north of the injection well. Most importantly, the seismic anomalies were confirmed by well-log and core data. At the time postburn seismic data were collected, several cores and logs through the burned zone were collected.

In Figure 9, density and sonic traveltime logs showing the reservoir sand from a preburn observation well, W306, are compared to similar logs collected in two postburn boreholes. Within the zones of clean-burnt sand, the logs show substantial decreases in both density and velocity.

The combustion process was expected to increase the gas saturation with consequent changes in density and velocity. Comparison of log density values showed decreased density in burned zones averaging about 5 percent. This density decrease can be fully accounted for by a change from 100 percent fluid-filled pores to partial gas saturation. The sonic-log velocities measured in burned zones decreased 15 percent to 35 percent, averaging 25 percent. This decrease in velocity is much greater than can be accounted for by increased gas saturation in the original pore space.

Ultrasonic measurements made on preburn core showed a 3 to 4 percent decrease in velocity, going from 100 percent water saturation to 100 percent gas saturation. Although this is a smaller decrease than that reported by Domenico (1976), a similar result was reported by Frisillo and Stewart (1980). In our case, the larger effect on velocity was due to permanent alteration of the rock matrix by the very high formation temperatures. Ultrasonic measurements showed a 25 percent decrease in velocity for cores heated to 700°F. The velocity decrease may be due to weakening of the rock by oxidation of organics and an alteration of clays. Therefore, the observed velocity decrease is the combined effect of changes in fluid saturation and of damage to the rock matrix.

The effect of increasing gas saturation on the seismic response is a nonlinear relation. As shown by Domenico (1974), it is the first few percentage increases in gas saturation (up to about 10 percent) that affect the seismic impedance the most. Further increases in gas saturation change the impedance very little. Therefore, if combustion gas is forced ahead of the burn zone in sufficient volume to increase gas saturation even a few percent, what is observed as a bright spot is both the clean-burnt zone and the zones ahead of the combustion point reached by steam and combustion gas.

Certainly the phenomenon of attenuation is even more complex. Similar to seismic impedance, it is the initial change from 100 percent water saturation that increases the attenuation substantially. However, unlike impedance, attenuation decreases as the rock reaches 100 percent gas saturation (Frisillo and Stewart, 1980). Therefore, in the initial stages of the process, the dim spot will reflect both the clean-burnt zone and the zone reached by combustion gases. However, as 100 percent gas saturation is reached, the dim spot will more likely be an indication of just the clean-burnt zone.

Propagation

Interpretation of the positions of the bright-spot and dim-spot anomalies over calendar time provides a reasonable description of the combustion propagation. First, it is quite clear from Figures 8a and 8b that the area around well W101 (to the north of the injection well) was not affected by the combustion process. This well was the only production well in which large quantities of gases were not observed. Therefore, this well was either too far downdip to be reached by the combustion process or was isolated by permeability barriers.

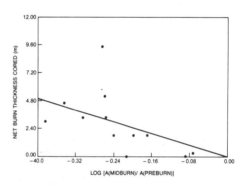

FIG. 10a. Net burn thickness from postburn cores versus natural logarithm of the ratio of midburn over preburn dim-spot (Palo Pinto) amplitude. The line is a least-squares fit to the data points.

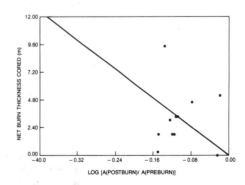

FIG. 10b. Net burn thickness from postburn cores versus natural logarithm of the ratio of postburn over preburn dim-spot (Palo Pinto) amplitude. The line is a least-squares fit to the data points.

One can also see from Figure 8 that the process did propagate to the southwest of the injection well, probably guided by the fracture system. The location of the strongest dim spot at midburn time shown in Figure 8a was verified by an observation well W306, which recorded the highest formation temperature at the time of the midburn survey.

The postburn dim spot (Figure 8b) decreased in amplitude and lateral extent compared to midburn time. It appears that the combustion process reached its maximum lateral propagation within the production area by midburn time or soon after. In a full-scale EOR project, this knowledge would be crucial in adjusting the program to sweep the reservoir more efficiently.

Although the midburn bright spot (Figure 6a) also shows that the process moved to the southwest, it does not extend back to the injection well. Therefore, the combustion gas most likely propagated laterally within the reservoir until it encountered a vertical permeability pathway which allowed the gas to stream to the top of the reservoir. Once established, this pathway also allowed the burn to move to the top of the reservoir. Up to that point, the successful part of the burn was contained in the middle of the reservoir.

The postburn bright spot (Figure 6b) increased in area from midburn time. A major fault (providing the southern closure to the field) is located approximately 300 m (1000 ft) to the south of the test. This fault probably blocked further southward propagation of the combustion gases, forcing them back along the top of the reservoir toward the injection well. If sustained injection of gas from midburn to postburn time had continued to fuel the combustion process, that process would have moved out beyond the production area and the area of seismic coverage.

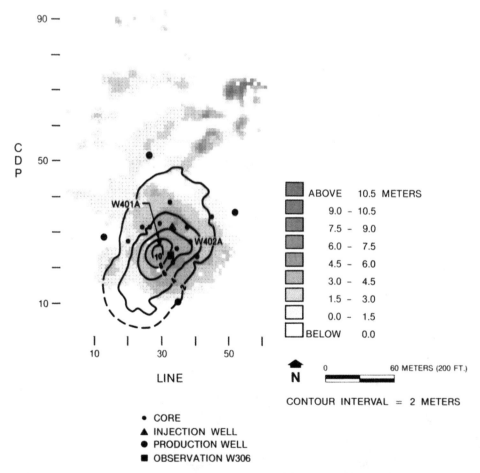

FIG. 11. Burn thickness calculated from midburn dim-spot amplitudes using equation (2) and the slope of the line in Figure 10a for the calibration constant. Overlain is a line contour map of net burn thickness observed in cores.

Burn volume

The final objective of this study was to estimate the volume of reservoir swept by the combustion process. Although the data do not have the spatial resolution to map the detailed distribution of the process, we have attempted to interpret the decreased amplitudes in the Palo Pinto reflection as estimates of burn thickness. The mechanisms of the attenuation are not separable, but several factors are certainly important: pore fluid state and interaction with the rock matrix, formation temperature, matrix velocity and density, and the increased reflectivity due to gas saturation in the reservoir.

A simple mathematical approach was chosen (after Waters, 1978) in which amplitude is expressed as

$$A = A_0 e^{-\alpha z}, \tag{1}$$

where A_0 is the initial amplitude of the propagating wavelet, α is the attenuation parameter, and z is the propagation distance. If two seismic waves are considered identical except that one has passed through a zone Δz where the attenuation is different, then the reflection amplitude from a level past the zone of attenuation can be compared directly to find Δz. For this study, observed seismic waves are the before-burn and after-burn data traces, and Δz is the estimate of burn thickness. After taking the natural logarithms and accounting for two-way propagation, the equation becomes the linear relation

$$\Delta z = \frac{1}{2(\alpha_B - \alpha_A)} \ln\left(\frac{A_A}{A_B}\right). \tag{2}$$

The reservoir was cored in twelve locations within the test pattern at postburn time. The cores confirmed that the burn had occurred in somewhat vertically isolated zones within the reservoir. The net burn thickness observed in each core was compared to the logarithm of the ratio for appropriate seismic amplitudes at the CDPs corresponding to the bottom-hole location of each core. Figure 10a shows the comparison of net burn thickness to the logarithm of midburn amplitude over preburn amplitude, and Figure 10b shows the same relationship for postburn data. Using least-squares estimation, a line was fit through the data points (lines are forced through the origin). The slope of either of these lines could be used in equation (2) to estimate burn thickness at all other CDPs. Since the midburn data (Figure 10a) appeared to fit the simple model better, the midburn data were used to estimate burn thickness. This reemphasizes the belief that burn propagation, at least within the production area, had ceased by midburn time. The postburn dim spot is more likely a map of formation damage only, suggesting that a more complicated model is needed to explain the attenuation due to alteration of the rock fabric.

Using the slope of the line in Figure 10a and equation (2), the midburn data were converted to an estimate of net burn thickness (Figure 11). Overlain on that estimate is a computer-generated contour map based on core data. A good correlation is observed, and the correlation could have been even better if there had been core data to the southwest. This also implies that even without the core data for calibration, a good estimate of relative burn thickness could have been made using only the seismic data. The observation of seismic attenuation is a useful approach in mapping certain recovery processes. Resolution could be improved utilizing borehole-to-borehole techniques.

CONCLUSIONS

Reflection seismic surveying can be used to monitor the progress of some EOR processes. In this case study, a fireflood process was detected, its propagation direction and extent were determined, and an estimate of net burn volume was made.

The 3-D seismic data detected the burn zone and showed that the gas propagated predominantly updip to the southwest. A dim spot observed in a reflector just below the reservoir level was interpreted as a map of areal extent of the burned zone. A region of maximum net burn thickness was located about 30 m (100 ft) from the initiation point of the burn. Comparison of the midburn and postburn dim spots led to the conclusion that the majority of the reservoir swept by the combustion process occurred in the first few months after ignition. The shape, orientation, and volume of the burn interpreted from the seismic data were confirmed by temperature monitor wells and postburn coring.

It was concluded that the attenuation increase, due to high-temperature alterations of the reservoir rock and pore fluid changes, was the best seismic indicator of the combustion process.

The subtraction of the baseline (preburn) data from the midburn and postburn data for interpretation of dynamic anomalies proved to be a very powerful technique. The subtraction technique has great potential for detecting anomalous seismic responses related to active reservoir processes.

ACKNOWLEDGMENT

The authors would like to thank ARCO Oil and Gas Company for allowing us to publish these results. Production Engineering Research provided the financial support for this project. In Geophysical Support, L. F. Konty and D. R. Paschal were instrumental in survey planning. L. J. Hix and P. W. Wise added their expertise in data acquisition. Data processing was designed by R. Chen and S. A. Svatek. M. L. Batzle made the petrophysical measurements on core samples and provided core descriptions. J. D. Robertson contributed support and ideas during the interpretation of the data sets.

REFERENCES

Domenico, S. N., 1974, Effect of water saturation on seismic reflectivity of sand reservoirs encased in shale: Geophysics, **39**, 759–769.
―――― 1976, Effect of brine-gas mixture on velocity in an unconsolidated sand reservoir: Geophysics, **41**, 882–894.
Frisillo, A. L., and Stewart, T. J., 1980, Effect of partial gas/brine saturation on ultrasonic absorption in sandstone: J. Geophys. Res., **85**, 5209–5211.
Sheriff, R. E., 1980, Seismic stratigraphy: Internat. Human Res. Dev. Corp., 185–198.
Tadema, H. J., 1959, Mechanism of oil production by underground combustion: Proc. 5th World Petr. Congress, sec. II, 279–287.
Taner, M. T., and Sheriff, R. E., 1977, Application of amplitude frequency, and other attributes to stratigraphic and hydrocarbon determination: Am. Assn. Petr. Geol. Memoir **26**, 301–302.
Waters, K. H., 1978, Reflection seismology: John Wiley and Sons, 203–207.
Widess, M. B., 1973, How thin is a thin bed?: Geophysics, **38**, 1176–1180.

SHEAR WAVES

SPWLA TWENTY-FOURTH ANNUAL LOGGING SYMPOSIUM, JUNE 27-30, 1983

BASIS FOR INTERPRETATION OF Vp/Vs RATIOS IN COMPLEX LITHOLOGIES
by
Raymond L. Eastwood and John P. Castagna
ARCO Oil and Gas Company

Abstract

This paper explains a systematic basis for interpretation of sonic data, especially Vp/Vs ratios, from full waveform sonic logs in complex lithologies. The Toksoz model (Kuster and Toksoz, 1974a,b; Toksoz, Cheng and Timur, 1976) for calculating effective bulk modulus and rigidity of an aggregate provides this basis and yields Vp, Vs, and Vp/Vs ratios that compare favorably with established laboratory data and results from full waveform sonic logs. Thus, the Toksoz model can be used to compare results for which interpretation experience is lacking.

We calculated Toksoz models for binary and ternary mixtures of common minerals (quartz, calcite, etc.) and water for expected values of porosity, pore aspect ratio and volume percentage and aspect ratio of inclusions of a second mineral. Our computations considered the mathematical limitation of the model for porosities less than aspect ratio.

Modeled limestones (calcite-water) and dolomites (dolomite-water) having equant pores nominally agree with published results and full waveform sonic logs as having constant Vp/Vs ratios, independent of porosity. Modeled sandstones (quartz-water), and shaly sands (quartz-shale-water) generally compare favorably with published results and full waveform sonic logs, but imply further important interpretational criteria. Sandstones and shaly sands fall within an area delimited by quartz-water, quartz-shale and shale-water binary mixtures; Vp/Vs increases with shaliness and porosity. For quartz-water mixtures, Vp/Vs is seemingly independent of aspect ratio for values of aspect ratio greater than 0.05 but very dependent for values less than 0.05. Modeled results for Vp agree with the Wyllie time-average equation only for pore aspect ratios of about 0.1 which may be a good average effective pore aspect ratio for sandstones.

Introduction

Vp, Vs and Vp/Vs ratios are now obtainable by computer analysis of data from full waveform sonic logging tools for intervals which may contain varied and complex lithologies. Our interpretational experience for porosity and lithology relationships in these complex reservoirs is limited, especially for Vs and Vp/Vs ratios. Yet, we believe that data from full waveform sonic logs may provide significant clues to enhance our understanding of porosity/lithology relationships for reservoir rocks.

Several papers in logging literature discuss some aspects of interpretation of shear and compressional wave data (e.g., Morris, Grine and Arkfeld, 1964; Nations, 1974; Kithas, 1976; Leeth and Holmes, 1978). Nations (1974), Kithas (1976) and Leeth and Holmes (1978) discuss interpretation of lithology and porosity from log derived Vp and Vs data. Anderson and others (1973) and Tixier and others (1975) show how Poisson's ratio is related to shaliness. Petrophysical literature discusses relationships between Vp/Vs ratios, lithology and porosity based on laboratory data (Pickett, 1963; Gregory, 1977; Benzing, 1978; Johnson, 1978) for various rock types. Tatham (1982) and Minear (1982) discuss use of theoretical models of effective elastic moduli for interpreting various lithologic characteristics.

Our analysis and results differ in detail from previous workers in that we have used Toksoz models differently and have sought only to correlate them with aspects of the gross variation of sonic log data. It is the goal of this paper to point out ways in which the use of theoretical models, and Toksoz models in particular, can provide interpretative insight relative to lithology/porosity relationships in complex reservoir rocks. We shall compare these results with Vp/Vs ratios obtained from mixed lithologies.

Elastic Constants of Minerals

Necessary for the implementation of theoretical models with which to calculate elastic constants of an aggregate are the elastic constant data for pure, zero porosity constituents (minerals or shale). Compilations of these data have been made by Birch (1966) and by Christensen (1982). Shown on a crossplot of Poisson's ratio and Young's modulus (Figure 1) are various minerals having possible occurrences in complex reservoir rocks. The separation of minerals or mineral groups on this plot indicates generally the extent to which elastic properties can be used to distinguish mixtures of minerals in rocks of complex lithology. Notice that quartz has unique elastic properties, while those of phyllosilicates, carbonates, feldspars, halides and sulfates are similar. Magnetite, hematite, garnet and other heavy minerals are characterized by Young's modulus. Thus, mixtures of quartz sand with carbonates, phyllosilicates or pyrite ought to be discernable through interpretation of full waveform sonic logs.

Based on our analysis of literature tabulations (e.g., Birch, 1966; Christensen, 1982), we have used the values listed in Table I for our model calculations. Due to unreliable literature values of the Vs for dolomite, we have calculated it assuming a Vp/Vs for dolomite of 1.800. Experienced log analysts will recognize that the values in Table I are not those commonly cited as typical of sandstone, limestone or dolomite. Our experience has shown that the values for shale are reasonable but may be in error for some applications. The shale values fall within the range of those used by Minear (1982). The values in Table I are also shown in Figure 1 for reference.

Table I. Velocity Parameters Used in Model Calculations

Mineral	μsed/ft	μsed/ft	Vp/Vs	gm/cc
Quartz	52.11	77.5	1.487	2.65
Calcite	49.1	94.8	1.931	2.71
Dolomite	43.5	78.4	1.800	2.85
Shale	70.4	136.3	1.936	2.40
Water	204	--	--	1.00

Computation of Toksoz Models

Numerous theoretical models have been developed during the past 30 years with which to compute the effective elastic moduli of a mixture, given the elastic moduli of the individual components. Two models which seek to allow ellipsoidal inclusions rather than spherical ones have been created by O'Connell and Budiansky (O'Connell and Budiansky, 1974; Budiansky and O'Connell, 1976) and by Toksoz and coworkers (Kuster and Toksoz, 1974a,b; Toksoz, Cheng and Timur, 1976). Models by Walsh (1965) and by Mavko and Nur (1978) also consider possible effects of cracks. We have used the Toksoz model principally due to its convenience and debate about the validity of the O'Connell and Budiansky model (Bruner, 1976; Henyey and Pomphrey, 1982). However, for spherical inclusions (e.g., quartz-calcite mixtures); other models may give equally valid results.

We refer the reader to the papers by Toksoz and coworkers for the particulars of the algebraically complex equations. Suffice it to say that these equations relate the effective bulk modulus and rigidity of an aggregate to the bulk modulus and rigidity of a medium in which are embedded inclusions having bulk modulus and rigidity and of known concentration (porosity) and aspect ratio. The aspect ratio is the length of the minor axis of an ellipsoidal pore divided by that of its major axis. Although the Toksoz equations are capable of handling pores having a spectrum of aspect ratios, we have made calculations for pores of only a single aspect ratio. This is unrealistic but has resulted in distinguishing effects of equant pores of high aspect ratio from cracks having low aspect ratio. The Toksoz model also possesses a mathematical limitation that the concentration of pores having a particular aspect ratio must be less than that aspect ratio. This implies that the model may not be applicable to some porous rocks having pores of low aspect ratio; perhaps some shaly sands or tuffs. Another assumption made by the theoretical model is that the ellipsoidal pores are randomly oriented so that the modeled material is acoustically homogeneous and isotropic. This may limit application of the Toksoz model for some shaly rocks due to presumed alignment of clay minerals and pores of low aspect ratio.

Porosity and Pore Aspect Ratio Relationships

It is well known in the Rock Physics literature that Vp, Vs and Vp/Vs ratios are very dependent on porosity and pore aspect ratio, especially for cracks. The dramatic effect of pore aspect ratio on calculated values of Vp and Vs (e.g., Minear, 1982) implies that effects due to pore aspect ratio may be more important than those of porosity. To evaluate this idea, in Figure 2 are found logarithmic crossplots of porosity against pore aspect ratio for quartz-water and calcite-water models within the limits of the Toksoz model and on which are contoured values of Vp and Vs. It is clear that Vp and Vs are dependent on both porosity and pore aspect ratio. For values of aspect ratio above about 0.05, the Vp and Vs are more dependent on porosity than on pore aspect ratio. Below about 0.05, Vp and Vs are about equally dependent on porosity and pore aspect ratio. Tatham (1982) also came to this conclusion from the same type of plots for similar data.

In Figure 3 are found logarithmic crossplots of porosity against pore aspect calculated for quartz-water and quartz-shale-water systems within the mathematical limits of the Toksoz model and on which values of Vp/Vs are contoured. It is clear that the Vp/Vs ratio is dependent on porosity and relatively independent of pore aspect ratio for high aspect ratios but is very dependent on pore aspect ratio for low aspect ratios. This difference in behavior occurs as aspect ratio values of about 0.05.

This points the the conclusion that we must regard interpretation of Vp, Vs and Vp/Vs for rocks having equant pores as different from that for rocks dominated by cracks. Toksoz, Cheng and Timur (1976) and Cheng and Toksoz (1979) have shown that the sandstones and limestones they studied are dominated by equant pores, especially at pressures greater than about 0.5 kbar which corresponds to a depth of about 6000 feet. Moreover, the success of conventional sonic tools as porosity indicators suggests that for many applications reservoir rocks can be considered as dominated by equant porosity

Still, it is necessary to consider the possible occurrence of cracks in rock otherwise characterized by equant pores. We have no data that supports Tatham's (1982) conclusion that different lithologies have inherently different pore aspect ratios. There are two cases in which interpretation of velocities for reservoir rocks may be affected by pores of low aspect ratio. First, as it is the effective pressure that controls closure of cracks, overpressured formations, especially shaly formations, ought to be considered as potentially influenced by cracks. Accordingly, higher than "normal" Vp/Vs ratios ought to be indicative of overpressured zones. Secondly, it is not yet clear to us whether actual shaly sands should be considered as dominated by equant pores or by cracks.

Simple Binary and Ternary Models

We have computed the Vp, Vs, and Vp/Vs ratios for simple binary models (e.g., quartz-water, quartz-calcite) and simple ternary models (e.g., quartz-shale-water) within the limitations of the Toksoz model discussed above for rocks dominated by equant pores. The results for the binary models are shown in Figure 4. The modeling suggests that the hierarchical dependence of Vp and Vs is on porosity, pore aspect ratio, concentration of included minerals and aspect ratio of these included minerals (most important to least important). For mineral pairs, the calculated velocity parameters are not very sensitive to aspect ratio of the included mineral. The lines connecting mineral pairs have slight curvature and represent spherical inclusions of minerals. The P-wave traveltime is sensitive to variations of proportion for mixtures of quartz and dolomite. However, P-wave traveltime is not quite linearly proportional, as Nations (1974) assumed, because the bulk modulus of dolomite is larger than that of quartz. The S-wave traveltime is insensitive to variations of proportion for quartz-dolomite mixtures; while the P-wave traveltime is relatively insensitive to variations in quartz-calcite mixtures at small proportions of calcite. Generally, Vp/Vs ratios depend on mixing proportions for all binary models. For quartz-shale mixtures, the dependence of Vp/Vs (Poisson's ratio) on shaliness has been pointed out by Anderson and others (1973) and by Tixier and others (1975).

Mineral-water models for equant pores nominally exhibit two types of general behavior. For dolomite-, calcite- and shale-water models, the Vp/Vs ratios remain fairly constant with increasing porosity (increasing P-wave and S-wave travel time). The Vp/Vs ratios for quartz-water increase with increasing porosity. This behavior agrees well with results of Pickett (1963) which are shown in the inset of Figure 4 and with those of Nations (1974). Pickett found that limestone has Vp/Vs ratios of about 1.9 while dolomite has Vp/Vs of about 1.8. Comparison of Pickett's sandstone data with the results of quartz-water indicate that Vp/Vs ratios for sandstone increase faster with porosity than do Vp/Vs ratios for the modeled sandstone. Thus, comparison of the simple binary models for carbonates with the results of Pickett (1963) are very good, while those for sandstone are only qualitatively correct.

Our ternary quartz-shale-water model qualitatively agrees with the experimental work of Tosaya and Nur (1982). We believe that quantitative differences are due to using different values for elastic moduli for the clay component. The porosity-1/Vp shale volume crossplot used by some log analysts (Millard, 1982) has the same systematics as those calculated from Toksoz models and as the work of Tosaya and Nur (1982).

Comparison with Full Waveform Sonic Logs

Vp/Vs ratios have been obtained from analysis of full waveform sonic logs for a variety of lithologies and geographic areas. Data were obtained by means of short-spaced, long-spaced or experimental full waveform sonic tools by computer processing designed to take advantage of unique characteristics of each data set. In the comparisons which follow, we want to show chiefly how Vp/Vs ratios are indicators of lithology.

We are presenting Vp/Vs ratios for three different carbonate formations in two different wells, one in Edwards County, Texas, and the second in Lee County, Virginia. Vp/Vs is plotted against 1/Vs in Figure 5 for the limestone of Cambro-Ordovician age in the Appalachian Mountains. The major trend of the data is for a nearly constant Vp/Vs ratio with increasing porosity (increasing 1/Vs). This agrees with the calcite-water model discussed above and with Pickett's (1963) experimental data for limestones. The Appalachian limestone data falls to the right of the line which represents the quartz-calcite zero porosity Toksoz model. The average Vp/Vs ratio for the Appalachian limestone is about 1.87, which is slightly lower than Pickett's limestone data. This may be caused by inclusion of small amounts of quartz, dolomite, or possibly shale or by slightly lower pore aspect ratios in the Appalachian limestone.

Shown in Figure 6 is the variation of Vp/Vs with 1/Vs for the Strawn Limestone of Pennsylvania age from the West Texas well. In contrast to the Appalachian limestone, these data do not show a distinct trend at constant Vp/Vs ratio. Rather, the data are limited by the line which represents the quartz-calcite, zero porosity model and tend to be smeared along the line. We believe these data are consistent with the view that the Vp/Vs ratio is indicative of the proportion of quartz in this sandy limestone.

Another example of Vp/Vs for mixed lithologies is seen in Figure 7 which is data for the Ellenberger, a sandy dolomite and dolomitic limestone, from a well in Edwards County, Texas. The lines in Figure 7 represent the zero porosity quartz-dolomite and dolomite-calcite models. Also included for comparison are data for the underlying Bliss Sandstone. The data are limited on the left by the quartz-dolomite model. Data points for the Bliss Sandstone lie close to the quartz point and are strung out towards the dolomite point. We believe that Vp/Vs is indicative of the proportions of quartz and dolomite and that the bulk of the Ellenberger data points lie to the right of the model line according to their porosity. Less compelling is the possible mixture of dolomite and calcite and its correlation with the dolomite-calcite model. Three factors may affect this correlation; error in the Vp for dolomite, presence of cracks and analytical uncertainty in the Vp/Vs ratio. Still, we are encouraged that the correlation is reasonably good.

Interpretation and correlation of data from shaly sands and shales with results of Toksoz modeling is more tentative due to uncertainties about the possible effects of low pore aspect ratios. Still, there are several systematic relationships that are notable. Shown in Figure 8 are the data for Vp/Vs from the Wolfcamp shale from an Edwards County, Texas well. Surprisingly, the data have a rather limited distribution which is bounded on the left by the zero porosity quartz-shale model line. The variation of Vp/Vs may be due in part to either variation of the proportion of quartz in these shaly rocks or to effects of low pore aspect ratios. As these shales are Paleozoic and have densities averaging between 2.65 and 2.7 gm/cc, their porosities are very low. Therefore, at least some of the displacement of the trend away from the quartz-shale model line must be due to occurrence of cracks. Also, as elastic constants for shale are now well known, it is possible that we could have chosen better values for the elastic constants of shale in our model calculations.

Calculated and digitally recorded logs through the Frio Formation in a well in Brazoria County, Texas, are shown in Figure 9. The Vp/Vs ratios crossplotted against 1/Vs for sandstones, shaly sandstones, and shales of the Frio Formation in this well are seen in Figure 10. Here, Vp/Vs ratios for the shaly rocks trend from about 1.6 to more than 1.9 with increasing 1/Vs. This trend is seemingly continuous with the trend of increasing Vp/Vs ratio with increasing porosity for experimentally determined sandstone data of Pickett (1963). Extrapolation of this trend to higher transit times is consistent with the Vp/Vs ratio obtained for the Frio Formation by vertical seismic profiling (Lash, 1980).

The shaly sandstone data points in Figure 10 are generally bounded by lines for the quartz-water, quartz-shale, and shale-water binary models. The trend implies that shaliness is chiefly responsible for the major variation of Vp/Vs ratios. This is shown in Figure 11 on which average values of Vp/Vs are plotted and contoured; shaliness and porosity value were obtained from analysis of neutron-density log crossplots. As predicted by Toksoz models, Vp and Vs are more sensitive to variations of porosity than of shaliness. However, due to the greater natural variation in shaliness, the Vp/Vs in this example is primarily controlled by shaliness and, secondarily, by smaller porosity variations.

Comparison of Frio shaly sandstone data (Figure 10) with that of the Wolfcamp shale (Figure 8) provides some insight into the relative importance of porosity and pore aspect ratio in actual rocks. Model calculations indicate that changing pore aspect ratio from 1.0 to 0.1 for shaly sandstones increases the Vp/Vs ratio only very slightly (Figure 3). Increasing porosity and volume of shale increases Vp/Vs significantly. However, pores of low aspect ratio (i.e., cracks, aspect ratio less than 0.01) increase the Vp/Vs ratio markedly. The range of the Vp/Vs ratios for the Wolfcamp shale must, therefore, be due to the occurrence of a small number of cracks and/or to the variation in the proportion of quartz to clay. The much wider range of Vp/Vs for the Frio shaly sandstones may also be due to these causes, may be related to variation of porosity or to its overpressured condition.

The effect of gas saturation on sonic velocities for a sandstone can be seen in Figures 10 and 11. The cluster of data having Vp/Vs between 1.4 and 1.55 and 1/Vs between 90 and 110 microsec/ft (Figure 10) and the low average Vp/Vs values for shaliness less than 0.2 and porosity of about 10% are due to gas saturation of a sandstone. These observations are consistent with our modeling and with observations from the literature (e.g., Gregory, 1977). Gas saturation reduces Vp which, in turn, reduces the Vp/Vs ratio. Leeth and Holmes (1978) show the utility of the Vp/Vs ratio for finding gas in shaly formations.

Toksoz Model and the Wyllie Time-Average Equation

Like Minear (1982), we have found that agreement between the Toksoz models and the Wyllie time-average equation is for pore aspect ratios of about 0.1. Thus relative to the Toksoz model, average reservoir rocks would seemingly have average pore aspect ratios of 0.1. Similarly, we have considered the "improved" transform described by Raymer, Hunt, and Gardner (1980). A cursory examination suggests that some of the curvature of the "improved" transform may be caused by low pore aspect ratios which tend to increase P-wave travel time without affecting porosity. However, it is not evident that lithologic factors, such as shaliness, or effects of possible gas saturation on Vp have been adequately accounted for by Raymer, Hunt, and Gardner (1980). If modification of the Wyllie time-average equation is necessary, we believe it can be made consistent with theoretical models, such as that of Toksoz.

Summary and Conclusions

It is clear that Toksoz modeling provides an important basis for interpretation of Vp, Vs, and Vp/Vs ratios. This model:

1. provides a basis for understanding effects of mixed lithologies which are characteristic of complex reservoir rock and allows for quantitative estimates to be made.

2. permits an improved estimation of porosity to be made for these mixed lithologies, and

3. substantiates known effects of gas saturation.

We have shown that the results of Toksoz modeling are consistent with laboratory experimental data of rock physics and with Vp, Vs, and Vp/Vs ratios from full waveform sonic logs obtained in generally known lithologies. The Toksoz models provide an understanding of effects of cracks (low pore aspect ratio), but translation of this into practical interpretation requires further study. We anticipate that some fractured reservoirs and overpressured reservoirs may be indicated by high Vp/Vs ratios.

Acknowledgements

We wish to express our thanks to J. D. Robertson, E. P. Howell, M. M. Backus, E. S. Pasternack, and D. J. Lafferty for helpful comments and assistance during this study and to ARCO Oil and Gas Company for permission to publish this paper.

References Cited

Anderson, R. A., Ingram, D. S. and Zanier, A. M., 1973, Determining Fracture Pressure Gradients from Well Logs: J. Petrol. Tech., v. 25, pp. 1259-1268.

Benzing, W. M., 1978, Vs/Vp Relationships in Carbonates and Sandstones Laboratory Data: Presented at the 48th Annual Internation SEG Mtg., San Francisco.

Birch, F., 1966, Compressibility; Elastic Constants: Handbook of Physical Constants; Geol. Soc. America Memoir 97, pp. 97-173.

Bruner, W. M., 1976, Comment on 'Seismic Velocities in Dry and Saturated Cracked Solids' by Richard J. O'Connell and Bernard Budiansky: Jour. Geophys.Res., v. 81, pp. 2573-2578.

Budiansky, B., and O'Connell, R. J., 1976, Elastic Moduli of a Cracked Solid: Int. J. Solids Structures, V. 12, pp. 81-97.

Cheng, C. H., and Toksoz, M. N., 1979, Inversion of Seismic Velocities for the Pore Aspect Spectrum of a Rock: Jour. Geophys. Res., v. 84, pp. 7533-7543.

Christensen, N. I., 1982, Seismic Velocities: CRC Handbook of Physical Properties of Rocks, v. II, pp. 1-228.

Gregory, A. R., 1977, Aspects of Rock Physics From Laboratory and Log Data that are Important to Seismic Interpretation: AAPG Memoir 26, pp. 15-46.

Henyey, F. S., and Pomphrey, N., 1982, Self-Consistent Elastic Moduli of a Cracked Solid: Geophys. Res. Letters., v. 9, pp. 903-906.

Johnson, W. E., Relationship Between Shear-Wave Velocity and Geotechnical Parameters: Presented at the 48th Annual International SEG Mtg., San Francisco.

Kithas, B. A., 1976, Lithology, Gas Detection, and Rock Properties from Acoustic Logging Systems: SPWLA 17th Ann. Log Symp., R, 10p.

Kuster, G. T., and Toksoz, M. N., 1974a, Velocity and Attenuation of Seismic Waves in Two-Phase Media: Part I. Theoretical Formulations Geophysics, v. 39, pp. 587-606.

Kuster, G. T., and Toksoz, M. N., 1974b, Velocity and Attenuation of Seismic Waves in Two-Phase Media: Part II. Experimental Results: Geophysics, v. 39, pp. 607-618.

Lash, C. C., 1980, Shear Waves, Multiple Reflections, and Converted Waves Found by a Deep Vertical Wave Test (Vertical Seismic Profiling): Geophysics, v. 45, pp. 1373-1411.

Leeth, R. and Holmes, M., 1978, Log Interpretation of Shaly Formations Using the Velocity Ratio Plot: SPWLA 19th Ann. Logging Symp. CC, 9p.

Mavko, G. M., and Nur, A., 1978, The Effect of Nonelliptical Cracks on the Compressibility of Rocks: Jour. Geophys. Res., v. 83, pp. 4459-4468.

Millard, F. S., 1982, Personal Communication.

Minear, J. W., 1982, Clay Models and Acoustic Velocities: SPE 11031.

Morris, R. L., Grine, D. R., and Arkfeld, T. E., 1964, Using Compressional and Shear Acoustic Amplitudes for the Location of Fractures: Jour. Petrol. Tech., v. 16, pp. 623-632.

Nations, J. F., 1974, Lithology and Porosity from Acoustic Shear and Compressional Wave Transit Time Relationships: SPWLA 15th Ann. Log. Symp., Q, 16p.

O'Connell, R. J., and Budiansky, B., 1974, Seismic Velocities in Dry and Saturated Cracked Solids: Jour. Geophys. Res., v. 79, pp. 5412-5426.

Pickett, G. R., 1963, Acoustic Character Logs and Their Applications in Formation Evaluation: Jour. Petrol. Tech., v. 15, pp. 659-667.

Raymer, L. L., Hunt, E. R., and Gardner, J. S., 1980, An Improved Sonic Transit Time-to-Porosity Transform: SPWLA 21st Ann. Log. Symp. P, pp. 1-13.

Tatham, R. H., 1982, Vp/Vs and Lithology: Geophysics, v. 47, pp. 336-344.

Tixier, M. P., Loveless, G. W. and Anderson, R. A., 1975, Estimation of Formation Strength from the Mechanical-Properties Log: J. Petrol. Tech., v. 27, pp. 283-293.

Toksoz, M. N., Cheng, C. H., and TImur, A., 1976, Velocities of Seismic Waves in Porous Rocks: Geophysics, v. 41, pp. 621-645.

Tosaya, C., and Nur, A., 1982, Effects of Diagenesis and Clays on Compressional Velocities in Rocks: Geophys. Res. Letters, v. 9, pp. 5-8.

Walsh, J. B., 1965, The Effects of Cracks on the Compressibility of Rock: Jour. Geophys. Res., v. 70, pp. 381-389.

Raymond L. Eastwood is a Senior Research Geologist with ARCO Oil and Gas Company and has worked in log interpretation research since 1980. He was formerly Assistant Professor of Geology at Northern Arizona University (1970-1979) and Senior Research Mineralogist with Phillips Petroleum Company (1968-1978). He holds degrees from Kansas State University, B.S. in physics (1962) and M.S. in geology (1965), and from the University of Arizona, PhD. in geology (1970). He is a member of SPWLA (Dallas), AGU, and GSA.

John P. Castagna is a Senior Research Geophysicist with ARCO Oil and Gas Company. He has worked in sonic logging research from 1979 to 1981 at the University of Texas at Austin, and from 1981 to the present with ARCO Exploration and Production Research. He has a B.S. (1976) and M. A. (1980) in geology from City University of New York and is working towards a doctorate in geophysics at the University of Texas at Austin. He is a member of SPWLA and SEG.

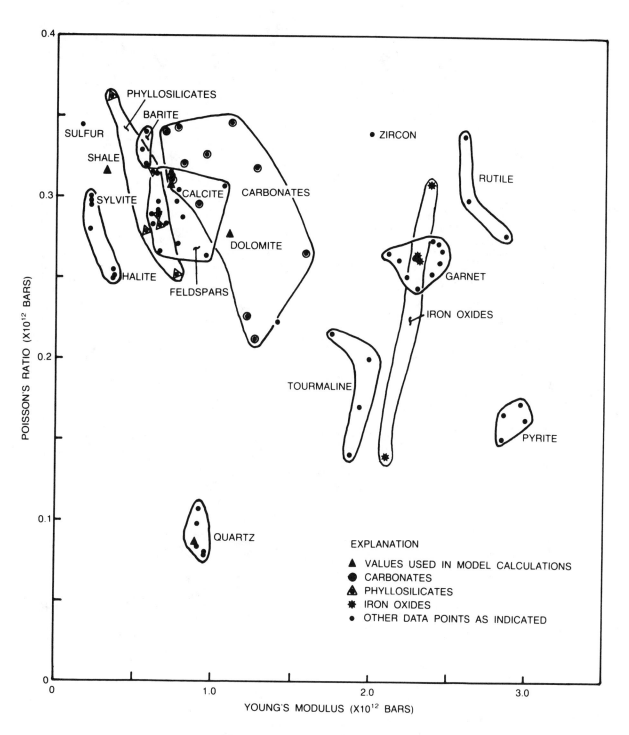

Figure 1. Crossplot of Poisson's Ratio and Young's Modulus for Possible Minerals in Complex Reservoir Rock (from compilations of Birch, 1966 and Christensen, 1982).

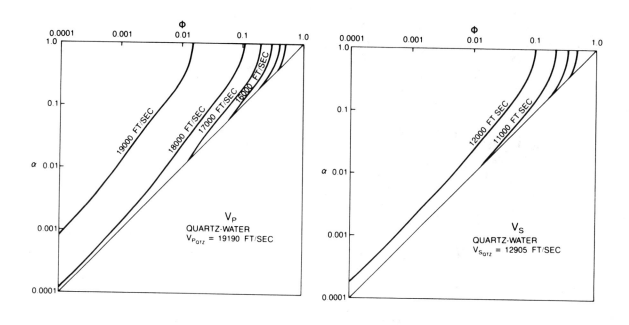
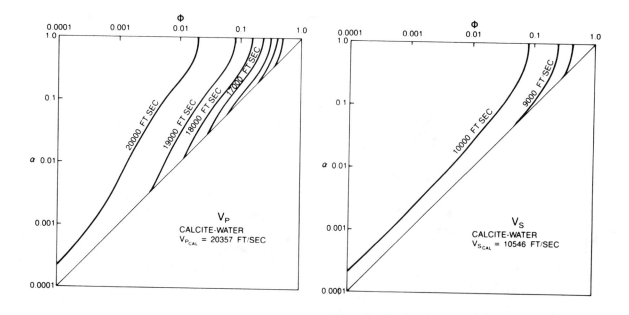

Figure 2. Logarithmic crossplot of porosity and pore aspect ratio; a) Vp contoured, quartz-water, b) Vs contoured, quartz-water, c) Vp contoured, calcite-water, d) Vs contoured, calcite-water.

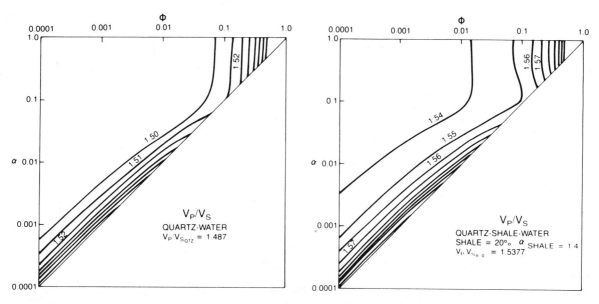

Figure 3. Logarithmic crossplot of porosity and pore aspect ratio, Vp/Vs contoured; a) quartz-water, b) quartz-shale-water for shale=20%.

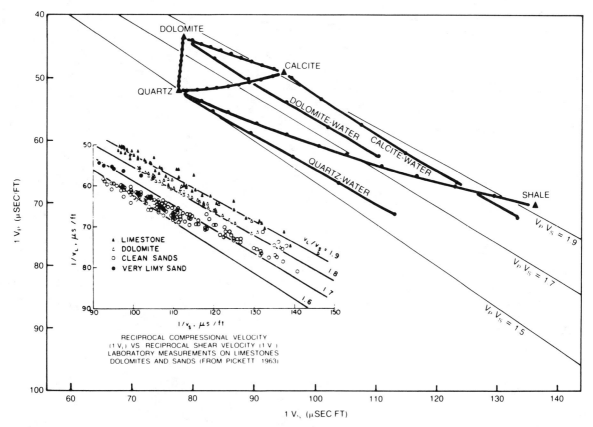

Figure 4. Modeled Binary Systems plotted as Dtp (1/Vp) versus Dts (1/Vs), inset is from Pickett (1963).

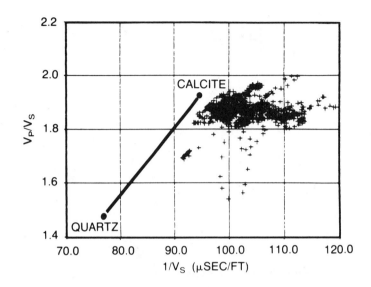

Figure 5. Crossplot of Vp/Vs and 1/Vs for the Appalachian Limestone.

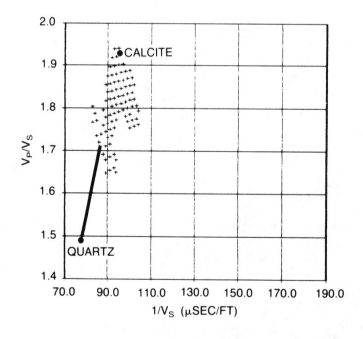

Figure 6. Crossplot of Vp/Vs and 1/Vs for the Pennsylvanian Strawn Limestone of Edwards County, Texas.

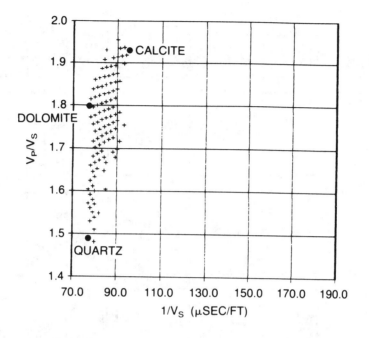

Figure 7. Crossplot of Vp/Vs and 1/Vs for the Ellenberger Dolomite of Edwards County, Texas.

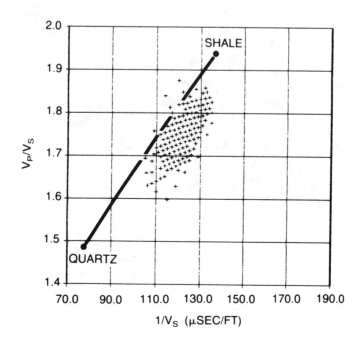

Figure 8. Crossplot of Vp/Vs and 1/Vs for the Wolfcamp Shale of Edwards County, Texas.

-15-

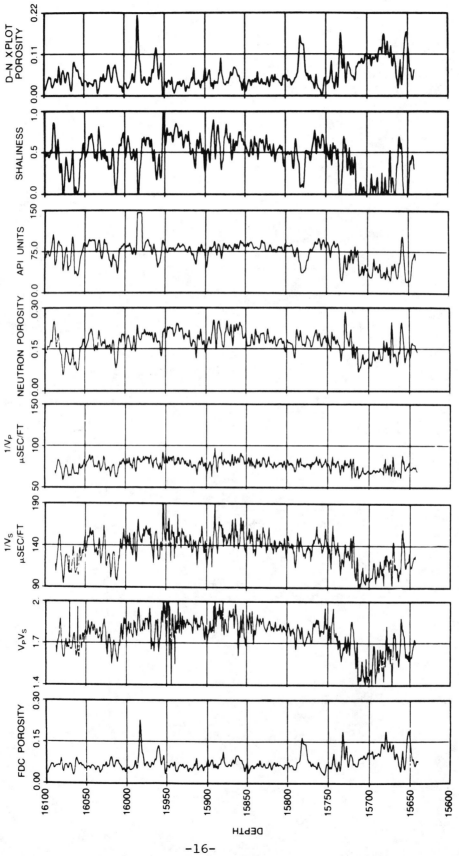

Figure 9. Logs through the Frio Formation, Brazoria County, Texas.

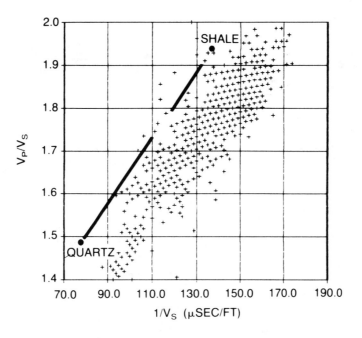

Figure 10. Crossplot of Vp/Vs and 1/Vs for the Frio Formation of Brazoria County, Texas.

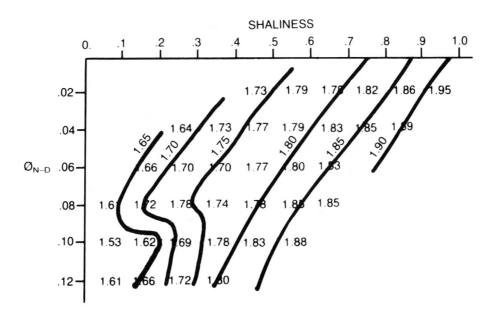

Figure 11. Crossplot of Shaliness and Porosity for the Frio Formation, Average Vp/Vs are contoured.

Evaluation of direct hydrocarbon indicators through comparison of compressional- and shear-wave seismic data: a case study of the Myrnam gas field, Alberta

Ross Alan Ensley*

ABSTRACT

Shear waves differ from compressional waves in that their velocity is not significantly affected by changes in the fluid content of a rock. Because of this relationship, a gas-related compressional-wave "bright spot" or direct hydrocarbon indicator will have no comparable shear-wave anomaly. In contrast, a lithology-related compressional-wave anomaly will have a corresponding shear-wave anomaly. Thus, it is possible to use shear-wave seismic data to evaluate compressional-wave direct hydrocarbon indicators. This case study presents data from Myrnam, Alberta which exhibit the relationship between compressional- and shear-wave seismic data over a gas reservoir and a low-velocity coal.

INTRODUCTION

A recent paper (Ensley, 1984) documented a new method of evaluating "bright spots" or other direct hydrocarbon indicators (DHIs). The technique involves the qualitative comparison of compressional (P) wave and shear (S) wave[1] seismic data. In practice, such a comparison offers a viable means of evaluating DHIs previously observed on P-wave data. Ensley (1984) describes the theory behind this technique, and demonstrates its feasibility with model data and a case history. This paper presents an interpretation of P- and SH-wave seismic data from the Myrnam field, Alberta as a second case history.

[1] S-wave is here understood to include both horizontally polarized shear waves (SH-waves) and vertically polarized shear waves (SV-waves). All of the S-wave seismic data discussed within this report are SH-wave data and will be referred to as such when appropriate.

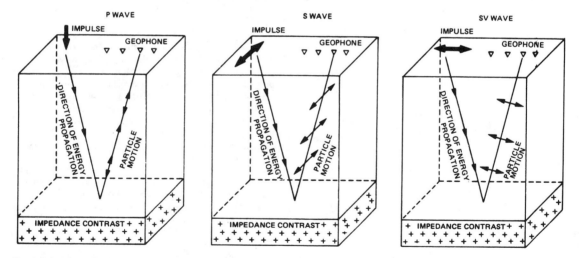

FIG. 1. Direction of energy propagation and particle motion for compressional (P) waves, horizontal shear (SH) waves, and vertical shear (SV) waves.

Manuscript received by the Editor June 4, 1984; revised manuscript received August 1, 1984.
*Exxon Production Research Company, P. O. Box 2189, Houston, TX 77001.
© 1985 Society of Exploration Geophysicists. All rights reserved.

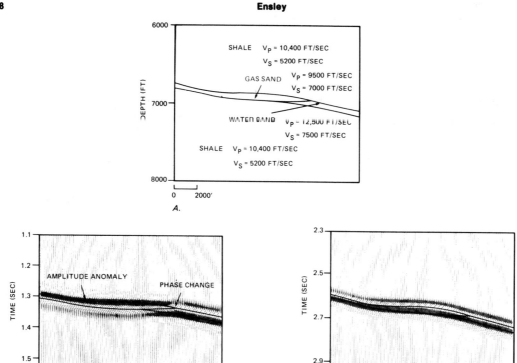

FIG. 2. Seismic models showing the differences between the P- and S-wave response to a gas-filled reservoir. The models are 2-D ray-tracing models and the velocities used are representative of moderately compacted sediments. (a) Depth model of a gas reservoir; (b) P-wave synthetic section showing a DHI characterized by a high-amplitude reflection along the top of the reservoir and a phase change at the edge of the reservoir; (c) S-wave synthetic section showing no anomalous response to the gas reservoir.

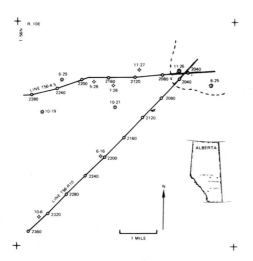

FIG. 3. Location map of Myrnam field.

Shear waves differ from compressional waves in both the direction of particle motion relative to the direction of wave propagation (Figure 1) and in the rock properties which control the wave velocity. A P-wave is an elastic wave in which the particle motion is parallel to the direction of wave propagation. In contrast, an S-wave is an elastic wave in which the particle motion is perpendicular to the direction of wave propagation. Because of this difference between P- and S-waves, the velocities of the two are functions of different rock properties.

Consideration of the elastic properties which control the velocity of P- and S-waves in a rock indicates that P-waves are sensitive to the type of pore fluid present within a rock while S-waves are only slightly affected by changes in fluid type. Thus, if the presence of gas within a reservoir rock gives rise to an anomalous seismic expression on P-wave data, a DHI, there should be no comparable expression on S-wave data. This relationship is illustrated by the hypothetical model shown in Figure 2. A P-wave anomaly generated by a lithological feature, however, should have a corresponding S-wave anomaly. One consequence of this relationship is that it is possible to evaluate the potential of P-wave DHIs through a comparison of P- and S-wave seismic data recorded over a prospect. This concept is more thoroughly discussed by Ensley (1984).

The application of SH-wave seismic data for evaluation of

Direct Hydrocarbon Indicators

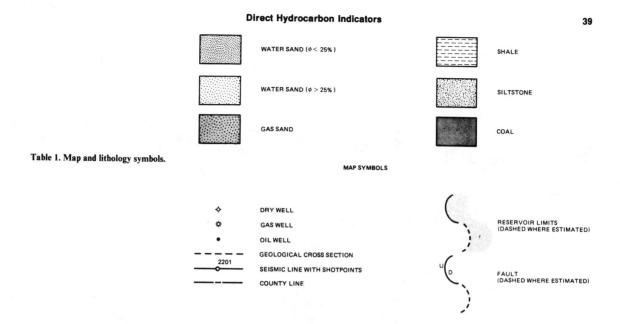

Table 1. Map and lithology symbols.

DHIs was previously documented with a case study of *P*- and *SH*-wave data from the Putah Sink field of central California (Ensley, 1983, 1984). Robertson (1983) reviewed the interpretation of data from the Putah Sink field and one other site in central California.

CASE STUDY

During 1979, Hudson's Bay Oil and Gas Company, acting as contractor for itself and several other companies, recorded seismic data at the Myrnam field in Canada to test the application of *SH*-wave data for gas detection. The Myrnam field is located 90 miles east of Edmonton, Alberta. *P*- and *SH*-wave seismic data were recorded along two lines (Figure 3). Sections from line T56-4.5 will be used for examples in this report, but the second line shows the same relationship between *P*- and *SH*-wave data. Table 1 is a key to the symbols used in this paper.

The Myrnam field produces gas from the Cretaceous Colony formation. The reservoir contains a maximum net pay of 22 ft within the study area and the gas/water contact is at a depth of 1 759 ft below ground level. The exact geometry and trapping

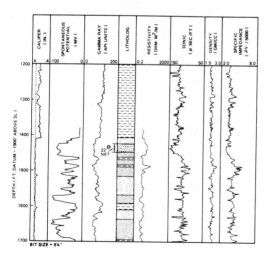

Fig. 4. Well logs from the zone of interest in the Duvernay 11–26 well.

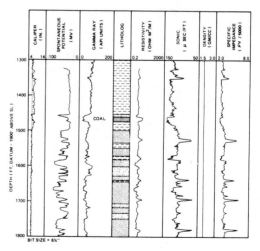

Fig. 5. Well logs from the zone of interest in the Duvernay 5–28 well.

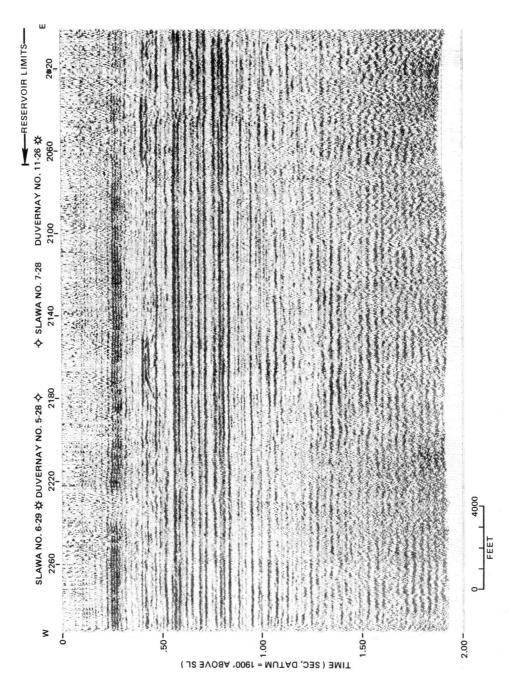

FIG. 6. P-wave seismic data from the Myrnam field, uninterpreted.

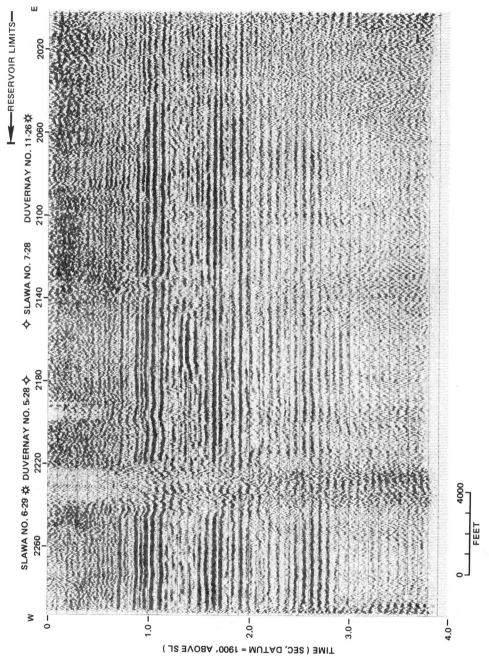

FIG. 7. *SH*-wave seismic data from the Myrnam field, uninterpreted.

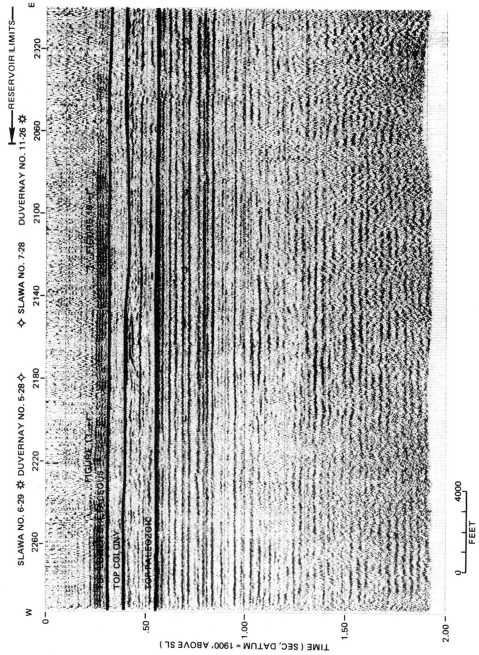

FIG. 8. *P*-wave seismic data from the Myrnam field, interpreted.

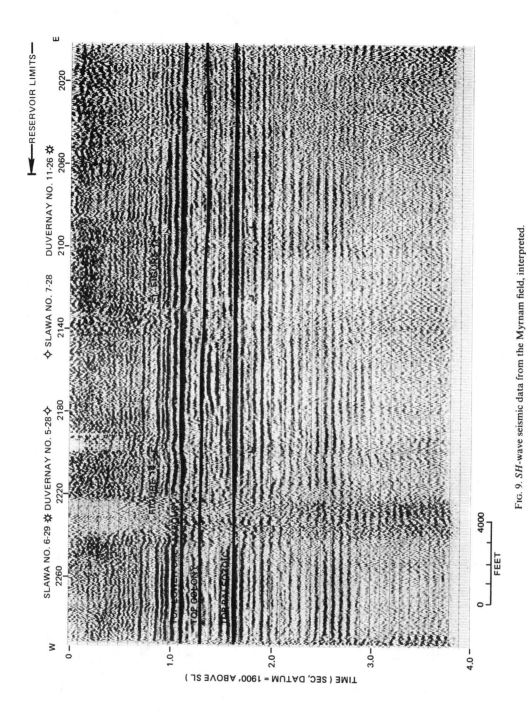

Fig. 9. SH-wave seismic data from the Myrnam field, interpreted.

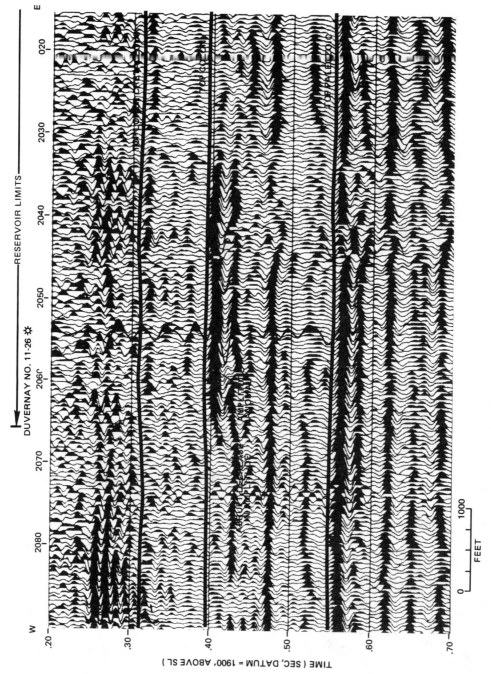

FIG. 10. P-wave seismic data from the Myrnam field, expanded section showing a gas-related DHI.

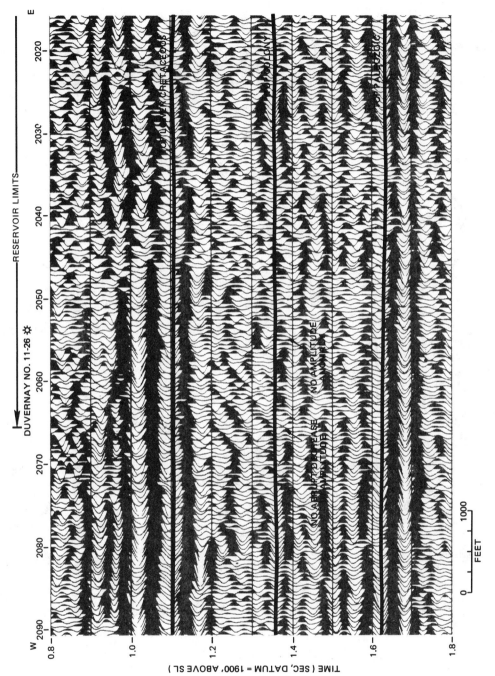

FIG. 11. *SH*-wave seismic data from the Myrnam field, expanded section showing no gas-related DHI.

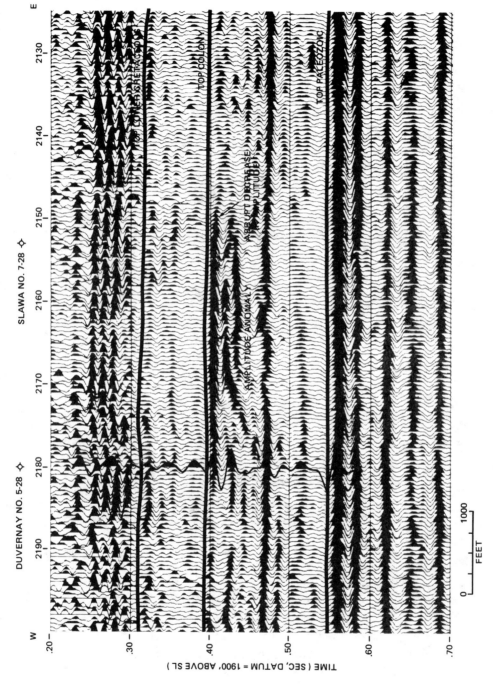

FIG. 13. *P*-wave seismic data from the Myrnam field, expanded section showing a false DHI.

Direct Hydrocarbon Indicators

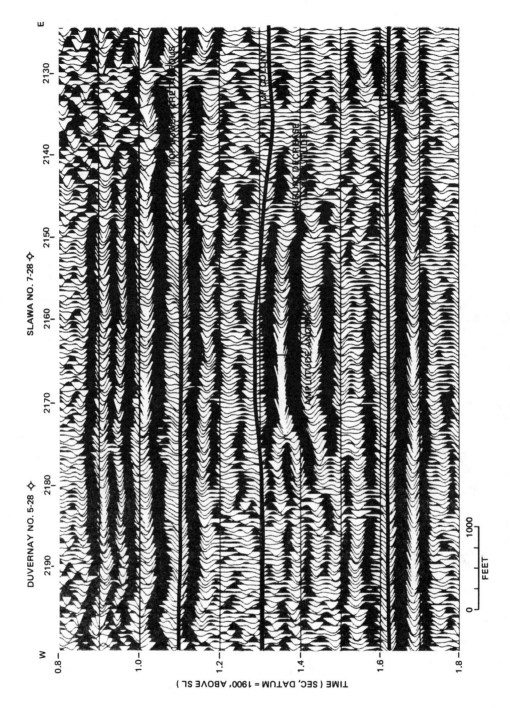

FIG. 14. *SH*-wave seismic data from the Myrnam field, expanded section showing a false DHI.

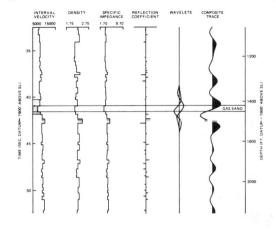

FIG. 12. Detailed plot of the synthetic seismogram from the Duvernay 11–26 well. The pulse used is a zero-phase wavelet with a peak frequency of 30 Hz.

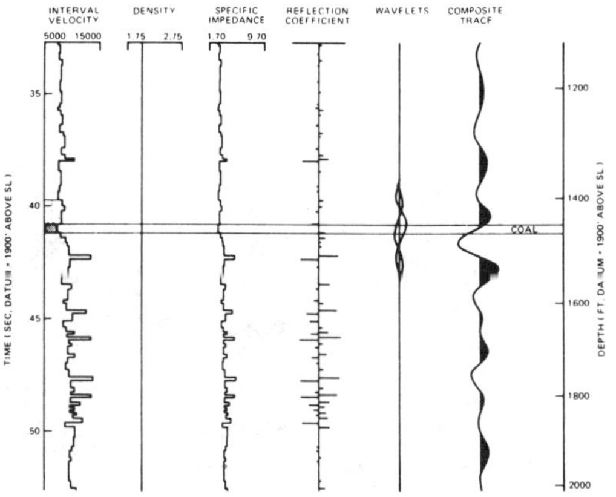

FIG. 15. Detailed plot of the synthetic seismogram from the Duvernay 5-28 well. The pulse used is a zero-phase wavelet with a peak frequency of 30 Hz.

mechanisms of the reservoir are unknown because of limited well control. In addition to the gas reservoir, the two lines also cross known coal deposits within the Colony formation. The coal beds range in thickness from 12 to 20 ft and are at approximately the same stratigraphic level as the gas reservoir.

Well logs from the Duvernay 11-26 well show that the gas sand has a slower velocity than the underlying water sand and the overlying shale (Figure 4). This results in a low impedance for the gas sand relative to the surrounding rocks. Logs from the Duvernay 5-28 well show that the coal has an impedance similar to the gas sand and is also surrounded by rocks of higher impedance (Figure 5).

The quality of the seismic data recorded at the Myrnam field is good and both the P- and SH-wave sections show comparable reflection continuity (Figures 6 and 7). This enabled the two sections to be correlated on the basis of reflection character (Figures 8 and 9).

In the primary zone of interest, over the gas reservoir, the P-wave data exhibit a DHI but there is no comparable SH-wave expression (Figures 10 and 11). The P-wave DHI is characterized by an amplitude anomaly and an abrupt change in amplitude laterally. The SH-wave data, though, show only low-amplitude continuous reflections. The dark trace on Figure 10 represents a synthetic seismogram from the Duvernay 11-26 well. A detailed plot of the synthetic seismogram confirms that the P-wave DHI is caused by the gas (Figure 12).

In the second zone of interest, over the coal deposit, both the P- and SH-wave data exhibit similar anomalies (Figures 13 and 14). The P-wave DHI is characterized by an amplitude anomaly and an abrupt change in amplitude laterally, much like the DHI over the gas reservoir. The SH-wave data exhibit a similar expression. A detailed plot of the synthetic seismogram from the Duvernay 5-28 well demonstrates the P-wave DHI is caused by the coal bed and is a false DHI (Figure 15).

CONCLUSIONS

This case history demonstrates that although a gas-related DHI and a false, lithology-related DHI may have similar appearances on P-wave seismic data, it is possible to distinguish between them through a comparison of P- and SH-wave data. This conclusion, in conjunction with similar results previously published (Ensley, 1983, 1984; Robertson, 1983), strongly suggests that the comparison of P- and SH-wave seismic data is a viable method for evaluating P-wave DHIs.

ACKNOWLEDGMENTS

I would like to thank Exxon Production Research Company for the opportunity to publish this paper. I would also like to acknowledge Don Hartman, Paul Tarantolo, and Gordon Weisser who were instrumental in organizing the shear-wave project at Exxon.

REFERENCES

Ensley, R. A., 1983. Direct hydrocarbon detection with P- and SH-wave seismic data (abs.): 53rd Ann. Inter. SEG Mtg., Abs. with Biographies, 349–351.
——— 1984. Comparison of P- and S-wave seismic data: A new method for detecting gas reservoirs: Geophysics, v. 49, p. 1420–1431.
Robertson, J. D., 1983, Bright spot validation using comparative P-wave and S-wave seismic sections (abs.): 53rd Ann. Inter. SEG Mtg., Abs. with Biographies, 355–356.

Relationships between compressional-wave and shear-wave velocities in clastic silicate rocks

J. P. Castagna*, M. L. Batzle*, and R. L. Eastwood*

ABSTRACT

New velocity data in addition to literature data derived from sonic log, seismic, and laboratory measurements are analyzed for clastic silicate rocks. These data demonstrate simple systematic relationships between compressional and shear wave velocities. For water-saturated clastic silicate rocks, shear wave velocity is approximately linearly related to compressional wave velocity and the compressional-to-shear velocity ratio decreases with increasing compressional velocity. Laboratory data for dry sandstones indicate a nearly constant compressional-to-shear velocity ratio with rigidity approximately equal to bulk modulus. Ideal models for regular packings of spheres and cracked solids exhibit behavior similar to the observed water-saturated and dry trends. For dry rigidity equal to dry bulk modulus, Gassmann's equations predict velocities in close agreement with data from the water-saturated rock.

INTRODUCTION

The ratio of compressional to shear wave velocity (V_p/V_s) for mixtures of quartz, clays, and other rock-forming minerals is significant in reflection seismology and formation evaluation. In this paper, we investigate the V_p/V_s ratio in binary and ternary mixtures of quartz, clays, and fluids.

The classic paper by Pickett (1963) popularized the use of the ratio of compressional to shear wave velocities as a lithology indicator. Figure 1, reproduced from Pickett's paper, shows the distinct difference in V_p/V_s for limestones, dolomites, and clean sandstones. Nations (1974), Eastwood and Castagna (1983), and Wilkens et al. (1984) indicated that V_p/V_s for binary mixtures of quartz and carbonates tends to vary almost linearly between the velocity ratios of the end members with changing composition.

Figure 2 shows compressional and shear wave velocities for minerals reported in the literature (e.g., Birch, 1966; Christensen, 1982). These velocities are calculated from single crystal data and represent isotropic aggregates of grains. Also plotted is an extrapolation of Tosaya's (1982) empirical relation for V_p and V_s in shaly rocks to 100 percent clay and zero porosity. The position of this "clay point" depends upon the particular clay mineral, and it is plotted only to indicate roughly the neighborhood in which velocities of clay minerals are to be expected. Interpretation of sedimentary rock velocities should be done in the context of these mineral velocities.

We establish general V_p/V_s relationships for clastic silicate rocks by comparing in-situ and laboratory data with theoretical model data. Available velocity information is examined for data from water-saturated mudrocks and sandstones. We examine laboratory data from dry sandstone and compare with simple sphere pack and cracked media theoretical model data. Data from water-saturated rocks are similarly investigated. The results of the relationships established between V_p and V_s are then applied to calculations of rock dynamic moduli. Finally, the general V_p-V_s trends versus depth are estimated for Gulf Coast clastics.

EXPERIMENTAL TECHNIQUES

We have combined a variety of in-situ and laboratory measurements for clastic silicate rocks that includes our data and data extracted from the literature. In-situ compressional and shear wave velocities were obtained by a number of sonic and seismic methods which are described in detail in the literature cited. Due to the development of shear wave logging, it is now possible to obtain shear wave velocities routinely over a wide range of borehole/lithologic conditions with few sampling problems and in quantities which were never before available (Siegfried and Castagna, 1982).

Both our laboratory data and laboratory data we assembled from the literature were obtained using the pulse transmission technique. Compressional and shear wave velocities are determined for a sample by the transit time of ultrasonic pulses (approximately 200 kHz to 2 MHz). The jacketed sample is placed in either a pressure vessel or load frame so that stresses can be applied. Temperature and pore fluid pressure can also be controlled. Details of the techniques are found in Gregory (1977) or Simmons (1965).

Manuscript received by the Editor March 12, 1984; revised manuscript received October 12, 1984.
*ARCO Oil and Gas Company, Exploration and Production Research, P.O. Box 2819, Dallas, TX.
© 1985 Society of Exploration Geophysicists. All rights reserved.

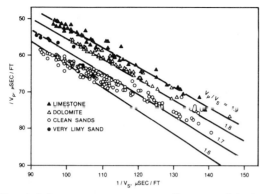

FIG. 1. Laboratory measurements on limestones, dolomites, and sandstones from Pickett (1963). V_p = compressional velocity, V_s = shear velocity.

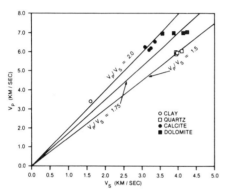

FIG. 2. Compressional and shear velocities for some minerals.

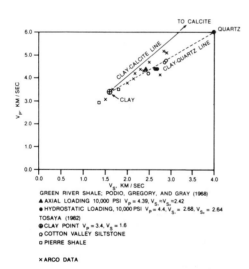

FIG. 3. Ultrasonic laboratory measurements for various mudrocks.

OBSERVATIONS IN MUDROCKS

We define mudrock as clastic silicate rock composed primarily of clay- or silt-sized particles (Blatt et al., 1972). Lithified muds are composed primarily of quartz and clay minerals. Owing to the difficulty associated with handling of most mudrocks, laboratory measurements on these rocks are not commonly found in the literature. Measurements that do exist are generally biased toward highly lithified samples.

Figure 3 is a V_p-versus-V_s plot of laboratory measurements for a variety of water-saturated mudrocks. For reference, lines are drawn from the clay-point velocities extrapolated from Tosaya's data ($V_p = 3.4$ km/s, $V_s = 1.6$ km/s) to calcite and quartz points. The data are scattered about the quartz-clay line, suggesting that V_p and V_s are principally controlled by mineralogy.

In-situ sonic and field seismic measurements in mudrocks (Figure 4) form a well-defined line given by

$$V_p = 1.16 V_s + 1.36, \qquad (1)$$

where the velocities are in km/s. In view of the highly variable composition and texture of mudrocks, the uniform distribution of these data is surprising. We believe this linear trend is explained in part by the location of the clay point near a line joining the quartz point with the velocity of water. We hypothesize that, as the porosity of a pure clay increases, compressional and shear velocities decrease in a nearly linear fashion as the water point is approached. Similarly, as quartz is added to pure clay, velocities increase in a nearly linear fashion as the quartz point is approached. These bounds generally agree with those inferred from the empirical relations of Tosaya (1982); the exception is for behavior at very high porosities. The net result is that quartz-clay-water ternary mixtures are spread along an

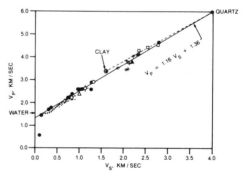

FIG. 4. Compressional and shear wave velocities for mudrocks from in-situ sonic and field seismic measurements.

elongate triangular region loosely defined by clay-water, quartz-clay, and quartz-water lines.

Figure 4 and equation (1) indicate that V_p/V_s for mudrocks is highly variable, ranging from less than 1.8 in quartz-rich rocks to over 5 in loose, water-saturated sediments. This is a direct result of the nonzero intercept at the water point.

Compressional and shear wave velocities obtained by sonic logging in geopressured argillaceous rocks of the Frio formation that exhibit clay volumes in excess of 30 percent, as determined by a neutron-density crossplot, are plotted in Figure 5. Most of the data are consistent with equation (1) and Figure 4. Figure 6 is a plot of sonic log data in shaly intervals reported by Kithas (1976). As with Figure 5, these data are well described by equation (1).

OBSERVATIONS IN SANDSTONES

The trend of Pickett's (1963) laboratory data for clean water-saturated sandstones (Figure 7) coincides precisely with equation (1) established for mudrocks. The correspondence of V_p/V_s for sandstones and mudrocks is not entirely expected. Figure 8 is a plot of sonic log V_p and V_s data in sandstones exhibiting less than 20 percent neutron-density clay volume in the Frio formation. Except for some anomalously low V_p/V_s ratios indicative of a tight gas sandstone (verified by conventional log analysis), the data again fall along the water-saturated line established for mudrocks. Also falling along this line are sonic velocities for an orthoquartzite reported by Eastwood and Castagna (1983). In-situ measurements for shallow marine sands compiled by Hamilton (1979) fall above the line. Sonic log velocities reported by Backus et al. (1979) and Leslie and Mons (1982) in clean porous brine sands tend to fall slightly below the line (Figure 9).

Figure 10 is a compilation of our laboratory data for water-saturated sandstones with data from the literature (Domenico, 1976; Gregory, 1976; King, 1966; Tosaya, 1982; Johnston, 1978; Murphy, 1982; Simmons, 1965; Hamilton 1971). To first order, the data are consistent with equation (1); however, they are significantly biased toward higher V_s for a given V_p. The location of some sandstone data on the mudrock water-saturated line, although other data fall below this line, is presumably related to the sandstone texture and/or clay content.

Following the lead of Tosaya (1982), we used multiple linear regression to determine the dependence of sonic waveform-derived compressional and shear wave velocity on porosity and clay content for the Frio formation. We applied conventional log analysis to determine porosity and volume of clay (V_{cl}) from gamma ray, neutron, and density logs. The resulting relationships for the Frio formation are

$$V_p \text{ (km/s)} = 5.81 - 9.42\phi - 2.21 V_{cl} \tag{2a}$$

and

$$V_s \text{ (km/s)} = 3.89 - 7.07\phi - 2.04 V_{cl}. \tag{2b}$$

The correlation coefficient r for both of these relationships is .96. The equations of Tosaya (1982) for laboratory data are

$$V_p \text{ (km/s)} = 5.8 - 8.6\phi - 2.4 V_{cl}, \tag{2c}$$

and

$$V_s \text{ (km/s)} = 3.7 - 6.3\phi - 2.1 V_{cl}. \tag{2d}$$

The similarity between our equations and Tosaya's is evident.

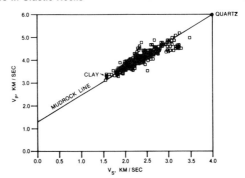

FIG. 5. Compressional and shear wave velocities for geopressured shaly rocks of the Frio formation from sonic logs.

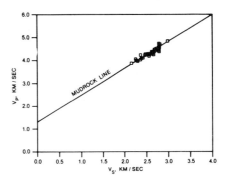

FIG. 6. Sonic data from Kithas (1976), mostly for shales.

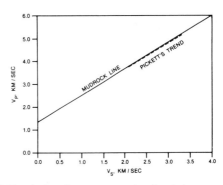

FIG. 7. Trend of sandstone compressional and shear wave velocities from Pickett (1963).

From Tosaya's equations the sonic properties of zero porosity clay are: P-wave transit time = 89.6 μs/ft, S-wave transit time = 190.5 μs/ft, and $V_p/V_s = 2.125$. The corresponding values for the Frio formation clay are: P-wave transit time = 84.7 μs/ft, S-wave transit time = 165.1 μs/ft, and $V_p/V_s = 1.95$.

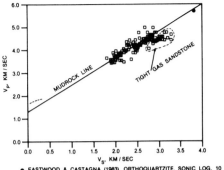

- ● EASTWOOD & CASTAGNA (1983), ORTHOQUARTZITE, SONIC LOG, 10 KHz
- □ FRIO FORMATION SANDSTONES, SONIC LOG (15 KHz)
- --- HAMILTON (1979), SANDS

FIG. 8. Sonic log velocities in sandstones.

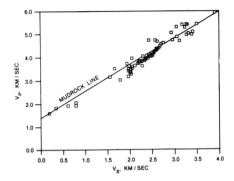

FIG. 10. Compilation of laboratory data for water-saturated sandstones, including ARCO and literature data.

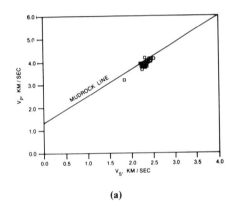

(a)

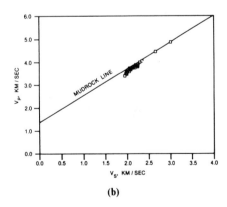

(b)

FIG. 9. Sonic log velocities for clean water-saturated sandstones. (a) Backus et al. (1979), (b) Leslie and Mons (1982).

Some algebraic manipulation yields equations which explicitly reveal the dependence of V_p/V_s on porosity and volume of clay. From equations (2a) and (2b) we get

$$V_p/V_s = 1.33 + .63/(3.89 - 7.07\phi) \tag{3a}$$

for clean sand and

$$V_p/V_s = 1.08 + 1.61/(3.89 - 2.04V_{cl}) \tag{3b}$$

for zero-porosity sand/clay mixtures. These equations reveal that increasing porosity or clay content increases V_p/V_s and that the velocity ratio is more sensitive to porosity changes.

DRY SANDSTONES: LABORATORY DATA AND IDEAL MODELS

According to Gregory (1977), Poisson's ratio is about 0.1 (corresponding to $V_p/V_s \approx 1.5$) for most dry rocks and unconsolidated sands, and it is independent of pressure. Figure 11 is a crossplot of V_p/V_s versus V_p for dry and water-saturated Berea sandstone. Note that the water-saturated points are reasonably close to the relationship defined by equation (1), whereas the dry points are nearly constant at a V_p/V_s of about 1.5. Figure 12 is a compilation of laboratory compressional and shear wave velocities for dry sandstones. The field data of White (1965) for loose sands also are included. The data fit a line having a constant V_p/V_s ratio of 1.5.

We gain some insight into the behavior of dry sandstones by considering various regular packings of spheres. The reader is referred to White (1965) and Murphy (1982) for a detailed discussion. The dry compressional (V_p^D) to shear (V_s^D) velocity ratio, as a function of Poisson's ratio (v) of solid spheres, is

$$V_p^D/V_s^D = [(2 - v)/(1 - v)]^{1/2} \tag{4a}$$

for a simple cubic packing (SC) of spheres;

$$V_p^D/V_s^D = [4(3 - v)/(6 - 5v)]^{1/2} \tag{4b}$$

for a hexagonal close packing (HCP) of spheres; and

$$V_p^D/V_s^D = \sqrt{2} \tag{4c}$$

for a face-centered cubic (FCC) packing. For HCP and FCC packings propagation directions are (1, 0, 0).

These relations are valid only at elevated confining pressures when the packings have sufficiently high bulk and shear moduli to propagate elastic waves. The velocity ratio cancels the pressure dependence; however, this pressure restriction should be kept in mind.

From equations (4a) through (4c) we see that dry V_p/V_s is constant or dependent only upon Poisson's ratio of the spheres (v). For quartz spheres (v ≈ 0.1), dry V_p/V_s is virtually the same for these packings (FCC = 1.41, HCP = 1.43, SC = 1.45).

Dry V_p/V_s ratios based on these regular packings of spheres (Figure 13) indicate that the elastic properties of the grains are of secondary importance. The maximum range of dry V_p/V_s is for SC, and varies only from $\sqrt{2}$ to $\sqrt{3}$ when Poisson's ratio of the spheres varies from 0 to 0.5. For all practical purposes, dry V_p/V_s for packings of common rock-forming minerals is from 1.4 to 1.5.

For real rocks, of course, a number of complicating factors must be considered. One of these is the presence of microfractures or pores of low aspect ratio. One means of studying the effects of microfractures is to make velocity measurements on samples before and after cracking by heat cycling. Figure 14 shows the results of heat cycling on dry sandstones obtained by Aktan and Farouq Ali (1975). The effect of the addition of microfractures is to reduce both compressional and shear wave velocities in a direction parallel to the dry line, maintaining a nearly constant V_p/V_s.

Further insight into the effect of microfractures is provided by modeling the elastic moduli of solids with various pore aspect ratio spectra utilizing the formulation of Toksöz et al. (1976). This theory assumes randomly oriented and distributed noninteracting elliptical pores and, therefore, may not be valid for highly porous or shaly rocks.

As an example, we computed the V_p/V_s relationship for the inverted Boise sandstone pore spectrum (Cheng and Toksöz, 1976). This was done by maintaining the same relative concentrations among the aspect ratio distribution but uniformly increasing the entire distribution to increase the total porosity. Our intent was to show that aspect ratio distributions thought to be characteristic of actual rocks produce behavior in close agreement with observed trends. We do not imply that these distributions are quantitatively applicable to real rocks. Except

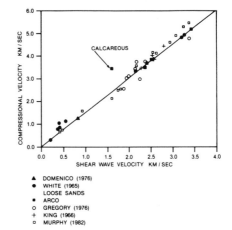

FIG. 12. Laboratory measurements for V_s and V_p for dry sandstones. Note that one sandstone with calcite cement plots well above the line.

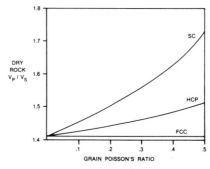

FIG. 13. Calculated V_p/V_s for regular packings of spheres versus Poisson's ratio (v) of the sphere material. SC = simple cubic packing, HCP = hexagonal close packing, FCC = face-centered cubic packing.

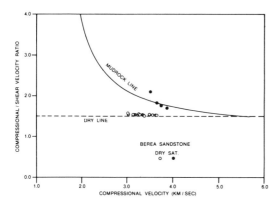

FIG. 11. Ultrasonic measurements of V_p/V_s for a dry and water-saturated Berea sandstone sample. The various points were obtained at different effective pressures.

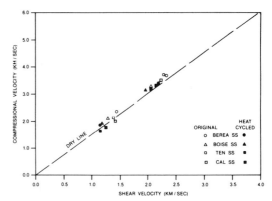

FIG. 14. The results of Aktan and Farouq Ali (1975) for several dry sandstones before and after heat cycling. Data are plotted for measurements at high and low confining pressures.

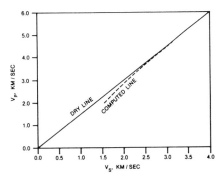

FIG. 15. Calculated V_s and V_p from the formulation of Cheng and Toksöz (1979) using the inverted pore aspect ratio spectrum of Boise sandstone.

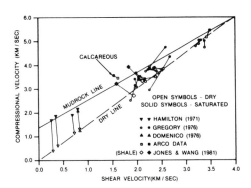

FIG. 16. Ultrasonic laboratory data for various sandstones under both dry and saturated conditions. Dry (open symbols) and saturated (solid symbols) data are plotted for the same effective pressure conditions and joined by tie lines.

for high crack concentrations (low velocities) where the limits of the model may be exceeded, the computed line for dry Boise sandstone is in excellent agreement with the observed dry sandstone line (Figure 15). It is interesting to note that an identical computed line is obtained when all cracks with pore aspect ratios less than 0.1 are closed. This is consistent with the results of Atkan and Farouq Ali (1975) shown in Figure 14. The addition of microfractures to a dry rock lowers V_p and V_s, but does not change the V_p/V_s ratio appreciably.

To summarize, we examined two extreme cases for dry rocks by two different models: (1) regular packings of spheres and (2) cracked solids. Both models predict a nearly constant V_p/V_s for dry sandstone, which is consistent with experimental observation.

SATURATED SANDSTONES:
LABORATORY DATA AND IDEAL MODELS

Figure 16 shows V_p/V_s relationships for dry and water-saturated sandstones. Lines join measurements made on single dry and water-saturated samples under the same effective pressure conditions. Qualitatively, the data from water-saturated sandstones are consistent with equation (1) but tend toward higher V_s for a given V_p.

Given the compressional and shear wave velocities obtained in the laboratory for dry sandstones, we use Gassmann's (1951) equations to compute velocities when these rocks are saturated with water. Gassmann's equations are:

$$K_W = K_S \frac{K_D + Q}{K_S + Q}, \qquad (5a)$$

$$Q = \frac{K_F(K_S - K_D)}{\phi(K_S - K_F)}, \qquad (5b)$$

$$\mu_W = \mu_D, \qquad (5c)$$

and

$$\rho_W = \phi\rho_F + (1 - \phi)\rho_S, \qquad (5d)$$

where K_W is the bulk modulus of the wet rock, K_S is the bulk modulus of the grains, K_D is the bulk modulus of the dry frame, K_F is the bulk modulus of the fluid, μ_W is the shear modulus of the wet rock, μ_D is the shear modulus of the dry rock, ρ_W is the density of the wet rock, ρ_F is the density of the fluid, ρ_S is the density of the grains, and ϕ is the porosity. Figure 17 is a plot of computed (from Gassmann's equations) and measured saturated compressional and shear wave velocities. Although the individual points do not always coincide, the computed and measured data follow the same trend.

We also can apply Gassmann's equations to the equations for dry regular packing arrangements of spheres given by Murphy (1982). Figure 18 shows the relationship between V_p and V_s for water-saturated SC, HCP, and FCC packings. Note that the packings of equal density (FCC and HCP) yield virtually the same curve. The curves obtained for these simple packings are qualitatively consistent with the low-velocity experimental data of Hamilton (1971) and Domenico (1976).

For water-saturated rock, the formulation of Toksöz et al. (1976) predicts that the addition of microfractures will change the V_p-V_s relationship (Figure 19). The inverted Boise sandstone pore aspect ratio spectrum yields a line close to equation (1). Closing all cracks with aspect ratios less than 0.1 yields a line which lies below equation (1). Thus, we might expect data for clean porous sandstones dominated by equant porosity to lie slightly below the line defined by equation (1), whereas tight sandstones with high concentrations of elongate pores would tend to lie along this line. Verification of this hypothesis is left as an objective of future research.

A good description of the water-saturated V_p-V_s relationship for sandstones is provided by the following formulation. The dry line established with laboratory data ($V_p/V_s \approx 1.5$) means that dry bulk modulus (K_D) is approximately equal to dry rigidity (μ_D)

$$\mu_D \approx K_D. \qquad (6)$$

These are exactly equal when

$$V_p^D/V_s^D = 1.53. \qquad (7)$$

From equation (5c) it follows that

$$K_D \approx \mu_D = \mu_W. \qquad (8)$$

Thus, the saturated shear velocity can be obtained from the dry bulk modulus by

$$V_s^2 \approx \frac{K_D}{\rho_w}. \qquad (9)$$

The wet bulk modulus is given by

$$K_W = \rho_W(V_p^2 - \tfrac{4}{3}V_s^2). \qquad (10)$$

Equations (9) and (10) and Gassmann's equations [(5a) through (5d)] allow computation of V_s given V_p, ϕ, and the grain and fluid densities and bulk moduli.

We computed shear-wave velocities for sandstone core porosities and sonic log compressional velocities given by Gregory et al. (1980) for depths from 2 500 to 14 500 ft in two wells 500 ft apart in Brazoria County, Texas. Figure 20 shows that the resulting V_p-V_s relationship is in excellent agreement with our sandstone observations.

Similarly, in Table 1 we compare the calculated shear velocities to the measured laboratory values. The differences are usually less than 5 percent, demonstrating excellent agreement with the theory. Since part of these differences must be due to experimental error and to our assumption that the matrix is 100 percent quartz, we consider the agreement remarkable. Additionally, Gassmann's equations are strictly valid only at low frequencies. Further corrections can be applied to account for dispersion. The Holt sand sample, which gives the largest discrepancy in Table 1, illustrates the importance of the assumption that dry bulk modulus is equal to dry rigidity [equation (6)]. The dry V_p/V_s for this Holt sand sample is 2.17 (Table 2), far different from the ratio of 1.53 required for equality of dry bulk and shear moduli. This variation is probably due to the high carbonate cement component of this rock. However, given this measured dry ratio, we can still apply Gassmann's equations to calculate the water-saturated values. As shown in Table 2, these predicted values are virtually identical to the measured velocities.

For clean sandstones at high pressures and with moderate porosities, porosity is often estimated by the empirical time-average formula (Wyllie et al., 1956),

$$\phi \simeq \frac{1/V_p - 1/V_p^s}{1/V_p^f - 1/V_p^s}, \qquad (11)$$

where V_p^s is the grain compressional velocity and V_p^f is the fluid velocity. Figure 21 shows the V_p-V_s relationship predicted by

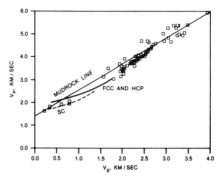

FIG. 18. Calculated V_s and V_p for saturated regular packings of quartz spheres using the calculated dry velocities and Gassmann's equations.

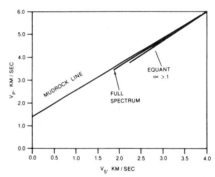

FIG. 19. Calculated V_s and V_p for Boise sandstone based on the formulation of Cheng and Toksöz (1979).

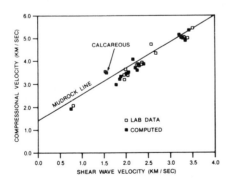

FIG. 17. Measured (open symbols) versus computed (solid symbols) V_s and V_p using the dry data from Figure 12 and Gassmann's equations.

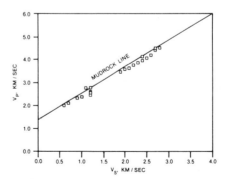

FIG. 20. Comparison of measured V_p with calculated V_s for sandstones from two wells as reported by Gregory et al. (1980). Calculations are based on measured wet V_p and porosity using Gassmann's equations and dry bulk modulus equal to dry rigidity.

Table 1. Holt sand: Comparison of observed water-saturated shear velocities with those calculated using the measured water-saturated compressional velocities and porosities. The calculations were based on Gassmann's equations and the assumption that $K_D \simeq \mu_D$.

Rock	Reference	V_p	Porosity	Predicted V_s	Observed V_s	Percent Error
Berea	Johnston (1978)	3.888	18.4	2.330	2.302	1.2
Berea	Johnston (1978)	4.335	18.4	2.700	2.590	4.2
Navajo	Johnston (1978)	4.141	16.4	2.520	2.430	3.7
Navajo	Johnston (1978)	4.584	16.4	2.890	2.710	6.6
Gulf Coast sand	Gregory (1976)	3.927	21.7	2.380	2.367	0.5
Gulf Coast sand	Gregory (1976)	3.185	21.7	1.730	1.975	−12.4
Boise	Gregory (1976)	3.402	26.8	1.970	1.960	.5
Boise	Gregory (1976)	3.533	26.8	2.080	2.073	.3
Travis peak	Gregory (1976)	4.732	4.45	2.860	2.581	10.8
Travis peak	Gregory (1976)	4.990	4.45	3.110	3.284	−5.3
Travis peak	Gregory (1976)	4.342	8.02	2.590	2.667	−2.9
Travis peak	Gregory (1976)	5.001	8.02	3.180	3.391	−6.2
Bandera	Gregory (1976)	3.492	17.9	1.970	2.032	−3.1
Bandera	Gregory (1976)	3.809	17.9	2.250	2.240	.4
Ottawa	Domenico (1976)	2.072	37.74	.740	.801	−7.6
Sample no. MAR	ARCO data	5.029	1.0	3.200	3.315	−3.5
Sample no. MAR	ARCO data	5.438	1.0	3.420	3.496	−2.2
Sample no. MDP	ARCO data	3.377	21.0	1.900	2.047	−7.2
Sample no. MDP	ARCO data	3.862	21.0	2.320	2.350	−1.3
Berea	ARCO data	3.642	19.0	2.120	1.992	6.4
Berea	ARCO data	3.864	19.0	2.310	2.267	4.3
Berea	ARCO data	3.510	19.0	2.000	1.680	19.0
Berea	ARCO data	3.740	19.0	2.200	2.130	3.8
St. Peter	Tosaya (1982)	5.100	6.6	3.250	3.420	−5.0
St. Peter	Tosaya (1982)	4.880	7.2	3.060	3.060	0.0
St. Peter	Tosaya (1982)	4.500	4.2	2.610	2.680	−2.6
St. Peter	Tosaya (1982)	4.400	7.5	2.630	2.600	1.2
St. Peter	Tosaya (1982)	4.400	5.0	2.540	2.600	−2.3
St. Peter	Tosaya (1982)	3.950	18.8	2.380	2.420	−1.6
St. Peter	Tosaya (1982)	3.600	19.6	2.090	2.070	1.0
St. Peter	Tosaya (1982)	3.170	14.5	1.580	1.560	1.3
Holt sand	ARCO data	3.546	16.3	1.990	1.539	29.3

Table 2. Water-saturated sandstones: A recalculation of velocities for the Holt Sand (Table 1) for which $K_D \neq \mu_D$. New water-saturated values are computed using both the V_p and V_s measured for the dry rock.

Holt Sand	
Porosity	16.3%
Water-saturated V_p (laboratory)	3.546 km/s
Water-saturated V_s (laboratory)	1.539 km/s
Water-saturated V_s (predicted from porosity and water-saturated V_p)	1.990 km/s
Percent error	29.3%
Dry V_p (laboratory)	3.466 km/s
Dry V_s (laboratory)	1.599 km/s
Dry V_p/V_s (laboratory)	2.17
Water-saturated V_p (predicted from dry data)	3.519 km/s
Percent error	−.8%
Water-saturated V_s (predicted from dry data)	1.540 km/s
Percent error	.1%

equations (5), (9), (10), and (11). This "time-average" line describes the laboratory data from water-saturated conditions presented in Figure 10 extremely well. Recalling the results of crack modeling shown in Figure 19, one explanation for the validity of this empirical formula would be the dominance of pores of high aspect ratio.

DISCUSSION OF RESULTS: DYNAMIC ELASTIC MODULI RELATIONSHIPS

Compressional and shear velocities, along with the density, provide sufficient information to determine the elastic parameters of isotropic media (Simmons and Brace, 1965). These parameters proved useful in estimating the physical properties of soils and characteristics of formations (see, for example, Richart, 1977). However, the relationships given by equation (1) for mudrocks or Gassmann's equations and equation (6) for sandstones fix V_s in terms of V_p. Hence, the elastic parameters can be determined for clastic silicate rocks from conventional sonic and density logs. This explains in part why Stein (1976) was successful in empirically determining the properties of sands from conventional logs.

In this discussion we assumed that equation (1) holds to first order for all clastic silicate rocks, with the understanding that Gassmann's equations might be used to obtain more precise results in clean porous sandstones if necessary. For many rocks, particularly those with high clay content, the addition of water softens the frame, thereby reducing the bulk elastic moduli. The following empirical relationships are, therefore, not entirely general but are useful for describing the wide variety of data presented here. As shown in Figure 22, bulk and shear moduli are about equal for dry sandstones. Adding water causes the bulk modulus to increase. This effect is most pronounced at higher porosities (lower moduli). Water-saturated bulk modulus normalized by density is linearly related to compressional velocity (Figure 23):

$$\frac{K_W}{\rho_W} \approx 2bV_p^W - b^2, \qquad b = 1.36 \text{ km/s.} \qquad (12)$$

Water-saturated rigidity normalized by density is nonlinearly related to wet compressional velocity:

$$\frac{\mu_W}{\rho_W} \approx \frac{3}{4}(V_p^W - b)^2. \qquad (13)$$

Dry bulk modulus and rigidity normalized by density are given by

$$\frac{\mu_D}{\rho_D} \approx \frac{K_D}{\rho_D} \approx \frac{3}{7}(V_p^D)^2. \qquad (14)$$

Poisson's ratio for water-saturated rock is approximately linearly related to compressional velocity (Figure 24). Poisson's ratio of air-saturated rock is constant.

DISCUSSION OF RESULTS: SEISMOLOGY AND FORMATION EVALUATION

In recent years there has been increased use of V_p, V_s, and V_p/V_s in seismic exploration for estimation of porosity, lithology, and saturating fluids in particular statigraphic intervals. The above analysis both complicates and enlightens such interpretation. It is clear that clay content increases the ratio V_p/V_s, as does porosity. The analyses of Tosaya (1982) and Eastwood and Castagna (1983) and equations (3a) and (3b) indicate that V_p/V_s is less sensitive to variation of clay content than to variation of porosity. However, the range of variation in clay content may be larger. Thus, V_p/V_s can be grossly dependent upon clay content.

Figure 25 shows V_p/V_s computed as a function of depth in the Gulf Coast for noncalcareous shales and clean porous sandstones that are water-saturated. The compressional velocity and porosity data given in Gregory (1977) are used to establish the variation with depth. Equation (1) is used to predict V_s for shales, and Gassmann's equations are used for sandstones. At a given depth, shale velocity ratios are on the order of 10 percent higher than sandstone velocity ratios.

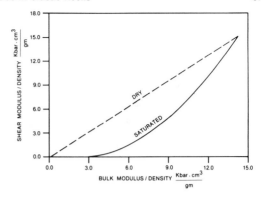

FIG. 22. The computed relationships between the bulk and shear moduli (normalized by density) based on the observed V_s and V_p trends.

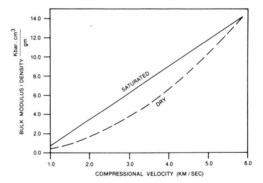

FIG. 23. The computed relationships between the bulk modulus (normalized by density) and V_p based on the observed V_s and V_p trends.

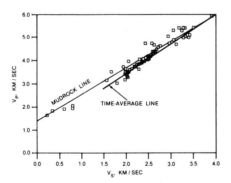

FIG. 21. Calculated V_s and V_p based on the time-average equation and Gassmann's equations.

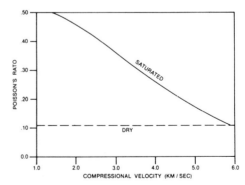

FIG. 24. The computed relationships between Poisson's ratio and V_p based on the observed V_s and V_p trends.

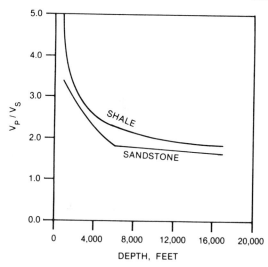

FIG. 25. V_p/V_s computed as a function of depth for selected Gulf Coast shales and water-saturated sands.

These conclusions are somewhat at odds with conventional wisdom that V_p/V_s equals 1.5 to 1.7 in sandstones and is greater than 2 in shales. Clearly, mapping net sand from V_p/V_s is not as straightforward as conventional wisdom would imply.

In clastic silicates, shear wave velocities for elastic moduli estimation or seismic velocity control may be estimated from equation (1) and/or Gassmann's equations and conventional logs. Alternatively, our relationships indicate that when V_p and V_s are used together, they can be sensitive indicators of both gas saturation and nonclastic components in the rock. Gas saturation will move the V_p/V_s toward the dry line in Figure 16. Nonclastic components can also move the ratio off the mudrock line, as shown by the Holt sand sample marked as calcareous in Figure 16.

The possibly ambiguous interpretation of lithology and porosity from V_p, V_s, and V_p/V_s in seismic exploration also applies to log analysis. However, because sonic logs are generally a part of usual logging programs, the interpretation of full waveform sonic logs should be made in the context of other logging information (as we did) to obtain equations (2a) and (2b). Other logs which are sensitive to clay content and porosity (e.g., neutron and gamma-ray) are needed to evaluate the details of how porosity and clay content separately affect clastic rocks. In gas-bearing and highly porous zones where the time-average equation fails, combined V_p and V_s information may potentially be used in porosity determination.

Assuming that the time-average equation works for compressional transit times and that dry incompressibility equals dry rigidity, Gassmann's equations yield a nearly linear relationship between shear wave transit time and porosity. This relationship is well described by a time-average equation

$$\phi = \frac{1\,V_s - 1\,V_s^s}{1.12 \text{ s/km} - 1\,V_s^s} \quad (15)$$

where V_s^s is the grain shear wave velocity. The constant 1.12 s/km is used in place of fluid transit time. This equation should be superior to the compressional wave time-average equation for porosity estimation when there is significant residual gas saturation in the flushed zone.

SUMMARY AND CONCLUSIONS

To first order, we conclude that shear wave velocity is nearly linearly related to compressional wave velocity for both water-saturated and dry clastic silicate sedimentary rocks. For a given V_p, mudrocks tend toward slightly higher V_p/V_s than do clean porous sandstones.

For dry sandstones, V_p/V_s is nearly constant. For wet sandstones and mudstones, V_p/V_s decreases with increasing V_p. Water-saturated sandstone shear wave velocities are consistent with those obtained from Gassmann's equations. The water-saturated linear V_p-versus-V_s trend begins at V_p slightly less than water velocity and $V_s = 0$, and it terminates at the compressional and shear wave velocities of quartz. The dry sandstone linear V_p-versus-V_s trend begins at zero velocity and terminates at quartz velocity. Dry rigidity and bulk modulus are about equal.

Theoretical models based on regular packing arrangements of spheres and on cracked solids yield V_p-versus-V_s trends consistent with observed dry and wet trends.

We believe that the observed V_p-versus-V_s-relationship for wet clastic silicate rocks results from the coincidental location of the quartz and clay points. The simple sphere-pack models indicate that the low-velocity, high-porosity trends are largely independent of the mineral elastic properties. As the porosity approaches zero, however, the velocities must necessarily approach the values for the pure mineral. In this case, clay and quartz fall on the extrapolated low-velocity trend.

REFERENCES

Aktan, T., and Farouq Ali, C. M., 1975, Effect of cyclic and in situ heating on the absolute permeabilities, elastic constants, and electrical resistivities of rocks: Soc. Petr. Eng., 5633.

Backus, M. M., Castagna, J. P., and Gregory, A. R., 1979, Sonic log waveforms from geothermal well, Brazoria Co., Texas: Presented at 49th Annual International SEG Meeting, New Orleans.

Birch, F., 1966, Compressibility; elastic constants, in Handbook of physical constants, Clark, S. P., Jr., Ed., Geol. Soc. Am., Memoir 97, 97–174.

Blatt, H., Middleton, G. V., and Murray, R. C., 1972, Origin of sedimentary rocks: Prentice-Hall, Inc.

Cheng, C. H., and Toksöz, M. N., 1976, Inversion of seismic velocities for the pore aspect ratio spectrum of a rock: J. Geophys. Res., 84, 7533–7543.

Christensen, N. J., 1982, Seismic velocities, in Carmichael, R. S., Ed., Handbook of physical properties of rocks, II: 2-227, CRC Press, Inc.

Domenico, S. N., 1976, Effect of brine-gas mixture on velocity in an unconsolidated sand reservoir: Geophysics, 41, 887–894.

Eastwood, R. L., and Castagna, J. P., 1983, Basis for interpretation of V_p/V_s ratios in complex lithologies: Soc. Prof. Well Log Analysts 24th Annual Logging Symp.

Ebeniro, J., Wilson, C. R., and Dorman, J., 1983, Propagation of dispersed compressional and Rayleigh waves on the Texas coastal plain: Geophysics, 48, 27–35.

Gassmann, F., 1951, Elastic waves through a packing of spheres: Geophysics, 16, 673–685.

Gregory, A. R., 1977, Fluid saturation effects on dynamic elastic properties of sedimentary rocks: Geophysics, 41, 895–921.

——— 1977, Aspects of rock physics from laboratory and log data that are important to seismic interpretation, in Seismic stratigraphy—application to hydrocarbon exploration: Am. Assn. Petr. Geol., Memoir 26.

Gregory, A. R., Kendall, K. K., and Lawal, S. S., 1980, Study effects of geopressured-geothermal subsurface environment of elastic properties of Texas Gulf Coast sandstones and shales using well logs, core data, and velocity surveys: Bureau of Econ. Geol. Rep., Univ. Texas, Austin.

Hamilton, E. L., 1971, Elastic properties of marine sediments: J. Geophys. Res., **76**, 579–604.

——— 1979, V_p/V_s and Poisson's ratios in marine sediments and rocks: J. Acoust. Soc. Am., **66**, 1093–1101.

Johnston, D. H., 1978, The attenuation of seismic waves in dry and saturated rocks: PhD. thesis, Mass. Inst. of Tech.

Jones, L. E. A., and Wang, H. F., 1981, Ultrasonic velocities in Cretaceous shales from the Williston Basin: Geophysics, **46**, 288–297.

King, M. S., 1966, Wave velocities in rocks as a function of changes of overburden pressure and pore fluid saturants: Geophysics, **31**, 50–73.

Kithas, B. A., 1976, Lithology, gas detection, and rock properties from acoustic logging systems: Trans., Soc. Prof. Well Log Analysts 17th Annual Logging Symp.

Koerperich, E. A., 1979, Shear wave velocities determined from long- and short-spaced borehole acoustic devices: Soc. Petr. Eng., 8237.

Lash, C. E., 1980, Shear waves, multiple reflections, and converted waves found by a deep vertical wave test (vertical seismic profiling): Geophysics, **45**, 1373–1411.

Leslie, H. D. and Mons, F., 1982, Sonic waveform analysis: applications: Trans., Soc. Prof. Well Log Analysts 23rd Annual Logging Symp.

Lingle, R., and Jones, A. H., 1977, Comparison of log and laboratory measured P-wave and S-wave velocities: Trans., Soc. Prof. Well Log Analysysts 18th Annual Logging Symp.

Murphy, W. F., III, 1982, Effects of microstructure and pore fluids on acoustic properties of granular sedimentary materials: PhD. thesis, Stanford Univ.

Nations, J. F., 1974, Lithology and porosity from acoustics and P-wave transit times: Log Analyst, November–December.

Pickett, G. R., 1963, Acoustic character logs and their applications in formation evaluation: J. Petr. Tech., **15**, 650–667.

Richart, F. E., Jr., 1977, Field and laboratory measurements of dynamic soil properties: Proc. DMSR 77, Karlsruhe, **1**, Prange, B., Ed., *in* Dynamical methods in soil and rock mechanics, 3–36.

Siegfried, R. W., and Castagna, J. P., 1982, Full waveform sonic logging techniques: Trans., Soc. Prof. Well Log Analysts 23rd Annual Logging Symp.

Simmons, G., 1965, Ultrasonics in geology: Proc. Inst. Electr. and Electron. Eng., **53**, 1337–1345.

Simmons, G., and Brace, W. F., 1965, Comparison of static and dynamic measurements of compressibility of rocks: J. Geophys. Res., **70**, 5649–5656.

Stein, N., 1976, Mechanical properties of friable sands from conventional log data: J. Petr. Tech., **28**, 757–763.

Toksöz, M. N., Cheng, C. H., and Timur, A., 1976, Velocities of seismic waves in porous rocks: Geophysics, **41**, 621–645.

Tosaya, C. A., 1982, Acoustical properties of clay-bearing rocks: PhD. thesis, Stanford Univ.

White, J. E., 1965, Seismic waves: Radiation, transmission and attenuation: McGraw-Hill Book Co.

Wilkens, R., Simmons, G., and Caruso, L., 1984, The ratio V_p/V_s as a discriminant of composition for siliceous limestones: Geophysics, **49**, 1850–1860.

Wyllie, M. R. J., Gregory, A. R., and Gardner, L. W., 1956, Elastic wave velocities in heterogeneous and porous media: Geophysics, **21**, 41–70.

Direct hydrocarbon detection using comparative P-wave and S-wave seismic sections

James D. Robertson* and William C. Pritchett‡

ABSTRACT

Two field experiments in the Sacramento basin and one in the Green River basin demonstrate that comparative P-wave and S-wave CDP seismic sections can be used to detect gas directly in sandstone reservoirs. The lines in the Sacramento basin were shot over producing gas fields known to correlate with amplitude anomalies on P-wave sections. Reflections of comparable strength are present on the P-wave and S-wave sections at lithologic boundaries, but gas-saturated zones correlating with P-wave bright spots show no equivalent S-wave amplitude anomalies. The responses are consistent with laboratory observations that P-wave velocity is more sensitive to the introduction of gas into liquid-saturated pore space than S-wave velocity. The line in the Green River basin was shot over a relatively deep gas field producing from an overpressured reservoir not associated with a conventional P-wave bright spot. The P-wave reflection strength of the reservoir is about 50 percent greater than the S-wave reflection strength, whereas the P and S strengths of other major reflectors are comparable. The three field tests show that an S-wave section validates a P-wave bright spot attributed to gas saturation when there is no anomalous amplitude at the equivalent S-wave event and that the technique is useful for verification of subtle as well as strong amplitude anomalies.

INTRODUCTION

Numerous theoretical and laboratory studies of elastic wave propagation in sedimentary rocks have shown that the ratio of P-wave velocity, V_p, to S-wave velocity, V_s, can be diagnostic of specific matrix and pore fluid properties. Several papers in the recent geophysical literature review these studies (Tatham and Stoffa, 1976; Domenico, 1977; Gregory, 1977; Mavko et al., 1979; Nur, 1982; Murphy, 1982). One result of the research has been the realization that V_p/V_s characteristically drops 10 to 20 percent when gas is introduced into the pore space of a liquid-saturated sandstone. The drop occurs because the change in fluid saturation substantially lowers the effective bulk modulus of the rock while minimally changing the density and effective shear modulus. Thus, V_p decreases sharply while V_s is not appreciably affected. It appears, unfortunately, that only the first few percent of gas are needed to produce most of the decrease in V_p and V_p/V_s, so the behavior of the velocities is more useful for detecting than quantitatively measuring significant partial gas saturation.

Laboratory observations have produced predictions that comparative P-wave and S-wave reflection seismic sections can be used in hydrocarbon exploration to distinguish gas-liquid contacts from lithologic interfaces (Gregory, 1976; Tatham and Stoffa, 1976; Domenico, 1977; Meissner and Hegazy, 1981; Nur, 1982). The idea is that an S-wave section validates a P-wave bright spot attributed to gas saturation when there is no anomalous amplitude at the equivalent S-wave event. If there is a roughly equivalent S-wave bright spot, the P-wave amplitude anomaly is a false alarm indicating lithologic change.

We present here the results of three field experiments conducted to test whether comparative P-wave and S-wave sections can directly detect gas. Two of the three were part of the 1977–1978 Conoco P-Wave/S-Wave Group Shoot (Rice et al., 1981). The lines come from the southern part of the Sacramento basin and were shot over producing gas fields known to correlate with bright spots on P-wave sections. The third experiment was conducted in 1975 over a producing gas field in Wyoming

---

Presented in part at the 53rd Annual International SEG Meeting September 13, 1983, in Las Vegas. Manuscript received by the Editor June 18, 1984; revised manuscript received September 7, 1984.
*ARCO Oil and Gas Company, P.O. Box 2819, Dallas, TX 75221.
‡ARCO Exploration Company, P.O. Box 2819, Dallas, TX 75221.
© 1985 Society of Exploration Geophysicists. All rights reserved.

by an ARCO/Chevron/Conoco/Shell consortium. This gas field is not an obvious amplitude anomaly on P-wave sections. The producing formation is highly overpressured, and the experiment was a test of whether P/S data can detect gas when the effect on P and S amplitudes is subtle.

WILLOW SLOUGH EXPERIMENT

Coincident P-wave and S-wave common-depth-point (CDP) data were acquired across the Willow Slough field, Yolo County, California (Figure 1). The seismic line, known to the Group Shoot participants as Yolo Line 2, is about two and one-half miles long, runs close to three producing gas wells, and ties to a dry hole off the field to the southwest (Figure 2).

A generalized geologic cross-section of the southern Sacramento basin abstracted from Drummond et al. (1976) is shown in Figure 3. The Tertiary section contains both marine and nonmarine clastics and is cut by several major unconformities. The horizons of interest for the P/S analysis are the sandstone members of the Upper Cretaceous Starkey and Winters formations. The Starkey, Delta shale, and Winters are separate lithofacies deposited as a regressive sequence across a basinward migrating shelf (Drummond et al., 1976). The Starkey sandstones constitute a shelf facies, the Delta is a silt and shale foreslope sequence, and the Winters sandstones form a basinal facies deposited by gravity flow down distributary channels in the foreslope. Both the Starkey and Winters sandstones can be prolific gas producers in the southern Sacramento basin; the Starkey is the main producing formation in the Willow Slough field.

The gas well near shotpoint 41 illustrates the vertical trapping sequence that typically occurs in the field (Figure 4). The main pay in this well is the third Starkey sandstone, which

FIG. 1. Location of Willow Slough and Putah Sink fields, northern California.

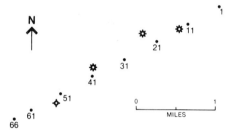

FIG. 2. Shotpoint map, Willow Slough seismic line.

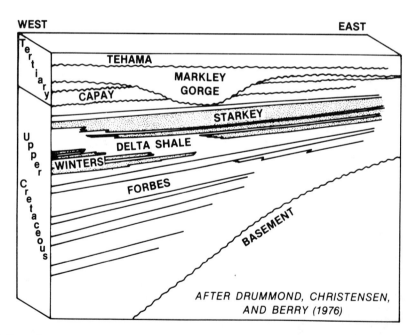

FIG. 3. Generalized geologic cross-section, southern Sacramento basin (modified from Drummond et al., 1976).

produces from several zones between 4 600 and 4 725 ft (Figure 5). The gas zones are sealed above by shale layers which stand out clearly on the spontaneous potential (SP) log. The pore fluid in the sandstone changes from gas to brine at about 4 620 ft and again at about 4 725 ft. The change at 4 725 ft is the base of the pay zone and represents a major boundary across which lithology is constant and pore fluid changes. The gas zones themselves are marked by low sonic velocities. The shales and water-saturated sandstones in the Starkey generally have similar sonic velocities, so one would expect strong P-wave reflections from the gas zones and weak P-wave reflections from boundaries between the shales and brine sands. Gas in the field is trapped by pinch outs and/or small faults.

The sonic and density logs were used to compute a P-wave synthetic seismogram (Figure 6). The synthetic is plotted between strips of traces from the P-wave surface seismic section. The match between the synthetic and the section is reasonably good, and major boundaries such as the base of the Tehama formation and the basal gas-water contact in the third Starkey are easy to correlate. The 20 Hz Ricker wavelet used to compute the synthetic seismogram was selected by comparing the amplitude of the stacked P-wave trace at shotpoint 41 to amplitude spectra of analytic wavelets until a reasonable fit was obtained (Figure 7). A piece of the synthetic seismogram displayed at expanded scale illustrates that the strongest events in the Starkey interval are directly related to the presence of gas and are true seismic bright spots (Figure 8).

The comparative P-wave and S-wave seismic sections across the field (Figures 9 and 10) were processed to preserve true amplitudes. In addition, the relative gains of the P and S sections were balanced on the basis of events above and below the Starkey sequence. This procedure minimizes any overall amplitude shift between the data sets resulting from different source strength, receiver sensitivity, near-surface attenuation, etc. Events on the S-wave section were identified using gross character correlation with the P-wave data away from the producing gas intervals. The event labeled "Forbes" has been so named on the basis of regional geology since no wells along

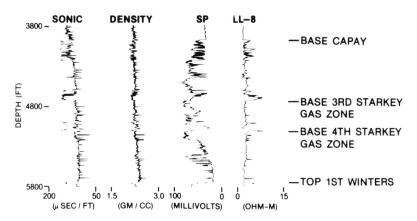

FIG. 4. Well logs near shotpoint 41, Willow Slough seismic line.

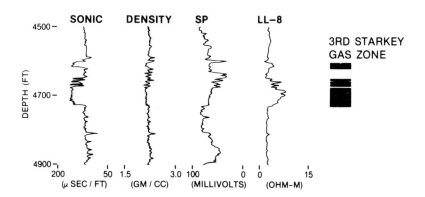

FIG. 5. Well logs near shotpoint 41 at expanded scale, Willow Slough seismic line.

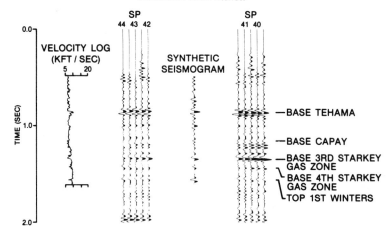

Fig. 6. *P*-wave synthetic seismogram near shotpoint 41, Willow Slough seismic line.

the line penetrate the Forbes. The name is not important for our purposes; what is relevant is that the event is probably a lithologic boundary and can be used to balance amplitudes and to correlate between sections.

Once horizons are identified and amplitudes are properly and fairly scaled, the remainder of the interpretation is straightforward. One searches for events in potentially productive intervals which possess anomalously high amplitudes on the *P*-wave section compared to the *S*-wave section. The *S*-wave amplitudes form, in effect, the baseline set against which *P*-wave amplitudes are measured. Similar *P* and *S* responses probably are indicative of lithologic interfaces, while grossly different responses where *P* is greater than *S* are diagnostic of gas-liquid contacts or contacts between shale and gas-bearing sandstone.

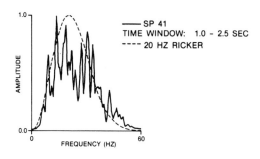

Fig. 7. Amplitude spectrum of *P*-wave seismic trace at shotpoint 41, Willow Slough seismic line.

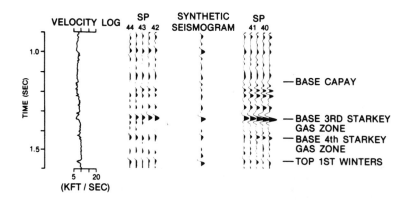

Fig. 8. *P*-wave synthetic seismogram near shotpoint 41 at expanded scale, Willow Slough seismic line.

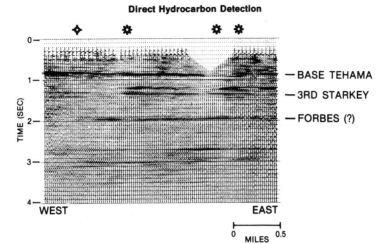

FIG. 9. *P*-wave section, Willow Slough seismic line.

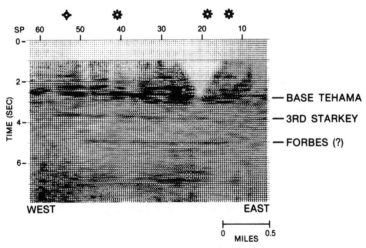

FIG. 10. *S*-wave section, Willow Slough seismic line.

The *P*-wave and *S*-wave sections across Willow Slough are displayed at expanded scale in Figures 11 and 12. The wells in Figure 2 are projected into the sections along strike at the Starkey level. The thicker gas zones in the Starkey stand out as anomalously high amplitudes on the *P*-wave display relative to the amplitudes of the equivalent zones on the *S*-wave display. The base of the third Starkey pay, for example, is the strong event just below 1.3 s. Multiple pays in the well at the right of the displays are obvious. This well produces from all four Starkey sandstones. The *P* and *S* sections define an important change in the third Starkey between shotpoints 40 and 50. The strong *P*-wave amplitude in the third Starkey diminishes to the left of shotpoint 40 and becomes roughly the same as the amplitude of the third Starkey *S*-wave event by about shotpoint 50. There is also a difference in the phase responses of the *P* and *S* events between these shotpoints. The seismic data show that the producing interval is changing from a gas-saturated to a fully brine-saturated sandstone, an interpretation confirmed by the well near shotpoint 53 which flowed only water from the third Starkey.

PUTAH SINK EXPERIMENT

Comparative *P* and *S* lines were also acquired over the Putah Sink field (Figure 1). These lines are about six miles long and cross the field's discovery well near shotpoint 99 (Figure 13). The well produces gas from the first Winters sandstone, which is about 100 ft thick at a depth of approximately 6 500 ft (Figure 14). The producing zones are sealed and underlain by shales at the well location (Figure 15). Owing to the presence of

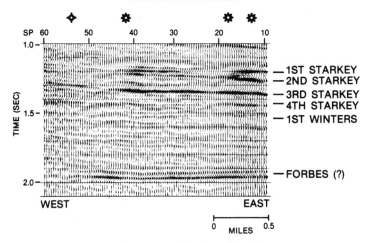

FIG. 11. *P*-wave section at expanded scale, Willow Slough seismic line.

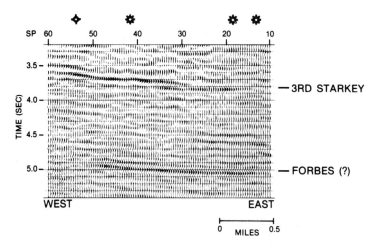

FIG. 12. *S*-wave section at expanded scale, Willow Slough seismic line.

the gas, the *P*-wave velocity of the first Winters is appreciably lower than that of the shales. In contrast, the sandstones in the overlying Starkey formation are water-saturated and possess velocities similar to those in surrounding shales. One would expect a strong, anomalous, *P*-wave seismic response from the first Winters.

Events on the *P*-wave section were identified by correlation with a synthetic seismogram (Figure 16) computed after an appropriate wavelet had been selected (Figure 17). As in the Willow Slough experiment, the synthetic seismogram correlates well with the *P*-wave section and verifies that the first Winters event is a true seismic bright spot (Figure 18).

The *P*-wave and *S*-wave stacked sections are fully displayed in Figures 19 and 20; the portions of the sections covering the gas field are plotted at expanded scale in Figures 21 and 22. These data were analyzed in the same way as were the Willow Slough data. True amplitude stacks were scaled such that lithologic boundaries like the base of the Markley Gorge (the discontinuous event at 1.0 s at the right of the *P*-wave section) have approximately the same amplitude. The first Winters pay zone stands out clearly as a *P*-wave amplitude anomaly at about 1.75 s. There is no corresponding *S*-wave amplitude anomaly. As in Figures 11 and 12, this difference is diagnostic of the presence of gas saturation.

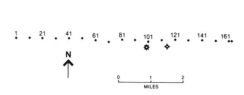

Fig. 13. Shotpoint map, Putah Sink seismic line.

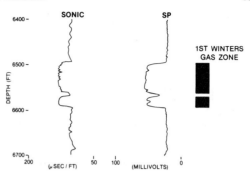

Fig. 15. Well logs near shotpoint 99 at expanded scale, Putah Sink seismic line.

CHURCH BUTTES EXPERIMENT

The Willow Slough and Putah Sink experiments were purposely conducted at locations where strong *P*-wave amplitude anomalies due to gas saturation were known to exist. The results are consistent with laboratory experiments and clearly demonstrate the concept that *S*-wave sections can be used to validate *P*-wave bright spots. In modern practice, one really would like to use *S*-wave data to verify *P*-wave amplitude anomalies when the anomalies are not obvious, since the obvious ones have probably already been drilled. The Church Buttes experiment was just such a test in a geologic setting not identified with conventional bright spot exploration.

The Church Buttes gas field is a low relief anticlinal trap

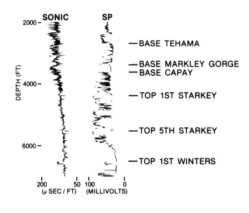

Fig. 14. Well logs near shotpoint 99, Putah Sink seismic line.

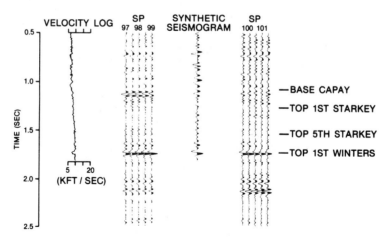

Fig. 16. *P*-wave synthetic seismogram near shotpoint 99, Putah Sink seismic line.

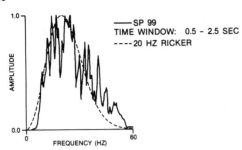

FIG. 17. Amplitude spectrum of P-wave seismic trace at shotpoint 99, Putah Sink seismic line.

the stacked traces is weighted toward near-offsets on the west and far-offsets on the east. Since the lines are very short and never get off the field, the experiment is essentially one-dimensional.

The events on the P-wave section (Figure 24) were identified using a synthetic seismogram computed from a sonic log in a well about a mile from the line. The log is of marginal quality, and the correlation is only good enough to match major packets of seismic energy with major formations. Shear-wave events (Figure 25) were identified by converting the P and S sections to depth using P and S stacking velocity functions. This procedure was adequate for correlating the major energy bands, but the depth conversion was too coarse to produce a reliable detailed correlation.

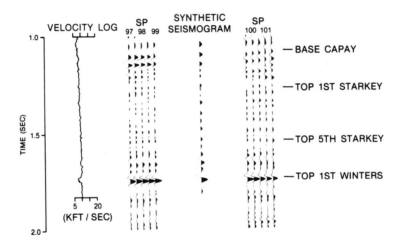

FIG. 18. P-wave synthetic seismogram near shotpoint 99 at expanded scale, Putah Sink seismic line.

located in the Green River basin of Wyoming along an arcuate fold known as the Church Buttes arch (Figure 23). The main producing horizon is the Cretaceous Dakota formation, which is 200 to 300 ft thick at a depth of 12 500 to 13 000 ft (Gras, 1968). The reservoir rocks in the Dakota are believed to be lenticular channel sandstones which are highly overpressured (Thomaidis, 1973, 1974). The sandstones produce gas with condensate under a gas-expansion drive. There is also some production from the Pennsylvanian Morgan formation (Thomaidis, 1973, 1974; Picard, 1977), though not in the northwestern part of the field where the seismic data were acquired.

Comparative P-wave and S-wave lines were shot in 1975 using P and S vibrators. The shear vibrators were predecessors of the more powerful units used by the 1977-1978 Conoco Group Shoot. Only a limited number of horizontal geophones was available, so the CDP data were acquired in an expanding spread configuration (sources and receivers each moving one group interval in opposite directions between shots). The result is that the P and S sections are only one-quarter mile long, but each stacked trace is 30-fold (Figures 24 and 25). A peculiarity of the expanding spread geometry is that the mix of offsets in

The interpretation of the data proceeded as follows. The P and S reflection strengths (instantaneous amplitudes) of the eight traces to the west were computed; maximum values were read for the Green River, Mesaverde, Dakota, and Morgan/Madison envelopes; the eight values for each event were averaged; and the P and S averages were normalized, respectivley, to the P and S averages for the Green River. Final results are displayed in Figure 26. Only the western eight traces were used in order to minimize offset-dependent amplitude variations. Normalization to the Green River was performed to remove variations due to source strength, near-surface attenuation, etc. similar to the California experiments.

The relative P and S reflection strengths of the Mesaverde and Morgan/Madison events are not significantly different after normalization. There is a significant difference at the Dakota event where P-wave reflection strength is about 50 percent higher than S-wave reflection strength. The results are consistent with a hypothesis that the overpressured gas in the Dakota lowers P velocity but minimally affects S velocity relative to P and S velocities in the overlying Aspen and underlying Morrison formations, thereby producing a relatively stronger P

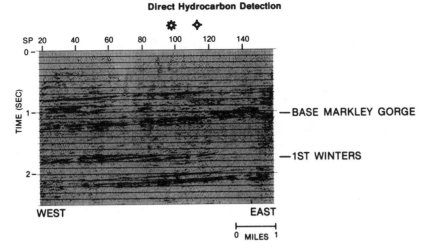

FIG. 19. *P*-wave section, Putah Sink seismic line.

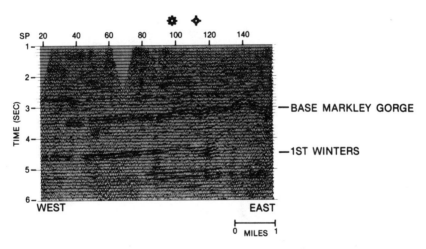

FIG. 20. *S*-wave section, Putah Sink seismic line.

reflection. Similar to the California tests, the *S*-wave data form the baseline which verifies that the *P*-wave amplitude response is in part an effect of gas saturation. The difference here is that the reflection strength is only tens of percent over baseline rather than the many hundreds of percent characteristic of conventional bright spots. The experiment indicates that it is possible to detect gas saturation directly using comparative *P*-wave and *S*-wave CDP data even when the effect on amplitudes is subtle.

SUMMARY

Direct detection of gas saturation in sandstones using *P*-wave bright spots is an established exploration technique in some geographic areas, most notably where there are reasonably shallow gas accumulations in Tertiary sand-shale sequences. A blanket correlation between bright spots and gas reservoirs can be misleading since there are complications such as interference patterns, thin beds, and unexpected lithologic changes that produce apparent bright spots. Comparative *S*-wave data constitute additional information which can help resolve these complications. This paper demonstrates through three case histories that *S*-wave sections can validate *P*-wave amplitude anomalies attributed to gas saturation. The field results are consistent with expectations from laboratory experiments. One of the case histories suggests that the comparative *P/S* technique may have application in searching for older,

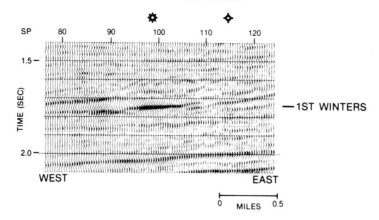

FIG. 21. *P*-wave section at expanded scale, Putah Sink seismic line.

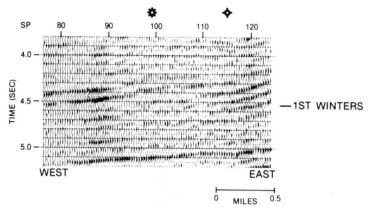

FIG. 22. *S*-wave section at expanded scale, Putah Sink seismic line.

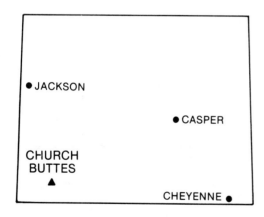

FIG. 23. Location of Church Buttes field, Wyoming.

more deeply buried gas reservoirs not normally considered targets for direct hydrocarbon detection.

ACKNOWLEDGMENTS

We thank W. T. Gant, V. J. McCullor, and W. W. Sharp for their comments and assistance during the course of this work, and ARCO Oil and Gas Company for permission to publish this paper.

The authors wish to note that Ensley (1983) presented a paper on *P/S* detection of gas saturation at the same SEG technical session at which this paper was first presented. Ensley discussed the Putah Sink experiment and reached the same general conclusions as are reached here.

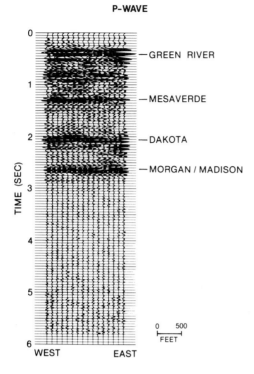

FIG. 24. *P*-wave section, Church Buttes seismic line.

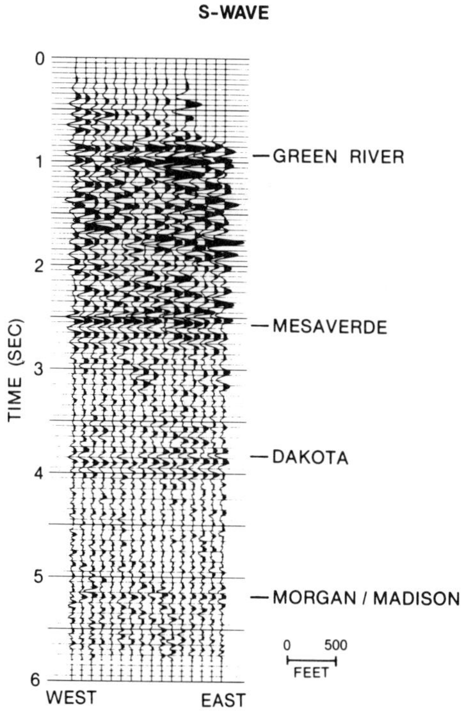

FIG. 25. *S*-wave section, Church Buttes seismic line.

REFERENCES

Domenico, S. N., 1977, Elastic properties of unconsolidated porous sand reservoirs: Geophysics, **42**, 1339–1368.
Drummond, K. F., Christensen, E. W., and Berry, K. D., 1976, Upper Cretaceous lithofacies model, Sacramento Valley, California: Am. Assn. Petr. Geol. Pacific Section Misc. Pub. no. 24, 76–88.
Ensley, R. A., 1983, Direct hydrocarbon detection with *P* and *SH*-wave seismic data: Presented at the 53rd Annual International SEG Meeting, September 13, Las Vegas.
Gras, V. B., 1968, Church Buttes gas field, Wyoming, *in* Natural gases of North America, **1**, Beebe, B. W., Ed.: Am. Assn. Petr. Geol., 9, 798–802.
Gregory, A. R., 1976, Fluid saturation effects on dynamic elastic properties of sedimentary rocks: Geophysics, **41**, 895–921.
——— 1977, Aspects of rock physics from laboratory and log data that are important to seismic interpretation, *in* Seismic stratigraphy—applications to hydrocarbon exploration: Payton, C. E., Ed., Am. Assn. Petr. Geol., Memoir 26: 15–46.
Mavko, G., Kjartansson, E., and Winkler, K., 1979, Seismic attenuation in rocks: Rev. Geophys. and Space Phys., **17**, 1155–1164.
Meissner, R., and Hegazy, M. A., 1981, The ratio of the *PP*- to the *SS*-reflection coefficient as a possible future method to estimate oil and gas reservoirs: Geophys. Prosp., **29**, 533–540.
Murphy, W. F., 1982, Effects of microstructure and pore fluids on the acoustic properties of granular sedimentary materials: Ph.D. thesis, Stanford Univ.
Nur, A., 1982, Wave propagation in porous rocks: Stanford Rock Physics Project, **13**, 2–77.
Picard, M. D., 1977, Petrography and stratigraphy of productive beds in the Morgan Formation, Church Buttes Unit No. 19, Southwest Wyoming: Wyoming Geol. Assn. Guidebook, 29th Field Conf., 179–196.

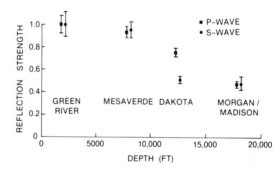

FIG. 26. Average *P*-wave and *S*-wave reflection strengths, Church Buttes seismic line.

Rice, R. B., Allen, S. J., Gant, O. J., Jr., Hodgson, R. N., Larson, D. E., Lindsey, J. P., Patch, J. R., LaFehr, T. R., Pickett, G. R., Schneider, W. A., White, J. E., and Roberts, J. C., 1981, Developments in exploration geophysics, 1975–1980: Geophysics, **46**, 1088–1099.
Tatham, R. H., and Stoffa, P. L., 1976, V_p/V_s—A potential hydrocarbon indicator: Geophysics, **41**, 895–921.
Thomaidis, N. D., 1973, Church Buttes Arch, Wyoming and Utah: Wyoming Geol. Assn. Guidebook 25th Field Conference, 35–39.
——— 1974, Structural-strat traps top targets for Church Buttes: Oil and Gas J., **72**, September 23, 213–216.

V_p/V_s and lithology

Robert H. Tatham*

ABSTRACT

Published laboratory investigations suggest an association exists between the ratio of seismic compressional and shear-wave velocities (V_p/V_s) and sedimentary rock lithology. Comparisons of some theoretical models with these laboratory studies suggest that crack, or pore, geometry has a stronger effect on observed V_p/V_s values than elastic constants of the minerals comprising the matrix. Further, it can be inferred that the observed association between lithology and V_p/V_s is a result of an association between lithology and distribution of pore and crack shapes. Direct observation of crack shapes for a variety of lithologies is a next step in strengthening these inferences. The present study reviews the empirical relations between V_p/V_s and lithology and examines two published theoretical models of cracked elastic media. The models suggest that seismic velocities of sandstones may be controlled by cracks and pores with aspect ratios in the range of 10^{-1} to 1, dolomite in the range of 10^{-2} to 10^{-1}, and dense limestones, of generally low porosity, in the range of 10^{-3} to 10^{-2}. Direct observations of the aspect ratio of cracks and pores in sedimentary rocks would test these inferences and offer a basis for physical and geologic insight into lithologic interpretations of V_p/V_s ratios.

INTRODUCTION

Recent industry interest in applying seismic shear-wave data in the search for hydrocarbons has led to techniques for gathering such reflection data in the field (Cherry and Waters, 1968; Tatham and Stoffa, 1976; Polskov et al, 1980). Suggested interpretation techniques (e.g., Erickson et al, 1968; Tatham et al, 1977) include correlation of P- and S-wave reflections to obtain the interval time ratio t_s/t_p within a given stratigraphic unit. Assuming the correlations are correct, this dimensionless ratio is equivalent to the velocity ratio V_p/V_s, and hence related to Poisson's ratio, a characteristic property of elastic solids.

Sedimentary rocks of interest in hydrocarbon exploration generally possess some finite porosity, and thus represent a two-phase system of matrix and pore saturant. In describing the bulk physical properties of such sedimentary rocks, we must consider not only the physical properties of the matrix material and the pore saturant, but also the shapes and distributions of the cracks and pores that compose the porosity. This pore geometry may vary with, among other things, grain geometry of detrital sediments and solution patterns of some carbonates.

The V_p/V_s ratio is especially sensitive to the pore fluid found in sedimentary rocks. In particular, the V_p/V_s value is much lower (10–20 percent) for gas saturation than for liquid saturation. This encouraging result has been suggested in the theoretical work of Gardner and Harris (1968), Kuster and Toksöz (1974a, b), and is consistent with the theory of Biot (1956a, b). Using the laboratory data of King (1966), Tatham and Stoffa (1976) showed the characteristic drop in V_p/V_s for gas-saturated sandstones. Gregory (1976) established the same observation for a wide range of well-consolidated sedimentary rocks, and Domenico (1976) observed such effects in unconsolidated sands.

Significantly, both laboratory and well log studies (Pickett, 1963; Nations, 1974; Kithas, 1976; Benzing, 1978) suggest a correlation between rock type (lithology) and observed V_p/V_s values. In particular, the data of Pickett (1963) yield V_p/V_s values, for a fairly wide range of porosity, near 1.9 for limestone, 1.8 for dolomite, and in the range of 1.6 to 1.75 for sandstones. This suggested sensitivity of V_p/V_s to whole-rock lithology is of great interest to explorationists, and the underlying rock properties controlling this observed sensitivity is my subject here.

DEFINITION OF THE PROBLEM

The relation between these properties—observed V_p/V_s, postulated crack geometry, and lithology—of sedimentary rocks is summarized in Figure 1. These properties are depicted as corners of a triangle, with sides of the triangle representing a concept relating the properties. The laboratory studies mentioned earlier suggest an empirical association between V_p/V_s and lithology, while consideration of theoretical models (examined in greater detail later) suggests an association between V_p/V_s and crack geometry. Several variables, such as flatness of cracks, shapes of crack ends, and general shape of the void (e.g., ellipsoidal or penny-shaped) are required to describe the geometry of a crack fully. For the present considerations, it is the aspect ratio, or flatness of cracks and pores, that describes the relevant effects of crack geometry. The remaining side of the triangle, direct observation of a correlation between pore shape and lithology, remains to be accomplished.

Presented at the 50th Annual International SEG Meeting November 19, 1980 in Houston. Manuscript received by the Editor October 14, 1980; revised manuscript received May 4, 1981.
*Formerly Texaco Inc., Bellaire, TX; presently Petty-Ray Geophysical, P.O. Box 36306, Houston, TX 77036.
0016-8033/82/0301—336$03.00. © 1982 Society of Exploration Geophysicists. All rights reserved.

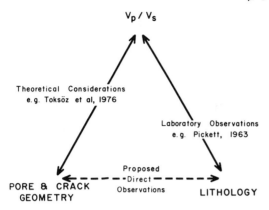

FIG. 1. Conceptual connection between sedimentary rock properties and the evidence supporting the connection. Direct observation of crack geometry for various lithologies would eliminate ambiguities in using both empirical observations and mathematical models to establish a strong V_p/V_s-lithology correlation.

An understanding of the underlying physical properties which control the V_p/V_s - lithology association is essential to the evolution of intuitive insights for interpretation of shear-wave data. Without some fundamental facts or models, the empirical data cannot be extrapolated beyond the actual laboratory observations.

The question arises as to whether or not the reported variations in V_p/V_s could be controlled by variations in the elastic properties of the matrix material alone. We will show that the more reasonable assumptions require adoption of the distributions of aspect ratios as the physical quantity controlling V_p/V_s variations.

Relative variations between any two independent elastic constants, however, can be related to variations of V_p/V_s. In considering a homogeneous and isotropic solid, only two elastic constants are required to describe the system. For our purposes, we will use K, the incompressibility (bulk modulus) and μ, the rigidity (shear modulus). Like Poisson's ratio, there is a correspondence between the ratio K/μ and the velocity ratio V_p/V_s. Such a correspondence, which can be expressed by $(V_p/V_s)^2 = 2(1 - \sigma)/(1 - 2\sigma) = k/\sigma + 4/3$, is plotted in Figure 2.

For a Poisson solid, Poisson's ratio σ is 0.25, V_p/V_s is $\sqrt{3}$ (1.732), and k/μ is 4/3 (1.67). Figure 2 shows a range in V_p/V_s from 1.3 to 4.0. A V_p/V_s ratio less than $\sqrt{2}$ requires a negative Poisson's ratio, which is found only in anisotropic material (Love, 1927). Since such values are observed for real sedimentary rocks (e.g., Gregory, 1976), they are included here.

In examining the bulk elastic properties K and μ of a sedimentary rock, we must keep in mind that both the elastic constants K and μ of the matrix and K of fluid materials, as well as the pore geometry, affect the final results. Using the relations described in Figure 2, we could substitute the bulk K/μ, or σ, for V_p/V_s at the apex of our conceptual triangle in Figure 1. Since V_p/V_s, or the identically equivalent t_s/t_p, is the interpretive quantity actually observed, we will maintain it in the present discussion. The relation between observed V_p/V_s and bulk K/μ shown in Figure 2 does leave some ambiguity in differentiating between the effects of K/μ of the matrix material and the crack distribution within the mineral matrix when interpreting observed V_p/V_s values for whole-rock samples. This ambiguity (to be discussed in greater detail later) stresses the need for direct observations and comparisons of pure geometry and lithology.

EFFECTS OF PORE SHAPE ON V_p/V_s

Numerous theoretical models have been proposed to describe elastic wave propagation in two-phase sedimentary rocks of finite porosity. Perhaps the most successful of these is the model of Biot (1962a, b), part of which is conveniently summarized and organized by Geertsma and Smit (1961). The Biot model requires knowledge of the elastic, or acoustic, properties of the matrix material, pore fluid saturant, and the elastic properties of the "frame" or "skeleton" composed of the matrix material. The elastic properties of the frame are determined empirically, with little theoretical consideration given to the pore geometry creating the nonuniform porosity other than its inclusion in such parameters as porosity and permeability. This limitation in Biot's theory limits the use of the model, in its present form, for implicitly including pore and crack geometry in the current discussion.

Brace et al (1972) discussed cracks and pores and how the presence of very flat cracks, even though they contribute little to the total porosity, can have a pronounced effect on bulk elastic properties. Further, there appears to be two general categories of rocks: cracked, generally crystalline rocks, and porous, generally sedimentary rocks. The strong effects of flat cracks, even in predominantly porous sedimentary rocks, may contribute significantly to their bulk elastic properties. Keep in mind that flat cracks have large surface area relative to their volume, and for large contrasts in elastic constants of the matrix material and fluid saturant, the surface area of the pores and cracks may be

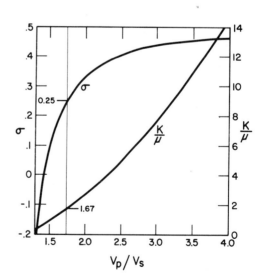

FIG. 2. Relation between the velocity ratio V_p/V_s and two other parameters, σ (Poisson's ratio), and K/μ (the ratio of incompressibility and rigidity). Note the relatively linear relation between V_p/V_s and K/μ, especially for larger values of V_p/V_s.

more important than the total crack volume (porosity). For these reasons, we will assume that crack and pore geometry may be crucial to the understanding of bulk seismic properties of sedimentary rocks.

Models that do consider pore geometry, or aspect ratio (e.g. Kuster and Toksöz, 1974a; O'Connell and Budiansky, 1974; Toksöz et al, 1976), have limitations in the maximum porosity that may be considered. This limitation is the result of the assumption of noninteraction between the individual pores or cracks. Even with this limitation, the models do provide some insight into the effects of pore shapes and will be considered here. Other investigators (e.g., Mavko and Nur, 1978, 1979) who have considered aspect ratio and actual crack shape have also considered the effect of the degree of fluid saturation on the attenuation of seismic waves. Such considerations are important in terms of fluid mobility and perhaps permeability, but they are not within the limits of the present discussion.

In the published theoretical models, the crack shape is summarized in a single parameter, the aspect ratio, which is the ratio of the minimum dimension to the maximum dimension. For flat, penny-shaped cracks, the aspect ratio is the thickness of the crack divided by its diameter, and thus less than unity. For spherical pores the aspect ratio is 1. For very flat cracks, the aspect ratio is much less than unity. Sprunt and Brace (1974), from direct observations of cracks in crystalline rocks, concluded that aspect ratios of cracks in granites range down to 10^{-3}, and they report observations of some cracks in Westerly granite with aspect ratios as small as 10^{-4}. Toksöz et al (1976), from theoretical studies, suggested that igneous rocks, where cracks are generally along intergranular boundaries, exhibit aspect ratios as small as 10^{-5} and, for sedimentary rocks, the aspect ratio may be in the range of 10^{-4} to 1. The larger values of aspect ratio are usually associated with voids between grains, and smaller values represent cracks at intergranular contacts. Together, the entire spectrum of cracks and pores, with varying aspect ratios, constitutes the total porosity of the rock.

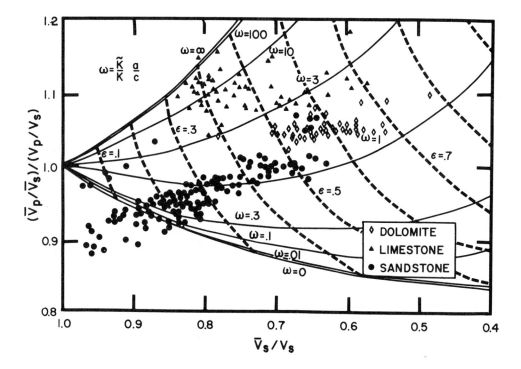

FIG. 3. Normalized values of V_p/V_s and V_s for cracked solids. Grid of ω (as defined in the figure) and ε (the crack density) are from O'Connell and Budiansky (1974). $\omega = (\tilde{K}/K)(a/c)$ where $\tilde{K}/K$ is the ratio of incompressibility ($\tilde{K}$) of the fluid saturant to the solid matrix material (K). $a/c = 1/\alpha$, where α is the aspect ratio of the cracks. ε is the crack density (number of cracks per unit volume). $\overline{V}_s$ and $\overline{V}_p/\overline{V}_s$ are S-wave velocity and velocity ratio for the cracked medium, while V_s and V_p/V_s, the normalizing quantities, are similar quantities for uncracked matrix material. Lithologic data plotted on the grid are from Pickett (1963). Assumed values of V_p/V_s and V_s, required to plot the Pickett data, are given in the text. For aid in interpretation of the data, ω for various α and $\tilde{K}/K$ are listed in Table 1. Note that for $\tilde{K}/K$ in the range of 10^{-2} to 10^{-1} (reasonable for brine-saturated rocks), velocity ratios appear to be dominated by aspect ratios in the range of $\alpha = 10^{-1}$ to 1 for sandstones, $\alpha = 10^{-2}$ to 10^{-1} for dolomite, and $\alpha = 10^{-3}$ to 10^{-2} for limestone.

Model of O'Connell and Budiansky

One model that addresses the effect of both aspects ratio and degree of saturation was introduced by O'Connell and Budiansky (1974). Their study described bulk elastic constants of cracked solids by applying potential energy considerations and a self-consistent assumption. The model fails, for dry cracks, at large crack density, and hence at high porosity. For *dry*, penny-shaped cracks with an aspect ratio of 10^{-1}, the maximum porosity that can be considered is about 18 percent. Fortunately, the model is better behaved for liquid saturation, even though the actual upper limits of crack density and porosity for which the model remains valid are the subject of some controversy in the literature (Bruner, 1976; Chatterjee and Mal, 1978). For the present purpose of examining gross effects and brine saturations, however, we will assume the model of O'Connell and Budiansky to be useful for a preliminary comparison of published data.

Some of the useful theoretical results of the work of O'Connell and Budiansky (1974) include examination of variations in crack density ε and a quantity $\omega = (\bar{K}/K)(1/\alpha)$ which is affected by both aspect ratio ($\alpha = c/a$) and relative incompressibility ($\bar{K}/K$) of the fluid ($\bar{K}$) to the matrix material (K). The background grid shown in Figure 3 is adapted from their paper. It represents a plot of ω and ε with respect to the velocity ratio V_p/V_s and the shear-wave velocity of V_s. The quantities ω and ε are plotted as families of curves on a grid of V_p/V_s and V_s. For plotting purposes, values of $\bar{V}_p/\bar{V}_s$ and $\bar{V}_s$, calculated for the cracked medium, are normalized against the values of V_p/V_s and V_s for the uncracked solid, which yields dimensionless quantities on the graph. Note that if V_p/V_s and V_s are known, an experimental determination of $\bar{V}_p/\bar{V}_s$ and $\bar{V}_s$ will yield values for ω and ε. If $\bar{K}/K$ is known, or can be reasonably assumed, then the model yields a value of aspect ratio α. To aid in the interpretation of the graph, values of ω, varying with $\bar{K}/K$ and α, are given in Table 1.

For a comparison of laboratory data with these theoretical results, the data of Pickett (1963), indicated by circles, diamonds, and triangles, have been plotted directly on the O'Connell and Budiansky graph. For normalization of the Pickett data, V_p of uncracked matrix material was assumed to be 19,500 ft/sec for sandstone, 22,000 ft/sec for limestone, and 24,500 ft/sec for dolomite. Further, the uncracked matrix material, for all lithologies, was assumed to be a Poisson's solid (i.e., $V_p/V_s = \sqrt{3}$).

As can be seen in Figure 3, for constant values of $\bar{K}/K$ we have different ranges of α for different lithologies. Recall that it is the Pickett data which support the original suggestion of an empirical association between V_p/V_s and whole-rock lithology. The applicability of the Pickett data, probably stretched beyond its limits, as well as the assumptions of both matrix velocity and the matrix being a Poisson solid, limit the reliability of the following conclusions. Further, applying a single aspect ratio for the curves on the O'Connell-Budiansky plot (Figure 3) to a real rock, as well as the low crack-density assumption, places even greater limits on the comparison. However, if we accept this model for a preliminary correlation of real data and take a value of $\bar{K}/K$ in the range of 10^{-2} to 10^{-1} (a reasonable range for most brine-saturated rocks), we tentatively conclude that the dominant aspect ratio for sandstone appears in the range $\alpha = 10^{-1}$ to 1; for dolomite $\alpha = 10^{-2}$ to 10^{-1} and; for limestone $\alpha = 10^{-3}$ to 10^{-2}. Thus, this rather crude comparison gives us encouragement to delve deeper into the relationship between V_p/V_s, aspect ratio, and lithology.

Model of Toksöz et al

Toksöz and his students (Kuster and Toksöz, 1974a, b; Tosköz et al, 1976; Cheng and Toksöz, 1976) have addressed the problem of aspect ratio and its effect on elastic wave velocities. They consider an aspect ratio spectrum which consists of defining a percentage volume concentration of pores for each aspect ratio considered, as well as the saturation state. Since we are presently concerned with only brine saturation, we will consider only the aspect ratio as a significant quantity.

Using the Toksöz et al (1976) model, calculations were completed with the goal of comparing aspect ratios with other parameters. Specifically, they give the expressions

$$\frac{K^* - K}{3K^* + 4\mu} = \frac{1}{3} \frac{K' - K}{3K + 4\mu} \sum_{m=1}^{M} c(\alpha_m) \cdot T_{iijj}(\alpha_m), \quad (1)$$

and

$$\frac{\mu^* - \mu}{6\mu^*(K + 2\mu) + \mu(9K + 8\mu)} = \frac{\mu' - \mu}{25\mu(3K + 4\mu)} \sum_{m=1}^{M} c(\alpha_m) \cdot \left[T_{ijij}(\alpha_m) - \frac{1}{3} T_{iijj}(\alpha_m)\right]. \quad (2)$$

Equations (1) and (2) are simply equations (5) and (6) of Toksöz et al (1976) and are based on the theoretical work of Kuster and Toksöz (1974a). In the above equations, K^* is the bulk modulus (incompressibility) of the composite medium, K is the bulk modulus of the matrix, K' is the bulk modulus of the fluid saturant, μ^* is the rigidity (shear modulus) of the composite medium, μ is the rigidity of the matrix and μ', the rigidity of the inclusions, is zero for a fluid saturant. The quantity $c(\alpha_m)$ is the volume concentration of pores and cracks with an aspect ratio α_m, where M is the total number of values considered in the aspect ratio spectrum. Thus, the porosity is given by

$$\phi = \sum_{m=1}^{M} c(\alpha_m) \quad (3)$$

which is equation (7) of Toksöz et al (1976). Further the quantities T_{iijj} and T_{ijij} are scalar functions of K, μ, K', μ', and α. These quantities are determined by rather involved algebraic combinations of the above five (in our case four) parameters, and are given in Appendix A of Toksöz et al (1976).

For the purposes of the present discussion, we want to consider the effects of individual values of aspect ratio α, so we will include each individual value inside the summations in equations (1) and (2). Thus, for each value of α and c inside the summation, we can calculate a value of K^* and μ^*. Knowing ρ and ρ', the densities of the matrix material and fluid, and applying equation (3) for the porosity, we can also determine ρ^*, the bulk density of the composite material. By this means, we can calculate values of the seismic velocities V_p and V_s for each α and c that contributes to the aspect ratio spectrum. Note that to obtain results for the entire spectrum, the sums to obtain K^* and μ^* must be

Table 1. Values of ω for various $\bar{K}/K$ and α, for use in interpretation of Figure 3. $\omega = (\bar{K}/K)(1/\alpha)$.

	10^{-1}	10^{-2}	10^{-3}
1	.1	.01	.001
10^{-1}	1	.1	.01
10^{-2}	10	1	.1
10^{-3}	100	10	1

performed over M values of α_m and c_m, the complete aspect ratio spectrum.

Basic elastic constants, expressed in megabars, used in the calculations were $K = 0.5$, $\mu = 0.3$, and density $\rho = 2.7$ g/cm^3 for the matrix material and $K' = 0.024$, $\mu' = 0.0$, and $\rho' = 1.03$ g/cm^3 for the pore fluid saturant. The matrix material is thus a Poisson solid, and the pore fluid is brine. Parameters calculated and displayed as a function of aspect ratio and volume concentration of pores are seismic velocities V_p and V_s, as well as the velocity ratio V_p/V_s. These quantities are plotted (in Figures 4 through 9) as a series of curves, representing contours on a three-dimensional (3-D) surface, with the vertical axis of the plot being aspect ratio and the horizontal axis the volume concentration c. Volume concentration is simply the porosity associated with a particular aspect ratio crack or pore. In the $\alpha - c$ space, the pore spectrum is described by a curve tracing $\alpha - c$ values, and the total porosity is the sum of volume concentrations c along this curve. This particular presentation was chosen to elucidate variations with respect to α.

Figure 4 shows V_p, the compressional wave velocity, as a surface in this $\alpha - c$ space. A noninteraction assumption in the model of Toksöz et al (1976) requires that $c(\alpha) < \alpha$, and thus only a portion of the $\alpha - c$ can be considered. Of course, the cracks and pores, in a rock of finite permeability, must be interconnected to some degree and thus do interact. The present models, however, assume noninteraction between cracks. This assumption should be kept in mind throughout the remaining discussion.

The surface shows how V_p, the seismic compressional wave velocity, varies with α and c. At a given α, V_p decreases with increasing porosity. Note that, except perhaps for small α, V_p is far more sensitive to porosity than to α.

Figure 5 is a similar plot for V_s, the seismic shear-wave velocity. Here we observe that V_s is somewhat more sensitive to α than V_p, but V_s is still quite dependent upon c, or porosity.

Figure 6 describes the velocity ratio V_p/V_s in the same context as V_p and V_s in Figures 4 and 5. Here, we observe the pronounced effects of both concentration of pores and the aspect ratio. We also see the constraints resulting from limiting $c(\alpha) < \alpha$, in that large concentrations of flat cracks or pores at a particular aspect ratio cannot be considered. Future models should address the problem of interactions between cracks. Before more sophisticated models are considered, however, direct observations of cracks and pores in real rocks may be in order.

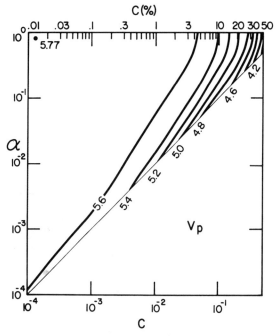

FIG. 4. Plot of calculated compressional-wave velocity V_p (km/sec) as a function of α (the aspect ratio) and c (the volume concentration of cracks with that aspect ratio). Velocities were computed using the model of Toksöz et al (1976). Assumptions requiring $c(\alpha) < \alpha$ limit the total range of concentrations to be considered. Total porosity is the sum of concentrations associated with a suite of aspect ratios, or pore geometry spectrum. V_p is shown as a surface with constant values of V_p. Note that V_p, especially at larger values of α, is more sensitive to pore concentration, or porosity, than to pore shape, or aspect ratio. Elastic constants in megabars and densities in g/cm^3 for these calculations are $K = 0.5$, $\mu = 0.3$, and $\rho = 2.7$ for the solid matrix material and $K' = 0.24$, $\mu' = 0.0$, and $\rho' = 1.03$ for the pore fluid saturant (brine). The matrix material is a Poisson solid.

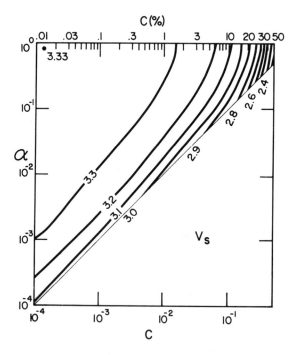

FIG. 5. Plot of shear-wave velocity V_s (km/sec) as a function of α and c. Model and elastic constants are the same as those considered in Figure 4. Note that, while V_s is more sensitive to pore concentration than to pore shape, the sensitivity of V_s to pore shape is stronger than was the case for V_p.

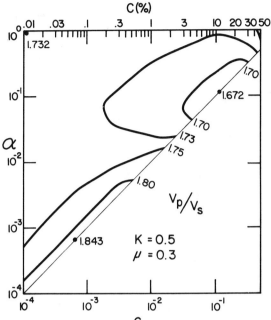

FIG. 6. Plot of velocity ratio V_p/V_s for model and elastic constants used in Figures 4 and 5. Note pronounced sensitivity of V_p/V_s to both pore shape α and pore concentration c. The lowest values of V_p/V_s, less than 1.70, are in the pore concentration, or porosity, range of 3–25 percent and aspect ratios range of 10^{-1} to 1. This is entirely consistent with laboratory observations for sandstones shown in Figure 10. At lower concentrations, higher V_p/V_s values generally result from flatter cracks. Flat cracks have a strong effect on V_p/V_s, even at very small concentrations.

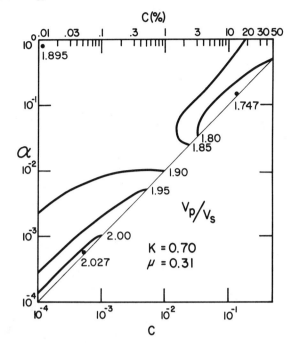

FIG. 7. Plot of V_p/V_s similar to Figure 6, the only difference being the elastic constants of the matrix used in the calculation. Constants of $K = 0.70$ and $\mu = 0.31$ megabars were chosen such that V_p/V_s for the non-Poisson solid matrix will be near 1.9, the whole-rock value observed by Pickett (1963) for limestone. All other parameters remain unchanged.

In Figure 6, we can make some interesting and relevant observations. For very low porosity, the matrix dominates and a V_p/V_s of 1.732 confirms that the matrix is a Poisson solid. For aspect ratios between about 0.3 and 1.0 (nearly equant voids), $V_p/V_s = 1.732$ at a porosity of 0.01 percent (upper left corner of plot), decreases to a value of less than 1.73 at 3–25 percent, consistent with known porosities and published V_p/V_s values for sandstones.

Further, the range in aspect ratios (10^{-1} to 1) for these low V_p/V_s values is consistent with the inferences drawn earlier from comparisons of the Pickett data and the O'Connell and Budiansky model. Higher V_p/V_s values are observed for flatter cracks in the lower porosity ranges. This suggests that, even at low concentrations, flat cracks can have strong effects on elastic constants. Also, the increase in V_p/V_s for aspect ratios less than 10^{-2} follows the earlier observations of the O'Connell and Budiansky model.

Figures 7, 8, and 9 show V_p/V_s plots similar to Figure 6 but for different elastic constants of the matrix material. Figures 7 and 8 show results with K and μ increased over those used in Figure 6. The values were chosen to yield V_p/V_s near 1.8 and 1.9 for a solid matrix. These calculations are attempts to duplicate solid-matrix values for observed whole-rock lithologies by choice of elastic constants rather than aspect ratios. Even in these rather extreme cases, the general shape of the V_p/V_s surface follows that for the Poisson solid in Figure 6. Further, locations of maximum and minimum V_p/V_s values in $\alpha - c$ space are not greatly affected, and the V_p/V_s values themselves for a particular $\alpha - c$ are between the matrix values and the values shown in Figure 6.

Figure 9 is a plot with K and μ values 20 percent less than those used for Figure 6, but where the solid matrix remains a Poisson solid. The differences between Figures 6 and 9 are slight, and the departures at higher porosities are a result of differences in the relative elastic constants between the pore fluid and the matrix.

From the above considerations, it is concluded that aspect ratio is a strong factor affecting observed V_p/V_s-lithology correlation.

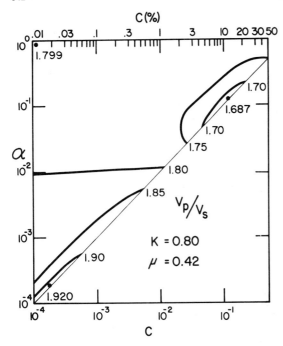

Fig. 8. Plot of V_p/V_s similar to Figures 6 and 7, the only difference being the elastic constants of the matrix used in the calculation. Constants of $K = 0.80$ and $\mu = 0.42$ megabars were chosen such that V_p/V_s for the non-Poisson solid matrix will be near 1.8, the whole-rock value observed by Pickett (1963) for dolomite. All other parameters remain unchanged. Note the similarity in the overall shape of the V_p/V_s surfaces shown in Figures 6, 7, and 8, especially the region of low V_p/V_s at crack concentrations in the range of 3–25 percent.

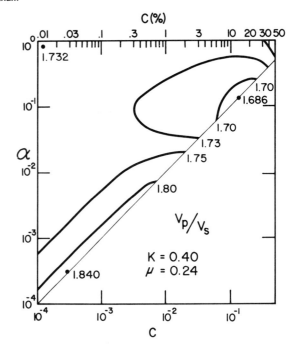

Fig. 9. Plot of V_p/V_s similar to Figure 6. The matrix material is assumed to be a Poisson solid, but the elastic constants of the matrix material are reduced 20 percent below those of Figure 6. Pore fluid (brine) is the same as that considered in all other calculations. Note the extreme similarity between Figures 6 and 9 with the only significant, but relatively minor, variations being at the higher crack concentrations. At these higher concentrations, the differences in $\bar{K}/K$ between Figures 6 and 9 are greatest.

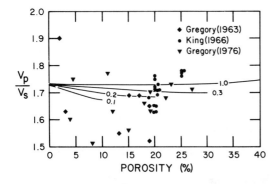

Fig. 10. V_p/V_s porosity plot for published sandstone velocities. Lines of constant aspect ratio α, from Figure 6, show V_p/V_s at a crack concentration for that particular α. All samples are either water- or brine-saturated, with a differential pressure of 5000 psi or greater. Sources of data are indicated by the shape of plotting symbols, and more detailed information is included in Table 2. Note the general trend of increasing V_p/V_s with increasing porosity, especially above a porosity of 15 percent. Significantly, all V_p/V_s values less than 1.7 are in a porosity range of 3–25 percent, entirely consistent with the model results shown in Figure 6.

DISCUSSION

Perhaps the most interesting observation in examining V_p/V_s variations is the zone of lower V_p/V_s seen in Figures 6 through 9. The aspect ratios and pore concentrations of this zone are suggestive of sandstones. Experimental confirmation of the existence of this low V_p/V_s zone would add strength to the suggestion of a V_p/V_s-α association for real rocks.

Table 2 lists V_p/V_s values, for a variety of porosities, for several sandstones. All data are from published laboratory studies of P- and S-wave velocities. The data represent either water- or brine-saturated sandstones, with experiments performed at a differential pressure (confining pressure – pore pressure) of 5000 psi or greater. The trends in the data, plotted in Figure 10, are in accordance with the trends suggested by the studies of Toksöz et al (1976). Also shown in Figure 10 are traces of calculated V_p/V_s values associated with constant aspect ratio α. For these traces, the porosity was taken as the pore concentration associated with that particular α. Recall that total porosity is the sum of pore concentrations for each α in the pore spectrum.

Significantly, Figure 10 shows that all observed V_p/V_s values less than 1.7 are in a porosity range of 3–25 percent, entirely consistent with the theoretical results of Toksöz et al (1976) plotted in Figures 6 through 9. Further, in a porosity range of

15–30 percent, there is a generally increasing trend in V_p/V_s with increasing porosity.

Tentative conclusions and speculations

Comparisons of theoretical models with published laboratory studies suggest that crack, or pore, geometry can have a stronger effect on observed V_p/V_s values than reasonable variations in elastic constants of the minerals comprising the matrix. Thus, I speculate that the observed association between lithology and V_p/V_s is likely to be a result of an association between lithology and distribution of pore and crack shapes. Direct observation of crack shapes for a variety of lithologies is a next step in confirming this speculation.

Comparisons of published data (Pickett, 1963) with the theoretical model of O'Connell and Budiansky suggest that the ratio of P- to S-wave velocities (V_p/V_s) for sandstones is controlled by cracks and pores with dominant aspect ratios in the range 10^{-1} to 1. Further, dolomite velocity ratios are controlled by dominant aspect ratios in the range 10^{-2} to 10^{-1}, and the velocity ratios of dense limestones considered by Pickett (1963), of generally low porosity, are controlled by aspect ratios in the range 10^{-3} to 10^{-2}.

Interpretive insights to be drawn

If pore shape controls the observed V_p/V_s lithology correlation, we have a basis for intuitively extrapolating observations to interpretation of rocks without establishing specific V_p/V_s-lithology correlation for every possible rock type. For example, siltstones, with grains similar to but smaller than sandstones, might be expected to have V_p/V_s values more similar to sandstones (at the same porosity) than to shales. Further, V_p/V_s values of vuggy or arenaceous limestones, at large porosity, might be more similar to sandstones than to dense limestones.

Future directions

As mentioned earlier, the next logical stage of development for the understanding of V_p/V_s lithology correlations is probably in the area of direct observation of crack and pore distributions. Further, V_p and V_s measurements on the same samples would establish correlations for all aspects of the problem discussed.

Direct observation of cracks and pores has been studied, to some degree, by Weinbrandt and Fatt (1969), Timur et al (1971), Brace et al (1972), and Sprunt and Bruce (1974). Some of their results, discussed earlier, are not extensive enough to address the present problem of examining a possible correlation between pore shape and lithology in sedimentary rocks. The experimental techniques discussed, however, do define some anticipated problems. They consider careful sample preparation by ion thinning, and direct observation of cracks with either an optical or scanning electron microscope. Small cracks in some denser rocks may require the use of a scanning electron microscope for direct observations. The preparation of undamaged samples, however, appears to be the most crucial aspect of successful observation of small cracks in rocks.

The goal of this paper has been to establish the need for direct observation of crack distributions for different lithologies of sedimentary rocks. Further, obtaining V_p and V_s observations on the same samples is highly desirable. Such an experimental study, directed toward soundly establishing fundamental correlations of rock porperties useful in hydrocarbon exploration, should be given high priority.

ACKNOWLEDGMENTS

I wish to express my gratitude to Jack Caldwell, Jay Dreves, and Ralph Knapp for critically reading the manuscript and to Texaco Inc. for permission to publish this paper.

Table 2. V_p/V_s values for a variety of sandstones with different porosities. All data are from published laboratory studies, and all measurements were made with differential pressures of 5000 psi or greater. References: 1—Gregory (1963); 2—King (1966); 3—Gregory (1976).

Sandstone	ϕ (%)	V_p/V_s	Pressure (psi)	Reference
Sundance	2	1.90	7,000	1
Oriskany	3	1.63	7,000	1
Travis Peak	4	1.60	5,000	3
Travis Peak	5	1.75	5,000	3
Travis Peak	8	1.51	5,000	3
Chugwater	11	1.77	5,000	3
Green River	12	1.63	5,000	3
Chugwater	13	1.55	6,000	1
Tensleep	15	1.69	6,000	1
Tensleep	15	1.53	5,000	3
Torpedo	17	1.69	6,000	1
Gulf Coast	18	1.66	5,000	3
Bandera	18	1.70	5,000	3
Berea (water)	19	1.65	6,000	1
Berea (brine)	19	1.52	6,000	1
St. Peter	19	1.68	5,000	2
Berea	19	1.63	5,000	3
Torpedo	20	1.65	5,000	2
Torpedo	20	1.65	6,000	2
Torpedo	20	1.63	8,000	2
Torpedo	20	1.63	10,000	2
Bandera	20	1.71	5,000	2
Bandera	20	1.72	6,000	2
Bandera	20	1.75	8,000	2
Bandera	20	1.76	10,000	2
Berea	21	1.73	5,000	2
Berea	21	1.71	6,000	2
Berea	21	1.71	8,000	2
Berea	21	1.69	10,000	2
Gulf Coast	22	1.68	5,000	3
Nichols Bluff	23	1.73	5,000	3
Boise	25	1.76	5,000	2
Boise	25	1.77	6,000	2
Boise	25	1.78	8,000	2
Boise	25	1.78	10,000	2
Boise	27	1.71	5,000	3

REFERENCES

Benzing, W. M., 1978, V_s/V_p relationships in carbonates and sandstones—Laboratory data: Presented at the 48th Annual International SEG Meeting, November 1, in San Francisco.

Biot, M. A., 1956a, Theory of propagation of elastic waves in a fluid-saturated porous solid. I. Low-frequency range: J. Acoust. Soc. Am., v. 28, p. 168–178.

―――― 1956b, Theory of propagation of elastic waves in a fluid-saturated porous solid. II. Higher frequency range: J. Acoust. Soc. Am., v. 28, p. 179–191.

―――― 1962a, Generalized theory of acoustic propagation in porous dissipative media: J. Acoust. Soc. Am., v. 34, p. 1254–1264.

―――― 1962b, Mechanics of deformation and acoustic propagation in porous media: J. Applied Phys., v. 33, p. 1482–1498.

Brace, W. F., Silver, E., Hadley, K., and Goetze, C., 1972, Cracks and pores: A closer look: Science, v. 178, p. 162–164.

Bruner, W. M., 1976, Comments on "Seismic velocities in dry and saturated cracked solids" (R. J. O'Connell and B. Budiansky): J. Geophys. Res., v. 81, p. 2573–2576.

Chatterjee, A. K., and Mal, A. K., 1978, Elastic moduli of two-component systems: J. Geophys. Res., v. 83, p. 1785–1792.

Cheng, C. H., and Toksöz, M. N., 1976, Seismic velocities in porous rocks, direct and inverse problem: Presented at the 46th Annual International SEG Meeting, October 27, in Houston.

Cherry, J. T., and Waters, K. H., 1968, Shear wave recording using continuous signal methods, Part I: Geophysics, v. 33, p. 229–239.

Domenico, S. N., 1976, Effect of brine-gas mixture on velocity in an unconsolidated sand reservoir: Geophysics, v. 41, p. 882–894.

Erickson, E. L., Miller, D. E., and Waters, K. H., 1968, Shear wave recording using continuous signal methods, Part II—Later experimentation: Geophysics, v. 33, p. 240–254.

Gardner, G. H. F., and Harris, M. H., 1968, Velocity and attenuation of elastic waves in sand: Trans 9th Annual Log. Symp., p. M1–M19.

Geertsma, J., and Smit, D. C., 1961, Some aspects of elastic wave propagation in fluid-saturated porous solids: Geophysics, v. 26, p. 169–181.

Gregory, A. R., 1963, Shear wave velocity measurements of sedimentary rock samples under compression, in Rock mechanics: p. 439–471; Proc. 5th symp. on rock mechanics, C. Fairhurst, Ed.: New York, Macmillian and Co.

——— 1976, Fluid saturation effects on dynamic elastic properties of sedimentary rocks: Geophysics, v. 41, p. 895–921.

King, M. S., 1966, Wave velocities in rocks as a function of changes in overburden pressure and pore fluid saturants: Geophysics, v. 31, p. 50–73.

Kithas, B. A., 1976, Lithology, gas detection, and rock properties from acoustic logging systems: SPWLA 17th Ann. Logging Symp., p. R1–R10.

Kuster, G. T., and Toksöz, M. N., 1974a, Velocity and attenuation of seismic waves in two-phase media: Part I. Theoretical formulation: Geophysics, v. 39, p. 587–606.

——— 1974b, Velocity and attenuation of seismic waves in two-phase media: Part II—Experimental results: Geophysics, v. 39, p. 607–618.

Love, A. E. H., 1927, A treatise on the mathematical theory of elasticity: Cambridge, Cambridge Univ. Press, p. 104.

Mavko, G. M., and Nur, A., 1978, The effect of nonelliptical cracks on the compressibility of rocks: J. Geophy. Res., v. 83, p. 4459–4468.

——— 1979, Wave attenuation in partially saturated rocks: Geophysics, v. 44, p. 161–178.

Nations, J. F., 1974, Lithology and porosity from acoustic shear and compressional wave transit-time relationships: SPWLA 15th Ann. Logging Symp., p. Q1–Q16.

O'Connell, R. J., and Budiansky, B., 1974, Seismic velocities in dry and saturated cracked solids: J. Geophys. Res., v. 79, p. 5412–5426.

Pickett, G. R., 1963, Acoustic character logs and their application in formation evaluation: J. Petr. Tech., p. 659–667.

Polskov, J. K., Brodov, L. Ju., Mironova, L. V., Michon, D., Garotta, R., Layotte, P. C., and Coppens, F., 1980, Utilisation conbinne des ondes longitudinales et transversales en sismique reflexion: Geophys. Prosp., v. 28, p. 185–207.

Sprunt, E. S., and Brace, W. F., 1974, Direct observation of microcracks in crystalline rocks: Int. J. Rock Mech. Min. Sci. and Geomech. (abstr.), v. 11, p. 139–150.

Tatham, R. H., and Stoffa, P. L., 1976, V_p/V_s—A potential hydrocarbon indicator: Geophysics, v. 41, p. 837–849.

Tatham, R. H., Danbom, S. H., Tyce, R. C., and Omnes, G., 1977, Seismic parameters and their estimation: Progress in estimation and interpretation: SEG Continuing Education School, "The stationary convolutional model of the reflection seismogram and progress in measurement of subsurface seismic parameters: velocity, density, specific attenuation" November 28–29, in Houston.

Timur, A., Hempkins, W. B., and Weinbrandt, R. M., 1971, Scanning electron microscope study of pore systems in rocks: J. Geophy. Res., v. 76, p. 4932–4948.

Toksöz, M. N., Cheng, C. H., and Timur, A.; 1976, Velocities of seismic waves in porous rocks: Geophysics, v. 41, p. 621–645.

Weinbrandt, R. M., and Fatt, I., 1969, A scanning electron microscope study of the pore structure of sandstone: J. Petr. Tech., May, p. 543–548.

V_p/V_s—A POTENTIAL HYDROCARBON INDICATOR

ROBERT H. TATHAM* AND PAUL L. STOFFA‡

Theoretically and experimentally, the shear-wave velocity of a porous rock has been shown to be less sensitive to fluid saturants than the compressional wave velocity. Thus, observation of the ratio of the seismic velocities for waves which traverse a changing or laterally varying zone of undersaturation or gas saturation could produce an observable anomaly which is independent of the regional variation in compressional wave velocity.

One source of shear-wave data in reflection seismic prospecting is mode conversion of P waves to shear waves in marine areas of high water bottom P-wave velocity. A relatively simple interpretative technique, based on amplitude variation as a function of the angle of incidence, is a possible discriminant between shear and multiple compressional arrivals, and data for a real case are shown. A normal moveout velocity analysis, carefully coupled with this offset discriminant, leads to the construction of a shear-wave reflection section which can then be correlated with the usual compressional wave section.

Once such a section has been constructed, the variation in the ratio of the seismic velocities can be mapped, and potentially anomalous subsurface regions observed.

INTRODUCTION

Recent developments in petroleum exploration have led to the use of "bright spots" (Craft, 1973) as a direct hydrocarbon indicator. Bright spots, or reflection amplitude anomalies, are a consequence of large seismic P-wave velocity and density contrasts at the *boundaries* of subsurface gas reservoirs caused by differences between the gas-saturated reservoir rock and the surrounding material.

Recent research in earthquake seismology, on the other hand, has led to the prediction of an earthquake in the Blue Mountain Lake region of New York State (Aggarwal et al, 1973b, 1975). The geophysical observations which led to this prediction, and others reported from the Garm region of Central Asia (Nersesov et al, 1969), tentatively have been explained by a model based on dilatancy in the zone of the impending earthquake. Basically, the dilatancy model suggests that cracks form, or existing cracks open further, prior to the ultimate mechanical failure of rocks under stress. The formation of this new porosity, or increase in porosity, leads to undersaturation of the cracks within the rock. The result is an observed (Aggarwal et al, 1975) reduction in the seismic P- and S-wave velocity in the vicinity of the impending earthquake. O'Connell and Budiansky (1974) theoretically have calculated similar variation in both P- and S-wave velocity for dry and saturated cracked media. The important conclusion is that the undersaturation causes P-wave velocity to decrease more than the S-wave velocity, resulting in an observable decrease in the velocity ratio V_p/V_s.

In the exploration problem, the incorporation of shear-wave data may make it possible to detect zones of gas saturation where reflection amplitudes are not reliable for this purpose, or to distinguish between gas saturation and anomalous P-wave velocity contrasts where velocities vary laterally for other reasons. For the first case, we require the presence of P- and S-wave reflections from reflectors which are at the same levels above

---

Lamont-Doherty Geological Observatory Contribution no. 2368. Paper presented at the 44th Annual International SEG Meeting, November 12, 1974, Dallas, Tex. Manuscript received by the Editor December 11, 1974; revised manuscript received March 8, 1976.
* Formerly Lamont-Doherty Geological Observatory, Palisades, N.Y., and Columbia University, New York, N.Y.; currently, Texaco Inc., Bellaire, Tex. 77401.
‡ Lamont-Doherty Geological Observatory, Palisades, N.Y. 10964.
© 1976 Society of Exploration Geophysicists. All rights reserved.

and below the zone in question, and that the reflectors remain the same distance apart for P and S waves. If these reflections are present, then the ratio V_p/V_s through the zone can be determined and may serve as a direct hydrocarbon indicator.

In this paper, we consider the use of the S-wave velocity as a normalizing quantity for the P-wave velocity, a suggestion based on published theoretical and experimental velocity data. We are attempting to derive an S-wave interval velocity function from conventional common-depth-point data by determining a moveout velocity function appropriate for S waves. Finally, we suggest the potential advantages of incorporating this type of analysis (when possible) into the usual bright-spot interpretation.

PROPOSED USE OF SHEAR WAVES

Decreases in P-wave velocity of approximately 10 percent to 20 percent have been observed (Whitcomb et al, 1973; Aggarwal et al, 1975) for the subsurface region surrounding impending earthquakes. These observations have been tentatively explained by undersaturation of pores or cracks (Nur, 1972; Aggarwal et al, 1973a; Whitcomb et al, 1973; Scholz et al, 1973). Similarly, Gardner et al (1974), using Gassman's theory, computed differences in P-wave velocity of approximately 20 percent between brine and gas saturated portions of a given sediment at depths less than 5000 ft. Thus, even in physical different media, similar P-wave velocity anomalies are expected (i.e., in preearthquake zones and subsurface gas reservoirs). Regional and lithologic variations in P-wave velocity, however, may be greater even than these anomalies. Hence, observation of P-wave velocity alone may not be sufficient to identify zones of undersaturation or gas saturation. Since S-wave velocity is less sensitive to the fluid saturant than the P-wave velocity, it can be used as a normalizing quantity with which to compare P-wave velocity. It is the observation of the velocity ratio V_p/V_s that has been applied successfully in earthquake prediction, and similar observations also might be applied in petroleum exploration. Erickson et al (1968) proposed that variations in V_p/V_s could be employed in identifying lateral variations in lithology. Further, they demonstrated the procedure with a specific quantitative example on actual reflection seismic data.

Kuster and Toksöz (1974) recently calculated P- and S-wave velocities for two-phase media

Table 1. Laboratory determinations (after King, 1966) for several sandstones and pore fluid saturants at a confining pressure of 5000 psi.

Rock (Sandstone)	Porosity %	Perm. (Md)	V_p (ft/sec)			V_s (ft/sec)			V_p/V_s		
			2N NaCl	Kerosene	Dry	2N NaCl	Kerosene	Dry	2N NaCl	2N Kerosene	Dry
Boise	25.0	1400	11,500	11,300	10,900	6,500	6,700	6,900	1.77	1.69	1.58
Bandera	20.0	3.5	11,600	12,000	11,350	6,800	7,500	7,500	1.71	1.60	1.51
Berea	20.5	250	13,500	13,400	13,200	7,800	8,300	8,350	1.73	1.61	1.58
Torpedo	19.8	6	12,700	12,800	12,600	7,700	8,250	8,350	1.65	1.55	1.51
St. Peter	18.6	130	15,200*	15,100	14,500	9,050*	9,400	9,200	1.68*	1.61	1.57

*Pore fluid—distilled water.

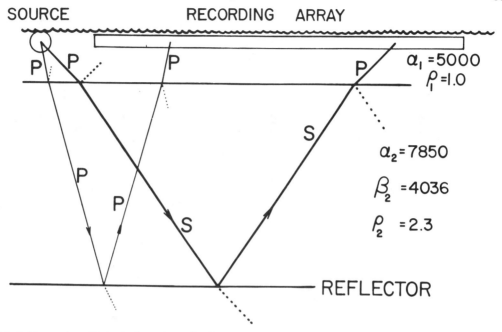

Fig. 1. Geometry of a typical marine seismic recording array, showing a *PSSP* ray at a large angle of incidence (from the vertical) being detected at a large offset trace.

such as porous and cracked rocks. For a model of a typical cyrstalline rock, with low porosity resulting from cracks rather than pores, V_p/V_s is 1.68 for water saturation and 1.50 for air saturation. Further, the theoretical work of O'Connell and Budiansky (1974), based upon different physical assumptions from those of Kuster and Toksöz, is also consistent with undersaturation leading to V_p/V_s anomalies. Domenico (1974) and Gardner et al (1974) have used Gassman's (1951) formula and available data to show that large differences in V_p should also be seen between water- and gas-saturated regions of a typical sedimentary rock. Combining this with the assumption that the shear modulus of a porous rock should not depend significantly upon fluid content, one concludes that V_p/V_s should change with pore fluid content. Kuster and Toksöz have proposed models which lead to similar conclusions.

Additional evidence of the diagnostic value of the ratio V_p/V_s is provided by actual laboratory measurements of seismic velocities. King (1966) determined both *P*- and *S*-wave velocities for various saturants in several different sandstone samples. His data, with corresponding V_p/V_s ratios, are given in Table 1. The velocities are those determined by King at a differential pressure of 5000 psi, corresponding to a depth of about 10,000 ft.

These results not only experimentally confirm diagnostic variations in V_p/V_s with gas saturation, but also suggest that V_p/V_s may be sensitive to the presence of liquid hydrocarbons. Moreover, for some samples his data show that V_s increases slightly with undersaturation, or with the addition of liquid hydrocarbons, rather than remaining completely unaffected by the pore fluid. King (1966, p. 63) points out that this result is qualitatively consistent with Biot's (1962a, b) theory. The slight increase in V_s, due in part to a decrease in density, serves to further enhance the diagnostic variations in V_p/V_s. Thus, we believe that the incorporation of shear-wave observations into seismic reflection interpretations may provide a useful tool for the direct detection of hydrocarbons.

ARTIFICIAL GENERATION OF SHEAR WAVES

Artificial generation of shear waves received considerable attention during the mid-1960s with experiments including both explosive (e.g., White and Sengbush, 1963; Geyer and Martner, 1969) and nonexplosive (e.g., Cherry and Waters, 1968) sources. One encouraging development was the construction of horizontal vibrators (Cherry and Waters, 1968), where observations of both reflections and refractions (Miller and Dunster, 1974)

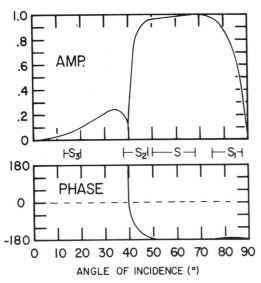

FIG. 2. Amplitude, as a fraction of the original P-wave amplitude, and phase of doubly mode-converted wave (PSSP) at all angles of incidence (after Choy, 1976). Note large amplitude of P wave, after two mode-conversions, at angles of incidence (from the vertical) beyond the critical angle (39°). The water layer has a P-wave velocity of 5000 ft/sec and density of 1.0. Sediments have P-wave velocity 7850 ft/sec, S-wave velocity 4036 ft/sec, and density 2.3 gm/cc. S_1, S, S_2, and S_3 show range of angles of incidence for rays to the long 12 traces (5600–8700 ft) of the recording array for shear-wave reflections indicated in Figure 7.

have been reported. Explosion generated shear waves observed by Geyer and Martner (1969) involved normal exploration-type explosions near-horizontal velocity discontinuities, and the results of White and Sengbush experimentally confirmed some of the theoretical considerations of cylindrically shaped charges (Heelan, 1955). Hazebroek (1966) considered theoretically the case of a finite line source and concluded that a charge shape deviating from spherical symmetry will generate shear waves. Significantly, his results suggest that shear waves will be more efficiently generated in media where Poisson's ratio is greater than 0.25.

MODE-CONVERTED SHEAR WAVES

An indirect source of shear-wave data, especially for marine areas, results from the conversion of P waves to S waves at a sharp refracting boundary. Such a boundary exists at the water bottom in regions of relatively high-velocity sediments or rock. This potential source of shear-wave data is especially attractive because much of this type of information may be present in existing marine common-depth-point (CDP) reflection data.

Mode conversion occurs only when the angle of incidence is nonvertical. Since mode conversion becomes more efficient as the angle of incidence increases from the vertical, we would expect to observe mode-converted arrivals primarily on the longer-range traces of multifold CDP data. We shall assume that such arrivals undergo conversion from P to S waves (SV) at only one interface (possibly the water bottom), are reflected from depth as S waves, and then undergo mode conversion from S to P upon reentering the upper (water) layer (Figure 1). Nafe and Drake (1957) have reported observations of S waves that originated by mode conversion within the sedimentary section. The principal observational difference between the usual P-wave reflection and the mode-converted S-wave reflection will be the reflection velocity across the recording array.

Figure 2 shows the anticipated amplitude and phase (Choy, 1976; Ergin, 1952) for all angles of incidence, for plane P waves converted to S waves at the water bottom, and another mode con-

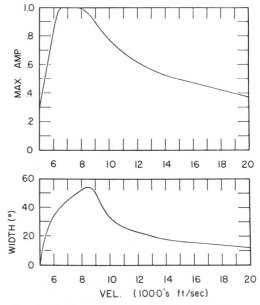

FIG. 3. Maximum amplitude of twice mode-converted waves and width of window (angle between first and second critical angle) for sediment velocities from 5000 to 20,000 ft/sec (after Choy, 1976). Note that the most efficient mode conversion occurs for a water-bottom velocity of about 6500–9000 ft/sec.

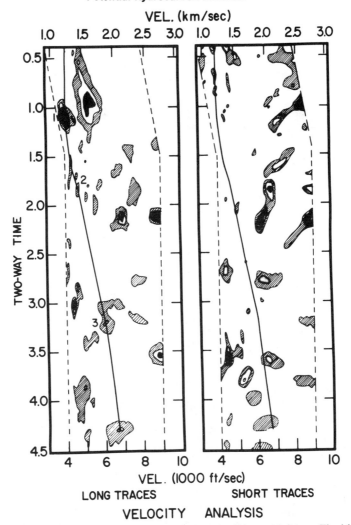

FIG. 4. Velocity analyses of seismic reflection data gathered offshore Alabama-Florida border area in the Gulf of Mexico. Analyses utilizing both long and short 12 traces of 24-fold CDP data are shown. Shear reflections are expected to be strongest on the set of long traces, while low-velocity multiples are expected to be of about equal intensity on both sets. Solid line indicates the interpreted S-wave normal moveout velocity function. Dashed lines define limits of velocity analysis. (Contours represent equal coherency; darker shading shows relatively higher coherency.) Numbers in the long-trace analysis identify S-wave reflections that are well correlated with P-wave reflections.

version back to a P wave upon reentering the water layer. The water-bottom parameters used in this calculation were determined by velocity analysis of the data in a later example. The amplitude is the fraction of the incident P wave for just the two transmission-mode conversion coefficients, ignoring attenuation and the shear-wave reflection coefficient at the subsurface reflector. Note that at angles of incidence (from the vertical) less than the critical angle of P waves (39°), the result-

ing amplitude is about 5 to 20 percent of the incident P-wave amplitude. Beyond the critical angle, however, the S-wave reflection, with two-mode conversions, has an amplitude nearly equal to the reflection coefficient of the S wave; i.e., most of the energy is mode-converted rather than reflected at the water bottom. Recall that beyond the critical angle, no P-wave energy penetrates the subsurface. This efficient mode conversion, as shown in Figure 3, persists for water-bottom P-

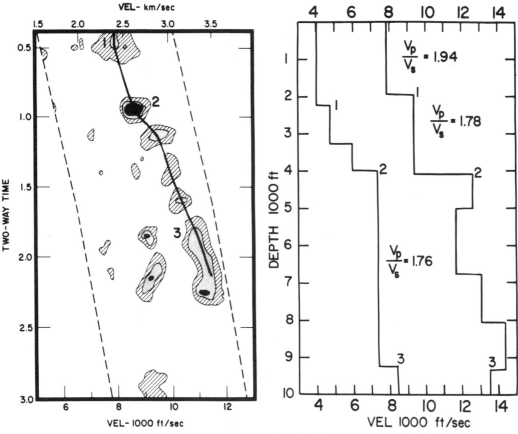

FIG. 5. Velocity analysis of same data used in Figure 4, but employing the full spread of 24 traces and setting limits consistent with P-wave reflections. Solid line indicates interpreted P-wave normal moveout velocity function, and numbers identify reflections correlating with S-wave reflections. Dashed lines define limits of velocity analysis.

FIG. 6. Velocity-depth profile for P and S waves interpreted from velocity analyses. Numbers identify correlated reflectors, and V_p/V_s is computed for intervals between these reflectors. The S-wave profile was interpreted with the constraint that the V_p/V_s ratio maintains reasonable values (e.g., 1.5 to 2.3).

wave velocities ranging from about 6500 ft/sec to 9000 ft/sec.

We searched for mode-converted shear wave reflections in existing CDP data, and selected an area off the U.S. Gulf Coast near the Alabama-Florida border. This area was chosen because the high-velocity water bottom provided the necessary interface for effective mode conversion. In a first attempt to observe shear-wave reflections, velocity analyses based on the coherence estimates (e.g., Taner and Koehler, 1969) were constructed on the 24-fold CDP reflection data, Figure 4, with the velocities limited to the range expected for shear waves. Since mode-converted S waves are anticipated principally for the off-vertical angles of incidence, the velocity analyses were made utilizing only the 12 longest-range traces (5400–8700 ft). Because multiple reflections also may occur in this low-velocity range, an identical velocity analysis utilizing only the 12 shortest range traces (2100–5100 ft) was also constructed for comparison. Generally, multiples should be equally well-observed on both the long and short range traces, but shear reflections should be recorded primarily on the long range traces. It must be kept in mind that in a particular geologic province the reflection coefficients for many seismic waves, both primaries and multiples, may increase as the angle of incidence deviates from vertical. For this data set, however, we feel this effect is less pronounced than the increase in mode conversion through the same deviation from vertical in-

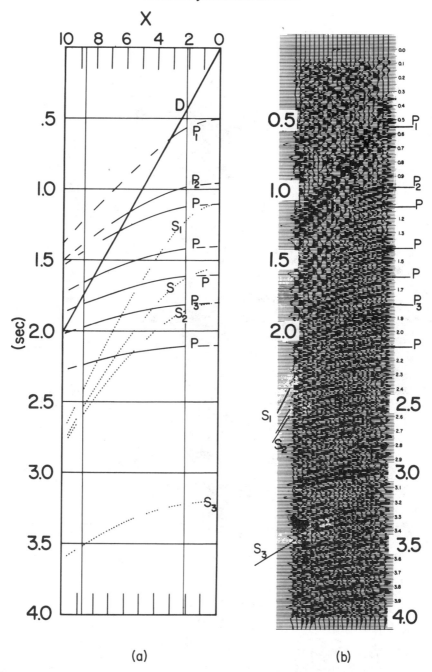

FIG. 7. (a) Time-distance plot showing positions of computed reflection hyperbolas corresponding to interpreted normal moveout velocities and observed arrival times. Offset distance (X) is in thousands of feet. Solid hyperbolas indicate interpreted P-wave reflections and dotted hyperbolas are the interpreted S-wave reflections. Solid sections of the hyperbolas indicate portions of the total array used in the velocity analyses. Note that all S-wave reflections arrive at least 0.5 sec after the computed direct water (D) arrival, and that there is no interference between wave types. (b) One CDP gather that was part of the input into the original velocity analyses. Positions of interpreted S- and P-wave reflection hyperbolas are indicated. Note the destructive effect of S_1 and S_2 upon the higher velocity compressional wave arrivals they encounter.

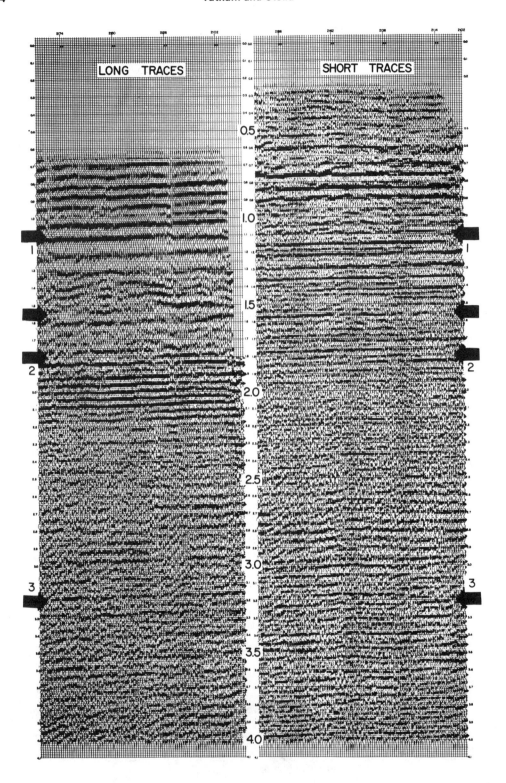

cidence (see Figure 2). Hence, the comparison of arrivals on the two data sets offers a possible means of discriminating between shear-wave reflections and multiple P-wave reflections of low apparent velocity.

An interpreted normal-moveout velocity function for the S waves is shown on both the long and short range velocity analyses in Figure 4. This function was interpreted in the following manner:

First, a trial S-wave moveout velocity function was determined by noting where large amounts of coherent energy appeared on the far-trace velocity analysis, but were not observed on the near-trace velocity analysis (see Figure 4).

Second, the interval P-wave velocity function was derived from the velocity analysis shown in Figure 5, and the interval S-wave velocity function was derived from the trial S-wave moveout function.

Third, the ratio of the interval P- to S-wave velocity was constrained to a broad but reasonable range (e.g., 1.5 to 2.3). That is, somewhat realistic Poisson's ratios must be maintained through gross portions of the section.

Fourth, we redetermined the S-wave moveout velocity function rejecting the points which gave unreasonable V_p/V_s ratios, while selecting additional points of coherent energy on the far-trace velocity analysis where no coherent energy was present on the near-trace analysis.

Thus, all the data available to us were used. Our requirements that the S-wave energy be concentrated in the far-trace range (large angles of incidence), and that the ratio V_p/V_s maintain a broad but reasonable range were maintained. The iterative interpretative method outlined above is necessary, since the largest coherent events on the far-trace velocity analyses may not necessarily be attributed to S waves.

In determining the interval velocity function, the Dix (1955) formulation could not be applied to the S-wave data because its applicability is limited to near vertical angles of incidence—an assumption that is necessarily violated for mode-converted S waves. Therefore, we computed all interval velocities by a procedure of ray-tracing and stripping of upper layers; that is, each subsequent layer velocity was computed after removing the effects of shallower layers (see Appendix). The data required for this procedure include not only the zero-offset times and apparent moveout velocity, but also the offset distances to the traces employed in the velocity analysis. The velocity-depth profiles for both P and S waves are shown in Figure 6 with the ratio V_p/V_s shown for three large intervals. Note that the moveout velocity of the first S-wave reflection is less than that for any anticipated multiple P-wave reflections. The numbered discontinuities are interpreted as being from the same reflector for both P and S waves. The arrival times of these reflections offer a means of correlating reflections between the P- and S-wave record sections.

Having identified both P- and S-wave reflections, we examine their reflection hyperbolas to insure that the interpreted shear reflections do not result from some coherent noise source or portions of other events. Figure 7a is a plot of computed reflection hyperbolas for both the P- and S-wave reflections interpreted from the velocity analyses. The position of the direct water-wave arrival (at 5000 ft/sec) is also shown. For this data set, any refractions would arrive earlier than this line at the ranges used to compute S-wave velocity and with an apparent phase velocity considerably faster than that associated with the nearly straight-line portions of the S-wave hyperbolas. Note that the hyperbolas associated with the S-wave reflections all occur at least 0.5 seconds after the direct arrival, and, thus, are probably not the result of coherent noise in the early part of the record. It is interesting to note that while distinct arrivals are not easily observed for the first two S-reflection hyperbolas on a common-depth-point gathered record (Figure 7b), they are more readily observed from their destructive interference on the strong P-wave arrivals they encounter. Since marine cables are designed to discriminate against low apparent phase velocities, consistent S-wave detection may require the redesign of marine

←

FIG. 8. CDP stacked record sections using the interpreted S-wave velocity function shown in Figure 4. Left section is stacked utilizing the 12 long-range traces, and the right-hand section stacked with 12 short-range traces. The only difference in the processing of these two sections and that of Figure 9 is the moveout velocity function applied. (Original data were recorded as 24-fold CDP.) Arrows indicated positions of interpreted S-wave reflections. Recall that S-wave reflections should be strongest on the long-range set, while low-velocity multiples should appear on both sections. Apparent low-frequency content at shallow levels on the long-trace section results from pulse distortion due to the large normal moveout correction required by the low stacking velocity and large offsets. Numbered reflectors correlate with numbered reflectors on P-wave section.

FIG. 9. Twenty-four fold CDP section stacked with the interpreted P-wave velocity function shown in Figure 5. This is the same original data set as that stacked in Figure 8. Arrows identify reflectors, and numbered arrows correspond to reflectors on the S-wave section.

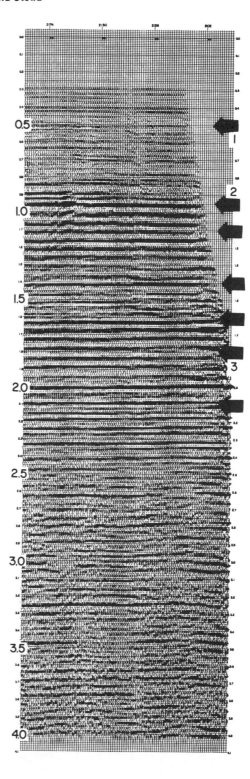

streamers for optimum observation of shear-wave reflections.

Using the interpreted velocity functions, we are able to apply normal moveout corrections and construct CDP stacked sections for both the S-wave (Figure 8) and conventional P-wave (Figure 9) reflections. The computed interval velocities allow the correlation of reflectors between these two sections. Since we are interested only in reflections that appear stronger on the long-trace section, simultaneous interpretation of both long-trace and short-trace S-wave sections provides a discriminant between S-wave reflections and possible multiples.

INTERPRETATION OF SHEAR-WAVE REFLECTION DATA

Once a stacked shear-wave reflection section is constructed, whether the shear waves are generated by mode conversion, horizontal vibrators, or special explosions, the same interpretative techniques can be applied. Since a structural interpretation is possible with the P-wave section alone, integration of the S-wave section into the total interpretation may provide expanded subsurface information while minimizing some possible ambiguities resulting from the use of a limited data set.

Bright-spot interpretations

Direct comparison of the P- and S-wave reflection sections could yield additional lithologic insight in bright-spot interpretations. For example, a bright spot which results from anomalously low P-wave velocity associated with the presence of interstitial gas should be absent or very small for the same reflector on the S-wave section. This difference in response occurs because shear-wave velocity is less sensitive to the effects of pore fluids than the P-wave velocity. Alternatively, if the amplitude anomaly results from the effects of anomalous subsurface geology, such as conglomeritic zones, it should be at least as strong, and quite possibly stronger, on the S-wave section than on the P-wave section. This is because the constrast between S-wave velocities for such materials as unconsolidated sediments and surrounding well-

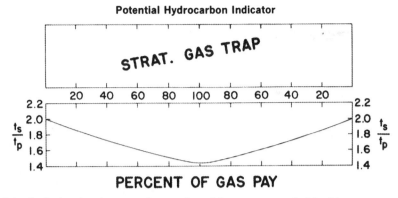

Fig. 10. Model calculation for the case of a stratigraphic gas trap sandwiched between two reflectors a fixed distance apart. The model stratigraphic trap is within a homogeneous sandstone of about 30 percent porosity, with gas pay thickness increasing from left (0 percent) to center (100 percent), and returning to no gas present (right). V_p for water saturation is assumed to be 8000 ft/sec, V_p for gas saturation is 5700 ft/sec, and V_s is assumed constant for both water and gas saturation at 4000 ft/sec. The upper portion illustrates the trap geometry, and the lower part shows variations in t_s/t_p (V_p/V_s) with indicated changes in gas pay thickness. For no gas present, $t_s/t_p = 2.00$, and as thickness of the nonreflecting gas layer increases, t_s/t_p decreases. For 100 percent gas pay, t_s/t_p attains a minimum of 1.43, a value inconsistent with changes in elastic parameters alone. Hence, this value is diagnostic of the presence of gas. As pay thickness decreases, t_s/t_p returns to the value associated with water saturation.

consolidated rock may be greater than the contrast in P-wave velocities for the same materials. Thus, shear-wave reflections respond differently to differing subsurface anomalies. This integration of both P- and S-wave data sets should minimize the potential ambiguities in bright-spot interpretations and provide a more reliable direct hydrocarbon indicator.

V_p/V_s interpretations

The use of the ratio of P-wave velocities to S-wave velocity offers considerable potential for exploration applications because it depends only upon reflecting boundaries above and below the zone (e.g., gas reservoir) in question. Hence, it may be applicable to gas in stratigraphic traps as well as structural traps. It should be noted, however, that the reflecting boundaries must be the same for both P- and S-wave reflections (with laterally consistent pulse shape), or they must be at known separations. Moreover, for reasonable sensitivity to V_p/V_s in the reservoir, the upper and lower reflections should not be far from the reservoir. Of course, these reflections must also be relatively undisturbed by laterally varying adjacent reflections. If these conditions are met, the combination of P- and S-wave seismic reflection sections offers a simple means of measuring and mapping this diagnostic parameter. Consideration of the reflection times between the two reflectors on both the P- and S-wave sections yields a traveltime ratio t_s/t_p which can be related to V_p/V_s for the zone in question. An interpretation very similar to the construction of an isochron or isopach map can then be made.

Figure 10 shows the expected nature of the t_s/t_p anomaly over a model stratigraphic gas reservoir. Two reflectors a fixed distance apart, with a sandstone of about 30 percent porosity between the interfaces, are considered. It is assumed that there are no additional reflectors observed between the two reflectors being mapped. The section starts with no gas present and $t_s/t_p = 2.00$. Proceeding to the right of the diagram, the thickness of the gas zone, located somewhere between the reflectors, increases; hence, t_s/t_p decreases. This results from the reduction in V_p while a constant V_s is maintained. For a gas thickness of 25 to 30 percent of the interval considered, t_s/t_p drops to 1.80, an anomaly of 0.2, which may be diagnostic of the presence of gas. For a gas thickness of 45 to 50 percent of the interval, t_s/t_p drops to less than 1.70. Values of t_s/t_p less than about 1.7 often require the existence of some anomalous condition. Thus, the observation of such values can be indicative of the presence of hydrocarbons, especially when changes in the parameter can be traced laterally from normal to anomalous regions. As the thickness of the gas zone increases, t_s/t_p continues to decrease, and a highly diagnostic and readily observed anomaly develops. Once the S-wave section has been constructed, the reliability of a t_s/t_p interpretation depends only upon the continuity of both the P- and S-wave reflections which have traversed the anomalous region. The relative amplitude data of high quality required for reliable bright-spot or attenuation interpretations are not necessary. Hence, the V_p/V_s technique may be applicable in regions of only moderately good data quality.

ACKNOWLEDGMENTS

We wish to express our gratitude to Texaco Inc. for arranging, and to A. Pollet of Teledyne and to Teledyne Exploration for supplying and processing the marine data crucial to the observations of mode-converted S-wave reflections. We wish to thank Professors John Kuo, L. R. Sykes, Paul G. Richards, and J. Ewing of Lamont-Doherty for critically reading the manuscript and Prof. C. H. Scholz for his useful discussions on the dilatancy model. The research was supported by the U.S. Geological Survey under contract USGS no. 14-08-0001-G-113; the Earth Science Section of the National Science Foundation, NSF grants GA-43295 and GX-34410; and the Office of Naval Research contract N00014-75-C-0210.

REFERENCES

Aggarwal, Y. P., Simpson, D. W., Sbar, M. L., Pomeroy, P. W., and Sykes, L. R., 1973b, First successful prediction of an earthquake in the United States: EOS, Trans. AGU, v. 54, p. 1134.

Aggarwal, Y. P., Sykes, L. R., Armbruster, J., and Sbar, M. L., 1973a, Premonitory changes in seismic velocities and prediction of earthquakes: Nature, v. 241, p. 101–104.

Aggarwal, Y. P., Sykes, L. R., Simpson, D. W., and Richards, P. G., 1975, Spatial and temporal variations in t_s/t_p and in P-wave residuals at Blue Mountain Lake, New York: Application to earthquake prediction: J. Geophys. Res., v. 80, p. 718–732.

Biot, M. A., 1962a, Generalized theory of acoustic propagation in porous dissipative media: J. Acoust. Soc. Am., v. 34, p. 1254–1264.

———, 1962b, Mechanics of deformation and acoustic propagation in porous media: J. of Applied Physics, v. 33, p. 1482–1498.

Cherry, J. T., and Waters, K. H., 1968, Shear-wave recording using continuous signal methods, Part I. Early development: Geophysics, v. 33, p. 229–239.

Choy, G. T., 1976, Theoretical seismograms calculated by Langer's uniformly asymptotic method: Application to core phases: Ph.D. dissertation, Columbia University (in preparation).

Craft, C., 1973, Detecting hydrocarbons: For years the goal of exploration geophysics: Oil and Gas J., Feb. 19.

Dix, C. W., 1955, Seismic velocities from surface measurements: Geophysics, v. 20, p. 68–86.

Domenico, S. N., 1974, Effect of water saturation on seismic reflectivity of sand reservoirs encased in shale: Geophysics, v. 39, p. 759–769.

Ergin, K., 1952, Energy ratio of the seismic waves reflected and refracted at a rock-water boundary: Bull. SSA, v. 42, p. 349–372.

Erickson, E. L., Miller, D. E., and Waters, K. H., 1968, Shear wave recording using continuous signal methods Part II—Later experimentation: Geophysics, v. 33, p. 240–254.

Gardner, G. H. F., Gardner, L. W., and Gregory, A. R., 1974, Formation velocity and density—The diagnostic basis for stratigraphic traps: Geophysics, v. 39, p. 770–780.

Gassman, F., 1951, Elastic waves through a packing of spheres: Geophysics, v. 15, p. 673–685.

Geyer, R. L., and Martner, S. T., 1969, SH waves from explosive sources: Geophysics, v. 34, p. 893–905.

Hazebroek, P., 1966, Elastic waves from a finite line source: Proc. Roy. Soc. London, Series A, v. 294, p. 38–65.

Heelan, P. A., 1953, Radiation from a cylindrical source of finite length: Geophysics, v. 18, p. 685–696.

King, M. S., 1966, Wave velocities in rocks as a function of changes in overburden pressure and pore fluid saturants: Geophysics, v. 31, p. 50–73.

Kuster, G. T., and Toksöz, M. N., 1974, Velocity and attenuation of seismic waves in two-phase media: Part I. Theoretical formulations: Geophysics, v. 39, p. 587–606.

Miller, D. E., and Dunster, D. E., 1974, High-resolution "Vibroseis" data—compressional and shear modes: submitted to Geophysics.

Nersesov, I. L., Semmova, A. N., and Simbivera, I. G., 1969, The physical basis of foreshocks: Moscow, Nauka.

Nafe, J. E., and Drake, C. L., 1957, Variation with depth in shallow and deep water marine sediments of porosity, density and the velocities of compressional and shear waves: Geophysics, v. 22, p. 523–552.

Nur, A., 1972, Dilatancy, pore fluids and premonitory variations of t_s/t_p traveltimes: Bull. SSA, v. 62, p. 1217–1222.

O'Connell, R. J., and Budiansky, B., 1974, Seismic velocities in dry and saturated cracked solids: J. Geophys. Res., v. 79, p. 5412–5426.

Officer, C. B., 1958, Introduction to the theory of sound transmission: New York, McGraw-Hill Book Co., Inc.

Scholz, C. H., Sykes, L. R., and Aggarwal, Y. P., 1973, Earthquake prediction: A physical basis: Science, v. 181, p. 803–810.

Taner, M. T., and Koehler, F., 1969, Velocity spectra—digital computer derivation and applications of velocity functions: Geophysics, v. 34, p. 859–881.

Whitcomb, J. H., Garmany, J. D., and Anderson, D. L., 1973, Earthquake prediction: variation of seismic velocity before the San Fernando earthquake: Science, v. 180, p. 632–635.

White, J. E., and Sengbush, R. L., 1963, Shear waves from explosive sources: Geophysics, v. 28, p. 1001–1019.

APPENDIX

METHOD OF COMPUTING INTERVAL VELOCITIES

The normal moveout velocities V_0 and zero offset times T_0 determined by a velocity analysis are derived by fitting observed data to the hyperbolic relation

$$T^2 = T_0^2 + \frac{X^2}{V_0^2}. \quad (A\text{-}1)$$

The resultant values of V_0 and T_0 may then be converted to interval velocities, provided X is not large, by the use of the Dix [1955, equation (12)] formula. For the observation of mode-converted shear waves we consider only large X. Thus, the Dix formula is not applicable. Dix (1955, p. 73) does suggest, however, that the effect of the layers above the interval under consideration can be removed by ray tracing. Given V_0, we can determine the ray parameter p for each X and T on the hyperbola (A-1) by differentiating (A-1).

$$p = \frac{dT}{dX} = \frac{X}{T V_0^2}. \quad (A\text{-}2)$$

It is then a simple procedure to determine the traveltime T_c and horizontal distance X_c traveled by a ray, which is reflected from the base of the interval under consideration, to the top of the layer by employing the ray-tracing equations (e.g., Officer, 1958, p. 51):

$$\frac{X_c}{2} = \int_0^Z \frac{p V(Z) \, dZ}{\sqrt{1 - p^2 V^2(Z)}}; \quad (A\text{-}3a)$$

$$\frac{T_c}{2} = \int_0^Z \frac{dZ}{V(Z)\sqrt{1 - p^2 V^2(Z)}}. \quad (A\text{-}3b)$$

Here, Z is the depth to the top of the layer under consideration and $V(Z)$ is the interval velocity in each layer, of thickness dZ, above Z. X_c and T_c are then subtracted from the X and T values for the reflection hyperbola described by (A-1) for each X used in the original determination of V_0 associated with the reflection from the base of the interval. New values of V_0 and T_0, associated solely with the interval being considered, can then be determined by a linear least-squares fit of the squares of the corrected ranges and times, i.e., a T^2-X^2 velocity determination for the source and receivers corrected to the top of the layer can be made.

Amplitude Variation with Offset

Plane-wave reflection coefficients for gas sands at nonnormal angles of incidence

W. J. Ostrander*

ABSTRACT

The *P*-wave reflection coefficient at an interface separating two media is known to vary with angle of incidence. The manner in which it varies is strongly affected by the relative values of Poisson's ratio in the two media. For moderate angles of incidence, the relative change in reflection coefficient is particularly significant when Poisson's ratio differs greatly between the two media.

Theory and laboratory measurements indicate that high-porosity gas sands tend to exhibit abnormally low Poisson's ratios. Embedding these low-velocity gas sands into sediments having "normal" Poisson's ratios should result in an increase in reflected *P*-wave energy with angle of incidence. This phenomenon has been observed on conventional seismic data recorded over known gas sands.

INTRODUCTION

During the past decade, the use of "bright spot" type analysis in petroleum exploration has become increasingly common. Oil companies, both large and small, are making use of the fact that high-intensity seismic reflections may be indicators of hydrocarbon accumulations, particularly gas. Bright spot exploration has significantly increased the recent success ratio for wildcat gas wells. Nonetheless, problems still do exist. Many seismic amplitude anomalies are not caused by gas accumulations, or they are caused by gas accumulations which are subcommercial. The latter problem is difficult to resolve. However, amplitude anomalies caused by nongaseous, abnormally high- or low-velocity layers may have distinguishing characteristics. This paper proposes a method which potentially may distinguish between gas-related amplitude anomalies and nongas related anomalies. Notable observations contained herein are (1) the somewhat surprising effects of Poisson's ratio on *P*-wave reflection coefficients; and (2) the existence of these effects in seismic amplitude anomalies related to gas accumulations.

BACKGROUND

Poisson's ratio

Poisson's ratio, sometimes denoted by the Greek letter small sigma (σ), is a somewhat neglected elastic constant. It is related to other elastic constants by a simple set of equations. In particular, Poisson's ratio for an isotropic elastic material is simply related to the *P*-wave (V_p) and *S*-wave (V_s) velocities of the material by

$$\sigma = \frac{(V_p/V_s)^2 - 2}{2[(V_p/V_s)^2 - 1]}. \qquad (1)$$

This equation indicates that Poisson's ratios may be determined dynamically using field or laboratory measurements of both V_p and V_s.

Poisson's ratio also has a physical definition. If one takes a cylindrical rod of an isotropic elastic material and applies a small axial compressional force to the ends, the rod will change shape. The length of the rod will decrease slightly, while the radius of the rod will increase slightly. Poisson's ratio is defined as the ratio of the relative change in radius to the relative change in length. Common isotropic materials have Poisson's ratios between 0.0 and 0.5. Incompressible materials such as liquids will have Poisson's ratios of 0.5, while "spongy" materials might have ratios closer to zero.

Reflection coefficients

In 1940, a classic article was published by Muskat and Meres showing the variations in plane-wave reflection and transmission coefficients as a function of angle of incidence. Since then, several additional articles on the subject have appeared in the literature, including those by Koefoed (1955, 1962) and Tooley et al. (1965). Using the simplified Zoeppritz equations given by Koefoed (1962), one can show that four independent variables exist at a single reflecting/refracting interface between two isotropic media: (1) *P*-wave velocity ratio between the two bounding media; (2) density ratio between the two bounding media; (3) Poisson's ratio in the upper medium; and (4) Poisson's ratio

---

Presented at 52nd Annual International SEG, Meeting in Dallas, Texas, on October 21, 1982. Manuscript received by the Editor February 15, 1983; revised manuscript received April 30, 1984.
*Chevron U.S.A., Inc., 2003 Diamond Boulevard, Concord, CA 94520.
© 1984 Society of Exploration Geophysicists. All rights reserved.

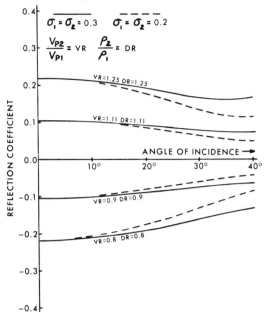

FIG. 1. Plot of P-wave reflection coefficient versus angle of incidence for constant Poisson's ratios of 0.2 and 0.3.

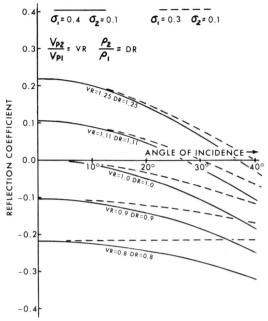

FIG. 2. Plot of P-wave reflection coefficient versus angle of incidence for a reduction in Poisson's ratios across an interface.

in the lower medium. These four quantities govern plane-wave reflection and transmission at a seismic interface.

Since Muskat and Meres (1940) had very little information on values of Poisson's ratios for sedimentary rocks, they used a constant value of 0.25 in all their calculations, i.e., Poisson's ratio was the same for both media. Results similar to theirs are shown in Figure 1 for various velocity and density ratios and constant Poisson's ratios of 0.2 and 0.3. One would conclude from these results that angle of incidence has only minor effects on P-wave reflection coefficients over propagation angles commonly used in reflection seismology. This is a basic principle upon which conventional common-depth-point (CDP) reflection seismology relies.

The work of Koefoed (1955) is of particular interest since his calculations involved a change in Poisson's ratio across the reflecting interface. He found that by having substantially different Poisson's ratios for the two bounding media, large changes in P-wave reflection coefficients versus angle of incidence could result. Koefoed showed that under certain circumstances, reflection coefficients could increase substantially with increasing angle of incidence. This increase occurs well within the critical angle where high-amplitude, wide-angle reflections are known to occur.

Figures 2 and 3 illustrate an extension of Koefoed's initial computations. Figure 2 shows P-wave reflection coefficients from an interface, with the incident medium having a higher Poisson's ratio than the underlying medium. The solid curves represent a contrast in Poisson's ratio of 0.4 to 0.1, while the dashed curves represent a contrast of 0.3 to 0.1. One may

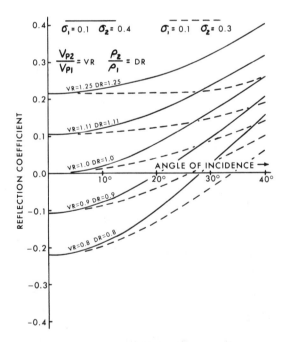

FIG. 3. Plot of P-wave reflection coefficient versus angle of incidence for an increase in Poisson's ratios across an interface.

Table 1. List of measured Poisson's ratios for various sedimentary rock types.

PODIO ET AL. (1968)	GREEN RIVER SHALE	0.22-0.30
HAMILTON (1976)	SHALLOW MARINE SEDIMENTS	0.45-0.50
GREGORY (1976)	CONSOLIDATED SEDIMENTS	
	BRINE SATURATED	0.20-0.30
	GAS SATURATED	0.02-0.14
DOMENICO (1976)	SYNTHETIC SANDSTONE	
	BRINE SATURATED	0.41
	GAS SATURATED	0.10
DOMENICO (1977)	OTTAWA SANDSTONE	
	BRINE SATURATED	0.40
	GAS SATURATED	0.10

conclude from these curves that if Poisson's ratio decreases going into the underlying medium, the reflection coefficient decreases algebraically with increasing angle of incidence. This means positive reflection coefficients may reverse polarity and negative reflection coefficients increase in magnitude (absolute value) with increasing angle of incidence.

Figure 3 shows the opposite situation to that shown in Figure 2. Here Poisson's ratio increases going from the incident medium into the underlying medium. In this case, the reflection coefficients increase algebraically with increasing angle of incidence. Negative reflection coefficients may reverse polarity, and positive reflection coefficients increase in magnitude with increasing angle of incidence.

The foregoing three illustrations point to a strong need for more information on Poisson's ratio for the various rock types encountered in seismic exploration. This is particularly important when one considers the long offsets commonly in use today and the resulting large angles of incidence. It will become evident later that this phenomenon has an important effect on bright spot analysis. For additional computations of reflection and transmission coefficients, the reader should refer to Koefoed (1962) and Tooley et al. (1965).

MEASUREMENTS OF POISSON'S RATIO

In the *Handbook of Physical Constants*, Birch (1942) lists Poisson's ratios for various materials, including many rock types. However, little significance can be placed on these values because of the methods and environments of measurement. As will become obvious later, any air or gas in cracks or pore spaces can severely alter measurement of Poisson's ratio. Until recently, other published measurements for sedimentary rocks were quite limited.

Many comprehensive measurements of Poisson's ratios for sedimentary rocks were reported in the literature during the 1970s. Hamilton (1976) presented a review of measurements made for shallow marine sediments including both sands and shales. His results showed that shallow, unconsolidated marine sediments to depths of 2 000 ft had Poisson's ratios between 0.45 and 0.50. Gregory (1976) gave results including fluid saturation effects for many consolidated sedimentary rocks. His samples included sandstones, limestones, and chalks ranging in porosity from 4 to 41 percent. The work of Domenico (1976, 1977) is of special interest because it applies to many of our shallow gas fields which have related seismic amplitude anomalies. In both a synthetic high-porosity glass bead and a high-porosity Ottawa sandstone mixture, Domenico found marked changes in Poisson's ratios between brine and gas saturations. In these unconsolidated 38 percent porosity specimens, the replacement of brine with gas reduced Poisson's ratio from 0.4 to 0.1. A summary of the foregoing results is shown in Table 1.

Several conclusions can be drawn from measurements of Poisson's ratios for sedimentary rocks. First, unconsolidated, shallow, brine-saturated sediments tend to have very high Poisson's ratios of 0.40 and greater. Second, Poisson's ratios tend to decrease as porosity decreases and sediments become more consolidated. Third, high-porosity brine-saturated sandstones tend to have high Poisson's ratios of 0.30 to 0.40. And fourth, gas-saturated high-porosity sandstones tend to have abnormally low Poisson's ratios on the order of 0.10.

The above conclusions result from the fact that the shear modulus μ of a rock does not change when the fluid saturant is changed. However, the bulk modulus k does change significantly (Gassmann, 1951). The bulk modulus of a fluid-saturated rock is a function of the bulk moduli of the fluid, the grains, and the dry rock framework. The bulk modulus of a brine-saturated rock is greater than that of gas-saturated rock because brine is significantly stiffer than gas. This results in the P-wave velocity (V_p) of the brine-saturated rock being considerably higher than that of a gas-saturated rock from equation (2). The S-wave velocity (V_s) defined in equation (3) is only affected by a small change in the density ρ. Since density is reduced by a gas saturant, the S-wave velocity is slightly increased with gas saturation. Equations (2) and (3) show the relationships among these parameters.

$$V_p = \left(\frac{k + \frac{4}{3} \times \mu}{\rho}\right)^{1/2}, \qquad (2)$$

and

$$V_s = \left(\frac{\mu}{\rho}\right)^{1/2}. \qquad (3)$$

Analysis of Gassmann's equation shows that the weaker the framework modulus, the greater the differences between brine and gas saturations. This explains the dramatic differences observed in poorly consolidated rocks such as those analyzed by Domenico.

Depth of burial and differential pressure also influence the elastic behavior of rocks. Differential pressure is the difference between the overburden pressure and the fluid pressure and is generally a monotonic function of depth. As differential pressure increases, the bulk and shear moduli of a rock increase, resulting in greater P- and S-wave velocities. Sediment consolidation and increased differential pressures tend to decrease fluid saturation effects with increased depth of burial.

Theoretical results also support the large reduction in Poisson's ratio as gas replaces brine as the saturant in high-porosity sandstones. Using the equations of Gassmann (1951), one can compute theoretical P-wave and S-wave velocities as a function

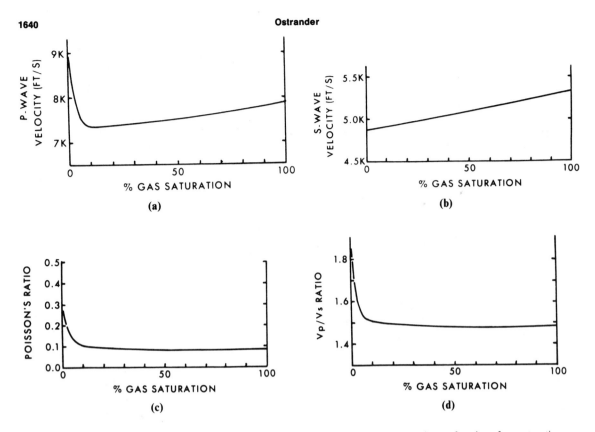

FIG. 4. Plots of (a) P-wave velocity, (b) S-wave velocity, (c) Poisson's ratio, and (d) V_p/V_s ratio as a function of gas saturation.

of percent gas saturation. Poisson's ratios can then be computed using equation (1). Theoretical results for a 35 percent porosity sandstone buried at 6 000 ft are shown in Figures 4a, 4b, 4c, and 4d. In Figure 4c, one sees that the major change in Poisson's ratio occurs with less than 10 percent gas saturation. From 10 to 100 percent gas saturation, Poisson's ratio changes very little around an average value of 0.09. These characteristic gas saturation curves have been supported by laboratory measurements (Domenico, 1976, 1977; Gregory, 1976).

GAS SAND MODEL

Using the foregoing review of physical parameters, one can now devise a hypothetical gas sand model. This model can be used to analyze plane-wave reflection coefficients as a function of angle of incidence. Calculations can be made for the reflections originating from both the top and base of the gas sand.

Figure 5 shows a three-layer gas sand model with parameters which might be typical for a shallow, young geologic section. Here, a gas sand with a Poisson's ratio of 0.1 is embedded in shale having a Poisson's ratio of 0.4. There is a 20 percent velocity reduction going into the sand, from 10 000 ft/s to 8 000 ft/s, and a 10 percent density reduction from 2.40 g/cm³ to 2.16 g/cm³. These parameters result in normal-incidence reflection

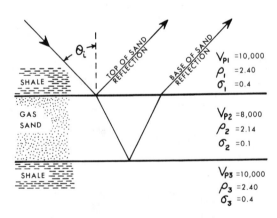

FIG. 5. Three-layer hypothetical gas sand model.

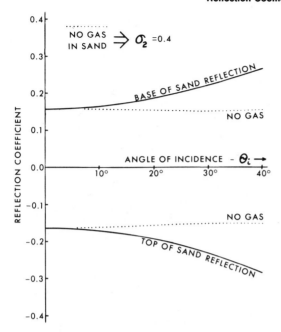

FIG. 6. Plot of P-wave reflection coefficients versus angle of incidence for three-layer gas sand model.

coefficients of -0.16 and $+0.16$ for the top and base of the gas sand, respectively.

Changes in plane-wave reflection coefficient as a function of angle of incidence for several cases are shown in Figure 6. The two solid curves are those reflection coefficients resulting from the gas sand model parameters shown in Figure 5. The effect of transmission and refraction on the base of sand reflection have been taken into account. The horizontal axis is the angle of incidence referenced to the top of the sand. Because of refraction, a 40-degree incident angle at the top of the sand represents only 31 degrees incident angle at the base of the sand. The top of sand reflection coefficient changes from about -0.16 to -0.28 over 40 degrees while the base of sand reflection coefficient changes from about $+0.16$ to $+0.26$. Thus, the amplitude of the seismic waveform resulting from this complex reflection would increase approximately 70 percent over 40 degrees.

The dotted curves in Figure 6 indicate what the reflection coefficients would be if Poisson's ratio in the sand were changed to 0.4. This would simulate the case of a low-velocity brine-saturated young sandstone embedded in shale. In this case, one sees only a slight decrease in the magnitude of the reflection coefficients as the angle of incidence increases.

DATA ANALYSIS

The obvious question at this point is how one can best observe and analyze changes in reflection coefficient with angle of incidence on today's conventionally recorded reflection seismic data. The answer lies in analyzing amplitudes on CDP-gathered traces prior to stacking. In this way, one can observe changes in reflection amplitude versus shot-to-group offset. As shot-to-group offset increases, the angle of incidence increases monotonically.

Angles of incidence

There are several ways to estimate angles of incidence from the depth to a reflector and the shot-to-group offset. The first and most simple is the straight-ray approach where the angle of incidence θ_i is given by

$$\theta_i = \tan^{-1}\left(\frac{X}{2 \cdot Z}\right), \qquad (4)$$

where X is the shot-to-group offset and Z is the reflector depth. If velocity increases with depth, which is most common, the angles computed from equation (4) will always be too small. In this situation, a better approach for estimating angles of incidence is illustrated in Figure 7. If the section interval velocity can be approximated in the form $V_I = V_0 + KZ$, then all ray-paths are arcs of circles whose centers are V_0/K above the

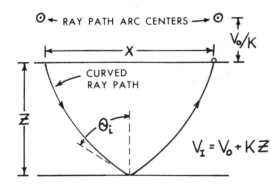

FIG. 7. Geometry for estimating angles of incidence for a velocity function of the form $V_I = V_0 + KZ$.

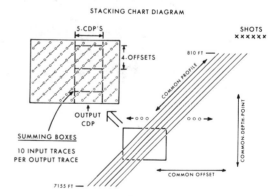

FIG. 8. Trace-summing technique to increase S/N ratios.

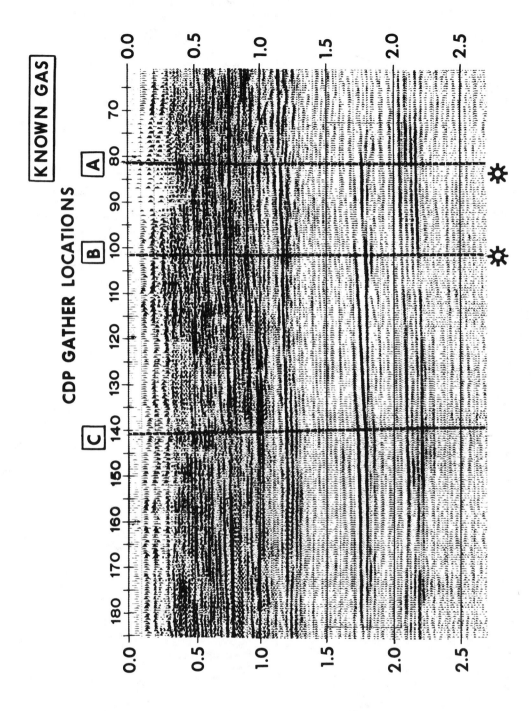

FIG. 9. Stacked seismic section for line SV-1.

ground surface. Using the resulting geometry, one can derive the following relationship:

$$\theta_i = \tan^{-1}\left(\frac{ZX + V_0 X/K}{Z^2 + 2V_0 Z/K - X^2/4}\right). \quad (5)$$

An example calculation for $V_I = 6\,000 + 0.6Z$, at a depth (Z) of 7 000 ft, and an offset (X) of 7 000 ft, gives an angle of incidence (θ_i) of 34 degrees. Thus, one sees that angles of 30 degrees or more are not uncommon in today's CDP recording and are in many cases unmuted during CDP stacking.

Trace summing—S/N improvement

In viewing single CDP-gathered traces for amplitude variations, one major drawback occurs: poor signal-to-noise (S/N) ratios. As a means of signal enhancement, trace summing can prove most worthwhile. A method of partial trace summing is illustrated in Figure 8. Shown in the figure is a stacking chart diagram on the right with an enlargement of the same in the upper left. The recording geometry is for 48-trace, single end, 24-fold CDP coverage with a near offset of 810 ft and a far offset of 7 155 ft. Here CDP traces lie on a vertical line, common-offset traces on a horizontal line, and common-shot profile traces on the diagonals as shown. In the enlargement, individual traces are shown as small circles. To form a partial sum trace, all traces fall within boxes which are 5 CDPs by 4 offsets in dimension and which are summed together forming a single output trace. This limited summing will produce a 10-fold sum trace, and thus improve the S/N ratio by a factor of about 3. Repeating the procedure for groups of 4 offsets will produce 12 traces, all with improved S/N ratios. Displaying these 12 partially summed traces in increasing average shot-to-group offset gives a desirable product for analyzing amplitude information. Variations on this type of limited offset summing are easily implemented for different recording geometries.

In the examples which follow, the reader will find that some CDP gathers are displayed as individual traces while others have been partially summed. The advantages of summing will become obvious.

EXAMPLES

Several examples will now be presented which illustrate apparent changes in reflection amplitude versus angle of incidence. All of the examples are in areas of well control, so the origin of the bright spot or amplitude anomaly is known. In two cases, the anomalies are caused by gas, while in the third, the seismic amplitude anomaly is caused by a high-velocity layer of basalt. Prior to drilling, all of these seismic anomalies were thought to be caused by gas-saturated sediments.

The illustrations presented for each example are (1) a conventionally stacked section showing the given amplitude anomaly; and (2) CDP gathers at locations indicated by vertical dashed lines and the letters A, B, and C on the stacked sections. The processing flow, prior to the displays shown, employed standard techniques. These included spherical divergence correction, exponential gain, minimum phase-spiking deconvolution, statics, velocity analysis, normal moveout (NMO) removal, time-invariant band-pass filtering, and single long-gated trace equalization. No wavelet processing was done on these data, so no implications as to reflector polarity can be made.

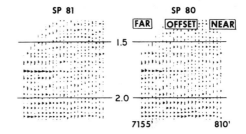

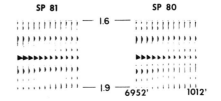

FIG. 10. CDP gathers for location A on line SV-1.

Line SV-1

Shown in Figure 9 is a 24-fold CDP stacked seismic line over a large gas field in the Sacramento Valley. The sand reservoir occurs at a depth of about 6 700 ft which corresponds to a seismic amplitude anomaly at about 1.75 s. The reservoir is a Cretaceous deep-sea fan deposit having a maximum net pay of 95 ft. The trap is both structural and stratigraphic with the reservoir being offset along a fault at about SP 95. The downthrown portion of the reservoir on the left is trapped against the fault, while the reservoir pinches out in the upthrown block at about SP 75. The velocity and density within the gas sand are substantially lower than the encasing shales, giving rise to strong seismic reflections at the top and base of the gas sand. A flat fluid contact reflection may be present at 1.8 s between SP 115 and SP 135. The discovery well is located at about SP 86 with the reservoir limits extending from about SP 75 to SP 130.

CDP gathers from three locations, A, B, and C, are shown in Figures 10, 11, and 12, respectively. Both single-fold and 10-fold summed gathers are shown for locations A and B, while only the summed gathers are shown for location C. Shot-to-group offset for all gathers increases to the left. These distances change on the summed gathers because the summing is done over four offsets. At the objective sand, the near-offset corresponds to about 5 degrees angle of incidence while the far-offset corresponds to about 35 degrees.

A strong amplitude increase with increasing offset is apparent in the gathers at locations A and B shown in Figures 10 and 11. The 10-fold summing obviously improves S/N ratios, and an amplitude increase by a factor of about three is indicated from the near offset to far offset. The gathers at location C, shown in Figure 12, show no indication of amplitude increase with offset and in fact show a decrease. This possibly indicates

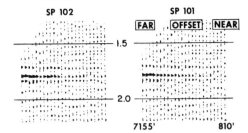
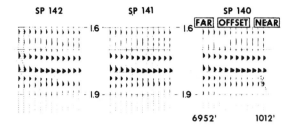

FIG. 12. CDP gathers for location C on line SV-1.

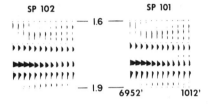

FIG. 11. CDP gathers for location B on line SV-1.

an absence of gas in the vicinity of location C. This possibility is also supported by the presence of a gas-water contact in a well which would structurally project in at about SP 120.

In this example and those which follow, no S-wave velocity data were available to model variations in reflection coefficient in shot-to-group offset. The P-wave sonic velocities within this gas reservoir may also be subject to error. The behavior of the observed seismic amplitude in shot-to-group offset is as expected from theory and laboratory measurements. However the magnitude of the change of amplitude may be influenced by

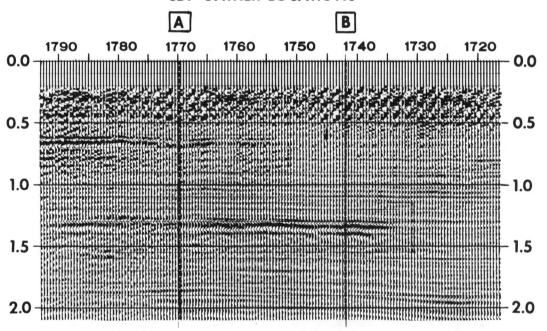

FIG. 13. Stacked seismic section for line GM-1.

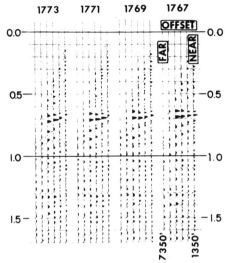

FIG. 14. CDP gathers for location A on line GM-1.

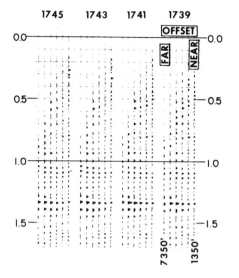

FIG. 15. CDP gathers for location B on line GM-1.

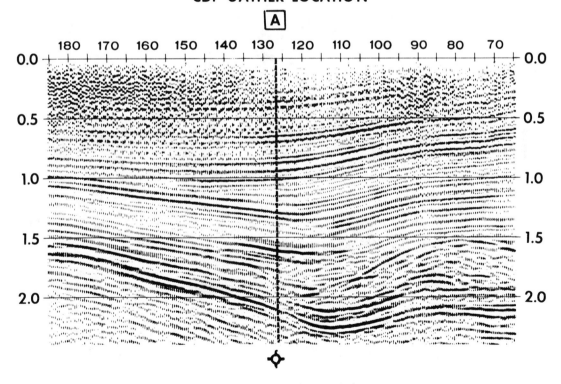

FIG. 16. Stacked seismic section for line FB-1.

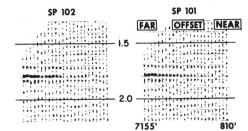
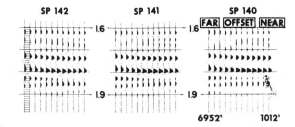

FIG. 12. CDP gathers for location C on line SV-1.

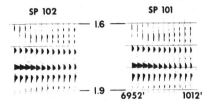

FIG. 11. CDP gathers for location B on line SV-1.

an absence of gas in the vicinity of location C. This possibility is also supported by the presence of a gas-water contact in a well which would structurally project in at about SP 120.

In this example and those which follow, no S-wave velocity data were available to model variations in reflection coefficient in shot-to-group offset. The P-wave sonic velocities within this gas reservoir may also be subject to error. The behavior of the observed seismic amplitude in shot-to-group offset is as expected from theory and laboratory measurements. However the magnitude of the change of amplitude may be influenced by

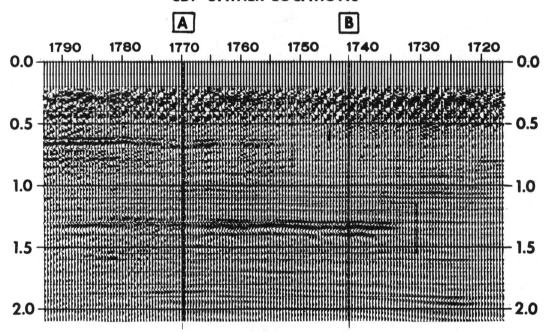

FIG. 13. Stacked seismic section for line GM-1.

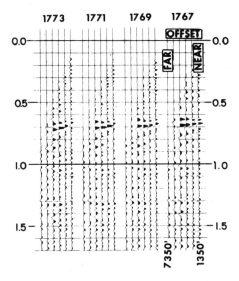

FIG. 14. CDP gathers for location A on line GM-1.

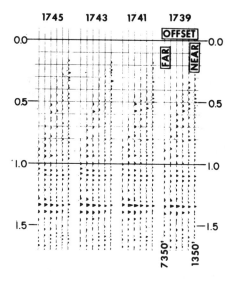

FIG. 15. CDP gathers for location B on line GM-1.

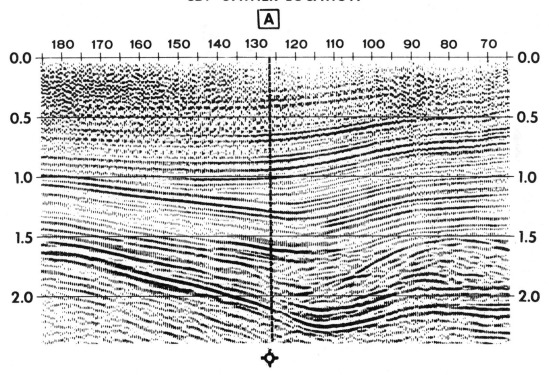

FIG. 16. Stacked seismic section for line FB-1.

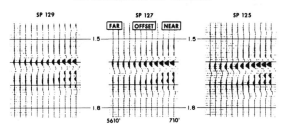

FIG. 17. CDP gathers for location A on line FB-1.

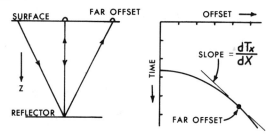

FIG. 18. Relationship between array attenuation, apparent reflector dip, and normal moveout.

factors to be discussed later. Because one has great difficulty in separating out "true" reflection amplitudes, the interpreter typically must rely on relative changes, concentrating on anomalous behavior of the amplitude.

Line GM-1

Figure 13 shows two gas-related seismic-amplitude anomalies on line GM-1 located in the Gulf of Mexico. The first of these anomalies is located on the left half of the seismic section at about 0.65 s. The second, deeper anomaly is toward the middle of the seismic section at about 1.35 s. Summed CDP gathers are displayed in Figure 14 for location A on the shallow anomaly and in Figure 15 for location B on the deeper anomaly. In both anomalies, it is quite apparent that reflection amplitude tends to increase with increasing offset. In the case of the shallower anomaly at location A, the effect of array attenuation and NMO stretch on the fifth offset trace is obvious.

Line FB-1

Shown in Figure 16 is a 24-fold CDP-stacked seismic line recorded in a virgin basin in Nevada. Several years ago, a well was drilled on this line at SP 127 (location A) to a depth below 2.0 s. A seismic amplitude anomaly is indicated on the stacked seismic data at this location and at a time of about 1.6 s. Upon drilling, the amplitude anomaly was found to originate from a high-velocity basaltic interval of about 160 ft in thickness. As has happened elsewhere, the apparent bright spot is not due to the presence of gas in the sediments.

The CDP gathers at the well location are shown in Figure 17. Here, there is a strong indication of a decrease in reflection amplitude with increasing offset or angle of incidence. This finding is consistent with a relatively uniform Poisson's ratio in the geologic section. Basalt is not expected to have an anomalous Poisson's ratio.

OFFSET AMPLITUDE ANALYSIS

At this point, the reader may wonder what type of amplitude balancing is desirable in order to analyze offset-dependent amplitude changes. One must then look at some of the major factors which affect the recorded amplitude of a reflection as a function of offset. Some of these factors are listed below.

(1) Reflection coefficient
(2) Array attenuation
(3) Event tuning
(4) Noise
(5) Spherical spreading
(6) Emergence angles
(7) Reflector curvature
(8) Spherical wavefronts
(9) Transmission coefficients
(10) Instrumentation/processing
(11) Inelastic attenuation

The first of these, the reflection coefficient, is the factor which one would like to observe. In actuality, one can only observe relative changes in reflection coefficient versus offset. If no other factors existed, one could simply observe the CDP gathers with a spherical divergence correction applied. However, because of the other offset-related amplitude factors listed above, simple observation of the reflection coefficient is not always feasible. Considered below are some of the other factors in more detail.

Array attenuation

Array attenuation arises because one generally does not have a point source and a point receiver. As the dip of the apparent reflector becomes large, geophone arrays tend to reduce amplitudes of reflections. The same is true for shot arrays. This effect is greatest for shallow reflectors at long offsets and diminishes with greater depths of reflectors and shorter offsets.

As illustrated in Figure 18, array attenuation is a result of NMO. For a flat-lying reflector, one can see that the apparent dip of the reflection across a geophone group array comes purely from NMO. This dip or slope is simply the derivative of the NMO with respect to offset, dT_x/dX. Using the NMO equation

$$T_x = (T_0^2 + X^2/V_{rms}^2)^{1/2}, \quad (6)$$

one finds that

$$\frac{dT_x}{dX} = X(T_0^2 V_{rms}^4 + X^2 V_{rms}^2)^{-1/2}. \quad (7)$$

In equations (6) and (7), T_x is the two-way event arrival time at shot-to-group offset X, T_0 is the zero-offset two-way time, and V_{rms} is the root-mean-square (rms) velocity to the reflector.

Using the above relationship and information about the

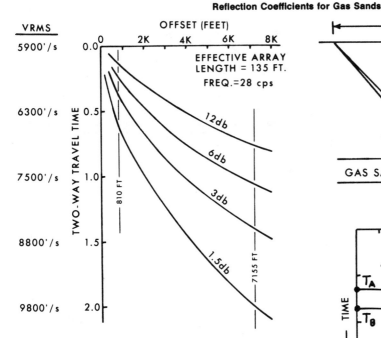

FIG. 19. Plot of array attenuation versus two-way traveltime and shot-to-group offset.

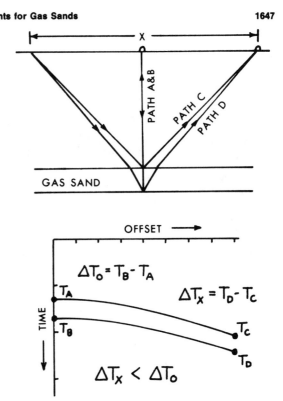

FIG. 20. Effect of differential traveltime on event tuning for thin gas sands.

recording geometry, one can obtain an array attenuation plot similar to the one shown in Figure 19. The recording parameters corresponding to this figure are for an effective shot and group array length of 135 ft. The plot is for a frequency of 28 Hz and for the velocity function shown along the vertical axis. The contours are in decibels (dB) and include the effects of both shot and group arrays. An overlay plot of this type is convenient in analyzing amplitudes on CDP gathers.

Event tuning

Event tuning is caused by differential traveltimes between two or more closely spaced reflections. This effect is illustrated in Figure 20 for a single gas sand with two reflectors, the top and base of the gas sand. The relationship between the two reflection arrival times and offset is shown by the two curves in the lower part of Figure 20. Because NMO naturally decreases with increased record time, the time difference between the top and base sand reflections will decrease with increasing offset, i.e., $\Delta T_x < \Delta T_0$.

As with array attenuation, this effect is best analyzed through the NMO equation given in (6). For small-event separations involved in tuning phenomenon, ΔT_0 is small and the ratio $\Delta T_x/\Delta T_0$ becomes important. As this ratio deviates from one, differential tuning effects occur with changes in offset. Letting $\Delta T_x/\Delta T_0$ approximate the derivative dT_x/dT_0, one has from (6)

$$\frac{dT_x}{dT_0} = (1 + X^2/T_0^2 V_{rms}^2)^{-1/2}. \quad (8)$$

Evaluation of equation (8), using realistic parameters for X, T_0, and V_{rms}, shows the derivative to have a minimum value of about 0.70 for extreme cases. This implies that interval time thicknesses for a thin interval such as a gas sand can decrease from near- to far-offset by about 30 percent. Differential tuning effects will therefore occur for two closely spaced reflections from near- to far-offset. Using the appropriate response curves, one can show that this differential tuning can cause amplitude changes of about 30 percent in the extreme cases. This effect can be either an increase or decrease in amplitude with offset depending upon bed thickness. For thin beds, such as many gas sands less than 50 ft thick, this effect results in a decrease in recorded amplitude with increasing offset.

Noise

Because noise can be strongly offset-dependent, its effect on analysis of offset amplitude can be significant. For example, vibrator data typically have noisy inside traces which after summing might be interpreted as lower reflection amplitude on the short-offset traces. Noise on the far-offset traces might have the opposite effect. This implies that any amplitude balancing which is data-dependent should be done on data which have the highest S/N ratio. Balancing of the noise energy will generally not balance low-energy signal.

Other factors

Spherical spreading gives rise to a very predictable decay in seismic amplitude with time and offset. Newman (1973) presented correction factors which account for this amplitude decay if one has detailed information on subsurface velocities. However, in typical seismic data processing, a simple zero-offset correction is applied which may not correct nonzero offset traces properly. Emergence angles may become large for nonzero offset seismic traces. Nonvertical emergence angles will cause attenuation in seismic amplitude at a vertically responding geophone. One can show from Newman's work (1973) that the emergence angle is directly coupled to spherical spreading. For anomalously low or high surface velocities, these two factors can be quite large with offset. However, they are of opposite sign and somewhat cancel each other.

Reflector curvature and nonplanar wavefronts at shallow depths can have effects on reflection coefficients as discussed by Krail and Brysk (1983). Transmission coefficients at nonvertical incidence angles are coupled with mode conversions and short path multiples and become difficult to analyze in general terms. Instrumentation, processing, and inelastic attenuation are also difficult to analyze in the shot-to-group offset domain. Surface-consistent processing can help reduce many of these distortions and also help with many noise problems.

Amplitude balancing

In all the examples presented here, the seismic data were simply trace-equalized over very long time gates. In using this method, relative changes in amplitude with offset are of principal significance. With a few exceptions, this method appears to work fairly well. Trace-equalizing data with a strong offset-dependent noise component has not been entirely satisfactory. Better techniques involving surface-consistent amplitude adjustment are currently under investigation and will undoubtedly improve our understanding of the phenomenon.

CONCLUSIONS

Two basic conclusions can be drawn from this paper: (1) Poisson's ratio has a strong influence on changes in reflection coefficient as a function of angle of incidence; and (2) analysis of seismic reflection amplitude versus shot-to-group offset can in many cases distinguish between gas-related amplitude anomalies and other types of amplitude anomalies.

The methods of analysis presented here have proven to be useful in many of the world's gas provinces. However, the methods are not foolproof and experience has shown them to fail on occasion. Other factors which affect observed reflection amplitudes versus offset need to be considered. Amplitude balancing during processing is quite important. Additional information on Poisson's ratios for other rock types needs attention as well as the effects of depth of burial and sediment consolidation on Poisson's ratio. Time and the drilling of additional wells will test the full utility of such methods in predicting the presence or absence of gas in the geologic section.

ACKNOWLEDGMENTS

The author wishes to thank Chevron, U.S.A. Inc. for permission to publish this paper. The author is also indebted to J. I. Foster, H. G. Lang, R. A. Seltzer, and D. D. Thompson for their advice, encouragement, and contributions to this project.

REFERENCES

Birch, F., ed, 1942, Handbook of physical constants: Geol. Soc. of Am., Special Paper 36.
Domenico, S. N., 1976, Effect of brine-gas mixture on velocity in an unconsolidated sand reservoir: Geophysics, **41**, 882–894.
—— 1977, Elastic properties of unconsolidated porous sand reservoirs: Geophysics, **42**, 1339–1368.
Gassmann, F., 1951, Elasticity of porous media: Vier. der Natur. Gesellschaft in Zurich, Heft I.
Gregory, A. R., 1976, Fluid saturation effects on dynamic elastic properties of sedimentary rocks: Geophysics, **41**, 895–921.
Hamilton, E. L., 1976, Shear-wave velocity versus depth in marine sediments: a review: Geophysics, **41**, 985–996.
Koefoed, O., 1955, On the effect of Poisson's ratios of rock strata on the reflection coefficients of plane waves: Geophys. Prosp., **3**, 381–387.
—— 1962, Reflection and transmission coefficients for plane longitudinal incident waves: Geophys. Prosp., **10**, 304–351.
Krail, P. M., and Brysk, H., 1983, Reflection of spherical seismic waves in elastic layered media: Geophysics, **48**, 655–664.
Muskat, M., and Meres, M. W., 1940, Reflection and transmission coefficients for plane waves in elastic media: Geophysics, **5**, 149–155.
Newman, P., 1973, Divergence effects in a layered earth: Geophysics, **38**, 481–488.
Podio, A. L., Gregory, A. R., and Gray, K. E., 1968, Dynamic properties of dry and water-saturated Green River Shale under stress: J. Soc. of Petr. Eng., **30**, 552–570.
Tooley, R. D., Spencer, T. W., and Sagoci, H. F., 1965, Reflection and transmission of plane compressional waves: Geophysics, **30**, 552–570.

VERTICAL SEISMIC PROFILING

Vertical Seismic Profiling—A Measurement that Transfers Geology to Geophysics

B. A. Hardage
Phillips Petroleum Company
Bartlesville, Oklahoma

> Vertical seismic profiling (commonly abbreviated to the shorter name VSP) is one of the rapidly developing areas of geophysical technology in the Western hemisphere. The measurement basically involves recording the total upgoing and downgoing seismic wavefields propagating through a stratigraphic section by means of geophones clamped to the wall of a drilled well. In most seismic measurements, both the energy source and the receivers are positioned on the earth's surface. What happens to the seismic wavelet as it propagates from the source to a subsurface reflector and back to the receivers is mostly a matter of inference based on the characteristics of the source and on the properties of the wavefield measured at the surface. Vertical seismic profiling replaces much of this inference with several closely-spaced direct physical measurements of the seismic wavefield in the real earth conditions that exist between the earth's surface and the subsurface reflector. These measurements are proving to be invaluable in structural, stratigraphic, and lithological interpretations of the subsurface and are particularly valuable when combined with surface-recorded seismic data covering a prospective area around a VSP well.
>
> VSP wavefield measurements provide two vital pieces of information needed in seismic stratigraphy: (1) they calibrate seismic signals in terms of the geology that exists at the depths where upgoing reflections are created; and (2) they provide an additional, and usually an improved, image of subsurface geology near a VSP well. Several examples of VSP data will be shown and discussed in order to illustrate how vertical seismic profiling accomplishes these two objectives of seismic calibration and geological imaging.

INTRODUCTION

Vertical seismic profiling (VSP) is a borehole seismic recording that allows the measurement of in situ behavior of seismic wavefields as they propagate through a stratigraphic section penetrated by a well bore. These subsurface geophone measurements provide seismic stratigraphers with information which allows subsurface stratigraphic relationships, lithological conditions, and rock properties to be correlated with the numerical characteristics of reflected seismic wavelets measured at the earth's surface. Consequently, good quality VSP data are now recognized by seismic stratigraphers as being an invaluable link between surface seismic data and subsurface geology. Surface-recorded seismic reflection data constitute the major investigative tool used by seismic stratigraphers. These seismic reflection signals provide images of the subsurface, which dependent on their quality, allow skilled interpreters to map subsurface distributions of stratigraphic sequences, depositional environments, and rock facies. However, the geological interpretations made from these seismic images are limited, since it is impossible to know exactly what frequency changes, amplitude losses, and phase shifts have been imposed on the seismic wavelet as it travels along its source-to-target-to-receiver path. This basic shortcoming of the seismic reflection method is emphasized by the simple raypath diagram in Figure 1.

Modern seismic processing procedures do an admirable job of manipulating surface-recorded reflection data so that the reflection signals exhibit numerical changes in amplitude, waveshape, and frequency content which often correlate quite well with the vertical and areal distributions of subsurface rock properties established by drilling. However, the interpretative value of these surface-recorded data can still be greatly enhanced if the seismic wavefield can be examined in detail as it moves through the earth, rather than being sampled only at the earth's surface. The recording of the seismic wavefield, deep in the earth as it moves through the rock conditions that we wish to infer from surface-recorded seismic responses, and the insights that these measurements provide to interpreters of seismic data, are major reasons for the rising interest in vertical seismic profiling.

Geological-Geophysical Relationships Provided by VSP Data

The seismic wavefield parameters that can be derived from VSP data provide several important relationships between the subsurface geological conditions penetrated by a VSP study well and the numerical attributes of seismic data recorded on the surface near that well. Specifically, VSP wavefield measurements provide two vital insights needed in seismic stratigraphy: (1) they calibrate seismic signals in terms of the geology that exists at the depths where the wavefields are observed; and (2) they provide an additional (and usually an improved) image of subsurface geology near a VSP well. Seismic stratigraphy applications resulting from these two VSP capabilities can include the following.

Geological Calibration of Seismic Data — Since VSP data measure seismic wavelet behavior within the earth, then in concept, any distinctive numerical property of seismic data recorded on the surface near a VSP well can be directly correlated to subsurface geological parameters by means of in situ numerical

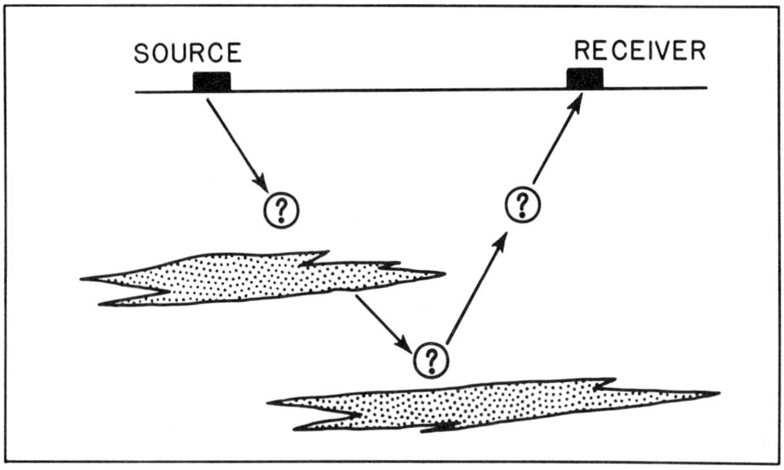

Figure 1. Seismic stratigraphy attempts to determine lithological and stratigraphic relationships in the subsurface from surface-measured seismic images of subsurface conditions. Unfortunately, one does not know precisely what happens to the seismic wavelet on its downward path to a target and on its upward path to a receiver, nor how the wavelet is altered at the reflecting interface. Conceptually, VSP data define these unknown seismic wavelet behaviors.

measurements of the seismic wavefield. These measurements are obtained by placing a geophone above, below, and within each stratigraphic unit that needs to be investigated. Specific examples of the geological calibration of seismic data which might be accomplished with VSP data could be (1) establishing precise depth-to-time conversion functions that allow geological data, measured as a function of depth, to be translated into functions of travel time that can be directly correlated with surface-recorded seismic responses; (2) using the measured amplitude decay of the expanding seismic wavefield to check the accuracy of the amplitude restoration function used when processing surface-recorded seismic data; (3) using seismic wavelet amplitudes, measured above and below subsurface impedance contrasts, to verify the accuracy of reflection coefficients derived from log data; (4) using the measured frequency losses that occur as a seismic wavelet progresses through a stratigraphic section to judge which portions of the seismic frequency band are diagnostic of subsurface conditions; (5) using in situ measurements of upgoing reflection wavelets to specify which surface-measured waveshapes define various stratigraphic relationships penetrated by a VSP well; or (6) establishing robust estimates of deconvolution operators from the strong, downgoing VSP wavefield that provide better vertical resolution of stratigraphic units than do operators determined from the weaker, upgoing wavefields recorded at the surface.

Detailed Imaging of Geological Conditions — In vertical seismic profiling, a seismic receiver is positioned deep in the earth, and thus is much closer to the target that is to be imaged than is a surface receiver. Because the target-to-receiver path is reduced in vertical seismic profiling, Fresnel zones are smaller, and the lateral resolution of the subsurface is improved. This technique of moving a receiver closer to a target in order to "see" it better is followed whenever close inspection of some object that is difficult to resolve is needed. For example, the technique will no doubt be used several times by people reading this book. If some interesting feature of an illustration is small and subtle, a reader will move his eye (the receiver) closer to the page (the target) in order to resolve the feature. The same principle works in seismic imaging. Because VSP recording geometry reduces the length of the seismic travel path, and consequently also reduces the diameter of Fresnel zones, then faults, angular unconformities, pinchouts, and weakly reflecting interfaces near a VSP well should be seen better with VSP data than with surface-recorded data. The lateral extent of this detailed horizontal imaging is limited, however, because the subsurface coverage of reflection points created by typical VSP recording geometries do not extend very far from a well bore.

Fundamental Relationships Between Vertical Seismic Profiling and Seismic Stratigraphy

Some explorationists prefer to divide the broad topic of seismic stratigraphy into two disciplines, depending on whether a seismic stratigraphy effort focuses on an exploration area, where subsurface control is limited, or on a known producing area, where well data are rather abundant. In Figure 2, the term "exploration stratigraphy" is used for the first type of investigation (little or no well data available); whereas, the term "development stratigraphy" is used to refer to the latter (where heavy emphasis is placed on correlating well data with surface seismic data).

In many subsurface interpretations, the type of seismic stratigraphy investigation that is done (exploration or development) is arbitrary and not especially critical for purposes of this discussion. However, this distinction in seismic stratigraphy efforts does help emphasize how vertical seismic profiling allows seismic measurements to be translated into meaningful subsurface geology.

In exploration stratigraphy, for example, there is interest in knowing such things as, "Which seismic reflection events define stratigraphic sequence A and which define sequence B?," or "What rock properties can be associated with reflection waveshape 1 and which with waveshape 2?" The first question can be answered by determining the depth at which each reflection event is created (depth calibration) and by establishing the upgoing reflection waveshapes associated with critical subsurface reflectors (sequence calibration). The second question can be answered by determining which amplitude, phase, and frequency changes in the reflected wavefield correspond to certain stratigraphic and lithological changes (physical properties calibration). The ability of VSP data to "calibrate" seismic data to geological conditions can provide many of these answers, so this attribute of VSP data is

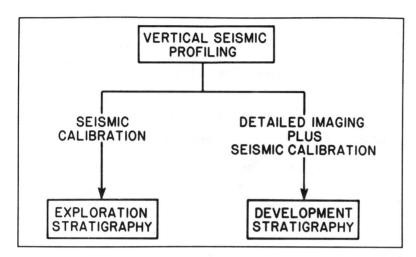

Figure 2. The principal ways by which vertical seismic profiling transfers geological information to seismic measurements are that VSP data allow the effects that subsurface conditions impose on propagating seismic wavelets to be measured (the concept of seismic calibration), and they allow the subsurface around a borehole to be resolved, both horizontally and vertically, with great resolution (the concept of detailed imaging).

emphasized in the left branch of the chart in Figure 2.

In exploration stratigraphy, one may be interested in seeing a specific part of the subsurface near a well imaged with the best possible resolution, but this need is usually secondary to that of creating a regional interpretation of a depositional model for selected stratigraphic intervals. Thus, the imaging capability of vertical seismic profiling is usually not as important in exploration stratigraphy as is the seismic calibration potential offered by VSP data.

On the other hand, when production is found, attention then focuses on improving the subsurface resolution of seismic data so that reserves can be estimated, and production wells can be sited. At this point, the second benefit of VSP data — its ability to provide detailed seismic images — plays a dominant role in transferring geology to seismology. Consequently, as seen in Figure 2, the full potential of VSP data, which includes both its detailed imaging as well as its wavefield calibration, is needed in development stratigraphy.

Classifying the end results of vertical seismic profiling into the two possibilities of seismic calibration and detailed imaging, and segregating seismic stratigraphy into two areas of emphasis, either exploration stratigraphy or development stratigraphy, leads to the relationships between seismic stratigraphy and vertical seismic profiling shown in Figure 2. The following material is organized according to the two conceptual approaches shown in this figure. The way by which VSP data transfer geological meaning to surface seismic measurements will be assumed to be either: (1) VSP data provide precise calibrations between seismic parameters and geological conditions, or, (2) VSP data provide improved seismic images of subsurface geology. Some of the specific topics that should be considered when evaluating the role of vertical seismic profiling in seismic stratigraphy are listed in Figure 3. Most of these VSP applications are described in the following sections. All of them are discussed in one or more of the references listed at the end of this paper.

The VSP Measurement Procedure

Vertical seismic profiling is performed by activating a seismic energy source on, or near, the earth's surface and recording the seismic response with receivers that are positioned at closely spaced depth levels in a drilled well. A typical geometrical arrangement of source, subsurface targets, and a single borehole receiver for a marine VSP experiment is shown in Figure 4. The recording geometry involved in an onshore VSP experiment would be essentially the same. However, a wider variety of energy sources, including shear wave sources, can be used in onshore VSP work. In onshore vertical seismic profiling, it is sometimes possible to record VSP data in a cased well after a drill rig has moved away. In marine VSP work, a drill rig (or a production platform) will always be present at the well site during a VSP experiment.

The term "wall-locked," is used in Figure 4 to describe the borehole geophone used to record VSP data because the geophone must be rigidly coupled to the borehole wall in order to record data suitable for seismic stratigraphy applications. A free-hanging receiver is adversely influenced by fluid-born tube waves, which may camouflage the seismic body wave signals that need to be analyzed. Perhaps more importantly, a geophone that is not coupled to the borehole wall cannot adequately detect subtle variations in the seismic wavelets moving in the surrounding formation. These small alterations in wavelet character must be known if detailed geological information is to be extracted from the seismic signals traveling upward through a stratigraphic section toward the earth's surface.

Presently, there are no borehole geophone systems available outside the Soviet Union which record data at more than one depth level simultaneously. Consequently, only one geophone is shown in the borehole in Figure 4. If data are recorded at only one depth, then the seismic source must be activated many times, and the geophone continually moved to new depth levels in order to acquire an appropriately sampled wavefield throughout the stratigraphic interval to be studied. A mobile energy source is shown in this situation, although a fixed source can often satisfactorily illuminate those parts of the subsurface that need to be studied.

It is important to note that both downgoing and upgoing seismic wavefields are shown arriving at the geophone position. Downgoing seismic wavelets cannot be analyzed as events that are distinct from upgoing wavelets in the responses measured by receivers located at the earth's surface, which is where receivers are located in order to record the common-reflection-point data that are so widely used in seismic stratigraphy analyses. Some of the unique geological and stratigraphic applications that become

EXPLORATION STRATIGRAPHY (CALIBRATION)
- **Establish depth-to-time conversion**
- **Identify seismic reflectors**
- **Adjust reflection coefficients**
- **Provide alternative to synthetic seismogram**
- **Define reflector dip**
- **Provide parameters for processing seismic data**
  Amplitude decay functions
  Deconvolution operators

DEVELOPMENT STRATIGRAPHY (IMAGING)
- **Locate faults**
- **Reduce Fresnel zone size**
- **Create high resolution horizontal stacks**
- **Provide detailed velocity behavior**
- **Create high signal-to-noise stacks in deviated wells**

Figure 3. Seismic stratigraphy objectives provided by VSP data.

available because the downgoing seismic wavefield is preserved in VSP data will be emphasized later.

The field geometry in Figure 4 is similar to that used when recording velocity check-shot data. However, the objectives of vertical seismic profiling are different from those of velocity surveying, specifically:

1. VSP data are recorded at vertically spaced depth intervals that are much smaller than those used when recording velocity check shot data. Successive VSP geophone stations should be no farther apart vertically than $\lambda/2$, where λ is the shortest spatial wavelength contained in a seismic wavelet being analyzed. If the geophones are farther apart than $\lambda/2$, then spatial aliasing exists in the data, and some of the numerical analyses of VSP data that need to be performed in seismic stratigraphy work cannot be properly done. Velocity filtering is an example of one important numerical process adversely affected by data which are improperly sampled in depth.

2. The only information that needs to be obtained in a velocity survey is the time-depth coordinates of the first break wavelets. However, in vertical seismic profiling, the weak downgoing and upgoing events that follow the first break wavelets are just as important as are the first breaks. Thus, the dynamic range, gain accuracy, and noise rejection properties of the downhole geophone, the logging cable, and the surface recording system must be superior to those which often exist in a velocity survey. Otherwise, these important weak amplitude events are not properly resolved.

VSP Field Procedures and Data Processing

The value of vertical seismic profiling in seismic stratigraphy investigations is strongly affected by the quality of the VSP data used. Very few VSP analyses capable of satisfying seismic stratigraphy requirements can be performed unless the VSP data exhibit recognizable reflection signals and contain minimal noise contamination. Consequently, the field procedures used to record VSP data must be carefully planned and executed so that a proper energy source and an adequate recording system are on-site, and so that the recording geometry correctly images the desired target. In addition, steps must be taken to insure that the downhole geophone is firmly coupled to the formation, and that all possible noise sources are eliminated during the data acquisition. Proper data processing techniques must then be used to analyze the data, or false geology may be introduced simply as an artifact of the processing software.

VSP field procedures and VSP data processing techniques are both subjects worthy of detailed discussion. However, neither topic will be considered here, since the intent is to focus attention on the benefits that VSP data offer to seismic stratigraphers after proper recording and processing; and not how VSP data can be "properly recorded" nor what constitutes "properly processed" VSP data. Common VSP noise sources that should be avoided and recommended field procedures that

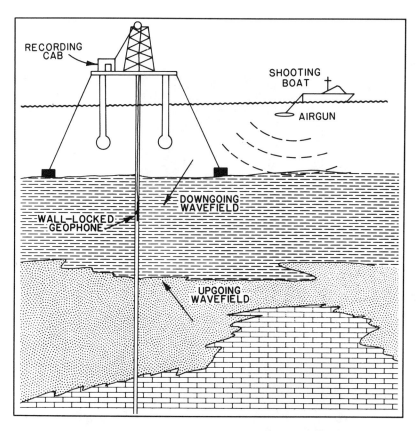

Figure 4. Equipment and field geometry involved in marine vertical seismic profiling.

can be employed in order to enhance VSP data quality are discussed by Balch et al (1982), Kennett and Ireson (1982), Hardage (1983), and Balch and Lee (1984). VSP data processing techniques are described by DiSiena, Byun, and Fix (1980), Gaiser and DiSiena (1982), Hardage (1983), Balch and Lee (1984), Kennett and Ireson (1971, 1982), Lee and Balch (1983), and Parrott (1980).

Vertical seismic profiling has been an active area of geophysical investigation in the Soviet Union for many years, and several novel field procedures and data processing techniques have been developed by Soviet geophysicists. An overview of the Soviet perspective of these topics, but which does not include marine VSP analyses, has been published by Gal'perin (1974, 1984) and Karus et al (1975).

The Basis of Most VSP Seismic Stratigraphy Applications - Dividing VSP Data into Downgoing and Upgoing Wavefields

The VSP recording geometry in Figure 4 indicates that both downgoing and upgoing seismic wavefields arrive at a subsurface borehole geophone position. This important fact forms the basis of most of the calibrating and imaging applications of VSP data that are used in seismic stratigraphy.

Consider the onshore VSP data shown at the top of Figure 5. These data represent only the vertical component of particle motion at each subsurface recording point. The vertical distance between recording levels is 50 ft (15.2 m). The energy source was a compressional vibrator which stayed at a fixed position 1000 ft (304.9 m) from the wellhead as the geophone was raised up the vertical borehole. In order to see subtle features of the data, a dynamically changing numerical gain function has been applied which makes all events have the same amplitude. Thus, no interpretations of seismic events in terms of their amplitude behavior should be made from this display. A sonic log recorded in the VSP well is shown at the left in order to demonstrate some of the impedance changes that create the upgoing reflection events.

The principal concept illustrated in Figure 5 is that the downgoing and upgoing seismic wavefields that are generated in this stratigraphic section by a surface energy source can be separated from each other by applying appropriately designed velocity filters to the original VSP data. These two segregated wavefields are shown at the bottom of the illustration.

The downgoing wavefield consists of a high amplitude first arrival followed by many downgoing multiples, which are generated at various depths in the stratigraphic section. In this instance, the majority of the downgoing multiples are created in the shallowest part of the earth above the topmost VSP trace. These multiples have reduced amplitudes, compared to the amplitude of the first arrival, but the gain function applied to the data camouflages this amplitude difference.

The upgoing wavefield consists of primary reflections, which by definition are events created only by the downgoing first arrival wavelet, together with numerous upgoing multiples, which by definition are created by downgoing multiples that follow the downgoing first arrival.

Now that the VSP data are segregated into these two wavefields, some of the

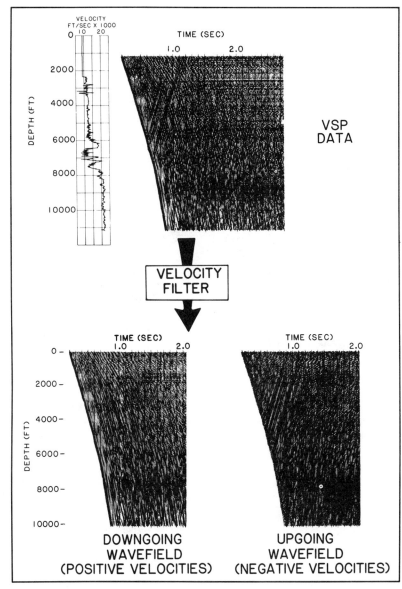

Figure 5. VSP data contain both downgoing and upgoing seismic wavefields. Applying a velocity filter which preserves events traveling in the $+Z$ direction and attenuates events traveling in the $-Z$ direction yields the downgoing wavefield. Applying a velocity filter which preserves events traveling in the $-Z$ direction and attenuates events traveling in the $+Z$ direction yields the upgoing wavefield. *(Courtesy K. D. Wyatt, Phillips Petroleum Company).*

applications listed in Figure 3 become more obvious. For instance, the depths of subsurface reflecting interfaces can be identified by extrapolating a selected phase point (peak, trough, zero-crossing) of each upgoing primary reflection downward until it intersects that same phase point on the downgoing first arrival. This procedure also identifies the one-way travel times to these reflectors, so that we know the two-way travel times at which the corresponding reflection events should arrive at the surface. An upgoing VSP wavefield is thus an alternative to the synthetic seismogram for defining the arrival times and waveshapes of surface-recorded reflections. An additional application results from studying the numerical amplitude decay of the downgoing first arrivals. This amplitude decrease confirms whether or not satisfactory approximations are being used to convert surface-recorded data near the VSP well into correct relative-amplitude form. Examples of these applications will be shown in the following sections.

A particularly critical insight into the advantages of VSP data in seismic stratigraphy analyses is the realization that the same multiple pattern that exists in the upgoing wavefield also exists in the downgoing wavefield, thus, deconvolution operators used to attenuate multiples can be determined from either wavefield. This option does not exist when processing surface-recorded reflection data.

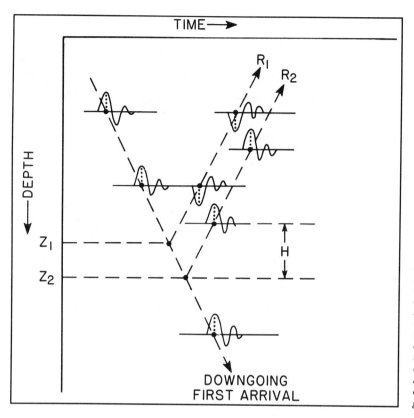

Figure 6. The location of a subsurface reflector separating two thick beds can be determined by extrapolating the trend of a selected phase point on an upgoing reflection wavelet downward until that trend coincides in space and time to the trend of its equivalent phase point on a downgoing first arrival wavelet. In some thin bed sequences, the phase properties of the upgoing reflected wavelet may be altered so that the reflection waveshape differs from the waveshape of the downgoing incident wavelet.

Identifying Seismic Reflectors

If VSP data are to be considered as an alternative to a synthetic seismogram, then the data must reliably identify the depths of all critical seismic reflectors penetrated by a study well. This requirement can be satisfied only if the VSP data have adequate signal-to-noise character, which they usually do if a properly sized energy source is used and if the downhole geophone is firmly coupled to the formation at all recording depths. In fact, the identification of seismic reflectors is one of the fundamental "calibration" properties of vertical seismic profiling that allows subsurface stratigraphy to be tied to surface-recorded seismic data.

One technique by which reflector depths can be identified with VSP data is illustrated in Figure 6. A prominent, but arbitrary, phase point on the downgoing first arrival wavelet is selected as a timing reference. In this example, only a few downgoing wavelets are drawn, and the apex of the first peak is chosen as the phase point along which propagation paths will be measured. The time-depth stepout trend of this reference point is shown by the down-to-the-right dashed line.

The same phase point must then be interpreted on all upgoing reflection events. Two such events, R_1 and R_2, are shown. Since the R_1 wavelets are inverted relative to the downgoing wavelet, this event must be created at an interface for which the reflection coefficient for a downgoing particle velocity pulse is negative. Event R_2 is generated at an interface where the reflection coefficient is positive. Both reflecting interfaces, Z_1 and Z_2, are assumed to be bounded by thick beds, thus the reflected waveshape is identical to the incident waveshape. In thin bed sequences, the upgoing reflected waveshape may be different from that of the downgoing wavelet, and in such cases, it may be more difficult to identify equivalent phase points on the downgoing and upgoing wavelets.

The trajectories of the equivalent phase points of R_1 and R_2 are shown by the up-to-the-right dashed lines. Extrapolating these lines downward until they intersect the down-to-the-right trajectory of the downgoing reference phase point identifies the two depths, Z_1 and Z_2, where the downgoing and upgoing phase values become coincident in time and space. These depths are, by definition, the positions at which these upgoing reflection events are created.

An upgoing wavelet cannot be recognized as an event distinct from the high amplitude downgoing first arrival until it has propagated upward a distance H above the interface where it is generated as shown in Figure 6. For propagation distances less than H, the weak upgoing wavelet is difficult to recognize because it is still convolved within the stronger downgoing first arrival and the short-period multiples that follow the first arrival. The distance H is a function of both the propagation velocity at the reflector depth and the time extent of the wavelet. Typically, H is 300 to 600 ft (91.5 to 183 m) or more. Thus, upgoing phase points must be projected over a sizeable distance in which they cannot be interpreted before they intersect the downgoing trend of the reference phase.

For simplicity, the travel paths in this hypothetical example are shown as straight lines. Ordinarily, the propagation velocity increases with depth so that the

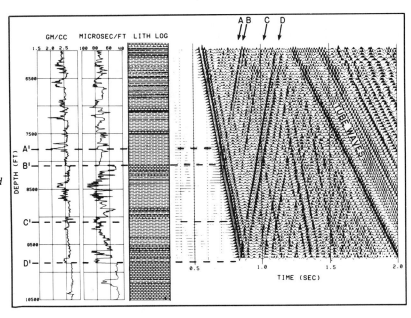

Figure 7. An example of the reliability with which VSP data can often identify primary seismic reflectors. Four upgoing primary reflections are shown by the lineup of black peaks labeled A, B, C, D. The subsurface depth of the interface(s) that generated each reflection can be defined) by extrapolating the apexes of the black peaks downward until they intersect the trend followed by the first trough of the downgoing compressional first arrival. These depths are labeled A', B', C', D'. These are raw field data. No processing has been done other than a numerical AGC function has been applied to equalize all amplitudes (from Hardage, 1983).

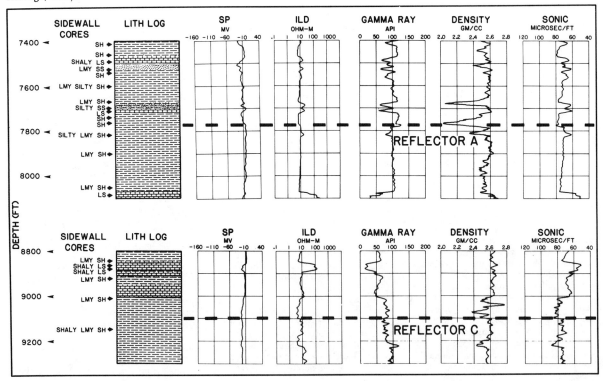

Figure 8. Subsurface data describing the physical rock properties creating seismic reflections A and C in Figure 7 (from Hardage, 1983).

time-depth stepout trends are curved rather than linear.

An example of VSP data recorded in a well where the stratigraphic and lithological conditions that create seismic reflections can be identified from a variety of borehole information is shown in Figure 7. This borehole is vertical, and vertically oriented, velocity-sensitive geophones were used in the downhole geophone tool. The lithological control in this VSP well consists of continuous mud log cutting samples, 160 sidewall cores, and a comprehensive set of commercial well logs. Therefore, the lithological description is about as complete as one can achieve from borehole data.

The basic seismic wavelet ingrained within the lengthy downgoing VSP first arrival is a short, symmetrical, zero-phase wavelet generated by a compressional vibrator. At any depth, the one-way travel time for the direct arrival is the time position of the apex of the first trough of the first arrival wavelet measured at that depth. Each upgoing reflection event should also be a symmetrical zero-phase wavelet with its travel time being the time coordinate of the apex of the central leg of the reflection wavelet. Using these wavelet criteria for measuring timing relationships, four reflection events, labeled A, B, C, and D, are traced downward to the depths A', B', C', and D' at which they originate. The depth A' is defined as the subsurface position where the projected trend of the black peaks, A, intersects the apex of the first trough of the downgoing first arrival. Depths B', C', and D' are defined by similar projections of the black peaks B, C, and D. The sonic and density logs are shown to verify that acoustic impedance contrasts do occur at depths A', B', C', and D', and that these contrasts can be defined in terms of negative reflection coefficients for a downgoing particle-velocity pulse.

A key observation to make from the data in Figure 7 is that the lithological log, which was made from continuous cutting samples and a drilling rate log, shows that events B and D mark lithostratigraphic boundaries between shale and limestone. However, events A and C appear to originate within thick shale units. If this lithological description of the subsurface is accurate, then reflections A and C would support the concept proposed by Vail, Todd, and Sangree (1977), and others, that seismic reflections are not necessarily lithostratigraphic markers.

Close examination of additional lithological control in this well shows that silty and lime-like components appear in the shale units at depths A' and C'. A detailed illustration of all the lithologically sensitive subsurface information recorded in the well at these reflector depths is shown in Figure 8. A sidewall core was obtained at each depth marked with a solid arrowhead. The lithology of each core sample is written beside the sampled depth. The depths A' and C', identified from the VSP data as the depths at which reflections A and C originate, are shown by the two dashed lines. These subsurface data imply that some seismic reflections in the vicinity of this well should be mapped as interfaces that are better characterized by slight changes in sediment grain size, or variations in cementation or compaction, than as boundaries where there is a change in basic lithology, since for both reflectors A and C, the basic lithology is shale.

In stratigraphic trap exploration, it is essential to know in this type of detail exactly what type of stratigraphic and lithological interpretations can be assigned to key seismic reflection events. Mapping a reflection as the boundary between a limestone and a shale, when actually it marks only a slight difference in grain size or a calcareous interval in a shale, will lead to numerous dry holes. It should be noted that although the slight change in grain size at depth A' is a mild alteration in lithology, the log data show that it is a facies change that is marked by significant changes in velocity and density, and thus is a strong seismic reflector. Because of experiments of this nature, many seismic stratigraphers have realized that a properly executed vertical seismic profile can accurately identify the depths and surface arrival times of critical seismic reflectors penetrated by a VSP well, and is therefore an invaluable interpretive aid that defines the geological significance of surface-recorded reflection events.

Calibrating Reflection Coefficients

The magnitude, polarity, and spatial separation of reflection coefficients occurring at sedimentary interfaces in a stratigraphic section establish the nature of the upgoing seismic wavefield measured at the earth's surface. A seismic stratigrapher needs to determine these three reflection coefficient attributes in order to reconstruct the impedance sequence occurring within a geological section and thereby understand the geological messages contained in the seismic response.

Forward and inverse seismic modeling are the principal interpretative procedures now used to construct the reflection coefficient sequence of the subsurface from surface-recorded seismic data. In forward modeling, the geometrical shapes and impedance values of earth layers are mathematically defined, and synthetic seismic responses are then calculated and compared to real seismic data to judge if the assumed earth model is sufficiently accurate. In inverse modeling, reflection coefficients and vertical impedance sequences are calculated from seismic trace data which are processed so as to preserve relative reflection amplitudes.

Both of these modeling processes concentrate on establishing relationships between the amplitude, phase, and frequency properties of seismic responses and the magnitude, polarity, and spacing of subsurface reflection coefficients. Once reflection coefficients are known, impedance sequences (that is, subsurface stratigraphy) can be inferred.

Sonic and density logs are the traditional data used to calculate reflection coefficients for forward modeling and to check the impedance estimates established by inverse modeling. VSP data can also be used for these same purposes (Balch et al, 1980; 1982), and in some instances, VSP data provide more accurate impedance and reflection coefficient values than do log data because logging tools measure rock properties in regions much smaller than the first order Fresnel zones associated with seismic measurements.

A comparison of log-derived and VSP-derived reflection coefficients is shown in Figure 9. A sonic log and two density estimates, a compensated density log (CDL) and borehole gravity meter responses, were recorded in this study well in addition to VSP data. A prominent shale-limestone interface occurs at a depth of 8,075 ft (2,461.9 m). The top arrowhead (labeled 1) at 7,950 ft (2,423.8 m) on the log curves is in the shale section, and the bottom arrowhead (labeled 2) is in the limestone unit. Two values of the shale-limestone reflection coefficient for a downgoing particle velocity pulse are calculated, using the log data at these two depths, and shown at the right.

VSP data were also recorded at depths 1 and 2, so an additional estimate of the shale-limestone reflection coefficient must be calculated from the expression

$$R = \frac{B}{A} - 1 \qquad (1)$$

where B is a measure of the amplitude of the first arrival wavelet that is transmitted through the shale-limestone interface, and A is a measure of the amplitude of the incident first arrival wavelet just above the shale-limestone boundary. In this example, A and B are arbitrarily chosen to be the amplitude of the first trough, but other wavelet amplitude measures could be used. The VSP estimate of the reflec-

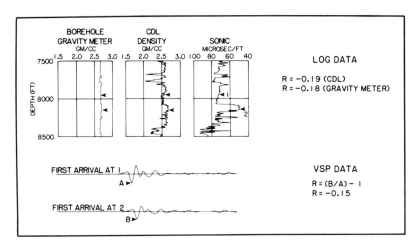

Figure 9. The arrowheads located beside each log curve indicate two positions, "1" and "2", above and below a prominent shale-limestone interface. The magnitude of the reflection coefficient, R, associated with this interface is traditionally estimated from log data. VSP data offer an alternative way to estimate reflection coefficients and calibrate reflection amplitudes. The VSP estimate is often more reliable since it results from a real geophone's response to an actual seismic wavefront. The value of R here is negative since the VSP geophone reacts to particle velocity, not to pressure (from Hardage, 1983).

Figure 10. Effect of acoustic impedance on seismic wavelet amplitude.

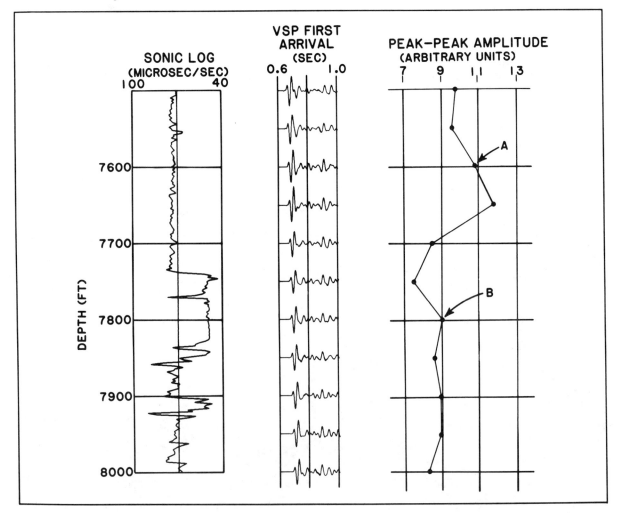

tion coefficient is shown at the lower right of Figure 9, and it differs from the log-derived estimates. Which value is the more accurate? Each data case must be considered on an individual basis in order to answer this question. Usually, log-derived estimates of seismic reflection coefficients will be reasonably accurate, but properly recorded VSP data, in which the borehole geophone is firmly bonded to the formation, should yield reflection coefficient values more like those experienced by the surface-recorded seismic wavefields that seismic stratigraphers use in their interpretations.

A second comparison between the subsurface impedance sequences defined by a sonic log and those defined by VSP data is shown in Figure 10. This example demonstrates some of the cautions that need to be exercised when using seismic wavelet amplitudes to estimate reflection coefficients. A peak-to-peak amplitude measure of the first arrival wavelet is plotted on the right. The objective is to determine the reflection coefficient associated with the impedance change shown at a depth of 7,740 ft (2,359.8 m), so that relative-amplitude surface data crossing this well can be properly calibrated to reveal impedance changes occurring along this interface as it is followed laterally away from the well.

No density log is available in this interval, so a log-derived reflection coefficient must be calculated by using only the sonic log velocities. In the interval from 7,600 to 7,700 ft (2,317.1 to 2,347.6 m), the VSP wavelet amplitudes vary in a way that does not correlate with the almost constant impedance defined by the sonic log. One logical explanation would be that some of these wavelet variations are caused by formation density changes since a density log was not recorded. However, they are likely due to inconsistent geophone couplings at these recording depths, which is a major caution that must be kept in mind when interpreting the geological significance of VSP amplitudes. Interestingly, the sonic log values at 7,600 and 7,800 ft (14,000 ft/sec and 20,000 ft/sec) and the VSP wavelet amplitudes at these same depths (10.8 and 8.9 units) yield the same reflection coefficient values for a downgoing particle velocity pulse, since:

$$R = \frac{V_1 - V_2}{V_1 + V_2} = \frac{14000 - 20000}{14000 + 20000} \quad (2)$$

$$= -0.176$$

and

$$R = \frac{B}{A} - 1 = \frac{8.9}{10.8} - 1 = -0.176 \quad (3)$$

However, the main point of emphasis is that when VSP data are used to calculate reflection coefficients, then some caution must be used to ensure that the wavelet amplitudes are not affected by noise, and that the borehole geophone is equally coupled to the formation at those depths where the wavelet analyses are made.

In Equations 1, 2 and 3, it is assumed that the quantities A and B describe the amplitudes of uncontaminated incident and transmitted wavelets, respectively. Depending on the relative separations between the depth where each wavelet is recorded and the depths to nearby reflecting interfaces, one or more upgoing reflection wavelet can be convolved within each of these first arrival wavelets. Thus, the amplitude estimates, A and B, are usually influenced to some degree by reflection wavelets, and are not pure measures of incident and transmitted seismic waves. In many instances, however, the numerical technique specified by Equation 1 is quite satisfactory for calibrating reflection coefficients.

VSP — An Alternative to Synthetic Seismograms

A synthetic seismogram is the traditional tool used to establish which geological features in the subsurface can be detected by seismic reflection wavelets. Scarcely any extensive seismic stratigraphy effort is completed without converting sonic and density log data in key study wells into reflection coefficient series so that synthetic seismic reflection responses can be calculated from them.

Synthetic seismograms will, and should, remain a cornerstone of seismic stratigraphy because of their great value in transferring geology to seismic responses. However, there are instances where a synthetic seismogram cannot be created which satisfactorily agrees with surface-recorded data. In these situations, there is uncertainty as to which data, the synthetic calculation or the surface-recorded data, are the better description of the subsurface geology near a study well.

There are numerous reasons why sonic and density log data may not be accurate measures of the acoustic impedance sequence experienced by a propagating seismic wavefront. Goetz, Dupal, and Bowler (1979) published an extensive analysis of the physical factors that cause sonic log velocities to differ from seismic propagation velocities, and all of their proposed mechanisms can cause synthetic seismograms to differ from real seismic responses. Most of the differences between log-derived and seismically-derived acoustic impedance properties of a layered rock system result from the fact that well-logging tools investigate a much smaller volume of the earth than does a seismic wavefront. In some stratigraphic sections, the rock properties around a borehole, even if accurately measured by logging tools, do not extend very far laterally away from the borehole.

An example of this type of stratigraphic situation is illustrated in Figure 11, where a well is shown penetrating a small sand body that is encased in a massive shale section. An important feature of this sand unit is its limited lateral extent. The horizontal dimension of the sand is shown to be less than a first order Fresnel zone, which depending on the depth of the sand, may have a diameter of several hundred feet. Because the sand is smaller than a Fresnel zone, the sand unit is a point diffractor, not a seismic reflector. Consequently, the sand cannot be imaged as a reflection event by surface-recorded seismic data. However, properly recorded sonic- and density-log data will indicate a change in acoustic impedance at the lithological interfaces, Z_1 and Z_2. A synthetic seismogram constructed from these log data will thus create a seismic reflection at these two sand-shale interfaces. As a result, we have the common situation where a synthetic seismogram and surface-recorded reflection data disagree as to what type of geology is being imaged, even though both the seismic data and the well-log data are accurate measures of the acoustic impedance in the respective volumes of the earth that they investigate.

In any study well, but particularly in those wells where synthetically calculated reflections do not agree with surface-measured reflections, vertical seismic profiling is a viable alternative for defining upgoing primary reflection events.

A comparison between a VSP's definition of upgoing primary reflections and a synthetic seismogram's definition of these same events is shown in Figure 12. The upgoing VSP wavefield is shown in data panel A after considerable processing has been done in order to eliminate all events except primary reflections moving upward with an apparent velocity between 0 and −4 msec per trace spacing. (An apparent velocity of 0 msec per trace spacing in this display would imply that a reflection event was generated at a flat, horizontal interface.)

The signal character of the upgoing VSP events can be further enhanced by summing the VSP traces in panel A into a single trace. This summed trace is shown repeated five times in panel C. The surface-recorded seismic cross section is

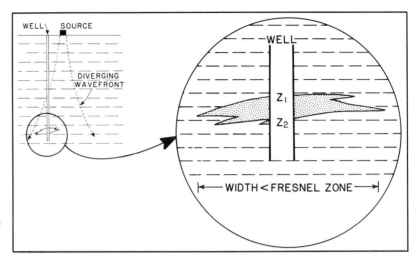

Figure 11. An example of impedance interfaces (at depths Z_1 and Z_2) at which a synthetic seismogram defines reflection events, but which generate no real seismic reflections since their lateral extent is less than the width of the first order Fresnel zone.

split at the well location, and this VSP's version of upgoing events together with the synthetic seismogram (panel B) made from the sonic log recorded in the well are inserted. The surface-recorded data are processed with state-of-the-art deconvolution and migration procedures and are displayed in a relative-amplitude format.

A critical stratigraphic interval in this prospect area is the section occurring between 2.1 and 2.2 sec. There is a rather serious mistie between the VSP data and the synthetic seismogram in this time interval, which is labeled as "1" in the display. In this instance, the VSP response is the better match with the surface-recorded data. The strong black peak in the synthetic trace at 2.15 sec is caused by a calcareous sand interval labeled "SD" on the sonic log above panel A. Evidently, this calcareous sand unit does not extend laterally very far from the wellbore, since neither the surface-recorded data nor the VSP data indicate it to be a reflector.

There are other features of the VSP data which collectively provide much more information about the geology around the borehole than can be obtained from the synthetic seismogram. For instance, an upgoing VSP event exhibits some amount of curvature if the reflector that generates the event has any dip (Hardage, 1983, Chapter 6). This reflection curvature will increase as the dip increases. Some reflection curvatures that extend upward for 5 or 6 trace spacings from their points of origin are labeled "2" in panel A. Numerical VSP modeling shows that these reflectors must have dips that vary from 5 to 10° in order to generate the curved VSP events shown between 2.1 and 2.4 sec.

This type of dip information cannot be obtained from a synthetic seismogram. Evidence of reflector dip in this interval can be seen in the surface-recorded data.

If the VSP recording geometry images the subsurface for some distance laterally away from a well, then faulted reflectors occurring within the imaged region are often revealed by VSP events that terminate, or abruptly change character, as they are followed upward from their point of origin. One such behavior is shown as feature "3". The interpreted fault is not shown in the surface-recorded data at this depth because a larger data window of the cross section must be examined in order to see its orientation.

This fault interpretation is another geological insight that cannot be provided by a synthetic seismogram. It is important also to note that the VSP's indication of a fault occurs at a recording time which is below the total depth of the well. Naturally, a synthetic seismogram cannot predict geological conditions below the bottom of a well since log data do not exist there. The ability to see below the bottom of a wellbore is another advantage of VSP data over a synthetic seismogram. Thus, late arriving primary reflections, such as those labeled "4", can be retrieved from the VSP data, but they are impossible to construct from these well-log data, since the logged interval ends at 2.41 sec.

Calibrating Reflection Amplitudes

Creating seismic cross sections that have accurate reflection amplitudes is a critical data processing requirement in seismic stratigraphy analyses. Several important stratigraphic and petrophysical relationships, such as lateral and vertical changes in rock type, amount of rock porosity, and type of pore fluid, can be inferred from relative-amplitude seismic cross sections. Relative-amplitude stacked data can also be used to construct color displays of seismic impedance estimates and pseudo sonic logs, which further help to identify potential reservoir facies, sealing units, and fluid contacts.

One of two types of gain functions can be applied to surface-recorded data in order to remove the amplitude decay caused by spherical divergence and other amplitude loss mechanisms. A data processor can calculate an amplitude-sensitive statistic, such as rms energy in successive time windows of surface-recorded data, determine the rate of amplitude decay of the upgoing wavefield arriving at the surface, and then design an analytical function that restores some, or all, of this amplitude decay in the field records. A second approach is to apply a V^2T gain function to each field record, where V is the rms velocity of the seismic wavefield being processed, and T is recording time. These V^2T functions restore the amplitude reduction caused by the spherical divergence of the seismic wavefield in a homogeneous, layered earth. Either of these gain functions creates valuable relative-amplitude cross sections if the seismic data have adequate signal-to-noise properties, and if there are no large variations in the receiver coupling or in the energy input from shot to shot.

One geology-to-seismic calibration property provided by VSP data which can be valuable in seismic data processing is the direct measurement of the amplitude decay of seismic wavefields within the

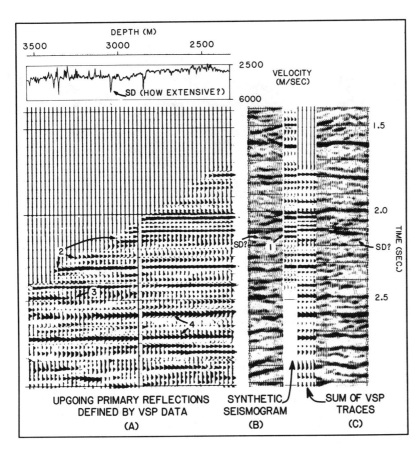

Figure 12. A comparison of VSP data and a synthetic seismogram with surface-recorded seismic data. The numbered features show some advantages of the VSP data over the synthetic data, which in this instance are: (1) the synthetic seismogram does not agree with the surface-recorded data in a particularly critical stratigraphic interval; (2) the curvature of the upgoing VSP reflections between 2.1 and 2.4 sec implies reflector dips in the range of 5 to 10°; (3) the termination of VSP reflections at 2.5 sec implies a fault is near the borehole at this depth; and (4) the VSP defines primary reflections below the total depth of the well.

earth. An example of such a measurement is shown in Figure 13. In this experiment, a compressional vibrator was stationed 685 ft (208.8 m) from the wellhead, and the internal earth response to this energy source was recorded at intervals of 50 ft (15.2 m) in the vertical, cased well bore. The rms amplitude of the first arrival wavelet is plotted as a function of depth in the bottom chart, together with a V^2T function calculated by measuring the first break time of the direct arrival wavelet every 500 ft (152.4 m) through the drilled stratigraphic section. The general shape of the V^2T curve is similar to that of the measured amplitude decay, but its magnitude is smaller. The measured amplitude decay is larger than what the V^2T function suggests because the measured value includes amplitude losses other than spherical divergence, such as dispersion, S-wave conversions, and scattering.

One can establish the accuracy of amplitude processing near a VSP well by comparing the measured amplitude decay of vertically traveling VSP wavelets recorded in the borehole with the analytical functions used to create relative-amplitude versions of the surface-recorded data crossing the well. Some caution must be used when interpreting the measured amplitude behavior of the VSP wavelets, since erroneously high amplitudes occur whenever a borehole geophone is not firmly coupled to the formation. Some anomously high amplitudes caused by poor geophone coupling are labeled A, B, C, and D in this figure.

Removing Multiple Images of Subsurface Interfaces

Surface-recorded seismic data always contain some amount of multiply reflected energy in addition to primary reflection signals. These multiple events overlay later-arriving primary reflections and prevent an optimum interpretation of the geological information contained in the seismic images of the deeper parts of the subsurface. Ideally, a stratigrapher needs a seismic cross section that contains only one image of each reflecting interface.

One of the significant advances made in recent years in seismic data processing has been the development of numerical procedures that deconvolve, or unwrap, multiple reflections from seismic data and leave only primary reflection events. Subsurface seismic imaging in most prospect areas, and particularly in marine environments, is greatly improved as a result of determining these deconvolution operators from the upgoing wavefields recorded by surface-positioned geophones and hydrophones, and then applying these operators to the surface-recorded data in order to attenuate multiples.

Vertical seismic profiling offers an alternative approach by which these deconvolution operators can be designed, and VSP-derived operators often improve the image quality of surface-recorded data beyond that achieved with operators determined from the surface-recorded data itself. The VSP approach determines deconvolution operators from the downgoing seismic wavefield; whereas, the traditional approach in CDP data process-

Figure 13. Comparison between the measured amplitude decrease of VSP first arrival wavelets and the decrease predicted by a (V^2T) function determined from the VSP first break times. The amplitude increases labeled A, B, C, D result because the geophone is not properly coupled to the formation (modified from Hardage, 1983).

Figure 14. The time delays T_1' and T_2' by which upgoing multiples M_1' and M_2' follow each primary reflection, P, are the same as the time delays T_1 and T_2 by which the downgoing multiples M_1 and M_2 follow the first arrival, FA. Thus deconvolution operators that attenuate M_1 and M_2 will also attenuate M_1' and M_2'.

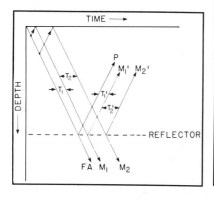

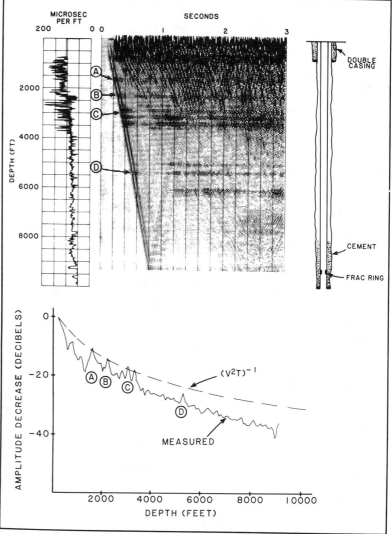

ing is to calculate these operators from the upgoing wavefield measured by surface-positioned receivers (Anstey, 1980; Hubbard, 1979; Kennett and Ireson, 1981).

The concept is based on the relationships between downgoing and upgoing seismic events illustrated in Figure 14. Only one downgoing event can generate primary reflections; that being the downgoing first arrival which is labeled FA. All subsequent downgoing events, such as M_1 and M_2 are multiples which create upgoing multiples M_1' and M_2'. Based on this fact, the following observations should be noted:

(1) The time delays by which multiples M_1 and M_2 follow the first arrival FA are the same as the time delays by which upgoing multiples M_1' and M_2' follow the primary reflection, P, created by FA.

(2) The amplitudes of M_1' and M_2' relative to each other and to the amplitude of P are the same as the amplitudes that M_1 and M_2 have relative to each other and to FA.

(3) The amplitudes of events FA, M_1, and M_2 are considerably larger than the amplitudes of P, M_1', and M_2'. Usually, downgoing seismic wavelets have amplitudes 10 to 100 times larger than the amplitudes of upgoing wavelets, depending on the magnitudes of the reflection coefficients involved. Consequently, the relative amplitudes and time delays measured in a downgoing seismic wavefield are less affected by noise than are these same attributes when measured in an upgoing wavefield. Thus, deconvolution operators determined from a downgoing wavefield should be statistically more robust than operators calculated from an upgoing wavefield.

A numerical procedure by which deconvolution operators can be derived from VSP data is illustrated in Figure 15. The VSP data are first segregated into

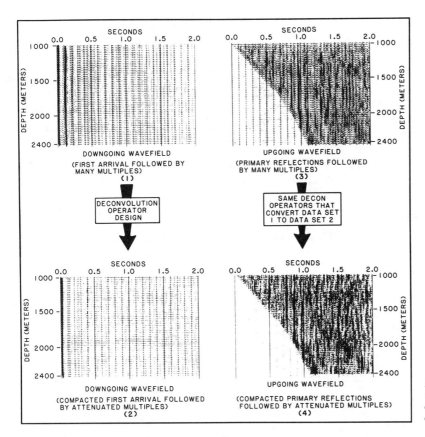

Figure 15. The sequence of operations by which robust deconvolution operators are determined from VSP data (data courtesy of Statoil; modified from Hardage, 1983).

downgoing and upgoing wavefields, and the traces are statically time-shifted so that an event not affected by structural dip is vertically aligned at the same record time for all recording depths (e.g. data sets 1 and 3). Deconvolution operators are then derived that attenuate all downgoing multiples that follow the first arrival, and which also reshape and shorten the first arrival wavelet if desired (data set 2).

These same operators are then used to remove the multiples in the upgoing VSP wavefield (data set 4), so that the origins of primary reflection events can be better interpreted. Perhaps more importantly, these operators can be used to attenuate multiples in the upgoing wavefield recorded at the surface near a VSP well.

The top panel in Figure 16 shows a seismic line crossing a VSP study well in which a technique similar to that described in Figure 15 was used to remove multiples from the upgoing VSP wavefield recorded in this well. The resulting primary reflection events defined by the VSP data are shown as the six repeated identical traces inserted in the seismic cross section at the well position. This repeated trace is usually created by vertically summing the deconvolved upgoing VSP wavefield (for example, data set 4, Figure 15) into a single composited trace. State-of-the-art deconvolution procedures were used when processing the surface-recorded seismic data crossing the well. The processing sequence used to create the cross section in the bottom panel is identical to that used to make the top display, except that the VSP-derived deconvolution operators were also used to shape the wavelets and attenuate the multiples in the surface-recorded data. The subsurface imaging is noticeably improved by the VSP operators in this instance, particularly in the zone immediately underneath the unconformity surface labeled A. An example is the improved definition of the location of the pinchout of interface B. The low frequency event enclosed in the rectangle centered at 3.3 sec also continues toward the right with an improved time registration relative to the upgoing low frequency event retrieved from the VSP data.

Reduction of Fresnel Zone Dimensions — A Key to Detailed VSP Imaging

In order to construct improved images of subsurface geology, seismic resolution must be increased in both the horizontal and vertical directions. One important consequence of vertical seismic profiling is that horizontal resolution is improved, relative to that achievable with surface-recorded seismic data, because VSP recording geometry reduces the dimensions of Fresnel zones.

The interplay between seismic Fresnel zones and horizontal resolution was discussed in AAPG Memoir 26 (Sheriff, 1977), and will not be repeated here. Basically, two subsurface features must be separated by a horizontal distance greater than or equal to the diameter of a first order Fresnel zone in order to be reliably interpreted as two distinct objects. Thus, if horizontal resolution is to be increased, the diameter of Fresnel zones must be reduced. Using some reasonable assumptions, the diameter, d, of the first order Fresnel zone associated

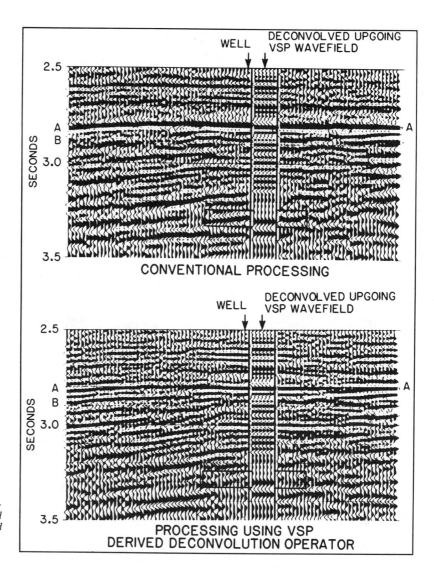

Figure 16. A comparison of surface-recorded data processed by conventional deconvolution techniques and by a deconvolution operator determined from a downgoing VSP wavefield (after Hubbard, 1979).

with vertical seismic travel paths can be expressed as:

$$d = 2\sqrt{\lambda \frac{ab}{a+b}} \qquad (4)$$

where λ is the wavelength of the radiation illuminating the subsurface target, a is the vertical distance from the source to the target, and b is the vertical distance from the target to the receiver (Hardage, 1983). For surface-recorded data, a and b are identical. The diameter of a Fresnel zone can be reduced by decreasing a, b, or λ. In exploration seismology, high frequencies cannot be propagated to great depths, so the wavelength λ cannot be significantly decreased except in shallow stratigraphic work. As long as the source and receiver are restricted to the earth's surface, neither can the travel paths, a and b, be reduced.

However, as shown by the recording geometry in Figure 17, the target-to-receiver distance in a vertical seismic profiling experiment is shorter than the path involved in surface recording, since the VSP receiver is deep below the surface and closer to the target that is being imaged. If the difference between travel paths SR_1X_1 and SR_4X_2 in Figure 17 is less than $\lambda/2$, then energy returning to the surface from any point between R_1 and R_4 is approximately in phase, and no portion of the interface between R_1 and R_4 can be distinguished from any other part of that interval. This distance R_1R_4 is then the diameter of the first order Fresnel zone at this reflecting interface as long as the receivers are located on the surface. If the receivers are arranged vertically beneath the surface, the horizontal separation between those reflection points which define source-to-interface-to-receiver travel paths that differ by $\lambda/2$ becomes smaller. In this example, raypaths SR_2Z_1 and SR_3Z_2 are drawn so that they also differ by the same amount as do raypaths SR_1X_1 and SR_4X_2. For this VSP recording geometry, the diameter of a first order Fresnel zone is R_2R_3. The resolution along the interface is thus increased in vertical seismic profiling because the Fresnel zone size is dimin-

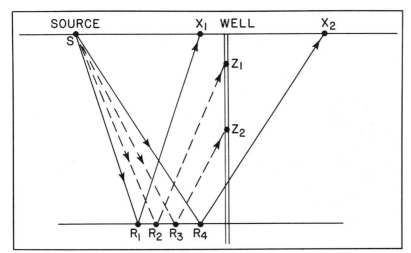

Figure 17. By ray tracing, it can be demonstrated that the diameter of a first order Fresnel zone decreases if a receiver moves vertically toward an imaged interface, rather than horizontally along the surface above the interface. In this example, the difference in path lengths SR_1X_1 and SR_4X_2 is the same as the difference in path lengths SR_2Z_1 and SR_3Z_2. If R_1R_4 is the diameter of a first order Fresnel zone for data recorded at surface positions X_1 and X_2, then R_2R_3 is the diameter of a first order Fresnel zone when the geophones are at Z_1 and Z_2.

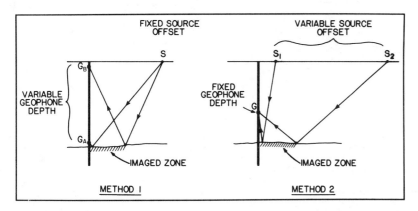

Figure 18. Two VSP field procedures by which one can look laterally away from a borehole (from Hardage, 1983).

ished, relative to the Fresnel zone size that exists for surface-positioned geophones.

Migration of seismic data is another process that increases horizontal resolution since it focuses energy into smaller spatial volumes. The fact that horizontal resolution increases as geophones are lowered toward subsurface targets is, in fact, the basis of wave equation migration. Thus, migrating VSP data, which inherently have small Fresnel zones, further improves the horizontal resolution of subsurface features.

Creation of High-Resolution Horizontal Stacks from VSP Data

High-resolution seismic cross sections can be created from VSP data if the source is not positioned directly above the subsurface geophone. Two common VSP recording geometries which result in subsurface reflection points being distributed laterally away from a vertical borehole are shown in Figure 18. In Method 1, the source remains at a fixed surface position while the geophone is moved up a vertical well bore. In Method 2, the geophone is clamped at a fixed recording depth, but the source is moved to various offset distances. In both instances, a flat, horizontal reflector can be imaged from the borehole out to a maximum distance of one-half the source offset.

By appropriately distributing NMO-corrected VSP data recorded with either of these geometries over the horizontal reflection point coordinate, X, and the vertical recording-time coordinate, t, a multi-fold seismic cross section of high resolution extending away from a VSP well can be created (Wyatt and Wyatt, 1981, 1982, 1984). Since the Fresnel zone involved in VSP recording is smaller than the Fresnel zone associated with a surface-positioned receiver, these stacked VSP data have a higher resolving power in the horizontal direction than do surface-recorded data.

An example of horizontally stacked VSP data is shown in Figure 19. In this experiment, a surface source was positioned 1,000 ft (304.9 m) away from a vertical borehole, and VSP data were recorded at intervals of 50 ft (15.2 m) over the bottom two-thirds of the well. Surface-recorded data crossing the well are shown on the left. These surface data have a trace spacing of 165 ft (50.3 m) between reflection points. A horizontal stack of the VSP data is shown in the center panel. The trace spacing of these data is 25 ft (7.6 m), and a more detailed image of the subsurface near the well is achieved.

A magnified view of part of the VSP stacked data is shown at the right. A small fault, which cannot be seen in the surface-recorded data, is interpreted in

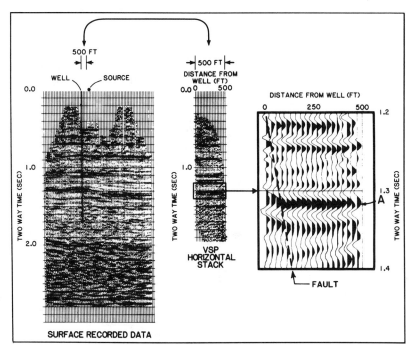

Figure 19. A comparison of the horizontal resolution obtained with surface-recorded seismic reflection data and a horizontal stack made from VSP data. Different time datums were used when processing the surface-recorded data and the VSP data. As a consequence, the VSP data should be moved approximately 40 msec earlier in time in order to tie with the surface data (after Wyatt and Wyatt, 1982).

the VSP data by these authors. At this window depth (1.2 to 1.4 sec), no VSP data can be stacked at an offset distance of 500 ft (152.4 m), so the rightmost trace maintains a zero value. Inspection of the center data panel shows that some VSP data begin to stack at this offset distance at about 1.9 sec.

This type of detailed imaging near a well can be crucial in some development stratigraphy problems.

Imaging Beneath Deviated Wells

An accurate detailed image of the subsurface is an urgent necessity when a stratigrapher is confronted with the problem of constructing a reservoir model for development purposes. Both production and injection wells must be properly sited in order to develop a field, and the positioning of these wells cannot be done confidently unless the highest quality seismic image of the reservoir is available. Offshore development can be particularly affected by inadequate seismic imaging because most offshore production facilities have only a limited number of well slots. Production engineers thus have a fixed number of opportunities to position wells properly, and in addition, the stepout distance of these wells from the production platform is limited. Onshore, there is usually more flexibility as to the number and location of wells that can be drilled to develop a field.

Consequently, the emphasis here will be on marine development stratigraphy in which most development wells are deviated holes. The concepts apply, of course, to onshore work, although deviated holes are not as common in onshore development.

A cross-sectional view of a single deviated marine well is drawn in Figure 20. Logging tools are seldom lowered down an uncased deviated hole since they tend to jam in washout zones or against borehole ledges. Thus, a significantly deviated well is always cased before any type of logging device is sent down-hole. The types of logs that can be recorded through casing are rather limited; in particular, sonic and density logs cannot be run. Thus, synthetic seismogram ties with surface-recorded data cannot be made at cased, deviated wells. Fortunately, VSP data can be recorded in cased wells, and the common unavailability of synthetic seismograms in deviated wells makes the recording of VSP data in these wells even more important. VSP tools can be lowered down most cased, deviated wells without modifying the tool. In extremely deviated wells, some weight may have to be added, sometimes wheels are attached to the tool barrel so that it moves with less resistance, and in rare cases, some tools are pushed down-hole with the drill string.

A common VSP data acquisition technique in a deviated well is shown in Figure 20. The objective here is to always position the source directly above the geophone as the VSP tool is brought up-hole. In this two-dimensional view, the horizontal and vertical coordinates are X and Z, respectively. When the geophone is at coordinates X_1, Z_1, the surface source is positioned at B, directly above the geophone. When the geophone is at a shallower level (X_2, Z_2), the source is moved to A in order to remain directly above the geophone position.

This shooting geometry allows the downgoing and upgoing seismic wavefields to be recorded along an oblique geophone trajectory through the subsurface, and two important consequences result. First, the vertical motion of the geophone causes downgoing and upgoing events to have opposite apparent velocities, so that the downgoing and upgoing wavefields can be separated by velocity filtering. This wavefield separation allows deconvolution operators to be determined from the downgoing wavefield, and these operators can in turn be used to create optimal vertical resolution in the upgoing wavefield. Second, the horizontal motion of the geophone allows this deconvolved, high-resolution upgo-

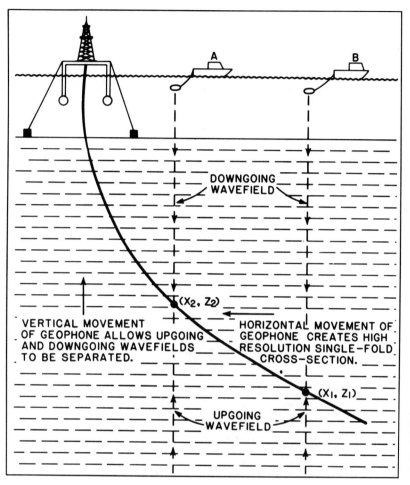

Figure 20. The detailed imaging capability resulting from vertical seismic profiling is probably best demonstrated when VSP data are recorded in a deviated borehole, such as would be encountered in a development stratigraphy problem.

ing wavefield to be displayed as a cross-sectional image of the stratigraphic interval below the well bore.

An example of VSP data recorded in this manner in a deviated well is shown in Figure 21. Only the deconvolved upgoing wavefield arriving from beneath the well bore is plotted. The trace spacing becomes less as the hole steepens, but on the average, the trace spacing is about 50 ft (15.2 m). Thus, the subsurface underneath the well is imaged with a rather precise resolution in both the horizontal and vertical directions. An interpretation of the structure and stratigraphy near the well is shown on the right (Figure 21). Event A is the bifurcated trend of troughs occurring immediately below 2.2 sec in the wiggle trace data, whereas, events B and C correlate with trends of black peaks in the VSP data.

A comparison between the imaging capabilities of VSP data recorded in a deviated hole and surface-recorded data that follow the track of the same well is shown in Figure 22. These data are from the Statfjord field in the North Sea. A deconvolved version of the upgoing VSP wavefield recorded in a deviated well in the field is shown on the right. On the left is a vertical slice chosen from a surface-recorded three-dimensional seismic grid which follows the track of the deviated well. State-of-the-art deconvolution and migration were used in processing the 3-D surface-recorded data.

The position of the Statfjord sand in the VSP data is defined from well data. The remaining structural and stratigraphic features noted in the VSP data are interpretations based strictly on the appearance of those data. These same features are translated to the surface-recorded data as accurately as possible.

There is an obvious difference between the resolution (image quality) of the reservoir interval as defined by the surface-recorded data and the VSP data. The VSP image is thought to be the more reliable picture of the stratigraphic relationships in this part of the field. Obviously, the improved definition of the fault relationships in the productive interval is invaluable for siting development wells.

A particularly thorough example of VSP imaging in several deviated development wells in the Piper field in the North Sea is described by Johnson, Riches, and Ahmed (1982).

Because of the improved subsurface imagery that can be obtained with VSP data, some companies are considering deliberately deviating onshore development wells in order to use this imaging technique in nonmarine work.

CONCLUSIONS

It is impossible to include within the scope of a single paper every possible

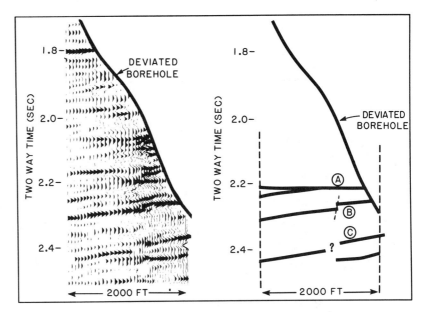

Figure 21. An example of the detailed imaging provided by an upgoing wavefield recorded in a deviated well (after Kennett and Ireson, 1981).

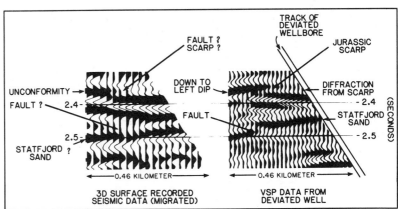

Figure 22. Comparison between surface-recorded reflection data and VSP data recorded in a deviated well bore (from Kennett and Ireson, 1981).

application of vertical seismic profiling that can aid seismic stratigraphers. Hopefully, the applications that have been discussed demonstrate the value of VSP data in seismic stratigraphy and serve to encourage expanded usage of VSP technology in stratigraphic and lithological analyses.

There may be no other geophysical measurement which allows as direct a tie between seismic behavior and geology as does vertical seismic profiling. Consequently, the present state of VSP technology will no doubt expand considerably as inventive minds continue to explore ways that additional lithological, stratigraphic, and structural information can be retrieved from this in situ seismic measurement.

One particular topic that should be emphasized in future seismic stratigraphy work is the measurement and analysis of three-component VSP data (DiSiena and Gaiser, 1983; Montmollin, 1983; Omnes, 1980; Turpening, Liskow, and Thomson 1980; Gal'perin, 1984). All of the VSP data shown in this chapter, and all of the applications that have been discussed, involve only the vertical component of particle motion. Some exciting possibilities concerning the detection of subtle stratigraphic traps arise when both P-wave and S-wave surface-recorded data are used in a seismic stratigraphy interpretation. Any seismic stratigraphy study utilizing shear waves can be greatly advanced if good quality three-component VSP data are recorded, so that fundamental relationships between propagating P and S body waves can be determined, and so that the geological effects on both of these types of wave modes can be documented.

REFERENCES

Cited

Anstey, N. A., 1980, Seismic delineation of oil and gas reservoirs using borehole geophones: Great Britain Patents 1,569,581 and 1,569,582; Canadian Patents 287,178 and 375,890-7. (According to Seismograph Service, Ltd., these patents were assigned to Seismograph Service Ltd., July, 1981.)

Balch, A. H., and M. W. Lee, 1984, Vertical seismic profiling - technique, applications, and case histories: Boston, International Human Resources Development Corporation, 488 p.

——— , et al, 1980, The use of vertical seismic profiles and surface seismic profiles to investigate the distribution of aquifers in the Madison Group and Red River Formation, Powder River Basin, Wyoming-Montana: Dallas, Society of Petroleum Engineers 55th Annual Fall Technological Conference Preprint, SPE Paper 9312.

——— , et al, 1982, The use of vertical seismic profiles in seismic investigations of the earth: Geophysics, v. 47, p. 906-918.

DiSiena, J. P., and J. E. Gaiser, 1983, Three component vertical seismic profiles - an application of Gal'perin's polarization position correlation techniques: Society of Exploration Geophysicists 53rd Annual International Meeting, Technical Program Abstracts, Paper S19.1, p. 522-524.

——— , B. S. Byun, and J. E. Fix, 1980, Vertical seismic profiling - a processing analysis case study: Houston, Society of Exploration Geophysicists 50th Annual International Meeting, Paper R-19.

Gaiser, J. E., and J. P. DiSiena, 1982, VSP fundamentals that improve CDP data interpretation: Society of Exploration Geophysicists 52nd Annual International Meeting, Technical Program Abstracts, Paper S12.2, p. 154-156.

Gal'perin, E. I., 1974, Vertical seismic profiling: Society of Exploration Geophysicists Special Publication No. 12, 270 p.

——— , 1984, The polarization method of seismic exploration: Hingham, MA, Kluwer Academic Publishers, 268 p.

Goetz, J. F., L. Dupal, and J. Bowler, 1979, An investigation into discrepancies between sonic log and seismic check-shot velocities: Australian Petroleum Exploration Association Journal, v. 19, pt. 1, p. 131-141.

Hardage, B. A., 1983, Principles of vertical seismic profiling: Amsterdam, Geophysical Press, 450 p.

Hubbard, T. P., 1979, Deconvolution of surface recorded data using vertical seismic profiles: Society of Exploration Geophysicists 49th Annual International Meeting, Preprint S-46.

Johnson, R., H. Riches and H. Ahmed, 1982, Application of the vertical seismic profile to the Piper field: European Petroleum Conference, Paper EUR 274, p. 39-47.

Karus, E. V., et al, 1975, Detailed investigation of geological structures by seismic well surveys: Ninth World Petroleum Congress, PD 9(4), v. 26, p. 247-257.

Kennett, P. and R. L. Ireson, 1971, Recent developments in well velocity surveys and the use of calibrated acoustic logs: Geophysical Prospecting, v. 19, p. 395-411.

——— , and ——— , 1981, The V.S.P. as an interpretation tool for structural and stratigraphic analysis: Venice, European Association of Exploration Geophysicists 43rd Annual Meeting. (Also available as a 42 page report from Seismograph Service, Ltd., Keston, Kent, England.)

——— , and ——— , 1982, Vertical seismic profiles: European Association of Exploration Geophysicists Continuing Education Short Course, 184 p.

Lee, M. W., and A. H. Balch, 1983, Computer processing of vertical seismic profile data: Geophysics, v. 48, p. 272-287.

Montmollin, V. de, 1983, Three-component vertical seismic profile-geometrical processing and wave identification: Society of Exploration Geophysicists 53rd Annual International Meeting, Technical Program Abstracts, Paper S19.2, p. 524-527.

Omnes, G., 1980, Logs from P and S vertical seismic profiles: Petroleum Technology, v. 32, p. 1843-1849.

Parrott, K. R., 1980, An investigation of the interior of a salt structure using the vertical seismic profiling technique: Colorado School of Mines, Master's thesis, 147 p.

Sheriff, R. E., 1977, Limitations on resolution of seismic reflections and geological detail derivable from them, in C. E. Payton, ed., Seismic stratigraphy - applications to hydrocarbon exploration: AAPG Memoir 26, p. 3-14.

Turpening, R. M., A. Liskow, and F. S. Thomson, 1980, Seismic investigations of Antrim shale fracturing: U.S. Department of Energy Publication FE-2346-90, 35 p.

Vail, P. R., R. G. Todd, and J. B. Sangree, 1977, Seismic stratigraphy and global changes in sea level, part 5 - chronostratigraphic significance of seismic reflections, in C. E. Payton, ed., Seismic stratigraphy - applications to hydrocarbon exploration: AAPG Memoir 26, p. 99-226.

Wyatt, K. D., and S. B. Wyatt, 1981, The determination of subsurface structural information using the vertical seismic profile: Society of Exploration Geophysicists 51st Annual International Meeting, Paper S5.2, p. 1915-1949.

——— , and ——— , 1982, Downhole vertical seismic profile survey reveals structure near borehole: Oil and Gas Journal, v. 80, no. 42, p. 77–82.

——— , and ——— , 1984, Determining subsurface structure using the vertical seismic profile, in M. N. Toksoz and R. R. Stewart, eds., Vertical seismic profiling, part B, advanced concepts: Amsterdam, Geophysical Press, p. 148–176.

Additional References

Balch, A. H., M. W. Lee, and D. C. Muller, 1980, A vertical seismic profiling experiment to determine depth and dip of the Paleozoic surface at drill hole U10bd, Nevada test site, Nevada: U. S. Geological Survey Open-File Report 80-847.

——— , et al, 1981, Seismic amplitude anomalies associated with thick First Leo sandstone lenses, eastern Powder River Basin, Wyoming: Geophysics, v. 46, p. 1519-1527.

——— , et al, 1981, Processed and interpreted U.S. Geological Survey seismic reflection profile and vertical seismic profile, Powder River and Custer counties, Montana: U.S. Geological Survey Chart OC-108.

DiSiena, J. P., J. E. Gaiser, and D. Corrigan, 1981, Three-component vertical seismic profiles - orientation of horizontal components for shear wave analysis: Society of Exploration Geophysicists 51st Annual International Meeting, Paper S5.4, p. 1990-2011.

Hardage, B. A., 1981, An examination of tube wave noise in vertical seismic profiling data: Geophysics, v. 46, p. 892-903.

Hauge, P. S., 1981, Measurements of attenuation from vertical seismic profiles: Geophysics, v. 46, p. 1548-1558.

Jolly, R. N., 1953, Deep-hole geophone study in Garvin County, Oklahoma: Geophysics, v. 18, p. 662-670.

Kennett, P., 1979, Well geophone surveys and the calibration of acoustic velocity logs; chapter 3, in A. A. Fitch, ed., Developments in geophysical exploration methods: London, Applied Science Publishers, Ltd., 311 p.

——— , R. L. Ireson, and P. J. Conn, 1980, Vertical seismic profiles - their applications in exploration geophysics: Geophysical Prospecting, v. 28, p. 676-699.

Lang, D. G., 1979, Downhole seismic technique expands borehole data: Oil and Gas Journal, v. 77, no. 28, p. 139-142.

——— , 1979, Downhole seismic combination of techniques sees nearby features: Oil and Gas Journal, v. 77, no. 29, p. 63-66.

Lee, M. W., et al, 1981, Processed and interpreted U. S. Geological Survey seismic reflection profile and vertical seismic profiles, Niobrara County, Wyoming: U.S. Geological Survey Oil and Gas Investigations Chart OC-114.

Levin, F. K., and R. D. Lynn, 1958, Deep hole geophone studies: Geophysics, v. 23, p. 639-664.

Miller, J. J., et al, 1981, Processed and interpreted U. S. Geological Survey seismic reflection profile and vertical seismic profile, Powder River and Carter counties, Montana: U. S. Geological Survey Oil and Gas Investigations Chart OC-110.

Mons, F., 1980, Vertical seismic exploration and profiling technique: Great Britain Patent No. 2,029,016.

———, 1980, Method and equipment for vertical seismic exploration: France Patent No. 2,432,177.

Omnes, G., 1978, Vertical seismic profiles - a bridge between velocity logs and surface seismograms: Houston, Society of Petroleum Engineers 53rd Annual Conference and Exhibition, Paper No. 53.

Riggs, E. D., 1955, Seismic wave types in a borehole: Geophysics, v. 20, p. 53-67.

Ryder, R. T., et al, 1981, Processed and interpreted U. S. Geological Survey seismic reflection profile and vertical seismic profile, Carter County, Montana, and Crook County, Wyoming: U. S. Geological Survey Oil and Gas Investigations Chart OC-115.

Steward, R. R., R. M. Turpening, and M. N. Toksoz, 1981, Study of a subsurface fracture zone by vertical seismic profiling: Geophysical Research Letters, v. 8, p. 1132-1135.

Wuenschel, P. D., 1976, The vertical array in reflection seismology - some experimental studies: Geophysics, v. 41, p. 219-232.

The use of vertical seismic profiles in seismic investigations of the earth

A. H. Balch*, M. W. Lee*, J. J. Miller*, and Robert T. Ryder*

ABSTRACT

During the past 8 years, the U. S. Geological Survey has conducted an extensive investigation on the use of vertical seismic profiles (VSP) in a variety of seismic exploration applications. Seismic sources used were surface air guns, vibrators, explosives, marine air guns, and downhole air guns. Source offsets have ranged from 100 to 7800 ft. Well depths have been from 1200 to over 10,000 ft.

We have found three specific ways in which VSPs can be applied to seismic exploration. First, seismic events observed at the surface of the ground can be traced, level by level, to their point of origin within the earth. Thus, one can tie a surface profile to a well log with an extraordinarily high degree of confidence. Second, one can establish the detectability of a target horizon, such as a porous zone. One can determine (either before or after surface profiling) whether or not a given horizon or layered sequence returns a detectable reflection to the surface. The amplitude and character of the reflection can also be observed. Third, acoustic properties of a stratigraphic sequence can be measured and sometimes correlated to important exploration parameters. For example, sometimes a relationship between apparent attenuation and sand percentage can be established.

The technique shows additional promise of aiding surface exploration indirectly through studies of the evolution of the seismic pulse, studies of ghosts and multiples, and studies of seismic trace inversion techniques.

Nearly all current seismic data-processing techniques are adaptable to the processing of VSP data, such as normal moveout (NMO) corrections, stacking, single- and multiple-channel filtering, deconvolution, and wavelet shaping.

INTRODUCTION

The U. S. Geological Survey (USGS) has been investigating new field techniques and new exploration applications of vertical seismic profiles (VSP) for 8 years. The basic idea—recording the seismic wave field at many levels in a deep borehole—is not new. McCollum and La Rue (1931) and Musgrave et al (1960) described the use of deep well geophones for detection and mapping of salt domes. Slotnick (1936a, b) and Dix (1939) described the use of borehole geophones to obtain time-depth curves and depth-velocity relationships. Jolly (1953), Levin and Lynn (1958), and Lynn (1963) discussed this technique, particularly to observe the evolution and attenuation of seismic pulses propagating into earth. Gal'perin (1973) included in his references a rich variety of articles on the topic from both the Russian and English technical literature, including several of his own, which date back to the early 1950s. Gal'perin's (1977) work on the polarization method of seismic investigations described some of his later work in three-component seismic recording and interpretation. Gal'perin's (1980) review paper provided a well-organized summary of petroleum industry applications of vertical seismic profiles. Gal'perin et al (1980) presented some recent Soviet results on the use of three-component VSP to estimate compressional/shear velocity ratios and Poisson's ratio.

A new series of articles and papers has recently started to appear in the English language, including Karus et al (1975), Balch et al (1977), Hague (1981), Hubbard (1979), Kennett et al (1980), Kennett and Ireson (1973), Lang (1979), Omnes (1980), Spencer et al (1982), Turpening et al (1980), S.S.L. (1980), Balch, Lee, and Muller (1980) and Balch, Lee, Miller, and Ryder (1980, 1981). Wuenschel's (1976) paper on vertical arrays is also highly relevant to this technique. As one might expect, the newer papers reflect the expanded potential of the method, which has been made possible by new developments in other areas of seismic exploration. The impact of computer processing is apparent in the more recent articles.

In the course of our investigations, we have run a total of 19 VSPs. Some of these profiles have been similar to those discussed in recent literature. Others have been quite different in field technique used, or in processing and interpretation. In the field, we have used a variety of sources, including explosives, surface air guns, downhole air guns, and vibrators, singly and in arrays. We have used both electromechanical and spring-actuated locking devices on the downhole detectors. We have occasionally used three-component downhole detectors. The deep-well observation levels usually have been very closely spaced (on the order of 15 ft) and digitally recorded. The close spacing of digitally recorded data has enabled us to perform extensive computer processing that is otherwise difficult or impossible.

Presented at the 33rd Annual Midwestern Meeting of SEG, March 23–25, 1980 in Tulsa. Manuscript received by the Editor February 10, 1981; revised manuscript received August 31, 1981.
*U.S. Geological Survey, Box 25046 Federal Center, Denver, CO 80225.
0016-8033/82/0601—906. This paper was prepared by an agency of the U. S. government.

We would like to share here some of our experiences and results in using VSP. We feel that by so doing, we may help stimulate further interest in and use of this powerful technique. We also hope this material will be of some value to others working in this area who have encountered problems similar to the ones we describe and may, therefore, be interested in our solutions to these problems.

A BRIEF DESCRIPTION OF THE USGS VERTICAL SEISMIC PROFILING METHOD

Ideal and realizable field configurations

Figure 1a is an idealized cross-sectional view of our typical VSP layout. A seismic source located near or at the surface is energized, and an array of seismometers R is clamped in a well. All downward- and upward-traveling (reflected) events are recorded in sequence at every level in the hole. Typical seismometer spacing would be 10–15 ft. With an ideal compressional source wavelet, no noise, and a single horizontal reflector, two events (a direct arrival and a single reflection) would be easily recognized on a plot of the recorded data and identified on the basis of their arrival times (Figure 1b). The depth to the reflector could be easily determined, as well as the character or shape of the reflected event. However, this ideal is never realized.

In actual practice, currently available technology has restricted us to one seismometer in the borehole. We can record at only one level at a time. Our system is usually constrained further to the measurement of the vertical component of motion or velocity.

Detector and recording systems

The well seismometer needs to have an electromechanical or electrohydraulic clamping arrangement. Typically, the tool has a locking arm driven by an electric motor that can be activated from the surface. With this arrangement, the arm is driven against the side of the hole with considerable force. A current surge results when the arm engages the hole wall, which gives a positive indication that the arm is engaged. Five to 10 ft of slack is then put in the cable, thus suppressing noise that would otherwise travel down a taut cable and interfere with the recording.

Other tool options include a spring-driven locking arm, a pressure-sensitive hydrophone, and a three-component detector. Any of these optional devices can be used, but all call for some compromise or trade-off with some other desired tool capability. For example, a three-component detector may preclude the use of a motor-driven locking arm, but a spring-driven locking arm may not keep the tool from slipping or creeping when recording in a cased hole.

All data are recorded digitally and given extensive digital processing necessary for interpretation. In investigations related to surface seismic exploration, it is desirable to use a recording system as nearly similar to a surface recording system as possible in order to correlate both types of data.

Seismic sources

Explosives, surface air guns, vibrators, and downhole air guns were used in the VSP investigations. To date, surface air guns have produced the best results. Explosives, in general, provided higher frequencies and a broader band of frequencies. Explosives yielded a higher output amplitude. There were two interrelated problems in the use of explosives: nonreproducibility of the source waveform and a vastly more complicated field operation.

The complexity of the field operation using explosives resulted from the need to record at several hundred levels, one level at a

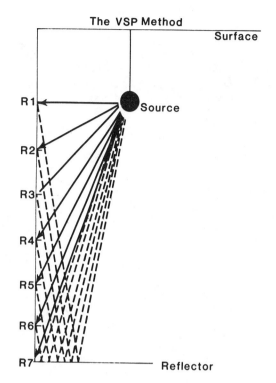

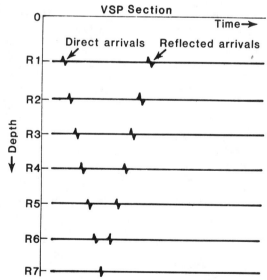

FIG. 1. (a) Schematic drawing of a typical vertical seismic profiling field configuration. (b) Idealized data set from the configuration shown in (a).

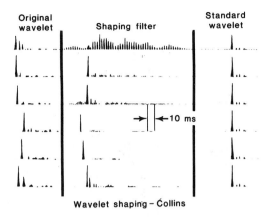

FIG. 2. Left, source monitor waveforms from the Coronado Collins no. W-1 well, Weston Co., Wyoming. Center, shaping filters designed to convert source waveform by convolution. Right, convolution product of source waveforms and corresponding shaping (conversion) filter.

time. Even with only one shot per level, hundreds of shots were required. We used relatively small charges (on the order of 1–2 lb per shot) to maintain a nearly constant source waveform and to obtain as many shots as possible from each hole. Even so, as many as 40 shotholes per well were required. This necessitated an expensive and time-consuming shothole drilling operation. In addition, hole caving often prevented the firing of subsequent shots at the same level in a given hole, resulting in source waveform changes. The problem was further complicated by difficulty in removing cap wire from the hole when small charges were used. Water tamping of the holes was difficult to carry out expeditiously during a profiling operation. When shifting the source to a different hole, additional source waveform changes resulted. Even cap wire disposal during shooting was laborious and time consuming. (For example, if a source depth of 100 ft is used for 400 shots, 40,000 ft of wire must be picked up.)

To expedite the recording operation, we tried shooting two or three holes alternately. This technique proved unsatisfactory and dangerous. With several cap leads lying on the ground, misunderstandings arose as to which hole was to be fired next. Because of these and other problems with explosive sources, we eventually resorted to surface air guns almost exclusively.

Surface air guns, such as the Bolt LLS-3 type, eliminated most of the problems with explosives mentioned above. No source holes were required, and the guns could be fired almost indefinitely at one surface location. The lower amplitude was compensated for by firing repeatedly at each level (as many as 20 times) and then compositing these shots. The narrower frequency band, inherent in the surface source, was considered an acceptable trade-off for the convenience of the source. Most of the high frequencies obtained from explosive sources are quickly attenuated in the earth. The absence of high frequencies in the air-gun source was not nearly so apparent in the air-gun data recorded at depth as first suspected.

Two major practical problems with surface air guns were reliable operation in very cold (subzero) weather and multiple gun synchronization. We found that skillful, experienced operators were the most essential factor in keeping the air guns operating in cold weather. Synchronization was attained by using a multiple-trace-monitor oscilloscope. Source accelerometer pulses were continuously observed with the oscilloscope, and the firing delays to the sources were frequently adjusted to ensure that the guns fired simultaneously.

A REVIEW OF SOME PROBLEMS ENCOUNTERED IN VERTICAL SEISMIC PROFILING AND SOME POSSIBLE SOLUTIONS

Changes in source waveform

As mentioned earlier, ideally we would energize the source only once and record simultaneously at all levels in the well, but equipment constraints restrict us to recording at only one level at a time. In principle, we can produce an equivalent set of data by energizing the source many times and recording at all desired levels, one at a time. The two data sets would be equivalent only if the source waveform remained constant. In practice, this is rarely the case, especially when using explosives.

Figure 2 shows an example of this phenomenon. The source waveforms were monitored with a geophone buried about 100 ft beneath the source; several of these waveforms are shown on the left-hand side of the figure.

Using a small explosive source, constant source depth, and water tamp, we can greatly reduce this variation in wave shape. Wuenschel (1976) described a more elaborate field procedure for maintaining source waveform consistency.

We could rarely reproduce any explosive source waveform more than 10 or 12 times. This change in source waveform created great difficulties, both in processing and in subsequent interpretation. For example, one effective processing technique to enhance desired events is velocity filtering. In our model of the multichannel process, used in the design of a multichannel velocity filter, we assume that the "desired events" have identical wave shapes (but different arrival times) on sets of adjacent traces. If the source waveform changes from level to level, this will not be the case, and the quality of the velocity-filtered data will deteriorate. In interpretation, the waveform may change from level to level. We cannot infer anything about the physical properties of the medium from these changes, unless we know that they are not merely due to source waveform changes.

To compensate for source waveform changes, a wavelet-shaping filter can be designed. One of the source waveforms, as measured on a monitor detector, is arbitrarily chosen as a "standard wavelet;" then a shaping filter is designed for every shot at every level. It converts the waveform observed on the monitor detector into the standard wavelet. This filter is then applied to the corresponding well recording. In this manner, a new set of VSP data is obtained that approximates what would have been recorded in the field if the source waveform had been constant.

The center portion of Figure 2 shows some sample shaping filters corresponding to the monitor detector recordings shown at the left. On the right of the figure, we see the convolution product of the shaping filter and the source waveform recorded by the monitor.

Even when the surface air gun is used for a seismic source, a change in waveform may occur. Gradual compaction of the soil beneath the source takes place as the gun is fired repeatedly. This causes a "drift" in the source waveform. The surface air-gun source location often becomes unusable after many firings and must be changed. This almost always causes a change in source waveform. Figure 3 shows an example of this, taken from the Pfister Fee well in Wyoming. The eighth trace from the bottom corresponds to a change in source location. The change in the first arrival, caused by this change in location, is clearly apparent.

Tube noise

We define tube noise as any seismic disturbance that is confined to the vicinity of the well bore. The most prevalent type of tube noise is thought to be a compressional wave traveling in the fluid column. Another mode can travel down the steel casing (casing breaks). Another is a mode confined to the boundary between the borehole and the surrounding rock. Others may exist also. The tube modes carry little information about the rock lithology and can badly obscure the desired body-wave recording.[1] Because they are confined to the well bore, there is little geometrical spreading. As a result, their amplitude with respect to body waves can be quite large at great depth.

Figure 4 shows an example of a VSP in which a tube-wave train (marked with an arrow in the figure) has an amplitude 5 to 10 times greater than the downgoing source pulse. While this is an extreme case, it is by no means rare.

One of the most effective ways to reduce tube wave interference is to keep the seismic source away from the immediate vicinity of the well head. Our experience indicates that most tube wave disturbances enter the well bore near the surface, either as the result of a direct arrival from the source or, in some cases, as a shallow refraction. Since this energy encounters the hole while traveling in a horizontal, or nearly horizontal direction, the noise can be reduced further by source patterns designed to suppress horizontally traveling energy. Additional suppression of this noise can often be accomplished by velocity filtering. In some cases, the tube noise amplitude can be so great that it persists in spite of all attempts to remove it. The ultimate resolution of the VSP can be controlled by tube wave noise.

Tool creep

It is always desirable to put slack on the cable, after locking the tool in the hole, to reduce cable noise. In some cases the weight of the tool, combined with part of the weight of the wire line bearing on top of the tool, causes minute slippage. The tool may gradually creep down the hole. This phenomenon can be especially troublesome in casing, although tool creep is not confined to cased sections of holes. Figure 5 shows an example of the effect of tool creep on the downhole recordings. The lower traces exhibit pronounced creep, whereas the upper traces were recorded while the tool was properly clamped and stationary.

As with the tube noise problem, the best solution is to eliminate the noise in the field instead of recording it. It may be necessary to replace or overhaul the clamping system; this involves a costly round trip for the tool. If the tool creeps in casing, it may be possible to bring the tool to rest at casing joints. Then the casing joints dictate the spatial sampling interval. This interval may be unacceptably coarse and frequently is not constant; the length of individual casing sections can vary considerably. Ultimately, one may be forced to record in the presence of creep noise and then suppress it in processing by compositing (stacking) and frequency filtering.

Weak signal

As mentioned earlier, one of the disadvantages of using the surface air gun, compared to explosives, is its lower amplitude

[1] An exception to this is described by Huang and Hunter (1981a, b).

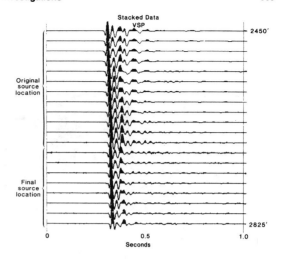

FIG. 3. Composited, edited VSP recordings from the Pfister Fee no. 1 well, Niobrara County, Wyoming. Note the change in first arrival waveform owing to surface air-gun relocation.

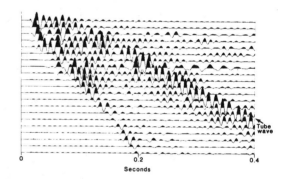

FIG. 4. VSP raw data set from the Coronado Collins no. W-1 well, Weston County, Wyoming, showing a high amplitude tube wave train.

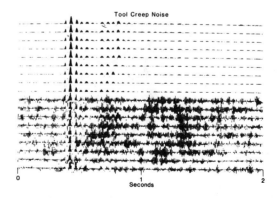

FIG. 5. Raw VSP data from the USGS Hydrologic Test Hole no. 1, Crook County, Wyoming. Lower traces exhibit creep noise; upper traces were recorded when the tool was stationary, and properly clamped.

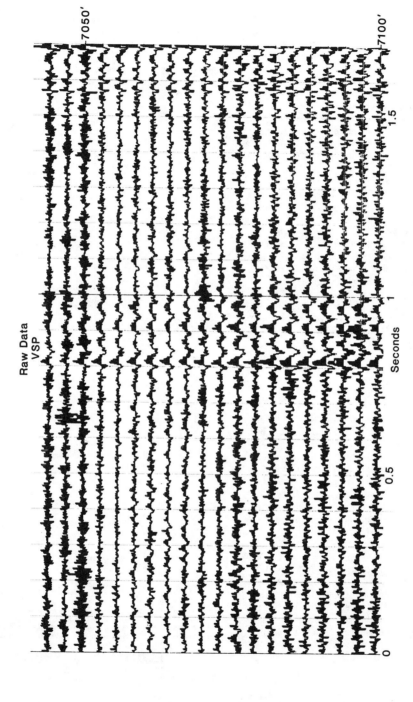

FIG. 6. Weak signals obtained from a single surface air-gun source near the 7000 ft level at the USGS Hydrologic Test Hole no. 2, Custer County, Montana.

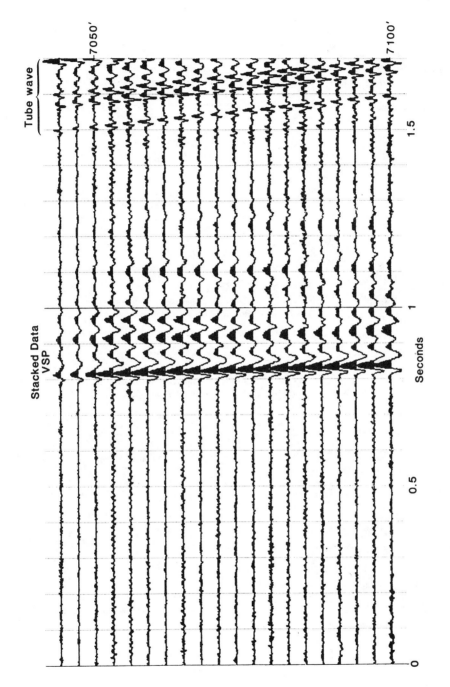

FIG. 7. Data corresponding to Figure 6 after 20-25 fold composite.

signal. This problem is especially troublesome at great depth where geometrical spreading, attenuation, and other transmission effects have substantially reduced the signal amplitude. Figure 6 illustrates this problem, encountered at the 7000+ ft level at the USGS Hydrologic Test Hole (HTH) no. 2 near Powderville, Montana. A single air gun was used for the source. To increase this low signal-to-noise ratio (S/N), we composited (or stacked) 20 to 25 shots at each level. The resulting composite is shown in Figure 7. The improvement is dramatic. Note that coherent noise is also enhanced. The low-velocity event at the right of the illustration, marked T in this figure, is coherent tube noise. Other means must be used to suppress or eliminate this phenomenon.

Multiple surface sources also serve to increase the source strength. Provided the sources can be properly synchronized, the S/N should be directly proportional to the number of sources, but only proportional to the square root of the number of shots in a composite or stack.

In order to produce the high S/N that we require in the VSP work, both improvement methods were frequently used. Ten shots per level with three synchronized sources comprised a typical configuration.

Near-surface reverberation and ghosting

An implicit assumption frequently made in seismic reflection interpretation is that the source generates a single, short duration pulse. In actual fact, this is rarely the case. This is demonstrated in Figure 8, taken from VSP at the Collins well. The recordings have been time-shifted to align downward-traveling events.

The highest amplitude event at all levels is the first arrival from the source, as might have been expected. However, this first arrival has a long "tail" and certainly cannot be considered a short duration pulse. There also is a second burst of seismic energy after the first arrival, and an additional third arrival. The presence of the later arrivals is confirmed by the plot in Figure 9 which shows the autocorrelation of selected traces from Figure 8. The consistent correlation trough marked "LVL ghost" and "surface ghost" correspond to the later arrivals shown in Figure 8.

We compensate for this effect, at least partially, by wavelet shaping. The procedure is similar to that described previously under source wavelet correction. If direct information about the source pulse(s) is not available, we can often estimate the source-pulse characteristic statistically from autocorrelations of the downhole data.

Reflections obscured by high-amplitude downward-traveling events and noise

The velocity-filtering techniques are described briefly here because they comprise an essential part of the vertical seismic profiling method. In Figures 1–8, few if any upward-traveling (reflected) events are observed. They are present, but much smaller in amplitude than the downward-traveling waves. Because of their small amplitude, they are difficult or impossible

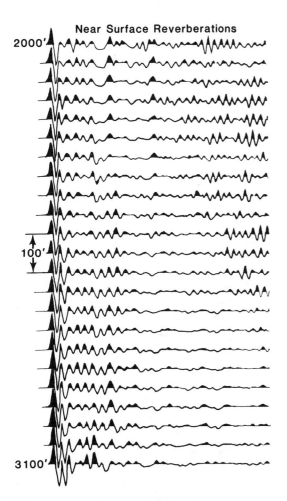

FIG. 8. First arrivals from part of the VSP data set obtained from the Coronado Collins no. W-1 well, Crook County, Wyoming, time shifted to align downward-traveling events. Note the long "tails" following the first arrivals, presumably owing to near-surface reverberations.

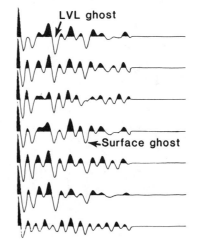

FIG. 9. Autocorrelations of the data shown in Figure 8.

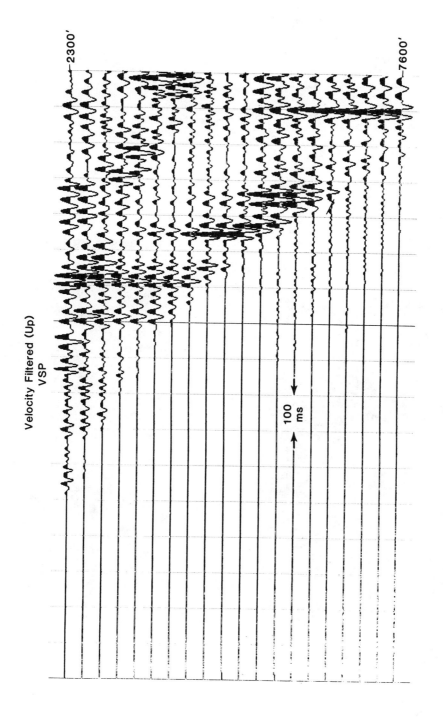

FIG. 10. Vertical seismic profile recordings from the USGS Hydrologic Test Hole no. 2, Custer County, Montana, after velocity filtering to enhance upward-traveling events. Traces have been time shifted to align these events.

to see in the presence of high-amplitude downward-traveling events and random noise.

In Figure 10, a set of data from the HTH no. 2 VSP has been velocity filtered to pass only events with negative (upward-traveling) velocities. The traces were subsequently time shifted to align these seismic events vertically.

The coherent reflections (upward-traveling events) are now readily apparent. In addition, the regions of origin of the reflections can be determined from the depth scale on the right of the figure.

COMPUTER PROCESSING OF DIGITAL VSP RECORDINGS—A SUMMARY

Nearly all computer-oriented data enhancement and interpretation techniques applicable to surface-recorded seismic data are useful in dealing with VSP recordings. In most cases, some modification of, or variation in, the techniques is required because of inherent differences in the nature of the data. A typical data processing sequence is outlined below:

1) Demultiplex.
2) Edit, sort, static time correct, dynamic time correct, amplitude balance, and adjust.
3) Stack.
4) Design and apply wavelet-shaping filter.
5) Multichannel velocity filter.
6) Make gain adjustment (if needed).
7) Do cumulative summation of upgoing waves (if well tie to surface seismic data is desired).
8) Merge upgoing and downgoing waves (usually with amplitude boost for upgoing waves).
9) Make target horizon transfer function estimates.
10) Inverse filter "trace inversion" if VSP-generated impedance log of target horizons is desired.

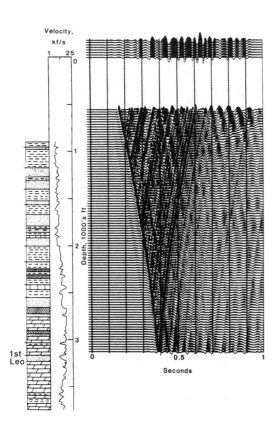

FIG. 11. Vertical seismic profile from the Mule Creek Oil Co., Fee no. 1-4235, Niobrara County, Wyoming, showing tie from lithologic log, on the left, to the surface seismic data, across the top. After edit, compositing, and wavelet shaping, the data have been separately velocity filtered to enhance upward- and downward-traveling compressional waves, then recombined with a gain boost in the upward-traveling waves.

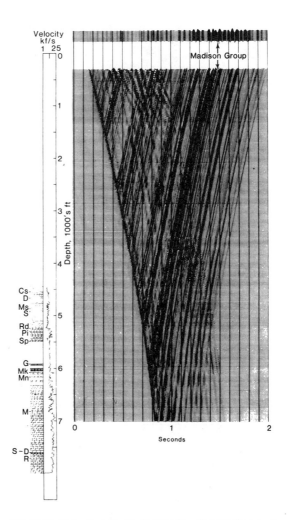

FIG. 12. Vertical seismic profile from the Sam Gary Madison no. 1 well, Bell Creek oil field, Montana, showing tie from lithologic log on the left to surface seismic data across the top. Processing sequence similar to that described in Figure 11.

APPLICATIONS TO SEISMIC EXPLORATION

Identification of reflected events on surface seismic profiles

Figure 11 shows a processed vertical seismic profile from the Pfister Fee no. 1 well, Niobrara County, Wyoming, and Figure 12 shows a similar profile from the Sam Gary Bell Creek Madison no. 1 well, Powder River County, Montana. A portion of the acoustic log and lithologic log is plotted beside the vertical-depth scale. A portion of a conventional surface seismic profile runs horizontally across the top of the figures. In both cases, the portion of the surface profile shown runs almost directly over the well itself.

If a major event on the surface profile is selected, for example the one labeled "Madison Group" in Figure 12, we can identify a corresponding event at the top of the VSP. We can then track that event on the VSP, level by level, to its zone of origin which starts at about 6800 ft. The Madison identification is reliable because this zone of origin corresponds to the top of the Madison on the lithologic log. It is also identified by the significant deviation shown on the velocity log just below that level.

Another event of considerable interest to us is from the First Leo horizon (labeled 1st L in Figure 11), which can be traced from its origin at about 3200 ft at the Pfister Fee well location to the surface profile run across the well. Other events on the surface recordings can be similarly identified.

Although other procedures are available to identify surface seismic events (such as synthetic seismograms, based on the theoretical response of a simplified model of the earth), we believe that actual observations of actual seismic events generated in the real earth give one a considerably higher degree of confidence in the identification of these events.

Figure 13 offers another type of surface-subsurface tie. It shows a portion of a surface seismic line across the Bell Creek field well. The seismic cross-section has been split in the middle, and a set of traces from the top of the VSP (Figure 12) has been inserted. The matching of events between the two different sets of real seismic data—recorded 6 months apart by different crews using different equipment and different seismic sources—is quite good.

Analysis of a surface seismic exploration problem

The Department of Energy's Nevada Test Site is a major testing ground for nuclear weapons. The depth to the "Paleozoic basement" surface (Tonopah formation) and the local structural configuration of this surface are two essential parameters used to select sites for the underground detonation of these weapons.

Efforts to map this surface with surface-to-surface reflection seismic surveys have had limited success. The thick attenuating cover of loose alluvium has frequently been blamed for the apparent absence of reflections. Other investigators have observed that the Paleozoic basement surface is erosional. It is, therefore, probably an irregular surface. Such surfaces are often poor reflectors of coherent, recognizable seismic energy.

In September 1978, a series of VSP investigations was carried out at the U-10-bd location of the Test Site (Balch, Lee, and Muller, 1980). One of the processed data sets is shown in Figure 14. These data have been velocity filtered to enhance upward-traveling events.

Two features are immediately apparent on these data. First, a first arrival (source pulse) appears on all traces. This pulse is so

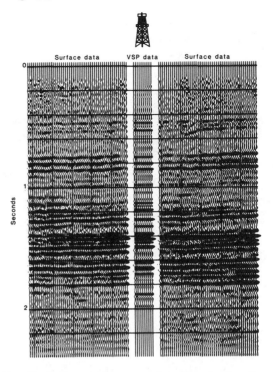

FIG. 13. Surface seismic cross section across the Sam Gary Madison No. 1 well, Bell Creek field, showing tie to vertical seismic profile data. Upper traces of the processed vertical seismic profile are spliced into the middle of the cross section.

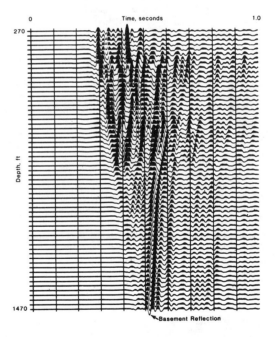

FIG. 14. Edited and processed vertical seismic profile from the UE-10-bd well, Nevada Test Site, Nevada. High amplitude coherent event at .5 sec on bottom trace is a Paleozoic basement reflection. After Balch et al. (1980).

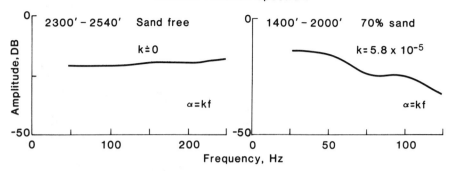

FIG. 15. (a) Amplitude spectrum plot (smoothed spectral ratio) for seismic waves traversing a nearly sand-free section of the Minnelusa Formation, derived from vertical seismic profile data. Attenuation α (nepers/ft) is assumed equal to the product of a constant k (neper-sec/ft), which is characteristic of the rock, and frequency (cycles/sec), or $\alpha = kf$. (b) Similar plot from a Minnelusa section containing approximately 70 percent sand.

strong that it appears in spite of the powerful attenuating effect of the multichannel velocity filter on downward-traveling events. Second, a strong Paleozoic basement reflection appears at about .510 sec on the lower trace in the figure. Clearly, neither of the previous explanations for the absence of Paleozoic reflections will stand up when examined in the context of the VSP data.

If we track the reflection upward in Figure 14, we observe that about halfway to the surface it becomes obscured by other events and finally disappears altogether. Our tentative explanation for this effect is that the top of the Paleozoic basement is block faulted close to the borehole. Block faulting is prevalent in the area. The Paleozoic reflections shown on the figure would then come from the plane of the fault. At the top of the fault, the reflections would become weaker and then disappear. One cannot construct a raypath emanating from the source, reflecting from the faulted Paleozoic layer, and arriving at the surface near the top of the U-10-bd well; the fault-plane geometry will not permit it. Only when the detector is deep in the well can such a raypath be constructed.

Other processing versions of these data show a pronounced reverberation effect in the upper 500 ft of the section. The reverberation from energy trapped between the surface and the alluvium-bedrock layer are so strong that they could completely obscure weaker, reflected events from the Paleozoic, if they were present.

Additional field investigations would be necessary to verify these interpretations. They have not yet been performed, so our conclusions remain tentative. However, results of the investigation demonstrate the value of VSP to diagnose and help prescribe a cure for a difficult, practical seismic field problem: to obtain Paleozoic basement reflections on a surface seismic profile at the Nevada Test Site.

Support of seismic-stratigraphic exploration

Vertical seismic profiles can be a valuable tool in stratigraphic exploration for petroleum and, in some cases, for water. A typical stratigraphic play involves one or more target (stratigraphic) horizons. VSP may be used to measure the seismic response of these key horizons under both porous-productive and the tight-

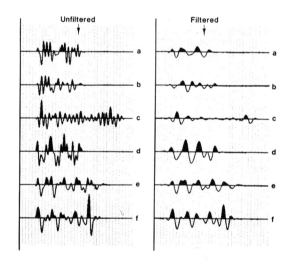

FIG. 16. Six reflection impulse responses from the Madison Group-Red River interval, eastern Powder River basin, Wyoming-Montana, calculated from vertical seismic profiles. The upper three response waveforms (a, b, and c) are from tight, barren sections; the lower three (d, e, and f) are from porous sections producing prolific quantities of water. Waveforms on the left are unfiltered; on the right, the waveforms have been zero-phase, band-pass filtered to approximate the typical surface seismic recording frequency band in the area. [After Balch, Lee, Miller and Ryder (1980)].

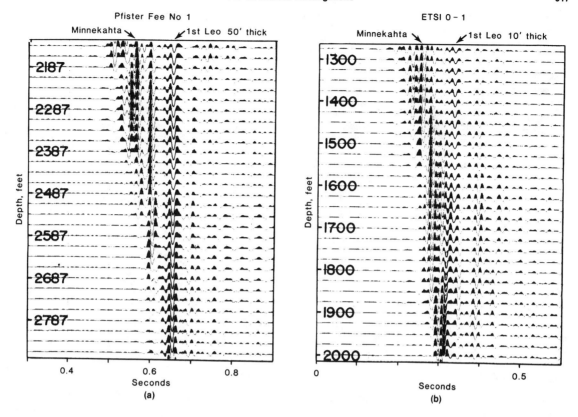

FIG. 17. First Leo sand reflections and Minnekahta reflections from two wells in the eastern Powder River basin, Wyoming, from vertical seismic profiles (a) Leo section 10 ft thick or less (b) 50 ft thick Leo section. Note the reduced relative amplitude of the Leo in (a). After Balch, Lee, Miller, and Ryder (1981).

barren conditions. If one can establish a difference in seismic response between these two "end-member" conditions, it may be possible to use this difference to detect the porous-productive end members on the basis of seismic reflections. We shall exhibit three types of seismic responses obtained with VSP by means of which end members might be identified: frequency-dependent attenuation, "character" of the impulse response of a seismic horizon, and amplitude of the impulse response of a seismic horizon.

In Figure 15, we show the result of two attenuation measurements made in the Minnelusa formation in the eastern Powder River basin of Wyoming. In this area, Leo sand buildups within the Minnelusa frequently contain commercial quantities of oil (Ryder, 1980). We assumed attenuation to be directly proportional to frequency (White, 1965, p. 93). When the spectral ratio of the transmitted seismic waveform with respect to the incident waveform is plotted to a logarithmic scale, the slope of the envelope is approximately equal to the factor of proportionality. In Figure 15a, the sand percentage in the Minnelusa, in the interval measured, is about 6. The observed attenuation proportionality factor is below 10^{-6} neper-sec/ft. Figure 15b shows a similar plot for a Minnelusa section containing about 70 percent sand.

The observed attenuation factor is approximately 5.8×10^{-5} neper-sec/ft. This suggests that attenuation may be a useful rock property to measure when exploring for (Leo) sand buildups in the Minnelusa. This type of measurement was discussed in considerable detail by Hague (1981) and Spencer et al (1982).

With VSP data we have been able to observe the downward-traveling (input) seismic wave train that enters the top of an interval, and also observe the resulting upward-traveling wave train reflected by the interval. Using these two wave trains, we calculate the reflection impulse response or transfer function of the interval. We take this result to be the reflection wave train that would have been generated if the input had been a short duration pulse. This impulse response is often zero-phase filtered to match the frequency band of surface seismic recordings made in the area.

Reflections from the Madison-Red River interval of the eastern Powder River basin, Wyoming and Montana, provide an example of how reflection character can help one distinguish between porous-productive and tight-barren members of a target stratigraphic horizon. Madison-Red River porosity in this area is often associated with prolific water production (on the order of 1000 gallons per minute). This water is currently in great demand for domestic use, agriculture, coal slurry pipelines, and secondary oil recovery activities. In other areas, for example, the Big Horn basin of Wyoming, Madison porosity is often associated with oil and gas reservoirs.

Six filtered impulse responses are displayed in Figure 16. The upper three (a, b, and c) are derived from tight-barren Madison-Red River sections. The lower three were obtained from porous-productive sections. We feel that the sets are readily distinguishable, even though it is difficult to describe the differences quantitatively. Interbed reflections from the intervals are rare in the tight-barren cases. Because much of the porosity is often confined to "zones" in the productive examples, interbed reflections would not be unexpected. The presence of reflections from within a porous Madison-Red River interval and the absence of interbed reflections when the interval is tight are also suggested on surface profiles (Balch, Lee, Ryder, and Miller, 1980).

We return to the Leo sand case as a final example of how we have used VSP data in support of our stratigraphic-seismic investigations. Figure 17a shows a processed VSP from the Pfister Fee no. 1 well, Niobrara County, Wyoming, which produces oil from a 50-ft section of Leo sand within the Minnelusa formation. Figure 17b shows a similarly recorded and processed VSP from the ETSI no. 0-1 well, Niobrara County, Wyoming, about 10 miles south of the Pfister Fee. In the ETSI well, the Leo is nonproductive and at most only 10 ft thick. Both VSPs have been amplitude adjusted to make the shallower Minnekahta reflection unit amplitude. Note the substantially greater Leo reflection amplitude relative to the Minnekahta at the Pfister Fee location. This suggests that commercially productive Leo sand buildups in the area may be detectable as amplitude anomalies. We believe this case for amplitude anomalies is a strong one. It is based on actually observed reflections (not synthetic data). We had control over the source amplitude and source conditions, so we can eliminate changes in near-surface conditions as a source of the amplitude anomaly. This amplitude anomaly was recently corroborated with a (horizontal) surface seismic profile (Balch et al, 1981).

SUMMARY AND CONCLUSIONS

Vertical seismic profiles have a wide variety of applications to reflection seismic exploration problems. We have cited examples from seismic-stratigraphic analysis, field exploration troubleshooting, and high confidence level identification of seismic events on surface-recorded seismic records. This is by no means an exhaustive list of current or potential applications. Kennett et al (1980) and Hubbard (1979) discussed applications of VSP to the deconvolution of surface-recorded (marine) seismic data, predicting geology ahead of the drill bit and other applications. Gal'perin (1973) listed 13 applications, some of which would include the nonstratigraphic examples given here. Many of Gal'perin's applications require the use of three-component detectors and/or long source offsets, two variations not considered here. Gal'perin had very little to say about computer processing of digitally recorded data.

We predict better equipment will become available and more individuals and organizations will become active in this area. We feel confident that in the near future, most of the applications cited, and others as well, will be widely utilized.

ACKNOWLEDGMENTS

The authors acknowledge with thanks the assistance and encouragement provided by many of their colleagues at the USGS. Special recognition is due Sam West, William Head, Elliott Cushing, and Richard Mast.

REFERENCES

Balch, A. H., Lee, M. W., and Ryder, R. T., 1977, Use of vertical seismic profiles in stratigraphic exploration (abstr.): SEG Midwestern Meeting, p. 54–55.
Balch, A. H., Lee, M. W., and Muller, D. C., 1980, A vertical seismic profiling experiment to determine depth and dip of the Paleozoic surface at Drill Hole U-10-bd, Nevada Test Site: U.S.G.S. Open-file rep. 80-847.
Balch, A. H., Lee, M. W., Ryder, R. T., and Miller, J. J., 1980, Use of vertical seismic profiles and surface seismic profiles to investigate the distribution of aquifers in the Madison Group and Red River Formation, Powder River Basin, Wyoming-Montana: Preprint 9312, 55th Ann. Conf. SPE, Dallas.
Balch, A. H., Lee, M. W., Miller, J. J., and Ryder, R. T., 1981, Seismic amplitude anomalies associated with First Leo sand buildups, eastern Powder River Basin, Wyoming: Geophysics, v. 46, p. 1519–1527.
Dix, C. H., 1939, Interpretation of well-shot data: Geophysics, v. 4, p. 24–32.
Gal'perin, E. I., 1973, Vertical seismic profiling: Tulsa, SEG Spec. Pub. no. 12, 270 p.
——— 1977, Polarization method of seismic investigations: Moscow, Nedra.
——— 1980, Vertical seismic profiles at the exploration and exploitation stage: Akad Nauk. SSSR Dokl., v. 253, no. 6.
Gal'perin, E. I., Merzoyan, U. D., and Oyfa, V. Ya., 1980, Results of the use of the polarization method of VSP in the Krasnodarsh Region, USSR: Ekspress Informatziya, Regional exploration and commercial geophysics, Min. of Geol., USSR, vyp 16.
Hague, P. S., 1981, Measurements of attenuation from vertical seismic profiles: Geophysics, v. 46, p. 1548–1558.
Huang, C. F., and Hunter, J. A. M., 1981a, The correlation of "tube wave" events with open fractures in fluid-filled boreholes, in Current research, Part A: Geol. Surv. Can., paper 81-1A, p. 161–376.
——— 1981b, A seismic tube wave method for in situ estimation of rock fracture permeability in boreholes: Presented at the 51st Annual International SEG Meeting, October, in Los Angeles.
Hubbard, T. P., 1979, Deconvolution of surface recorded data using vertical seismic profiles: Presented at the 49th Annual International SEG Meeting November 7 in New Orleans, 11 p.
Jolly, R. N., 1953, Deep-hole geophone study in Garvin County, Oklahoma: Geophysics, v. 18, p. 662–670.
Karus, E. V., Ryabinkin, L. A., Gal'perin, E. I., Teplitskiy, V. A., Demidenko, Yu. B., Mustafayev, K. A., and Rapaport, M. B., 1975, Detailed investigations of geological structures by seismic well surveys: 9th World Petr. Cong. PD 9(4), v. 26, p. 247–257.
Kennett, P., and Ireson, R. L., 1973, Some techniques for the analysis of well geophone signals as an aid to the identification of hydrocarbon indicators in seismic processing: Presented at the 43rd Annual International SEG Meeting October 23, in Mexico City.
Kennett, P., Ireson, R. L., and Conn, P. J., 1980, Vertical seismic profiles: Their applications in exploration geophysics: Geophys. Prosp., v. 28, p. 676–699.
Lang, D. G., 1979, Downhole seismic technique expands borehole data: Oil and Gas J., v. 77, no. 28.
Levin, F. K., and Lynn, R. D., 1958, Deep-hole geophone studies: Geophysics, v. 23, p. 639–664.
Lynn, R. D., 1963, A low-frequency geophone for borehole use: Geophysics, v. 28, p. 14.
McCollum, B., and La Rue, W. W., 1931, Utilization of existing wells in seismograph work: AAPG Bull., v. 15, p. 1409.
Musgrave, A. W., Woolley, W., and Gray, H., 1960, Outlining of salt masses by refraction methods: Geophysics, v. 15, p. 141–167.
Omnes, G., 1980, Logs from P and S vertical seismic profiles: J. Petrol. Tech., v. 32, p. 1843–1849.
Ryder, R. T., 1980, Depositional setting of the middle part of the Minnelusa formation (Middle and Upper Pennsylvanian) and its implication for stratigraphically trapped oil accumulations in the Powder River Basin, Wyoming (abstr.): AAPG-SEPM convention, Denver.
Seismograph Service Ltd., 1980, The VSP modeling atlas: Holwood, Holston, Kent, 48 p.
Slotnick, M. M., 1936a, On seismic computations, with applications I: Geophysics, v. 1, p. 9–22.
——— 1936b, On seismic computations, with applications II: Geophysics, v. 1, p. 299–305.
Spencer, T. W., Sonnad, J. R., and Butler, T. M., 1982, Seismic Q —Stratigraphy or dissipation: Geophysics, v. 47, p. 16–24.
Turpening, R. M., Liskow, A., and Thomson, F. S., 1980, Seismic investigations of Antrim shale fracturing: Dow Chemical Co., Midland, Mich. (U.S. Dept. of Energy publ. FE 234690), 35 p.
White, J. E., 1965, Seismic waves: Radiation, transmission, and attenuation: New York, McGraw-Hill Book Co. Inc., 302 p.
Wuenschel, P. C., 1976, The vertical array in reflection seismology: Geophysics, v. 41, p. 219–232.

PREDICTION OF OVERPRESSURE IN NIGERIA USING VERTICAL SEISMIC PROFILE TECHNIQUES

by S. Brun (Elf Nigeria), P. Grivelet (Schlumberger EPS) and A. Paul (Schlumberger Africa).

ABSTRACT

In traditional drilling practice, the real time overpressure indicators are generally of very short notice, especially in complex growth fault systems, which is the case in the Niger delta. The conventional prediction from regional correlation and surface seismic interpretation has some limitations which are difficult to overcome in case of dipping horizons, as well as in complicated fault systems.

The VSP iterative modelling inversion gives a more acurate predictive profile of acoustic impedance even when recorded at several hundred meters above the overpressured zones. The results are valid even for non-horizontal pressure fronts. This inversion procedure is independent of the low frequency content in the seismic signal, contrarily to other inversion methods based on integration. It gives the change of compaction trend (which is indicative of overpressure) as well as some other details (e.g. sand shale sequences).

A number of cases (three in all) in various geological conditions are presented. Results in every case are confirmed, by drilling, within less than a hundred meters. The method is valid assuming a continuity in the trend of acoustic velocities in the well, down to the top of overpressure.

Theory of VSP inversion by iterative modelling

VSP recording allows the acurate separation of the downgoing and the upgoing fields. The exact downgoing wave train is known at each level. In the case of a one dimensional vertical VSP, the up and downgoing waves are travelling in the same column of formations and the upgoing wavelet is identical to the downgoing wavelet except for the extra multiples generated below the recording point.

Fig.1 illustrates the inversion technique used:

A layered model below the recording point is built such that the incident downgoing wavelet produces, through the corresponding primary and multiple reflections, an upgoing synthetic trace as close as possible to the real one.

The first iteration states the time positions of all the considered boundaries. The total number of considered boundaries is a parameter which is usually limited to 50 , corresponding to a length of trace to be inverted of about 1 second.

The choice of these boundaries is done by stepwise regression. Using as reference the downgoing wavelet, the highest reflection coefficient is looked for on the upgoing trace and the corresponding wavelet is removed from the trace. the next highest reflection coefficient is then looked for on the resulting trace ; the process is repeated until the desired number of boundaries is obtained(fig.2). This process insures that only important constrasts are dealt with, and remains relatively immune to noise. This time series of events includes primaries and main multiples generated below the recording point.

The subsequent iterative process keeps the initial position of the boundaries, but computes the contrast accross these boundaries to obtain a synthetic trace which nearly coincides with the real upgoing trace.

The absolute value of the acoustic impedance is known at the recording point from the density and sonic logs. So the series of impedance contrasts will start from that reference value. The predicted series can take as additional input in its modelling some geological knowledge; this knowledge may be either absolutely sure(fixed constraint) or with some degree of uncertainty(constraint with a tolerance range). These constraints will be taken into account at each round of iteration (fig.3).

Each iteration computes the gradient of error for each of the reflection coefficients and modifies the corresponding time series. When the total error is found within a predefined threshold, the solution is finally output.

The solution is not unique, but this process ensures that it is the most probable one if the boundaries selection is kept above noise level, and the constraints used are justified.

Undercompaction and overpressure

A formation pore pressure far above normal hydrostatic can originate either from absence of compaction during burial or ,if normal compaction has occurred,from fluid communication with an undercompacted formation.
A well known undercompacted shale series in Nigeria is the Akata formation.The pressure level in this formation is at least 1.5 times the normal hydrostatic.Due to the plasticity of this type of formation,the overlaying sediments have been piling up over a non stable substratum and creating a fault sytem.The primary growth faults cut all the way down and the bottom part of these fault surfaces coincides with the top of the Akata formation.The top of these undercompacted formations is then displaced and dipping at the vicinity of these growth fault systems.(fig.4)
At the contact between the undercompacted formations and the growth fault,there is a possibility of propagation of the high pressure into adjacent reservoirs or,due to drainage along permeable fault surfaces to higher reservoirs.The overpressure is then communicated to a complete block of formations usually limited by two faults and an upper seal.The formations in this block are then "decompacted"leading to lower than normal acoustic impedance for these sand-shale series.
Depending on the type of pressure communication the overpressure front can have different shapes.
In the absence of a pressure drain the front can coincide with the top Akata (flat or dipping) usually with a progressive increase of pressure.
In presence of faults with drainage communication the pressure can change very abruptly when crossing a fault,or may vary progressively from the upper seal to the lower part of the overpressured block.
This implies that often the progressive decrease of acoustic impedance with depth corresponds to more or less horizontal reflectors,while a very sharp change of pressure system accross a fault corresponds to a dipping reflector.

Examples

1-overpressure detection in a flat series: well A

This well is drilled at the center of a wide anticline far away from the main growth fault.In that region the overpressure depth was correlated with nearby wells and was predicted at 3720m far above the Akata.

The VSP was recorded when the well had a depth of 3220m. The inverted trace presented in fig.5 is recorded higher up in the well at a depth of 2300m. The constraints used for computing the inverted profile of acoustic impedance were:
-the value at the recording point in the well: 22,000 ft/sec.g/cm3
-the value at the bottom of the well: 30,000 +/- 5,000 at 2.75 sec
this second value is taken with a wide tolerance as if it was extrapolated from the general compaction gradient and the well was not already drilled that deep.
50 coefficients are computed for 1 second of time between 2.0 and 3.0 seconds of TWT.
From the computed impedance profile,a sharp decrease is showing at 2.81 seconds.This corresponds to an extrapolated depth of 3580m.
With this indication drilling was continued while watching for overpressure at the indicated depth.Other indicators were monitored at the same time and the well was stopped at 3550m before going into overpressure.A second well was then drilled next to the first one in the same pressure system.The drilling program was this time designed to get down to the overpressure.
Fig. 5 shows the total profile of acoustic impedance as measured by sonic and den-

sity logs.

The predicted profile and the actual profile are very close. Of course all the details are not represented as we modelled only 50 boundaries, but most of the main sedimentary sequences are represented. The most interesting point is the prediction of the undercompacted zone.

The prediction of overpressure at 2.81 sec corresponds to the beginning of a transition zone shown by sonic and density logs in the interval from 2.8 to 2.85 sec.

The identification is quite easy in such a series where there is no sharp break which can be attributed to lithogical change. In the central area of the Niger delta the impedance trend is steady and only disturbed by the overpressure fronts.

Identification is even easier in this case as the pressure front and all reflectors are horizontal, which means they are positioned at the right time on the VSP trace.

2-overpressure detection for a slightly dipping major growth fault: well B

Fig. 6 shows well B was drilled down to a major growth fault which meets the top of the Akata shales tangentially.

The VSP survey was recorded when the well was drilled down to a depth corresponding to a two way time of 2.76 sec, to detect the transition to the Akata.

The upgoing traces of the VSP results of fig.6 are showing horizontal reflectors except for one.

This reflector is drifting to shorter time when the receiving geophone is travelling up the well. This indicates a dipping reflector which is identified as the top of undercompacted shales. The depth of this reflector can be predited by extrapolating it until it crosses the prolongated breaktime line below TD. This gives a prediction of 2.785 sec.

To refine this interpretation an inversion by modelling is performed on a trace recorded at a depth corresponding to a two way time of 2.5 sec in the well.

The starting impedance in the well is 23,000 ft/sec.g/cm3 and the general compaction gradient is simulated by a very loose constraint of 30,000 +/- 5000 at 3.3 sec. Over this interval from 2.5 to 3.3 sec 50 boundaries are modelled.

The results of the inversion is shown in fig. 6.

A steady compaction trend is first slightly decreasing at 2.7 sec and then completely stopped at 2.83 sec. This is interpreted as a transition zone, on top of a massive undercompacted series starting at 3.83 sec.

This prediction enabled stopping the exploratory drilling at the right depth.

3-overpressure prediction with a highly dipping pressure front: well C

Well C was drilled in the early part of the Niger delta and the sedimentary sequence does not show a very steady compaction trend. Besides, the well is going through a series of faults and the pressure system is probably linked to these faults.

The well was drilled to TD at 2100m and a VSP recorded from 2100 up to 1000m. the goal of the survey was to evaluate if it would have been possible to predict the overpressures which had already been encountered.

Fig. 7 shows the acoustic impedance profile given by the sonic and density logs. No steady compaction trend is seen, but knowing that the lithology is the traditional sand shale series of the Niger delta, the sharp decreases at 1.07 sec and 1.28 sec (1500m and 1850m) can be attributed to going into overpressures. This is confirmed by the dipmeter log of fig. 8 which shows two major fault breaks at the same depths.

The inversion process was performed on a trace recorded at 1060 m. Fifty boundaries were modelled between 0.8 and 1.8 sec, the starting impedance is 28,000 at 0.8 sec and one constraint is imposed: 30,000 +/- 15000 at 1.1 sec.

The inverted trace shows several major breaks.At the top a low impedance at 0.91 sec and a high impedance at 0.96 sec are correlating with the actual profile.Below the two major breaks at 1.02 sec and 1.22 sec are showing as shifted from the corresponding breaks in the actual profile at 1.07 and 1.28 .These breaks appear as highly dipping events on the VSP results and are strongly drifting to shorter time on the upper recorded levels.

On the inverted profile the horizontal reflectors appear at the right time,while the dipping ones are shifted to early times by an amount depending on the actual dip.Of course this type of one dimensional modelling cannot reposition correctly the dipping events.However,when these dipping events have very strong contrasts, they show as superimposed on the normal one dimensioned boundary series and can be picked out.The actual dipping of these events can be detected on the VSP upgoing traces plot,and the true amount of time drift can be approximated.

In this well,using the inverted trace as a prediction,it appears that the most probable overpressure front corresponds to the sharp decrease at 1.22 sec.Besides,the corresponding reflector shows a dipping trend which could be extrapolated to the first break line at 2.88 sec. The strong dip at that point was confirmed by the dipmeter which shows a fault surface at an angle of 40 degrees to the SE.The starting of the undercompaction at this point is also confirmed by a massive shale section just below the fault point.

J

REFERENCES

Grivelet P.A.,1983 53rd SEG meeting in Las Vegas,Inversion of Vertical Seismic Profiles by Iterative Modelling.

Schlumberger,1983 Well Evaluation Conference of West Africa, Chapter 5.

ABOUT THE AUTHORS

Serge Brun graduated from Pierre and Marie Curie University in Paris in Geophysics and received a geophysics engineering degree from French Petroleum Institute in 1979.He has been working with Elf in seismic processing and is presently with Elf Nigeria as geophysics interpreter.

Pierre Grivelet graduated from Ecole Centrale de Paris (France) in 1978 and has been working in the Interpretation Development Department at Etudes and Production Schlumberger in Clamart France.He is currently working in the interpretation of borehole seismic data.

Alain Paul graduated from Ecole Centrale de Paris (France) in 1966. He joint Schlumberger in 1968 and served in the Middle East and Africa as operating engineer,field manager and interpretation adviser. Presently he is a member of the Interpretation Development Group for the Schlumberger Unit of South Europe and Africa ,based in Paris.

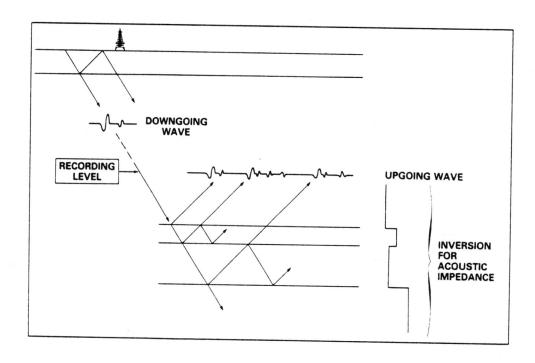

Fig. 1- *Primary and multiple reflections generated below the recording point by the downgoing wavelet.*

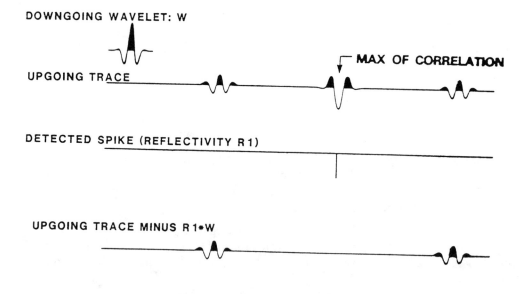

Fig. 2- *Selection of the reflective boundaries by stepwise regression.*

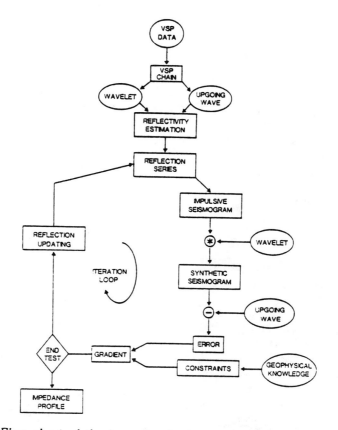

Fig. 3- Flow chart of the inversion by iterative modelling.

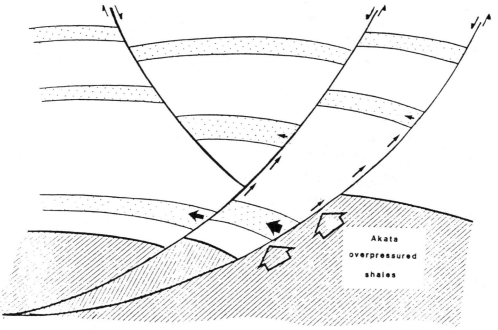

Fig. 4- Typical representation of overpressure communication in the Niger delta area.

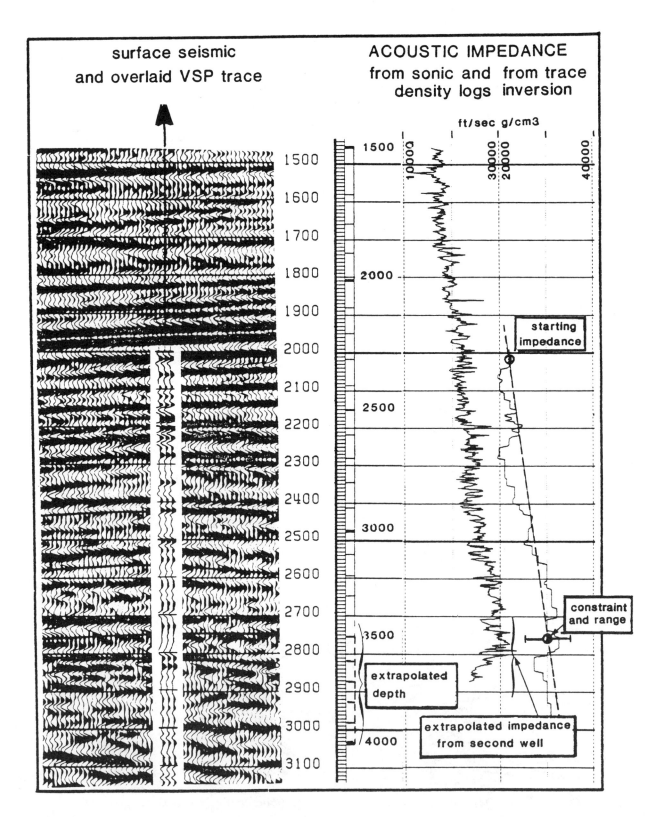

Fig. 5- Well A: overpressure prediction in a flat series.

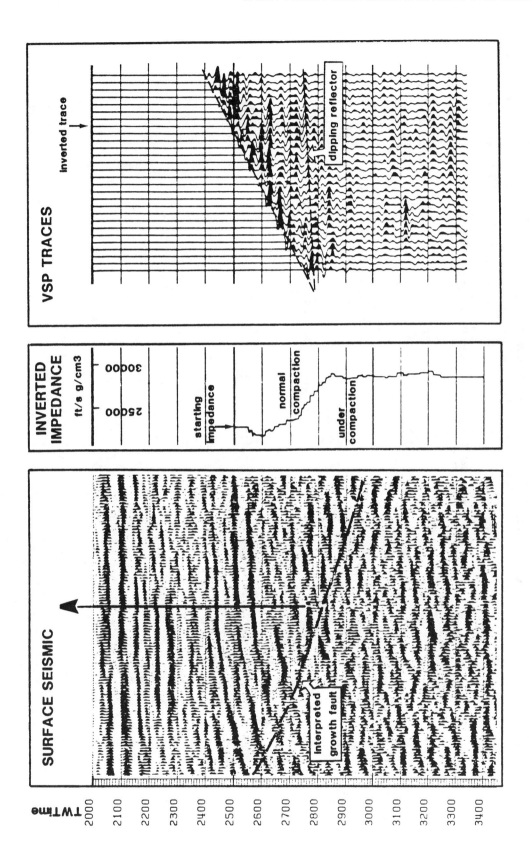

Fig. 6 - Well B: overpressure prediction in a slightly dipping transition to massive shales.

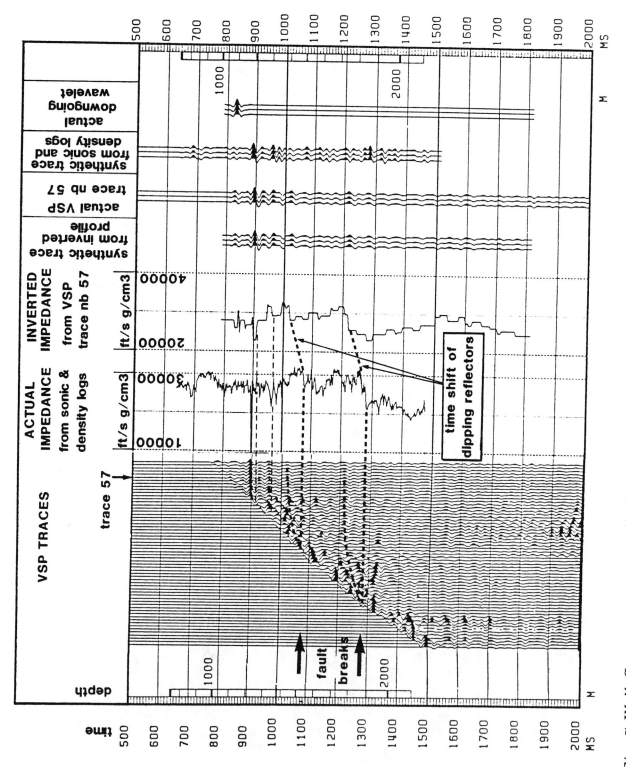

Fig. 7- Well C: overpressure prediction in a case of a highly dipping sharp transition through a fault surface.

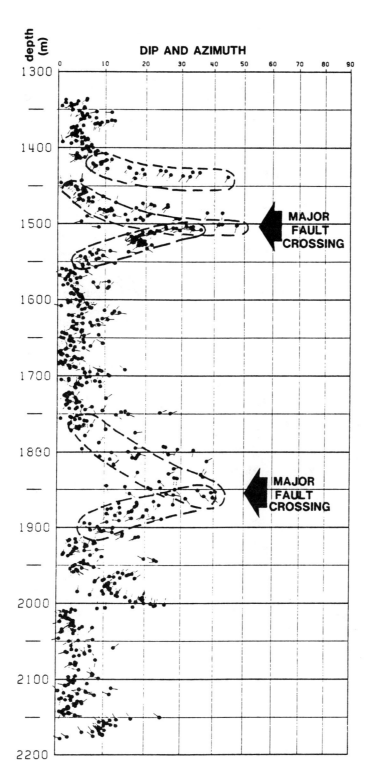

Fig. 8-Dipmeter results in well C showing sharp fault breaks.

OFFSET SOURCE VSP SURVEYS AND THEIR IMAGE RECONSTRUCTION*

P.B. DILLON and R.C. THOMSON**

Abstract

DILLON, P.B. and THOMSON, R.C. 1984, Offset Source VSP Surveys and their Image Reconstruction, Geophysical Prospecting 32, 790–811.

Zero-offset-source VSP surveys provide information about the subsurface only within the Fresnel zone centered at the well. Offsetting the source location moves the reflection zones away from the well thus providing lateral cover.

Conventional processing of this type of data gives rise to a distorted image of the subsurface. Using a simple ray-tracing scheme, this image may be reconstructed into the more familiar coordinate system of the surface seismic section. This simple data-independent mapping is based on the assumption of horizontal layering and requires a vertical velocity profile.

The technique of placing the source away from the borehole was first applied to the single-offset-source VSP survey. However, data from any survey geometry (such as deviated well with rig source, walkaway VSP, etc.) can be mapped to the coordinate system defined by the appropriate seismic section.

To obtain the best results from this type of survey the target area must be defined and simple modeling techniques used to optimize the source location(s). These pre-survey modeling methods may also be used to anticipate—and hence avoid a number of problem areas which experience has highlighted.

The data from any VSP survey is the result of a realizable experiment and as such obeys the wave equation. This implies that the wave equation may be used to migrate the data to its true subsurface location. Theoretically, such a process is more secure than ray-tracing techniques, although its practice presents many difficulties.

Introduction

The Vertical Seismic Profile (VSP) technique, as normally executed with a single source position near the well head, provides subsurface information only within the Fresnel zone centered at the well. Careful processing can yield quantitative information regarding, say, dipping or faulted reflectors (Kennett and Ireson, in a paper read at the EAEG meeting in Venice 1981), but in principle the survey is one dimensional and

---

* Based on a paper presented at the 45th meeting of the EAEG, Oslo, June 1983, revised version received February 1984.
** Seismograph Service (England) Ltd, Well Survey Division, Holwood, Westerham Road, Keston, Kent BR6 6HD, England.

hence insensitive to structure away from the borehole. The idea of altering the conventional survey geometry to allow illumination of subsurface structure away from the well is an attractive one, and is fairly new to the western world.

A major contribution towards developing new techniques has been the paper by Wyatt and Wyatt, read at the SEG meeting in Los Angeles 1981. This paper has shown that offsetting the source location with respect to a vertical array of downhole geophone stations can give sufficient information for the generation of "local" seismic sections. With a number of simplifying assumptions, the distorted image obtained from the VSP reflection response may be reconstructed into the familiar co-ordinate system of surface seismic sections. The mapping procedure is based upon raypath analysis and assumes that the Fresnel zone of each reflection from a subsurface interface is vanishingly small. Although the assumption is incorrect the mapping can lead to remarkably good results.

This arrangement, however, is only one of many possible survey geometries, all of which may be processed using the same basic mapping procedure. Here we shall consider, in addition to the far-offset source VSP, the deviated well VSP with a fixed source location and the so-called "walkaway" VSP with a traverse of source positions across the well.

Before proceeding, it should perhaps be noted that the established VSP processing techniques, originally developed to generate high-resolution reflection images for conventional (one-dimensional) zero-offset surveys, are still applicable to the two-dimensional geometries covered by this paper. The most important properties of VSP survey data which lead to effective processing are as follows:

(a) Geophones are located beneath the major multiple generators so that the downgoing wave varies little from geophone station to geophone station. This property, combined with the spatial arrangement of the geophones, leads to a very effective separation of the upgoing and downgoing wavefields.

(b) Each reflection from below the geophone is very nearly a replica of the downgoing wave. Since this is known, an accurate and completely deterministic deconvolution operator may be designed to collapse the multiples in the reflection response. Even very long term multiples of the order of a second can be accounted for. In contrast, the statistical deconvolution methods used on surface seismic data remove only the first few hundred milliseconds of multiple activity, leaving longer term multiples in the data. Because statistical deconvolution is based upon the assumption of randomly distributed reflection coefficients, a departure from true randomness may introduce artifacts into the final seismic section.

In addition to these properties, knowledge of the geophone location allows accurate determination of the depths of reflectors, especially those which intersect or approach that portion of the well traversed by the geophone.

Subsurface Illumination

By elementary ray tracing it is clear that offsetting the source with respect to the downhole geophone entails a movement of reflection points away from the well. What is not so clear is how these reflection points vary with depth and the velocity profile of the subsurface, for given source and receiver positions. The simplest model for investigation is that of a horizontally layered earth whose velocity is constant. Figure 1a traces the reflection ray paths through a 10-layer model from an offset source to a single receiver at depth. If the spatial sampling interval is reduced to increase the number of reflectors, then the points of reflection begin to describe a well-defined curve. This is the smooth curve which has been drawn through the points of reflection on the diagram and is the locus of reflection points for this particular source/receiver pair. As reflector depths increase, so the points of reflection approach the horizontal midpoint position between source and receiver. For the horizontally layered model, no reflection point may occur beyond this midpoint position.

When a nonuniform velocity profile is introduced into the model, the reflection point locus becomes more complex due to refraction effects. Figure 1b illustrates this situation and indicates that refraction must be taken into account for an accurate assessment of subsurface cover. In the following analysis, however, the velocity has been assumed constant to allow direct comparison of various VSP survey geometries.

Offset source VSP (OVSP)

A survey conducted with a regularly spaced vertical array of geophones and a single fixed source, offset from the well, generates a series of geophone traces which form an OVSP. The reflection point loci shown in Fig. 2a describe the subsurface illumination

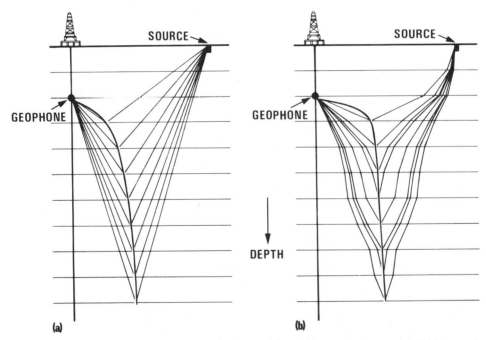

Fig. 1. (a) Ray path and reflection point locus plot for a uniform velocity model. (b) Ray path and reflection point locus plot for a non uniform velocity model.

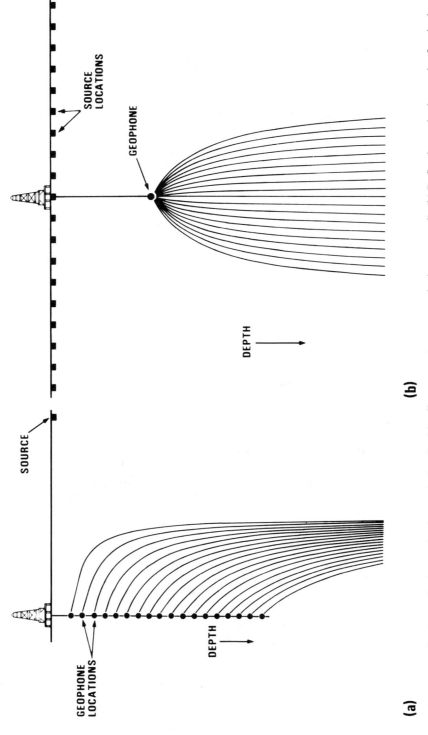

Fig. 2. (a) Reflection point locus plot for vertical well with offset source (constant velocity assumed). (b) Reflection point locus plot for single downhole geophone with a traverse of shots across the well.

obtainable with this survey geometry. Since the locus of reflection points for each source/geophone pair tends towards the midpoint as reflector depth increases, the spatial sampling of reflectors in this zone becomes increasingly concentrated. Such irregular spatial sampling may be considered a disadvantage but, conversely, it suggests the possibility of extremely detailed resolution of selected portions of the subsurface, provided the survey geometry has been correctly defined.

Note that this survey geometry is inappropriate for illumination of structure directly below the well and so its main application must be for investigation of the well environs.

Walkaway VSP

The walkaway VSP (also referred to as OSP) traditionally consists of a single downhole geophone recording the earth's response to a wide traverse of shots centered above the geophone. Figure 2b shows the reflection point loci which concentrate at the geophone but very quickly assume a well-behaved spatial separation with depth.

The major problem so far encountered with the walkaway VSP geometry is that of separating the upgoing and downgoing wavefields. With only a single downhole geophone station there is insufficient spatial information to achieve this. Ideally a string of regularly spaced geophones would give sufficient data for wavefield separation using multichannel filtering techniques. An equivalent but more time-consuming arrangement is to repeat the survey for a number of different geophone depths.

In contrast to the OVSP, the walkaway VSP survey illuminates the zone below the geophone, and this gives it the potential for prediction of structure ahead of the bit.

Deviated well VSPS

When a well has been significantly deviated, a moving source/moving geophone VSP survey may be performed wherein the source is maintained in a position vertically above the geophone (Kennett and Ireson, in a paper read at the SEG meeting, Calgary 1977). Such a survey gives rise to a mini seismic section over the lateral extent of the well, which has gained acceptance in the industry (Johnson, Riches and Ahmed 1982). Figure 3a illustrates the subsurface cover provided by the survey. The absence of "bend" in the reflection point loci indicates that, within the context of this presentation, no lateral "mapping" of the data is necessary. Simple time shifts, corresponding to the vertical traveltimes to the geophones, are all that is required to position the reflections at their required locations.

Data acquisition for marine wells requires mounting the source on a supply boat and steaming along the required traverse, firing the airgun at regular intervals. However, it may be inconvenient or impossible to place the source at the required locations. This is often true for land wells. Under these conditions, and if the path of the well is constrained approximately within a vertical plane, an alternative scheme is to use a fixed source at the well head. The subsurface covered by a source located at the well head is shown on fig. 3b and compares favorably with the moving source/moving geophone arrangement. Figure 4a shows the cover provided by a source directly above

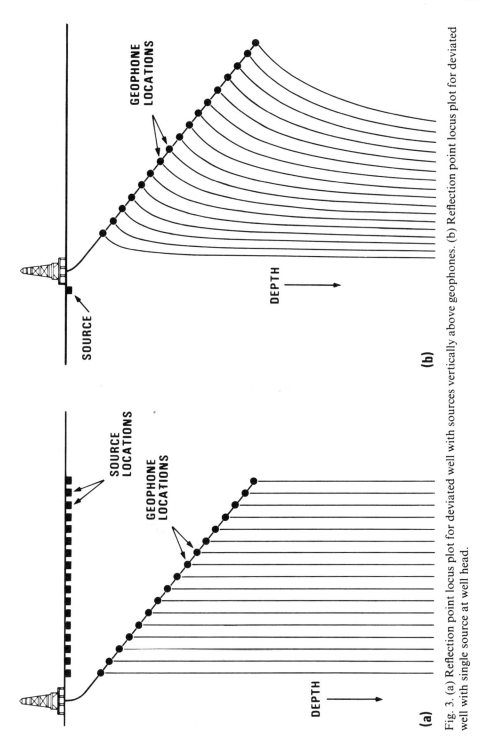

Fig. 3. (a) Reflection point locus plot for deviated well with single source at well head. (b) Reflection point locus plot for deviated well with sources vertically above geophones.

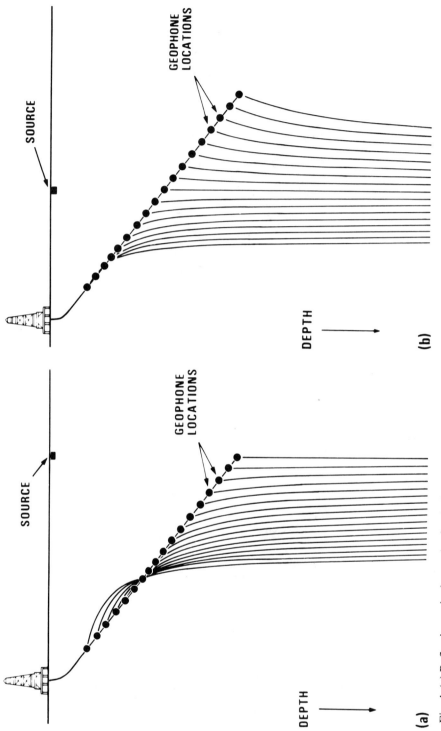

Fig. 4. (a) Reflection point locus plot for deviated well with source vertically above TD. (b) Reflection point locus point for deviated well with source vertically above mid-point position.

TD; in this case the extent of coverage tends to be limited with a much finer spatial sampling towards TD. The source position may be adjusted between these extremes to weight the coverage around a particular zone of interest; for example, see fig. 4b

Reconstruction of the VSP Reflection Image

Unlike true migration, which is a multichannel process, the simple mapping technique described here is a single-channel process and hence independent of the structure of the data: an event on a trace will be mapped without regard to its relation to events on adjacent traces. Because the layering is assumed horizontal, only reflections from horizontal and near-horizontal interfaces are correctly handled.

Figure 5 outlines the principle of the mapping technique for a constant velocity situation. The formulae simply transform a time on a trace to the x–y coordinates of a point in the subsurface. For real data with a complex velocity profile, ray-tracing techniques must be used to account for the refraction experienced by rays during their travels.

It is instructive to apply this simple mapping algorithm to the reflection response of a single downhole geophone in association with an offset source. The left-hand side of fig. 5 displays the reflected energy as recorded by the geophone and the right-hand side of the figure displays the same data after application of the simple mapping. By inspection it can be seen that the mapping program has effectively bent and stretched the trace to account for the moveout experienced by each sample of the reflection sequence. It follows that the mapping may be applied on a trace-by-trace basis to the complete reflection image of a VSP and the results superimposed to form the reconstructed image. Figure 7a shows the reflection image obtained from the data of an OVSP survey (after wavefield separation and deconvolution) and fig. 6 shows the image after this simple mapping procedure. This particular technique uses very little computer memory space because it takes advantage of the virtually unlimited storage capacity of an x–y plotter tape. Unfortunately, no further processing is possible on the data in this form and so it must be regarded as a final product.

A more versatile approach is to partition the subsurface so that it corresponds to a number of equally spaced traces, each trace representing some fixed offset distance from the well head. The mapping process is then the calculation of the x–y coordinates corresponding to each sample of the reflection sequence followed by its "binning" into the nearest trace. For comparison purposes, it may be useful to arrange the partitioning to be the equivalent to the processed surface seismic section.

This mapping procedure can be thought of as a trace-by-trace moveout correction followed by a species of CDP stacking. The moveout correction relocates a reflected event from its original recorded time to an estimate of its vertical two-way traveltime from the surface. The stacking procedure is of variable fold and gathers data into the nearest common-offset location. Figure 7b shows such a mapping of the same data as were used in fig. 6, but in this case, the velocity profile derived from the conventional zero-offset source VSP was used to establish the refracted ray paths.

Although the mapping procedure is not a true migration it has the advantage of being a very flexible algorithm which can be applied to the data from any survey

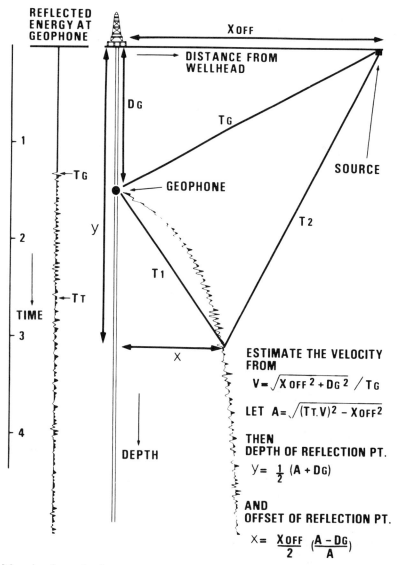

Fig. 5. Mapping formulae for constant velocity model and their application to a single OVSP trace.

geometry. Its flexibility is due to the very feature which precludes it from being classed as a true migration, i.e. the mapping is a single channel process. The only information required for each trace is the geophone location, the source location and a velocity profile. Thus the arrangement of geophone and source arrays is immaterial from a processing point of view.

For example, fig. 8a is a plot of the traces recorded during a walkaway VSP survey. As the shots are positioned further from the well, so the traces exhibit the expected

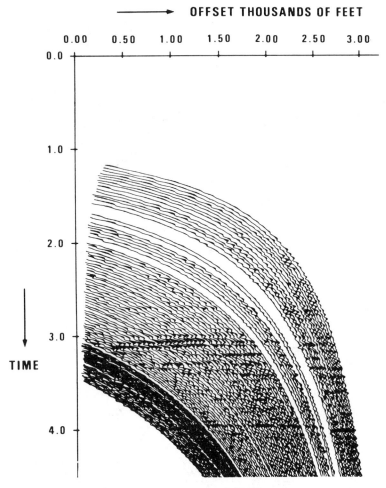

Fig. 6. Moveout corrected OVSP traces.

NMO distortion. Water multiples or near-surface multiples spend most of their propagation time in low-velocity media and as a result may be identified by their larger moveout distortion. Primaries and long-term multiples generated from deep reflectors propagate at higher velocities and hence show smaller moveout distortion. It is possible to exploit this difference in moveout in order to suppress the low-velocity multiples, using multichannel filtering techniques. Unfortunately, it is not possible in general to distinguish effectively between primaries and high-velocity downwave multiples for a survey conducted with a single downhole geophone station (as was used for this survey). However, high amplitude primaries can be seen; a rather obvious one occurs at 2.2 s in the well.

After applying the mapping algorithm to the traces, a local seismic section has been produced which clarifies the subsurface structure (fig. 8b). The primary identified in fig.

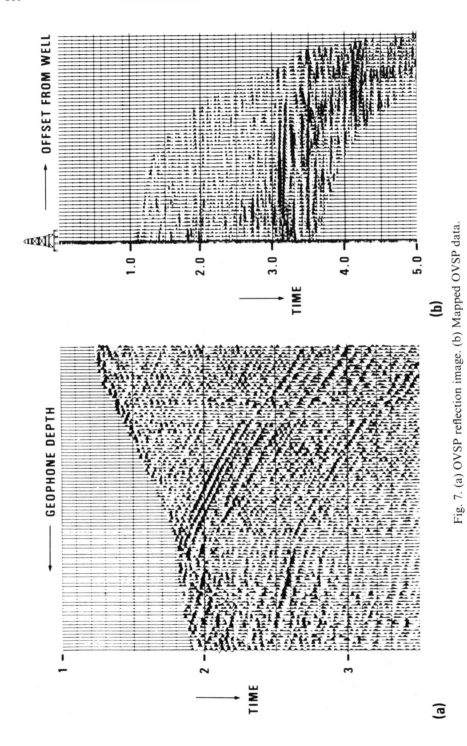

Fig. 7. (a) OVSP reflection image. (b) Mapped OVSP data.

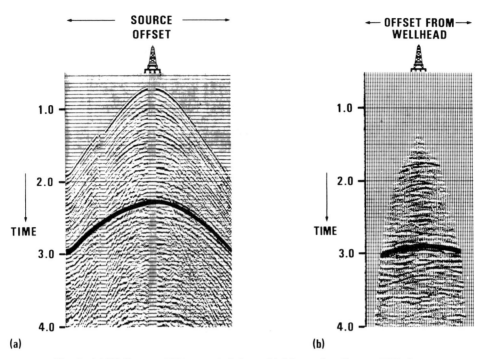

Fig. 8. (a) Walkaway VSP recorded data. (b) Mapped walkaway VSP data.

8a now occurs at 2.9 s (corrected two-way time) in this mapped section, and in addition a dipping event is also evident below this.

For deviated well surveys, the recorded traces have properties similar to those of the conventional zero-offset VSP in a vertical well. Figure 9 shows the traces as recorded during a deviated well survey with the source located at the rig. Broadly speaking, the incident wavefield is near enough parallel to the first arrival curve and the reflected wavefield is dipping in the opposite direction. After wavefield separation, deconvolution, and filtering, the VSP reflection image is available for inspection (fig 10a). To remove the spatial and temporal distortion, the mapping algorithm is applied to the data to produce the section of fig 10b. Only events from below the well are available from the survey, and so the plot represents a local seismic section from the well downwards.

Figure 10b indicates a major dipping reflector which intersects the well at approximately 2.64 s two-way vertical traveltime. This survey configuration, with the source positioned at the well head, has allowed complete illumination of the interface and shows it to be one continuous dipping structure. If the source had been positioned directly above the central portion of the reflector, then the structure would not have enjoyed complete illumination. This could have led to an erroneous interpretation of faulting where no fault exists. Thus attempts to improve illumination of a particular zone of interest by changing the source location may be defeated by unexpected dip.

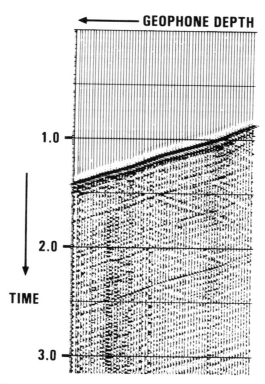

Fig. 9. Data from deviated well survey (single source at well head).

STRUCTURAL DISTORTION INTRODUCED BY THE MAPPING

The mapping procedure is secure for horizontal events but introduces errors for dipping events. Nevertheless, dipping reflectors are mapped correctly at and near to the geophones. The error is due to the assumption that each reflection comes from a horizontal interface. To account for the subsurface structure correctly, multichannel techniques based upon the wave equation should be used. Such techniques are not at present commercially available but are seen as a future development.

The accuracy of the mapping described in this paper may be established by simple modeling procedures. Figure 11a shows that part of an assumed structure which has been illuminated by an OVSP. This structure would be faithfully reproduced by perfect migration but the single channel mapping of fig. 11b has introduced some distortion. As expected, the horizontal faulted event has been mapped correctly, whereas the dipping events show a horizontal and vertical error away from the well. At and near to the well, all reflectors are correctly mapped because they are so close to the geophone array. The direction of dip has been preserved and, to a first approximation, the angle of dip is correct. However, the mapped image shows a migration down dip with respect to the assumed model and may be corrected by simple geometric considerations.

The same model has been assumed for a walkaway VSP. The illumination provided

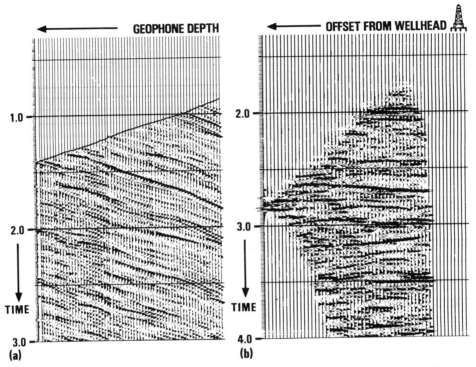

Fig. 10. (a) Deviated well survey (source at well head) reflection image. (b) Deviated well survey mapped data.

by the survey geometry is shown on fig. 12a. In this case, both sides of the well enjoy some cover. After mapping (fig. 12b), the image is seen to have similar distortions to the OVSP geometry. Again, events have been moved down dip and a correction is required to relocate them to their true positions.

This simple modeling technique is very usefully applied in the planning stage of a survey. Given the well location and the target structure, the parameters of the survey geometry may be adjusted until an optimum solution is found.

Practical Problems Encountered

High-frequency suppression

A noticeable feature of data acquired when the source is located a significant distance from the geophone is the loss of high frequencies, compared with the zero-offset VSP. The mechanisms for the loss of high frequencies may be absorption and scattering of the higher frequencies due to very local acoustic impedance changes. Lower frequencies are less susceptible to scatter because their wavefronts tend to "heal" fairly swiftly. The greater length of the propagation path in offset source surveys gives the loss mechanisms more opportunity to narrow the seismic pulse spectrum.

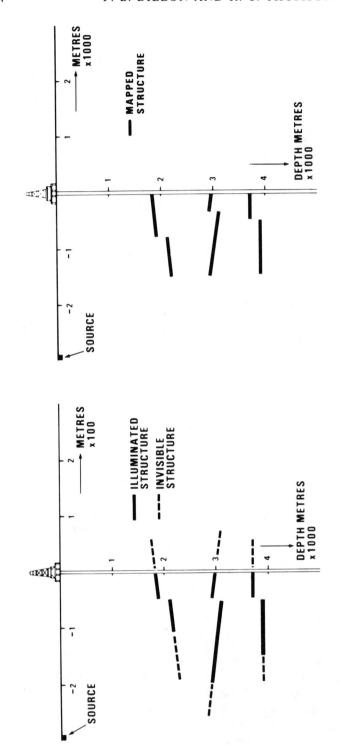

Fig. 11. (a) Illumination of structure provided by an OVSP geometry. (b) Image of illuminated structure after mapping (OVSP modeling).

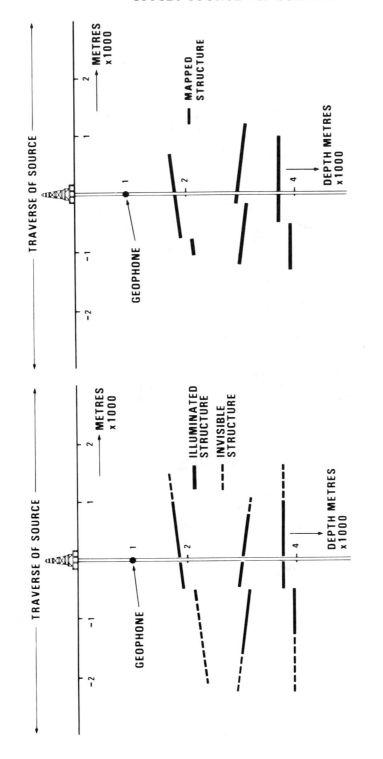

Fig. 12. (a) Illumination of structure provided by a walkaway VSP geometry. (b) Image of illuminated structure after mapping (walkaway VSP).

Mode conversion

Another problem which can be clearly seen on fig. 13a is that of mode conversion. Mode conversion is a direct consequence of high angles of incidence at large changes of acoustic impedance. At the low/high velocity interface, this section exhibits a significant mode-converted downgoing SV event recorded by geophones in the high-velocity medium (arrowed). As expected, the shear wave propagates at a lower

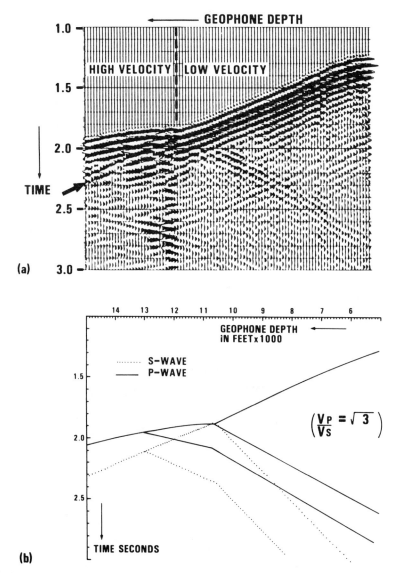

Fig. 13. (a) OVSP recorded data. (b) Kinematic estimate of P- and S-wave arrivals assuming a Poisson solid.

velocity than the P-wave first-arrivals. To verify that this is indeed shear energy, a simple two-layer model was defined assuming a Poisson solid for the lower layer ($V_p/V_s=\sqrt{3}$) and a kinematic plot of the expected geophone responses produced (fig. 13b). It is necessary to remove the shear agent with a reject dip filter because it interferes with the primary reflections recorded by the deeper geophones.

Refraction effects

When a geophone is locked into a low-velocity formation just above a high-velocity formation, it is possible for refracted "head" wave energy and reflected energy from within the high-velocity medium to arrive at the geophone before the incident energy whose path of propagation is restricted to the low-velocity medium. This apparently noncausal situation may lead to a false estimate of the propagation time. As the geophone descends into the high-velocity formation, refraction effects may also lead to timing inversions of the direct arrivals: that is, incident energy may reach deeper geophones before reaching neighboring shallower geophones. This can complicate the image reconstruction algorithm.

A useful parameter which gives a measure of this problem is the "Critical Source Offset" which we define as the maximum lateral distance the source may be placed from the downhole geophone before it becomes possible for reflected arrivals or refracted "head" arrivals to precede the direct arrival. During the design of a survey geometry it is good policy to obtain estimates of this distance for each geophone location—interval velocities derived from seismic stacking velocities are usually quite adequate for the calculation. The source should then be kept within this offset to minimize problems.

As an example, fig. 14 shows the variation of critical offset with geophone depth using the velocity profile derived from the OVSP survey.

Velocity mismatch

When a VSP survey is conducted with a far-offset source and a source at the rig, two sets of arrival times are obtained. Using these times it is possible to obtain two independent estimates of the velocity profile: one estimate from the vertical times and the other estimate (by ray-tracing techniques assuming horizontal layering) from the slant times. Experience has shown that the two estimates in general do not agree; fig. 15 shows two such estimates from the OVSP survey used previously and clearly indicates a difference between them.

The coarse spatial sampling provided by typical geophone separations may not account for fine layering of the true velocity profile. The route taken by the slant arrivals favors high-velocity layers in order to minimize the total traveltime in accordance with Fermat's principle. If the high-velocity layering is not resolved sufficiently then a mistie between the two estimates is inevitable.

The two sets of times allow estimation of velocity anisotropy within each layer bounded by two adjacent geophones, making the assumption that the layering is horizontal and the minor or major velocity axis is perpendicular to the surface (Byun

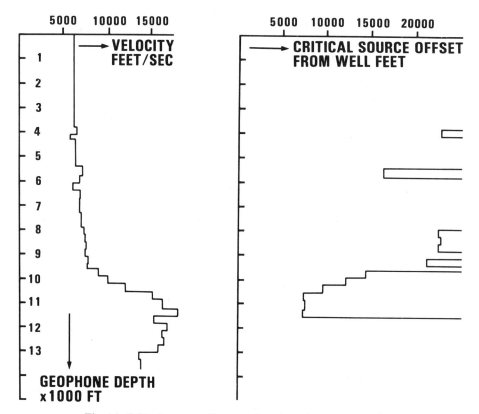

Fig. 14. Critical source offset as a function of geophone depth.

1983). Using ray-tracing techniques based upon Byun's formulae these two sets of times were processed. From the results (fig. 16) it would seem that reliable anisotropy estimates cannot be made from this data set.

Future Developments—Multichannel Migration

In this presentation only a straightforward mapping scheme has been applied to the VSP reflection image. That is, the VSP data have been transformed from the VSP coordinate system to the more acceptable surface seismic coordinate system. In order to apply the transformation, as described, certain assumptions have been made about the data:

(a) The reflecting interfaces are horizontal;
(b) All energy present is high frequency so that Fresnel zones are vanishingly small and ray-tracing techniques become valid;
(c) The velocity profile of the earth varies with depth but does not vary laterally.

The major advantage of this existing scheme is that it is very robust and hence

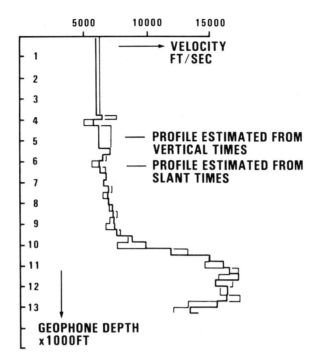

Fig. 15. Estimates of velocity profile derived from vertical and slant times.

tolerant of parameter variations. Its disadvantages are essentially defined by the assumptions (a) and (b) above. That is, dipping events are not correctly mapped and Fresnel zone effects are not collapsed. Despite these shortcomings the technique often leads to very good results.

However, for theoretically secure migration, a wave equation solution should be used to follow the wavefronts so that both P-wave and S-wave energy is accounted for. The S-wave energy is generated by mode conversion at each interface in the subsurface and is often seen on offset source VSP data. To quantify the mode conversions it is necessary to have detailed knowledge of the elastic properties of each rock layer (e.g., P-wave velocity and S-wave velocity). This information is not usually available for the unlogged shallow portion of the well where much of the mode conversion takes place. Even if this information were available, the complexity of calculating the stress displacement relationships at each interface would involve a high computing overhead.

It is, therefore, preferable to bypass the problem altogether by designing the survey so as to minimize mode conversion whilst meeting the survey objective. Any shear energy which is produced may be removed using dip filters and the remaining energy can then be assumed to be P-wave only. Thus the data should now be susceptible to a migration scheme using the scalar wave equation as for surface seismic data.

The differential form of the wave equation presents a number of difficulties for VSP data:

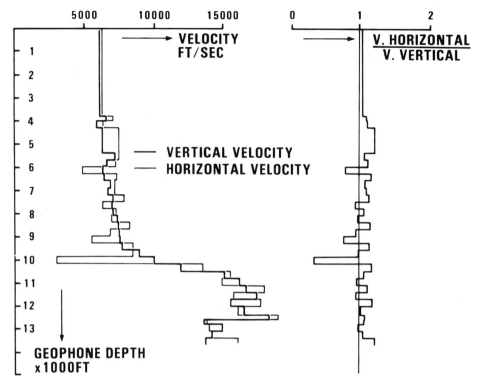

Fig. 16. An estimate of velocity anisotropy derived from two sets of arrival times.

(a) Both the incidence and reflected wavefields must be treated simultaneously so that the migrated data may be determined at their points of intersection. This is more complex than for the case of surface seismic data where an exploding reflector model can be assumed, that leads to one-way wave equation.

(b) For migrating an offset source VSP with a vertical array of downhole geophones "sideways continuation" of the phone array could be used. This would be analogous to the downward continuation methods as used with surface seismic data. A problem here is that the scalar wave equation is valid only for a constant velocity medium. This assumption is acceptable for surface seismic data where all the geophones can be assumed to be in the same medium during the downward continuation step. However, it is not a valid assumption for the VSP geophone array where the subsurface velocity may perhaps double over a few geophone positions.

(c) Edge effects are probably more important for VSP data than for surface seismic data. There are so few traces to be migrated that absorbing boundary conditions are essential. This problem must be addressed in the context of separate source and receiver in a varying velocity profile.

These problems are currently under investigation, and their solution should lead to an effective migration scheme to replace the ray-trace mapping technique at present being used. The final migration scheme is likely to differ markedly from existing seismic methods; the small number of VSP traces involved should allow more sophisticated or time-consuming processing techniques to become commercially viable.

Conclusions

Offsetting the source with respect to downhole geophones adds another dimension to the VSP image. Various survey geometries can be used, each with its own characteristics and pre-survey modeling techniques should be used to exploit these characteristics to best advantage with respect to the target zone.

The currently available single-channel mapping procedure, as discussed in this paper, may be used to reconstruct the image in the coordinate system of 2-D surface seismic sections. The mapping process is very flexible and efficient but the assumptions may restrict the accuracy of the final product. With this in mind, future developments must make use of the wave equation for the true migration of data. When this has been achieved, the industry will have a technique for the production of local high-resolution seismic sections. These sections will be used for the delineation of structure in the vicinity of the borehole and for detailed reservoir studies.

Acknowledgments

The authors wish to thank the Directors of SS(E)L for their permission to publish this paper. Thanks are also due to Amoco, Texaco North Sea UK Company, and other oil companies for their generosity in making their data available to us.

References

Byun, B.S. 1983, Seismic parameters for media with elliptical velocity dependencies, Geophysics 47, 1621–1626.

Johnson, R., Riches, H. and Ahmed, H. 1982, Application of the vertical seismic profile to the Piper field, Paper EUR 274, European Petroleum Conference 39–47.

OGJ REPORT

Drilling Technology Issue

About this report...

Can new drilling technology emerge from a recession-torn industry, strapped for cash and people?

This special Journal report leaves no doubt that it can.

In fact, new ideas presented here indicate that innovation and discovery have accelerated during hard times. Here's what you'll find in this report:

A new version of an old technique looks ahead of the bit for pressure, pay zones, or key formations.

More than a dozen important, practical uses have been discovered for sophisticated MWD. It's saving operators money now,
and the future looks bright.

How are computers coming? One new one answers a question near to the hearts and pocketbooks of drilling contractors and some operators. It creates a new rig count, with breakdowns by depth and contract, that tells where the rigs are, where the work is, and what the competition is like.

Another article may convince engineers to throw away their bit hydraulics slide rule.

And one article may prompt them to throw away their bit, too. A record 13,725 ft run is described, along with the near ideal conditions that made it possible.

Vertical seismic profiling technique emerges as a valuable drilling tool

R. J. Roberts
Consultant
Houston

J. D. Platt
Downhole Seismic Service Inc.
Houston

New developments in seismic profiling have created a useful drilling tool. The new techniques also are valuable exploration tools.

With adaptations of traditional seismic techniques, the drilling engineer can now learn more about how to drill the hole, while the geologist gathers his information on where to drill it. Types

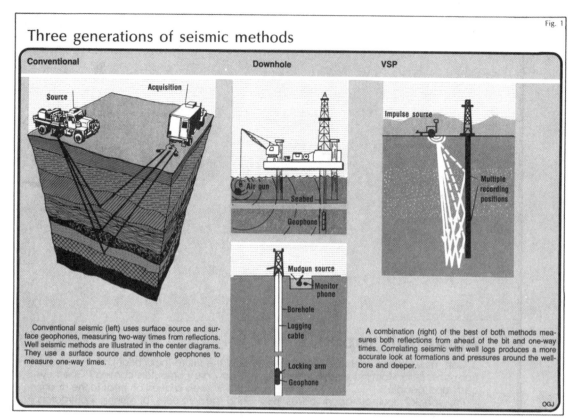

Fig. 1

Three generations of seismic methods

Conventional seismic (left) uses surface source and surface geophones, measuring two-way times from reflections. Well seismic methods are illustrated in the center diagrams. They use a surface source and downhole geophones to measure one-way times.

A combination (right) of the best of both methods measures both reflections from ahead of the bit and one-way times. Correlating seismic with well logs produces a more accurate look at formations and pressures around the wellbore and deeper.

of information now available to the man on the rig include:
- Pressure ahead of the bit.
- Tops of key formations or transition zones.
- Lateral distance from faults or salt domes.
- Relationship of the borehole to the target.

The general family of methods is known as vertical seismic profiling (VSP). It entails lowering a geophone downhole.

Measurements taken downhole greatly enhance data quality, allow more variables to be measured, and give a closer, more detailed look at the formations of greatest interest.

This article explains the new well seismic applications from a drilling standpoint and gives wellsite uses and methods.

The universal well seismic technique, the velocity survey, is being upgraded to a new status from its traditional exploration role in time-depth determination to the vertical seismic profile (VSP).

Today's VSP extends velocity survey theory to allow the driller to "see" below the drill bit and out into the strata laterally near the wellbore.

Introduction. Perhaps the most positive item to emerge from the current industry recession is the trend to employ advanced geoscience techniques in better evaluating the region around the borehole.

Now that the industry is working under new economic models the best way to save drilling dollars is by increasing drilling efficiency. The service industry has strived to reduce the cost per drilled foot, but the best method to reduce drilling costs is to drill no dry holes.

It is from this perspective that operators choose to utilize advanced geoscience techniques to better evaluate the wellbore before leaving location.

If a well is productive, then complete evaluation of the reservoir is desired to ensure the stepout is also a "keeper." If a well is dry, then complete evaluation of the drilled section is desired to reduce dry hole risk in the follow-up prospect.

Geophysicists analyze miles of surface-gathered seismic data to map the subsurface, but even the best geophysical acquisition, processing, and modeling techniques leave unanswered questions about the subsurface that only penetration of the strata by the bit can answer. Once a well is drilled, geologists run logs to evaluate the prospect and answer the questions of the geophysicists.

Traditionally, geophysicists look at "pictures" of the subsurface prior to drilling and geologists look at the subsurface prior to drilling.

What is needed is a combination effort of using logs with geophysical data prior to regional drilling to best recommend a prospect, to analyze logs of the prospect, to understand the well after drilling, and then return to the seismic domain and apply rock data from well logs to the adjacent seismic profiles.

The best method to extrapolate well-log-derived parameters into the seismic profile is by placing the geophysicist's geophone down the geologist's well.

This in-situ geophone is a technique the explorationist uses and it is called borehole geophysics or well seismics. By placing a geophone opposite the formation in the borehole, the seismic response of that formation can be calibrated with the physical characteristics of that formation as seen on well logs. In other words, by placing a geophone in a well, the explorationist can once and for all determine how that formation with its diagnostic properties (lithology, fluid content, porosity, etc.) has modified the received seismic signal.

This technique of recording the complete seismic profile in a well is termed a vertical seismic profile (VSP).

How VSP helps drill

Fig. 2

When it is known how the seismic signal of the VSP is related to log-derived rock parameters then the rock parameters can be extrapolated laterally into the surface seismic profile via the acoustic waveform. This will reduce dry hole risk in the next nearby well.

By calibrating the seismic profile with well log values the pre-drill direct detection of hydrocarbons is closer to reality. A VSP is the connection between the disciplines of geology and geophysics; it is a time series measurement at a known depth.

A by-product of VSP data is that it yields information beneath the bit that can increase the efficiency of a drilling program when the data are processed in a timely manner.

A VSP can tell the driller what hardness of rock is beneath the drill bit, what pressure it contains, and at what depth it will be reached.

It can also be used to determine geologic features laterally proximal to the wellbore. This allows one well to test multiple structures more accurately.

The techniques discussed here are all pioneer interpretive methods that will be commonplace within five years.

Fig. 1 puts the various types of seismic in perspective.

VSP contributions to drilling

Those who have performed a market or information analysis on the VSP industry have been frustrated in their results. No two clients report the same use of VSP data.

This has good and bad connotations. It guarantees a continued industry diversification in VSP development and growth, but it intimidates companies planning to enter the field since their market analysis shows no central area for profit in the VSP program.

The reason no two salesmen report the same client VSP interest is because a VSP has many applications. The uniqueness of a basic VSP allows the interpreter to use the data in three main areas:

• In applications to surface seismic interpretation and data processing

• For formation evaluation (reservoir property determination, borehole to borehole for production applications, and engineering parameter identification), and modeling into the seismic domain from the geologic domain

• For on-site decisions for the drilling program.

To complicate understanding of VSP markets even further, each of the three basic VSP uses has numerous subset applications.

This article focuses on the drilling VSP application.

Applications of VSP data to a drilling program consist of two subsets:

• "Seeing" beneath the drill bit, and

• "Seeing" laterally out from the borehole.

Looking beneath the bit has three main applications: in determining depth to reflectors (formation tops) not yet drilled, compaction curves (which yields mud weight and overpressure information) in strata not yet drilled, and hardness of rocks not yet drilled.

The last item will help determine rotary speed, weight on bit, and bit type. All three items help determine casing and mud programs, and for it to be contributory in an economic sense the data must be analyzed in time to yield an on-site decision.

The biggest technological problem to overcome in beneath-the-bit theory is determining the velocity of the strata below the bit. Reflections below the bit are analogous to surface seismic with its problem in evaluating Equation 2. That is because the only parameter out of three measured beneath the bit is two-way time. Velocity must be determined beneath the bit to convert two-way time to depth.

Seeing beneath the bit is absolutely no problem. Reflections readily emanate from acoustic basement which is typically 20,000 to 50,000 ft or more beneath the earth's surface. But the data are measured in time and must be converted to depth to be useful to the drilling program.

The current technique to obtain velocities beneath the bit is based on "backing out" the velocity function from the amplitudes of the VSP seismogram. The method is the reverse of calculating a synthetic seismogram from velocity and density functions obtained from standard well logs where the object is to create the amplitudes of a seismic trace.

A VSP is a real seismogram. Furthermore, it is not limited to the depth of the well like a synthetic is. Therefore, by calculating reflection coefficients (RCs) from VSP amplitudes beneath the well, and by removing the density effect from the RCs by recursive methods beginning with known densities in the well, then the residual will be formation velocities. With velocity and two-way time beneath the bit, then Equation 2 will yield depth.

For the future, look to see this problem also approached in the same manner as surface seismic finds stacking velocities, i.e., using moveout scans to plot power and amplitude correlation values with time on time vs. root mean square velocity plots. Any other method to find depth beneath the bit assumes, rather than calculates, a velocity function.

In practice, to determine depth beneath the bit, start with known nearby velocity functions, make assumptions appropriate for the spud site, and run a VSP and CVL/density log on the first logging run. Do not wait to run a single VSP at total depth, but run one at each logging interval. Statistics are needed to help extrapolate the known velocity function of the well to beneath the well.

Use nearby CVLs and seismic stacking velocities to model the pre-drill velocity assumption. Compare the assumption with logs of the well and accordingly modify the next assumption beneath the bit as it is compared to the 'backing out' technique.

Poisson's ratio

Table 1

Material	Poisson's ratio
Water	0.5
Gravel	0.47
River deposits	0.45
Dry loess	0.44
Stiff mud	0.43
Casing steel	0.37
Sandstone	0.34
Limestone	0.25
Concrete	0.21
Granite	0.20

(Heiland, 1951; Press, 1966; Sheriff, 1982)

Table 2

Common energy sources

Environment	Offshore	Onshore	Shallow water	Arctic
Energy source	Airgun Watergun Explosives	Airgun (in a tank) Mudgun (in a pit) Explosives Vibrator (P and S) Weight drop Land airgun	Airgun Mudgun Explosives	Airgun (in a tank) Explosives Vibrator (P and S)

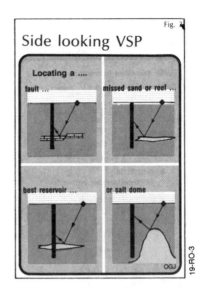

Fig.

Side looking VSP

Repeat this procedure at each logging run to build up statistics (and interpreter confidence) for beneath the bit depth predictions. This may become standard technique as soon as case histories are revealed.

Depth beneath the bit is particularly useful when combined with a compaction prediction program. In this application, repeat all the above procedures to determine the velocity function (including pre-spud analysis of nearby mud weight vs. transit time plots and transit time calculations from seismic stacking velocities) but display results as seen in Fig. 2. For the driller and petroleum engineer, display mud weight data in a familiar form. Assign mud weights to the transit time semi-log plots. This allows an engineer to draw conclusions based on the drilled section as it compares to prior predictions and present predictive values.

It is desirable to do this in any well where the mud weight function is in question or where overpressure is known to exist. Beneath the bit mud weight predictions should also become standard industry procedure. Pre-drill mud weight predictions from seismic velocities should already be standard technique.

To derive rock hardness beneath the bit also requires determination of velocities beneath the bit. This time both compressional (V_p) and shear (V_s) wave VSP data are needed. Technologically the problem here is acquiring quality V_s data. Shear wave data acquisition requires an appropriate shear wave source (so test this beforehand), the appropriate shear wave downhole geophones (check geophone response vs. hole deviation beforehand), and the elimination of transmission noise from the tool to the surface.

To calculate rock hardness use standard rock physics formulas for determination of V_p and V_s from elastic constants:

$$V_p = \sqrt{\frac{K + 4/3G}{D}} \quad K = \frac{E}{3(1 - 2\mu)}$$

$$V_s = \sqrt{\frac{G}{D}} \quad G = \frac{E}{2(1 + \mu)}$$

where K is compressibility, G is rigidity modulus, E is Young's modulus, d is density, and μ is Poisson's ratio.

Manipulate the equations until:

$$\mu = \left[\left(\frac{V_p^2}{2V_s^2} - 1\right)^{-1} + 2\right]^{-1} \quad (1)$$

so that Poisson's ratio is the quantity determined. Poisson's ratio is a unitless rock parameter that varies from 0.0 to 0.5 where a value of 0.25 indicates an isotropic elastic material. Table 1 shows Poisson ratio values for typical materials.

As a byproduct of obtained Poisson ratios, use the previously derived density function with the Poisson value to calculate the elastic constants K, G, and E. Here is a method to compare log-derived rock parameters to in situ rock measurements.

The contribution to drilling, exploration, development, and production programs using downhole V_p and V_s measurements are limited only by imagination and data quality. Look to see research groups concentrating more on downhole shear wave technology in the very near future.

Applications of VSP data to a drilling program by seeing laterally out from the borehole are shown graphically in Fig. 3. The technique involves imaging the subsurface and "bouncing" the data into the well geophone.

Remember that a more detailed picture of the target to be imaged is obtained with well seismics because the receiver is physically closer to the target.

Fig. 4 shows the two common VSP field procedures used to image a nearby event. Basically the technique involves a combination of stationary or multiple source positions with a combination of stationary or multiple downhole geophone positions. The movable source increases subsurface coverage of a suspected event. A multiple source VSP is often termed a "walkaway" VSP.

Fig. 5 shows why the term walkaway is often used. The acquisition object is to move from the well until the subsurface bounce point has passed the edge or boundary of the event in question.

In Fig. 5 the object is to walkaway from the well until the subsurface bounce point has crossed the boundary of the fault face.

Fig. 6 shows an idealized data set from the model in the preceding figure.

To determine where to position the well sonde and whether to gather data from stationary or multiple downhole stations, pre-job modeling should be performed.

Walkaway VSPs allow the interpreter to see both sides of a fault, to see both sides of a stratigraphic wedge, to see reservoir lateral extent, to delineate salt dome or shale mass or overthrust boundaries, to determine formation dip angles, and yield horizontal velocity control.

Actually the term walkaway should infer linear source location patterns. Often multiple source locations are used which are not in a line, and the proper term here is multi offset VSP, compared with a single offset VSP which entails only one source location throughout the survey.

For the driller and for on-site decisions, a walkaway VSP helps locate features of a nearby structure so that deviation instructions can be made to penetrate the target. For example, in a tight carbonate play with pay in vertical fractures or small relief faults not visible on surface seismic, a well near the target can be deviated into the potential reservoir after a walkaway VSP has imaged the nearby structure.

In a deviated well, a walkaway with the source positioned above the well phone will yield conventional or 100% seismic coverage. In this technique, the in situ geophone allows detailed imaging of strata not visible on surface-acquired seismic data. Deviated well VSP/walkaway combinations are a powerful tool for reservoir

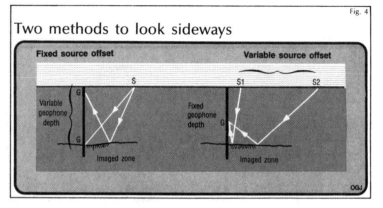

Fig. 4 Two methods to look sideways

delineation when working from an offshore platform.

Overall technical problems with support systems for VSP utilization in on-site drilling applications do exist. For any VSP contribution to the drilling program to be useful it must be processed in a timely manner. Major operators who regularly utilize advanced VSP techniques use their own computer hardware, particularly for quick data turnaround.

Hardware to handle the data volume of well seismics acquisition is not regularly available to the commercial user. In on-site applications, computer systems are either trucked to location or data is rushed to the computer system.

In regional work, a central location is optimal for advanced processing of high priority VSP data. However, on-site systems need to offer "quick look" VSP displays for acquisition quality control. Working with a well-coordinated drilling program, turnaround for VSP data should be sufficient for most applications if accomplished within 12 hr, and certainly no longer than 24 hr.

When possible, try to arrange for casing to be set, or a conditioning trip, or more logs to be run after VSP acquisition to allow processing time without incurring excessive rig costs. Also when feasible, try to run the VSP in sections, i.e., at each logging run. Analyze results in the upper section to ensure all parameters set for acquisition are best suited for that spud location.

Lateral changes in geology and near surface weathering may cause a field set-up from one location to be inadequate at the next location. Use data from the first VSP run to ensure data from TD is of the best quality. Quick looks are important because they show if the target has been imaged correctly or if the shooting plan needs to be altered.

On-site VSP quality control monitors are needed whether the data is high priority or not.

One hurdle that exists to on-site processing is the large volume of data accumulated in well seismic operations. A typical suite of electric logs will yield a single 7½ or 8-in. reel of digital tape.

In contrast, a basic check shot survey yields three to five reels of the same tape, and a VSP can easily generate 50 to 70 reels of data.

For wells not drilled near cultural centers, VSP data transmission can become a problem in quick processing applications. Current telephone and microwave remote data transmission systems often take as long to transmit as to record the raw data. Viable techniques include satellite link to a processing center and then microwave return to the rig.

Return data volume is insignificant compared to raw data volume because return data are conclusions.

Another processing method is to demultiplex raw data on-site, but this requires large hardware memory capacity on-site. One solution is to record in demuxed form. Once trace oriented displays are achieved, then manipulation for on-site use is greatly simplified. These problems will be more aggressively addressed in the near future. The rapid advancement in computer systems will help solve these problems.

Besides lacking hardware on site, software development available to the commercial user of VSPs is lagging significantly to that internal to major operators. Quality on-site VSP software is nonexistent. The only necessary on-site VSP software is that which either contributes to an on-site decision or to quality control. Current emphasis of a synthetic seismogram on-site, when a VSP is a real seismogram which also extends beneath the welbore, is an example. The object in on-site reduction of data is to aid in on-site applications. Lack of on-site reduction of VSP data is the item that keeps well seismics off of development and production wells.

The potential to every part of the drilling program offered by proper analysis of downhole seismic data is established in the professional literature, therefore it is only a matter of time before industry catches up to research. The next few years in the well seismic industry will be exciting.

VSP acquisition rules of thumb

It has been said that the only expensive VSP is a poorly acquired VSP. Sometimes it is false economics to gauge the value of a VSP by rig time and acquisition costs. Depending on the operator's interest in the area, no VSP and another dry hole or an unidentified high pressure zone blowout may be more costly than a detailed and properly planned well seismic program.

A VSP is not a panacea but when coupled with predrill objectives and predrill planning it can go a long way towards answering the geologic riddles seen on adjacent seismic profiles.

There has never been a method more accurate than the in situ geophone of a VSP to calibrate surface seismic with well geology. Some operators acquire minimal VSP coverage with no immediate data reduction in mind but, because the province is important to their program, they want a VSP for later analysis.

Once a well goes into production or is abandoned it may be too late to obtain well seismic measurements.

In planning a well seismics field program the first requirement is to define the survey objectives and then decide what energy source to use. Table 2 is a brief chart of energy sources. In choosing a source, pick as broad a frequency spectrum as is possible and choose a source that yields a repeatable energy pulse.

Research has shown that explosives will satisfy the above conditions but explosives have been used in VSP to a limited extent.

Downhole sources for borehole to borehole or borehole to surface measurements are a big research item. Technological problems to overcome include borehole destruction, signal repeatability, energy attenuation, and firing rate for production modes. The potential in downhole source applications is strong enough to warrant continued active research.

After source selection make sure survey objectives can be met with minimal cost in on-site time and equipment. It is strongly recommended that a computer model of the suspected geology near the borehole be made. This includes synthetic VSPs and ray tracing and conversion of nearby well data to the seismic domain.

From models, choose source and sonde locations to meet the objective. Consider whether to simultaneously record several surface lines with the VSP.

Fig. 7 shows a general star pattern surface geophone layout around the wellbore. Suggested spread patterns are in the direction of dip, or to prove closure nearby, or to look for reservoir limits, or to parallel an existing seismic profile, etc.

Work out several field models so as to have a back-up plan in case on-site quality control displays indicate the primary assumptions to be invalid. In other words, try to foresee the exploration, development, or production needs from an interpretive viewpoint, model the conclusions, and then see if a VSP program will help. If so, proceed with the models to establish imaging parameters and source and receiver parameters.

After deciding to proceed with the program, visit the well site (excepting deep water prospects) and check for permitting, topographic problems, etc, in laying out the surface equipment. Offshore, check for navigation problems (pipelines, well heads, etc.) and make arrangements with positioning companies for accuracy in multi offset and walkaway source locations.

On land, survey in as many locations, shotholes, etc., as possible beforehand. Also check acquisition parameters on nearby seismic sections to see if any need to be duplicated for the VSP program, i.e., vibrator parameters, geophone responses, recorder dynamic ranges, source signature monitor requirements. Line up contractors as necessary.

Where applicable, try to define potential problems that will affect the seismic signal, i.e., will the near surface inhibit energy penetration, how deep should shot holes be to avoid weathering zone problems, will the area generate excessive or unwanted near surface noise, will buried gas lines or railroads or autos or power lines transmit spurious signals to the geophones, etc.

Many of these items are simple common sense but, due to their simplicity, are often ignored. Some operators have check lists for VSP programs to ensure no common sense item is overlooked. All operators should satisfy themselves that contractor procedures are safe, efficient, and will meet job requirements.

Another critical aspect of VSP acquisition is the sonde. Some people consider the downhole geophone tool the weak link in well seismic systems. The tool must be able to support its own weight in the borehole and not

Walkaway locates faults, profiles reservoir
Fig. 5

slide when the wireline is slackened to isolate the tool from cable-borne noise.

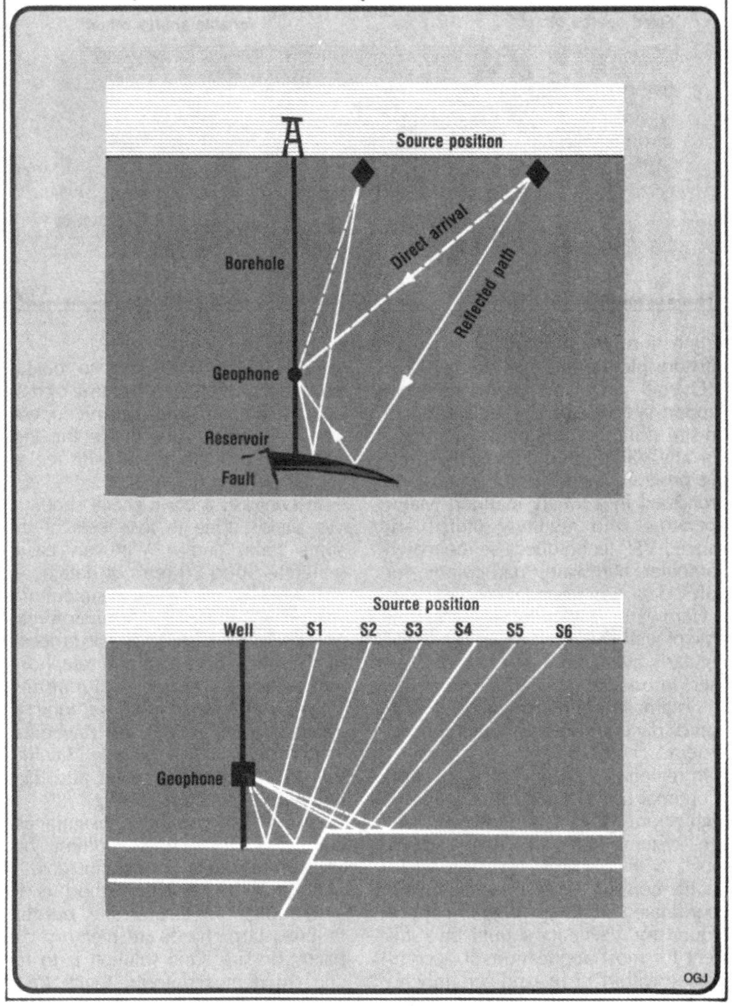

Temperature effects on downhole tools have been minimized in recent years but long borehole operating times always necessitate high temperature components.

Downhole amplification is another item that has improved recently. However, crossfeed into signal channels in uphole data transmission is a persistent problem that is compounded in tools with shear wave phones.

Three component tools, P-wave and the two orthogonal S-wave phones, present challenges to tool design personnel.

Look to see downhole telemetry emerge in VSP sondes.

Another item that has market appeal in 3-D VSP tools is an azimuth and inclinometer device for monitoring tool orientation. This will significantly aid in amplitude and attenuation studies.

Also needed is a depth monitoring device such as an SP button or radioactive unit or caliper/collar locator system. This ensures correct correlation to other logs.

All these design additions add requirements on the support weight of the locking arm, and often the arms must function in variant hole diameters.

Dual locking arms and hydraulic positive lock/quick release mechanisms are current VSP sonde research items.

It would reduce in-hole time if multiple recorders could be strung down hole but problems abound in this approach.

The most obvious problems include sonde wall coupling and cable tan-

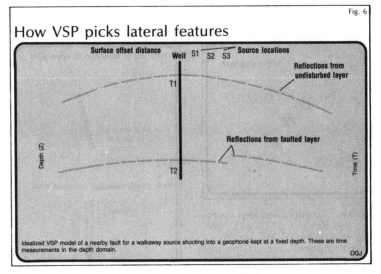

Fig. 6
How VSP picks lateral features

Idealized VSP model of a nearby fault for a walkaway source shooting into a geophone kept at a fixed depth. These are time measurements in the depth domain.

gling.

Last but not least, the VSP tool must withstand formation pressures.

Another neglected item very necessary to advanced VSP programs is on-site quality control systems.

Data must be monitored to ensure quality.

Downhole surface recorders compatible with conventional surface seismic systems are quickly becoming standard in well seismics. But full channel monitoring systems are lacking.

Real time displays (including summed and correlated data) in useful depth-oriented alignment need to be in VSP field systems. Velocity and time-depth analysis on-site are needed for VSP trace alignment input. These simple manipulations are required to ensure multioffset VSP data is imaging properly.

It is also important to decide when to run the VSP in relation to the sequence of rig events.

Remembering that rig time is money, the first consideration is the trade-off between multiple and single pass VSP log runs. The trend for multiple runs with overlapping sections is not one to be denied, and the author strongly recommends this where affordable for virtually every VSP application and especially for beneath the bit applications.

Next determine the position of the VSP in the logging sequence.

Run the VSP after a wiper trip and definitely never after dragging cores up the hole.

It is best to utilize the caliper and cemet bond logs to determine potential problem locations for optimal tool coupling (Fig. 8).

Logs that exhibit strata type should be viewed to firm-up preplanned borehole sampling intervals. A density or dipmeter log will aid in predicting at what depths to expect major reflections.

And definitely check the temperature log.

More general rig sequence decisions require placing the VSP in an order with casing/liner and cement programs.

In an open hole, the VSP sonde should need to stay less than eight minutes a level.

In a cased hole, level times should be less than five minutes. These figures are source dependent, but never stay more than fifteen minutes in one spot in open hole.

Open hole acquisition is sometimes better in that cased holes can generate fluid column borne noise at casing/collar points and at diameter changes.

A multiple run VSP that duplicates in-casing a run previously shot in open hole helps fathom noise problems.

Always allow time, where possible in rig sequencing, so the VSP can be processed prior to the resumption of drilling when on-site applications are utilized.

Before going in the hole, personnel involved should be provided with a list of depth stations to be shot, and of course all tools and equipment should be surface tested.

Test procedures for VSP systems include system voltage tests and geophone response tests.

Going into the hole, zero the depth indicator at the kelly bushing with the top of the tool. Confirm the zero mark when the tool comes out of the hole.

Like other logs, a VSP is recorded coming out of the hole, but for quality control purposes several stations going in the hole should be logged and compared with the repeat data when coming out of the hole.

Logging going down also proves-up the system and allows the determination of the number of energy source sums necessary for sufficient data recovery over background noise before TD is reached.

All components to the sum should go to tape separately in an unedited and unfiltered format. All on-site operations can use drilling crew help, particularly if unaccounted electrical disturbances from rig equipment finds its way to tape. Improperly grounded or isolated service equipment is most detrimental. The author remembers an electrical noise problem whose source turned out to be the vibration of a broken motor mount on a rig generator.

Another time it was the electric light bulb that kept the coffee pot warm.

If it had not been for midnight help by the rig electrician the whole survey might have been lost.

Well seismic history. The need for basic downhole geophone data acquisition (commonly called a velocity survey, check shot survey, or well shoot) initially arose when geologic structures were drilled based on maps contoured in time instead of in feet.

Time maps are made from reflection seismic data gathered at the earth's surface in opposition to depth maps which are made from well logs or from time maps that have been converted to depth.

To convert seismic time (t) to depth (z) and vice versa requires a velocity (v) such that (in its most basic form):

$$z = v(t/2) \qquad (2)$$

Note in Equation 1 that seismic time is two-way time, i.e., seismic time is the travel time for a surface-activated energy source to transmit a pulse down to a geologic horizon (termed a reflector in geophysics jargon) and then back to the surface where it is detected by a geophone (Fig. 1).

In surface seismic reflection data, only one measurement in Equation 1 is recorded—two-way travel time. Out of three variables in Equation 1, two are unknown. A geophysicist has no idea what the depth is to a reflecting horizon nor the average velocity of the earth from the surface to that reflector from preliminary measurements.

For a geophysicist to know at what depth his time structures are located, a velocity function must be known. Of course, drilling programs are always in feet, so how is the velocity function defined when drilling seismic prospects?

Two basic techniques exist to resolve the velocity function into depth: (1) Use a mathematical deterministic method to iteratively derive from two-way times a "best fit" velocity function and combine that with knowledge of regional geology to give a predrill depth estimate to structures; or (2) lower a geophone in an existing well to a known depth and measure the lapsed time for a surface-generated acoustic pulse to travel one-way to the downhole geophone (Fig. 1).

The latter method is a velocity survey and yields the unknowns to solve Equation 2.

Downhole geophysics began as a research idea developed in 1927 when the Geophysical Research Corp. (GRC) lowered the first geophone in a well.

That year GRC drilled the world's first prospects on reflection seismological data. They immediately discovered that time maps yielded no depth values unless a velocity function was available, so they invented the velocity survey.

Because explosives were the source most commonly used, the velocity survey was often called a well shoot. During 1927, GRC completed two velocity surveys, in Upton Co., Tex., to a depth of 2,150 ft, and in Crockett Co., Tex., to a depth of 2,624 ft. (The current depth record for a velocity survey is 31,000 ft near Elk City, Okla., shot by Seismic Reference Service for Lone Star in 1972.)

The basic velocity survey concept pioneered in 1927, and in use today, involves stopping the geophone sonde at any subsurface point (called a station or a level) of operator interest as predetermined from electric logs. Then the tool is forced against the casing or formation wall (Fig. 1) by activating a surface controlled motor driven downhole arm, typically located in the sonde immediately above or below the geophone elements. Early geophone tools did not have surface-controlled retractable arms but this has evolved to be the most satisfactory method to establish appropriate coupling of the sensors with the borehole wall.

The value of the first velocity survey to explorationists was immediately obvious and the skills developed by GRC were unique—by 1930, GRC had shot 22 velocity surveys. In 1932, GSI began shooting wells, Petty began in 1933, and Empire and SSC started in 1934.

By 1940, the following oil companies were recording their own surveys: Arkansas Fuel, Carter, Cities Service, Conoco, Gulf, Humble, Magnolia, Phillips, Shell, Stanolind, Superior, Texaco, and Texas.

And the following service companies were recording velocity surveys by 1940: Empire, GGC, GRC, GSI, IXC, NGC, Petty, SEI, SSC, and WGC.

On international scenes: Carter shot the first surveys in Canada in 1936 and in Sumatra in 1938. Petty was in Venezuela by 1936, SSC was in Iran in 1938, and Shell in Colombia in 1939.

From 1931 to 1935, 137 domestic well shoots were recorded; from 1936 to 1940, 300+ surveys were shot; and during the war years, 1941-1944, 345 surveys were recorded.

Today over 10,000 velocity surveys have been recorded in North America.

By the late 1940s, several oil company research groups were enthralled with velocity data to the point of having a station every 15 ft of hole depth. (Ironically, the industry is back to this detail of downhole geophone data acquisition for VSP surveys).

The tie-up of rig facilities was enormous.

It would take 30 or more men to handle shot holes for the explosives (weight drops and vibrators were not invented until the 1950s) and other acquisition tasks. Geophones were in the borehole from 24 to 48 hr at a time.

It was in this vein the research response in Magnolia, in 1949, was to invent the continuous velocity log (CVL) using acoustic techniques developed for submarine use in the previous war.

Actually, CVL invention was simultaneous at Magnolia, Humble, and Shell. Today the CVL has been improved by being borehole compensated.

The CVL concept avoided locking arms and stopping the tool and surface energy sources by using pressure sensitive geophones that could receive a signal through the well fluid column and by using a downhole transmitter in addition to the downhole receiver.

The object of CVL evolution was to

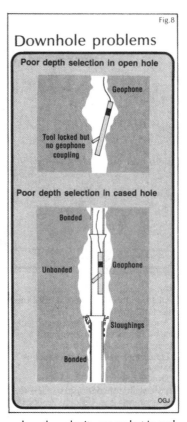

replace the velocity survey but in reality the velocity survey was now needed to check the CVL for borehole induced errors in acquired data. Thus the velocity survey, or well shoot, began to be known also as a check shot survey.

In 1956, Hicks and Berry published one of the first correlations between CVL velocity and lithology—that of velocity with rock type, fluid content, and porosity.

Also at that time, Wyllie, et al., began publishing their series of studies which led to the Wyllie time-average equation that empirically related formation transit times from the CVL to formation porosity.

Velocity as a function of rock parameters had been an item of study for some time, but the in situ measurements of the CVL gave data that no lab in rock physics could duplicate. With these types of studies, the CVL created a whole new market niche for its services.

Industry use of CVLs accelerated while less velocity surveys were run as each year passed. Velocity surveys became relegated to exploration wells only and by the early 1980s they were run in roughly one out of five exploration wells; whereas borehole compensated CVLs are run in virtually every well.

The authors...

Roberts

Platt

R.J. (Jeff) Roberts is a well seismic consultant in Houston. He started as a velocity engineer with Birdwell Division of Seismograph Service Corp. in 1976 and has held a variety of positions in wellsite seismic applications. He was a delegate to the Joint SEG/CGS meeting, Beijing, 1981. He recently completed a contract to develop and start up Dresser Atlas' entry into the field of well seismics, and is currently consulting to Texas Seismic. Roberts holds a BA in geology from University of Texas at Austin.

J.D. (Jack) Platt, Jr. is vice-president of Downhole Seismic Service Inc., Belle Chasse, La. He currently offices in Houston where he directs data acquisition and processing, and the firm's expansion into new well seismic markets. He has experience in drilling data collection and wireline logging. Platt worked for Geosource for six years including positions in Geosurvey and Seismic Reference Service before joining the management team founding Downhole Seismic Service.

In 1959, Russia began a well seismics program which later developed into a technique termed vertical seismic profile (VSP). They recognized that the best way to understand surface-gathered seismic reflection data was by intersecting the ray paths in situ or with a borehole geophone (Fig. 1).

They correctly surmised that by being near the source of the reflection (i.e., in the borehole), much more information could be obtained from the seismic signal than could be ascertained at the surface.

In actuality, all Russian VSP work takes off from two important American research projects of the 1950s. In 1952, Carter Co. theorized and proved that the best method to understand surface seismic reflection data was to record the same data with a geophone downhole.

They were interested in total waveform detection, not just the one-way time measurement of a basic velocity survey.

In 1958, Jersey Production Research Co. performed a continuation of the Carter project of 1952.

Its goal was to better understand acoustic pulse behavior as it passed through the earth into the well phone to better understand surface reflection data.

These two successful experiments were rediscovered in the U.S. during the early 1970s by studying Russian publications on VSP.

Of course the Russians contribute their own knowledge in peripheral subjects and subsequent experiments to the VSP concept, but the basis of VSP comes from the Carter and Jersey Co. tests of the 1950s.

While the Russians were becoming experts in well seismics, American geophysicists were going offshore, developing advanced computer processing techniques such as common midpoint theory, and converting to digital.

Well seismic research was not abandoned during the 1960s and 1970s but few significant research programs existed.

Two notable exceptions were a Century Geophysical well geophone prototype project begun in 1961 for the U.S. Air Force Office of Scientific Research, and a VSP test performed by Gulf Research and Development in the early 1970s. The results of these two studies were significant but not in the mainstream of contemporary geophysics.

It was in 1974 that borehole seismics under the name VSP began to be an outspoken topic in research circles.

In 1974, the Society of Exploration Geophysicists published a translated Russian-to-English text on VSP (Gal'pern, 1974) that drew immediate professional interest in the technique. By the mid 1970s, Europeans were acquiring VSP data routinely. Shortly thereafter VSP acquisition was in domestic price books of service companies, and vibrators and air guns in addition to explosives were offered as energy sources.

Major research groups embraced these concepts readily, they quickly began acquiring VSP in conjunction with surface seismic data, and moved into downhole shear wave in addition to downhole compressional wave VSP acquisition. By 1977, Gal'pern had been to the U.S. on a speakers tour at the request of American oil majors, most notably Phillips Petroleum.

Interest in well seismics was picking up.

Today, VSP applications have mushroomed.

Everyone seems to have a different use for VSP data. Use of basic VSP techniques have shifted from research groups to the independent operator interested in better evaluation of his prospect.

Downhole seismics is a concept back in vogue and here to stay.

Conclusion. Well seismic theory has come a long way since the first well shoot in 1927. It has evolved into a powerful method to accurately tie the geology of the well to the geophysics of the surface seismic profile, and this method is the vertical seismic profile (VSP).

Barely out of major company research programs, the basic VSP is a technique catching the eyes of the operator who wants to do everything possible to better evaluate his prospect.

Although the multitude of applications of a VSP survey seem confusing initially, further inspection shows it to be a tool of advantage for exploration, development, and production.

As more emphasis is placed on infield manipulation of VSP data, it will become a standard service for all drilling programs.

The best news concerning VSP utilization is that its wide range of applicability will spawn continued research and growth in all directions.

Acknowledgment

Special thanks for history information to Bob Audley, Ted Borne, Bob Broding, and Harry Mayne. The authors appreciate the help of Lilly Neale on the manuscript.

Bibliography

Anstey, N.A., "Simple Seismics," Inter. Human Resource Devl., Boston, 1982.
Broding, R.A., et al., "Final Report on Study of a Three-Dimensional Seismic Detection System," prepared for Air Force Office of Scientific Research Office of Aerospace Research, Washington, D.C., 1963.
Gal'pern, E.I., "Vertical Seismic Profiling," Society of Exploration Geophysicists Special Publication No. 12, 1974.
Heiland, C.A., Geophysical Exploration: 1951, Prentice-Hall Inc., New York, N.Y., p. 468.
Hicks, W.G., et al., "Applications of Continuous Velocity Logs for Determination of Fluid Saturation of Reservoir Rocks," Geophysics, 1956, Vol. 21, No. 3, pp. 739-754.
Jakosky, J.J., Exploration Geophysics, 1960, Trija Publishing Co., Newport Beach, Calif., p. 658.
Jolly, R.N., "Deep-Hole Geophone Study in Garvin, County, Oklahoma," Geophysics, 1953, Vol. 18, pp. 662-670.
Levin, F.K., et al., "Deep Hole Geophone Studies," Geophysics, 1958, Vol. 23, pp. 639-664.
Pennebaker, E.S., "The Use of Geophysics in Abnormal Pressure Applications," 43rd Annual Fall Meeting of SPE, Houston, Tex., 1968.
Method for Prediction of Abnormal Pressures from Routine or Special Seismic Records, U.S. Patent No. 3898610, 1975.
Press, F., "Seismic Velocities," Geological Society of America Memoir 97, 1966, p. 207.
Sheriff, R.E., et al., Exploration Seismology, Vol. 1: 1982, Cambridge University Press, New York, N.Y., p. 74.
Wuenschel, P.C., "The Vertical Array in Reflection Seismology—Some Experimental Studies," Geophysics, 1976, Vol. 41, No. 2, pp. 219-232.
Wyllie, M.R.J., et al., "Elastic Wave Velocities in Heterogenous and Porous Media," Geophysics, 1956, Vol. 21, pp. 41-70.
An Experimental Investigation of Factors Affecting Elastic Wave Velocities in Porous Media, Geophysics, 1958, Vol. 28, pp. 459-463.
Studies of Elastic Wave Attenuation in Porous Media, Geophysics, 1962, Vol. 27, pp. 569-589.
Zwart, W.J., et al., "Ed. Index of Wells Shot for Velocity," Society of Exploration Geophysics, 1983.